The Mitchell Beazley Joy of Knowledge Library

Science and The Universe

Scientiam non dedit natura semina scientiae nobis dedit
"Nature has given us not knowledge itself, but the seeds thereof."
Seneca

The Joy of Knowledge Encyclopaedia is affectionately dedicated to the memory of John Beazley 1932–1977, Book Designer, Publisher and Co-Founder, of the publishing house of Mitchell Beazley Limited, by all his many friends and colleagues in the company.

The Joy of Knowledge Library

General Editor: James Mitchell
With an overall preface by Lord Butler, Master of Trinity College, University of Cambridge

The Mitchell Beazley Joy of Knowledge Library

Science and The Universe

Introduced by Sir Alan Cottrell, FRS

Master of Jesus College, University of Cambridge; and

Sir Bernard Lovell, FRS

Professor of Radio Astronomy, University of Manchester

MITCHELL BEAZLEY

The Joy of Knowledge Encyclopaedia
© Mitchell Beazley Encyclopaedias Limited 1976

The Joy of Knowledge Science and the Universe
© Mitchell Beazley Encyclopaedias Limited 1977

Artwork © Mitchell Beazley Publishers Limited
1970, 1971, 1972, 1973, 1974, 1975 and 1976
© Mitchell Beazley Encyclopaedias Limited 1976
© International Visual Resource 1972

ISBN 0 85533 111 9

Typesetting by Filmtype Services Limited, England
Photoprint Plates Ltd, Rayleigh, Essex, England

Printed in England by Balding + Mansell

The Joy of Knowledge Library

Editorial Director	**Frank Wallis**
Creative Director	**Ed Day**
Project Director	**Harold Bull**

Volume editors	
Science and The Universe	John Clark
	Lawrence Clarke
The Natural World	Ruth Binney
The Physical Earth	Erik Abranson
	Dougal Dixon
Man and Society	Max Monsarrat
History and Culture 1 & 2	John Tusa
	Roger Hearn
Time Chart	Jane Kenrick
Man and Machines	John Clark
The Modern World	John Clark
Fact Index	Stephen Elliott
	Stanley Schindler
	John Clark

Art Director	Rod Stribley
Production Editor	Helen Yeomans
Assistant to the Project Director	Graham Darlow
Associate Art Director	Anthony Cobb
Art Buyer	Ted McCausland
Co-editions Manager	Averil Macintyre
Printing Manager	Bob Towell
Information Consultant	Jeremy Weston

Sub-Editors	Don Binney
	Arthur Butterfield
	Charyn Jones
	Jenny Mulherin
	Shiva Naipaul
	David Sharp
	Jack Tresidder
Proof-Readers	Jeff Groman
	Anthony Livesey
Researchers	Peter Furtado
	Malcolm Hart
	Peter Kilkenny
	Ann Kramer
	Lloyd Lindo
	Heather Maisner
	Valerie Nicholson
	Elizabeth Peadon
	John Smallwood
	Jim Somerville

Senior Designer	Sally Smallwood
Designers	Rosamund Briggs
	Mike Brown
	Lynn Cawley
	Nigel Chapman
	Pauline Faulks
	Nicole Fothergill
	Juanita Grout
	Ingrid Jacob
	Carole Johnson
	Chrissie Lloyd
	Aean Pinheiro
	Andrew Sutterby
Senior Picture Researchers	Jenny Golden
	Kate Parish
Picture Researchers	Phyllida Holbeach
	Philippa Lewis
	Caroline Lucas
	Ann Usborne

Assistant to the Editorial Director	Judy Garlick
Assistant to the Section Editors	Sandra Creese
Editorial Assistants	Joyce Evison
	Miranda Grinling
Production Controllers	Jeremy Albutt
	John Olive
	Anthony Bonsels
Production Assistants	Nick Rochez
	John Swan

Major contributors and advisers to The Joy of Knowledge Library

Fabian Acker CEng, MIEE, MIMarE; Professor Leslie Alcock; Professor H.C. Allen MC; Leonard Amey OBE; Neil Ardley BSc; Professor H.R.V. Arnstein DSc, PhD, FIBiol; Russell Ash BA(Dunelm), FRAI; Norman Ashford PhD, CEng, MICE, MASCE, MCIT; Professor Robert Ashton; B.W. Atkinson BSc, PhD; Anthony Atmore BA; Professor Philip S. Bagwell BSc(Econ), PhD; Peter Ball MA; Edwin Banks MIOP; Professor Michael Banton; Dulan Barber; Harry Barrett; Professor J.P. Barron MA, DPhil, FSA; Professor W.G. Beasley FBA; Alan Bender PhD, MSc, DIC, ARCS; Lionel Bender BSc; Israel Berkovitch PhD, FRIC, MIChemE; David Berry MA; M.L. Bierbrier PhD; A.T.E. Binsted FBBI (Dipl); David Black; Maurice I.F. Block BA, PhD(Cantab); Richard H. Bomback BSc (London), FRPS; Basil Booth BSc (Hons), PhD, FGS, FRGS; J. Harry Bowen MA(Cantab), PhD(London); Mary Briggs MPS, FLS; John Brodrick BSc(Econ); J.M. Bruce ISO, MA, FRHistS, MRAeS; Professor D.A. Bullough MA, FSA, FRHistS; Tony Buzan BA(Hons) UBC; Dr Alan R. Cane; Dr J.G. de Casparis; Dr Jeremy Catto MA; Denis Chamberlain; E.W. Chanter MA; Professor Colin Cherry D Sc(Eng), MIEE; A.H. Christie MA, FRAI, FRAS; Dr Anthony W. Clare MPhil(London), MB, BCh, MRCPI, MRCPsych; Professor Aidan Clarke MA, PhD, FTCD; Sonia Cole; John R. Collis MA, PhD; Professor Gordon Connell-Smith BA, PhD, FRHistS; Dr A.H. Cook FRS; Professor A.H. Cook FRS; J.A.L. Cooke MA, DPhil; R.W. Cooke BSc, CEng, MICE; B.K. Cooper; Penelope J. Corfield MA; Robin Cormack MA, PhD, FSA; Nona Coxhead; Patricia Crone BA, PhD; Geoffrey P. Crow BSc(Eng), MICE, MIMunE, MInstHE, DIPTE; J.G. Crowther; Professor R.B. Cundall FRIC; Noel Currer-Briggs MA, FSG; Christopher Cviic BA(Zagreb), BSc(Econ, London); Gordon Daniels BSc(Econ, London), DPhil(Oxon); George Darby BA; G.J. Darwin; Dr David Delvin; Robin Denselow BA; Professor Bernard L. Diamond; John Dickson; Paul Dinnage MA; M.L. Dockrill BSc(Econ), MA, PhD; Patricia Dodd BA; James Dowdall; Anne Dowson MA(Cantab); Peter M. Driver BSc, PhD, MIBiol; Rev Professor C.W. Dugmore DD; Herbert L. Edlin BSc, Dip in Forestry; Pamela Egan MA(Oxon); Major S.R. Elliot CD, BComm; Professor H.J. Eysenck PhD, DSc; Dr Peter Fenwick BA, MB, BChir, DPM, MRCPsych; Jim Flegg BSc, PhD, ARCS, MBOU; Andrew M. Fleming MA; Professor Antony Flew MA(Oxon), DLitt (Keele); Wyn K. Ford FRHistS; Paul Freeman DSc(London); G.S.P. Freeman-Grenville DPhil, FSA, FRAS, G.E. Fussell DLitt, FRHistS; Kenneth W. Gatland FRAS, FBIS; Norman Gelb BA; John Gilbert BA(Hons, London); Professor A.C. Gimson; John Glaves-Smith BA; David Glen; Professor S.J. Goldsack BSc, PhD, FINSTP, FBCS; Richard Gombrich MA, DPhil; A.F. Gomm; Professor A. Goodwin MA; William Gould BA(Wales); Professor J.R. Gray; Christopher Green PhD; Bill Gunston; Professor A. Rupert Hall LittD; Richard Halsey BA(Hons, UEA); Lynette K. Hamblin BSc; Norman Hammond; Peter Harbison MA, DPhil; Professor Thomas G. Harding PhD; Professor D.W. Harkness; Richard Harris; Dr Randall P. Harrison; Cyril Hart MA, PhD, FRICS, FIFor; Anthony P. Harvey; Nigel Hawkes BA(Oxon); F.P. Heath; Peter Hebblethwaite MA (Oxon), LicTheol; Frances Mary Heidensohn BA; Dr Alan Hill MC, FRCP; Robert Hillenbrand MA, DPhil; Catherine Hills PhD; Professor F.H. Hinsley; Dr Richard Hitchcock; Dorothy Hollingsworth OBE, BSc, FRIC, FIBiol,

FIFST, SRD; H.P. Hope BSc(Hons, Agric); Antony Hopkins CBE, FRCM, LRAM, FRSA; Brian Hook; Peter Howell BPhil, MA(Oxon); Brigadier K. Hunt; Peter Hurst BDS, FDS, LDS, RSCEd, MSc(London); Anthony Hyman MA, PhD; Professor R.S. Illingworth MD, FRCP, DPH, DCH; Oliver Impey MA, DPhil; D.E.G. Irvine PhD; L.M. Irvine BSc; E.W. Ives BA, PhD; Anne Jamieson cand mag(Copenhagen), MSc (London); Michael A. Janson BSc; G.H. Jenkins PhD; Professor P.A. Jewell BSc (Agric), MA, PhD. FIBiol; Hugh Johnson; Commander I.E. Johnston RN; I.P. Jolliffe BSc, MSc, PhD, ComplCE, FGS; Dr D.E.H. Jones ARCS, FCS; R.H. Jones PhD, BSc, CEng, MICE, FGS, MASCE, Hugh Kay; Dr Janet Kear; Sam Keen; D.R.C. Kempe BSc, DPhil, FGS; Alan Kendall MA (Cantab); Michael Kenward; John R. King BSc(Eng), DIC, CEng, MIProdE; D.G. King-Hele FRS; Professor J.F. Kirkaldy DSc; Malcolm Kitch; Michael Kitson MA; B.C. Lamb BSc, PhD; Nick Landon; Major J.C. Larminie QDG,Retd; Diana Leat BSc(Econ), PhD; Roger Lewin BSc, PhD, Harold K. Lipset; Norman Longmate MA(Oxon); John Lowry; Kenneth E. Lowther MA; Diana Lucas BA(Hons); Keith Lye BA, FRGS; Dr Peter Lyon; Dr Martin McCauley; Sean McConville BSc; D.F.M. McGregor BSc, PhD(Edin); Jean Macqueen PhD; William Baird MacQuitty MA(Hons), FRGS, FRPS; Professor Rev F.X. Martin OSA; Jonathan Martin MA; Rev Cannon E.L. Mascall DD; Christopher Maynard MSc, DTh; Professor A.J. Meadows; Dr T.B. Millar; John Miller MA, PhD; J.S.G. Miller MA, DPhil, BM, BCh; Alaric Millington BSc, DipEd, FIMA; Rosalind Mitchison MA, FRHistS; Peter L. Moldon; Patrick Moore OBE; Robin Mowat MA, DPhil; J. Michael Mullin BSc; Alistair Munroe BSc, ARCS; Professor Jacob Needleman; John Newman MA, FSA; Professor Donald M. Nicol MA PhD; Gerald Norris; Professor F.S. Northedge PhD; Caroline E. Oakman BA(Hons. Chinese); S. O'Connell MA(Cantab), MInstP; Dr Robert Orr; Michael Overman; Di Owen BSc; A.R.D. Pagden MA, FRHistS; Professor E.J. Pagel PhD; Liam de Paor MA; Carol Parker BA(Econ), MA (Internat. Aff.); Derek Parker; Julia Parker DFAstrolS; Dr Stanley Parker; Dr Colin Murray Parkes MD, FRC(Psych), DPM; Professor Geoffrey Parrinder MA, PhD, DD(London), DLitt(Lancaster); Moira Paterson; Walter C. Patterson MSc; Sir John H. Peel KCVO, MA, DM, FRCP, FRCS, FRCOG; D.J. Penn; Basil Peters MA. MInstP, FBIS; D.L. Phillips FRCR, MRCOG; B.T. Pickering PhD, DSc; John Picton; Susan Pinkus; Dr C.S. Pitcher MA, DM, FRCPath; Alfred Plaut FRCPsych; A.S. Playfair MRCS, LRCP, DObstRCOG; Dr Antony Polonsky; Joyce Pope BA; B.L. Potter NDA, MRAC, CertEd; Paulette Pratt; Antony Preston Frank J. Pycroft; Margaret Quass; Dr John Reckless; Trevor Reese BA, PhD, FRHistS; M.M. Reese MA (Oxon); Derek A. Reid BSc, PhD; Clyde Reynolds BSc; John Rivers; Peter Roberts; Colin A. Ronan MSc, FRAS; Professor Richard Rose BA(Johns Hopkins), DPhil (Oxon); Harold Rosenthal; T.G. Rosenthal MA(Cantab); Anne Ross MA, MA(Hons, Celtic Studies), PhD, (Archaeol and Celtic Studies, Edin); Georgina Russell MA; Dr Charles Rycroft BA (Cantab), MB(London), FRCPsych; Susan Saunders MSc(Econ); Robert Schell PhD; Anil Seal MA, PhD(Cantab); Michael Sedgwick MA(Oxon); Martin Seymour-Smith BA(Oxon), MA(Oxon); Professor John Shearman; Dr Martin Sherwood; A.C. Simpson BSc; Nigel Sitwell; Dr Alan Sked; Julie and Kenneth Slavin FRGS, FRAI; Professor T.C. Smout; Alec Xavier Snobel BSc(Econ); Terry Snow BA, ATCL; Rodney Steel; Charles S. Steinger MA, PhD; Geoffrey Stern BSc(Econ); Maryanne Stevens BA(Cantab), MA(London); John Stevenson DPhil, MA; J. Sidworthy MA; D. Michael Stoddart BSc, PhD; Bernard Stonehouse DPhil, MA, BSc, MInstBiol; Anthony Storr FRCP, FRCPsych;

Richard Storry; Charles Stuart-Jervis; Professor John Taylor; John W.R. Taylor FRHistS, MRAeS. FSLAET; R.B. Taylor BSc(Hons, Microbiol); J. David Thomas MA, PhD; D. Thompson BSc(Econ); Harvey Tilker PhD; Don Tills PhD, MPhil, MIBiol, FIMLS; Jon Tinker; M. Tregear MA; R.W. Trender; David Trump MA, PhD, FSA; M.F. Tuke PhD; Christopher Tunney MA; Laurence Urdang Associates (authentication and fact check); Sally Walters BSc; Christopher Wardle; Dr D. Washbrook; David Watkins; George Watkins MSc; J.W.N. Watkins; Anthony J. Watts; Dr Geoff Watts; Melvyn Westlake; Anthony White MA(Oxon), MAPhil(Columbia); Dr Ruth D. Whitehouse; P.J.S. Whitmore MBE, PhD; Professor G.R. Wilkinson; Rev H.A. Williams CR; Christopher Wilson BA; Professor David M. Wilson; John B. Wilson BSc, PhD, FGS, FLS; Philip Windsor BA, DPhil(Oxon), Roy Wolfe BSc(Econ), MSc; Donald Wood MA PhD, Dr David Woodings MA, MRCP, MRCPath; Bernard Yallop PhD, BSc, ARCS, FRAS Professor John Yudkin MA, MD, PhD(Cantab), FRIC, FIBiol, FRCP.

The General Editor wishes particularly to thank the following for all their support:
Nicolas Bentley
Bill Borchard
Adrienne Bowles
Yves Boisseau
Irv Braun
Theo Bremer
the late Dr Jacob Bronowski
Sir Humphrey Browne
Barry and Helen Cayne
Peter Chubb
William Clark
Sanford and Dorothy Cobb
Alex and Jane Comfort
Jack and Sharlie Davison
Manfred Denneler
Stephen Elliott
Stephen Feldman
Orsola Fenghi
Professor Richard Gregory
Dr Leo van Grunsven
Jan van Gulden
Graham Hearn
the late Raimund von Hofmansthal
Dr Antonio Houaiss
the late Sir Julian Huxley
Alan Isaacs
Julie Lansdowne
Professor Peter Lasko
Andrew Leithead
Richard Levin
Oscar Lewenstein
The Rt Hon Selwyn Lloyd
Warren Lynch
Simon macLachlan
George Manina
Stuart Marks
Bruce Marshall
Francis Mildner
Bill and Christine Mitchell
Janice Mitchell
Patrick Moore
Mari Pijnenborg
the late Donna Dorita de Sa Putch
Tony Ruth
Dr Jonas Salk
Stanley Schindler
Guy Schoeller
Tony Schulte
Dr E.F. Schumacher
Christopher Scott
Anthony Storr
Hannu Tarmio
Ludovico Terzi
Ion Trewin
Egil Tveteras
Russ Voisin
Nat Wartels
Hiroshi Watanabe
Adrian Webster
Jeremy Westwood
Harry Williams
the dedicated staff of MB Encyclopaedias who created this Library and of MB Multimedia who made the IVR Artwork Bank.

Science and The Universe/Contents

Science

The Universe

Keystone

Lord Butler, Master of Trinity College,
Cambridge, knocks on the great door of
the college during his installation
ceremony on October 7, 1965

Preface

I do not think any other group of publishers could be credited with producing so comprehensive and modern an encyclopaedia as this. It is quite original in form and content. A fine team of writers has been enlisted to provide the contents. No library or place of reference would be complete without this modern encyclopaedia, which should also be a treasure in private hands.

The production of an encyclopaedia is often an example that a particular literary, scientific and philosophic civilization is thriving and groping towards further knowledge. This was certainly so when Diderot published his famous encyclopaedia in the eighteenth century. Since science and technology were then not so far developed, his is a very different production from this. It depended to a certain extent on contributions from Rousseau and Voltaire and its publication created a school of adherents known as the encyclopaedists.

In modern times excellent encyclopaedias have been produced, but I think there is none which has the wealth of illustrations which is such a feature of these volumes. I was particularly struck by the section on astronomy, where the illustrations are vivid and unusual. This is only one example of illustrations in the work being, I would almost say, staggering in their originality.

I think it is probable that many responsible schools will have sets, since the publishers have carefully related much of the contents of the encyclopaedia to school and college courses. Parents on occasion feel that it is necessary to supplement school teaching at home, and this encyclopaedia would be invaluable in replying to the queries of adolescents which parents often find awkward to answer. The "two-page-spread" system, where text and explanatory diagrams are integrated into attractive units which relate to one another, makes this encyclopaedia different from others and all the more easy to study.

The whole encyclopaedia will literally be a revelation in the sphere of human and humane knowledge.

Butler

**Master of Trinity College,
Cambridge**

The Structure of the Library

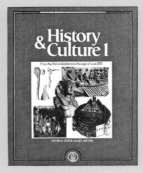

Science and The Universe

The growth of science
Mathematics
Atomic theory
Statics and dynamics
Heat, light and sound
Electricity
Chemistry
Techniques of astronomy
The Solar System
Stars and star maps
Galaxies
Man in space

The Physical Earth

Structure of the Earth
The Earth in perspective
Weather
Seas and oceans
Geology
Earth's resources
Agriculture
Cultivated plants
Flesh, fish and fowl

The Natural World

How life began
Plants
Animals
Insects
Fish
Amphibians and reptiles
Birds
Mammals
Prehistoric animals and
    plants
Animals and their habitats
Conservation

Man and Society

Evolution of man
How your body works
Illness and health
Mental health
Human development
Man and his gods
Communications
Politics
Law
Work and play
Economics

History and Culture

Volume 1 From the first
civilizations to the age of
Louis XIV

The art of prehistory
Classical Greece
India, China and Japan
Barbarian invasions
The crusades
Age of exploration
The Renaissance
The English revolution

Science and The Universe is a book of popular general knowledge about the physical sciences and astronomy. It is a self-contained book with its own index and its own internal system of cross-references to help you to build up a rounded picture of the subjects it covers.

It is one volume in Mitchell Beazley's intended ten-volume library of individual books we have entitled *The Joy of Knowledge Library*—a library which, when complete, will form a comprehensive encyclopaedia.

For a new generation brought up with television, words alone are no longer enough—and so we intend to make the *Library* a new sort of pictorial encyclopaedia for a visually oriented age, a new "family bible" of knowledge which will find acceptance in every home.

Seven other colour volumes in the *Library* are planned to be *Man and Society, The Physical Earth, The Natural World, History and Culture* (two volumes), *Man and Machines*, and *The Modern World. The Modern World* will be arranged alphabetically: the other volumes will be organized by topic and will provide a comprehensive store of general knowledge rather than isolated facts.

The last two volumes in the *Library* will provide a different service. Split up for convenience into A-K and L-Z references, these volumes will be a fact index to the whole work. They will provide factual information of all kinds on peoples, places and things through approximately 25,000 mostly short entries listed in alphabetical order. The entries in the A-Z volumes also act as a comprehensive index to the other eight volumes, thus turning the whole *Library* into a rounded *Encyclopaedia*, which is not only a comprehensive guide to general knowledge in volumes 1–7 but which now also provides access to specific information as well in *The Modern World* and the fact index volumes.

Access to knowledge

Whether you are a systematic reader or an unrepentant browser, my aim as General Editor has been to assemble all the facts you really ought to know into a coherent and logical plan that makes it possible to build up a comprehensive general knowledge of the subject.

Depending on your needs or motives as a reader in search of knowledge, you can find things out from *Science and The Universe* in four or more ways: for example, you can simply browse pleasurably about in its pages haphazardly (and that's my way!) or you can browse in a more organized fashion if you use our "See Also" treasure hunt system of connections referring you from spread to spread. Or you can gather specific facts by using the index. Yet again, you can set yourself the solid task of finding out literally everything in the book in logical order by reading it from cover to cover: in this the Contents List (page 6) is there to guide you.

Our basic purpose in organizing the volumes in *The Joy of Knowledge Library* into two elements—the three volumes of A-Z factual information and the seven volumes of general knowledge—was functional. We devised it this way to make it easier to gather the two different sorts of information—simple facts and wider general knowledge, respectively—in appropriate ways.

The functions of an encyclopaedia

An encyclopaedia (the Greek word means "teaching in a circle" or, as we might say, the provision of a *rounded* picture of knowledge) has to perform these two distinct functions for two sorts of users, each seeking information of different sorts.

First, many readers want simple factual answers to straightforward questions like "What is a rhombus?" They may be intrigued to learn that it is a four-sided plane figure with all of its sides equal and that a square is a rhombus with its interior angles right angles. Such direct and simple facts are best supplied by a short entry and in the *Library* they will be found in the two A-Z *Fact Index* volumes.

But secondly, for the user looking for in-depth knowledge on a subject or on a series of subjects—such as "What has man achieved in space?" short alphabetical entries alone are inevitably bitty and disjointed. What do you look up first—"space"? "astronautics"? "NASA"? "rockets"? "Skylab"? "Mariner"? "Apollo"?—and do you have to read all the entries or only some of them? You normally have to look up *lots* of entries in a purely alphabetical encyclopaedia to get a comprehensive answer to such wide-ranging questions. Yet comprehensive answers are what general knowledge is all about.

A long article or linked series of longer articles, organized

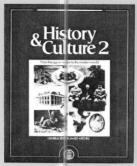    

History and Culture

Volume 2 From the Age
of Reason to the
modern world

Neoclassicism
Colonizing Australasia
World War I
Ireland and independence
Twenties and the
depression
World War II
Hollywood

Man and Machines

The growth of
technology
Materials and techniques
Power
Machines
Transport
Weapons
Engineering
Communications
Industrial chemistry
Domestic engineering

The Modern World

Almanack
Countries of the world
Atlas
Gazetteer

Fact Index A-K

The first of two volumes
containing 25,000 mostly
short factual entries
on people, places and
things in A-Z order. The
Fact Index also acts as
an index to the eight
colour volumes. In
this volume, everything
from Aachen to Kyzyl.

Fact Index L-Z

The second of the A-Z
volumes that turn the
Library into a complete
encyclopaedia. Like the
first, it acts as an
index to the eight
colour volumes. In this
volume, everything from
Ernest Laas to Zyrardow.

by related subjects, is clearly much more helpful to the
person wanting such comprehensive answers. That is why
we have adopted a logical, so-called *thematic* organization
of knowledge, with a clear system of connections relating
topics to one another, for teaching general knowledge in
Science and The Universe and the six other general knowledge
volumes in the *Library*.

The spread system
The basic unit of all the general knowledge books is the
"spread"—a nickname for the two-page units that
comprise the working contents of all these books. The
spread is the heart of our approach to explaining things.

Every spread in *Science and The Universe* tells a story
—almost always a self-contained story—a story on how
algebra works, for example (pages 34 to 35) or on the
nature of sound (pages 80 to 81) or on the evolution of
stars (pages 228 to 229) or on comets (pages 216 to 217).
The spreads on these subjects all work to the same discipline,
which is to tell you all you need to know in two facing
pages of text and pictures. The discipline of having to get in
all the essential and relevant facts in this comparatively
short space actually makes for better results—text that has
to get to the point without any waffle, pictures and
diagrams that illustrate the essential points in a clear and
coherent fashion, captions that really work and explain the
point of the pictures.

The spread system is a strict discipline but once you get
used to it, I hope you'll ask yourself why you ever thought
general knowledge could be communicated in any other way.

The structure of the spread system will also, I hope
prove reassuring when you venture out from the things you
do know about into the unknown areas you don't know,
but want to find out about. There are many virtues in
being systematic. You will start to feel at home in all sorts
of unlikely areas of knowledge with the spread system to
guide you. The spreads are, in a sense, the building blocks
of knowledge. Like living cells which are the building
blocks of plants and animals, they are systematically
"programmed" to help you to learn more easily and to
remember better. Each spread has a main article of 850
words summarising the subject. The article is illustrated
by an average of ten pictures and diagrams, the captions

of which both complement *and* supplement the
information in the article (so please read the captions,
incidentally, or you may miss something!). Each spread,
too, has a "key" picture or diagram in the top right-hand
corner. The purpose of the key picture is twofold: it
summarises the story of the spread visually and it is
intended to act as a memory stimulator to help you to
recall all the integrated facts and pictures on a subject.

Finally, each spread has a box of connections headed
"See Also" and, sometimes, "Read First". These are
cross-reference suggestions to other connecting spreads.
The "Read Firsts" normally appear only on spreads with
particularly complicated subjects and indicate that you
might like to learn to swim a little in the elementary
principles of a subject before being dropped in the deep
end of its complexities.

The "See Alsos" are the treasure hunt features of *The
Joy of Knowledge* system and I hope you'll find them
helpful and, indeed, fun to use. They are also essential if
you want to build up a comprehensive general knowledge.
If the spreads are individual living cells, the "See Alsos"
are the secret code that tells you how to fit the cells
together into an organic whole which is the body of
general knowledge.

Level of readership
The level for which we have created *The Joy of Knowledge
Library* is intended to be a universal one. Some aspects of
knowledge are more complicated than others and so readers
will find that the level varies in different parts of the
Library and indeed in different parts of this volume,
Science and The Universe. This is quite deliberate: *The Joy of
Knowledge Library* is a library for all the family.

Some younger people should be able to enjoy and to
absorb most of the pages in this volume on telescopes,
for example, from as young as ten or eleven onwards—
but the level has been set primarily for adults and older
children who will need some basic knowledge to make
sense of the pages on thermodynamics or biochemistry, for
example.

Whatever their level, the greatest and the bestselling
popular encyclopaedias of the past have always had one
thing in common—simplicity. The ability to make even

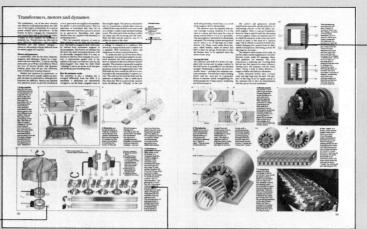

Main text Here you will find an 850-word summary of the subject.

Connections "Read Firsts" and "See Alsos" direct you to spreads that supply essential background information about the subject.

Illustrations Cutaway artwork, diagrams, brilliant paintings or photographs that convey essential detail, re-create the reality of art or highlight contemporary living.

Annotation Hard-working labels that identify elements in an illustration or act as keys to descriptions contained in the captions.

A typical spread Text and pictures are integrated in the presentation of comprehensive general knowledge on the subject.

Captions Detailed information that supplements and complements the main text and describes the scene or object in the illustration.

Key The illustration and caption that sum up the theme of the spread and act as a recall system.

complicated subjects clear, to distil, to extract the simple principles from behind the complicated formulae, the gift of getting to the heart of things: these are the elements that make popular encyclopaedias really useful to the people who read them. I hope we have followed these precepts throughout the *Library*: if so our level will be found to be truly universal.

Philosophy of the Library

The aim of *all* the books—general knowledge and *Fact Index* volumes—in the *Library* is to make knowledge more readily available to everyone, and to make it fun. This is not new in encyclopaedias. The great classics enlightened whole generations of readers with essential information, popularly presented and positively inspired. Equally, some works in the past seem to have been extensions of an educational system that believed that unless knowledge was painfully acquired it couldn't be good for you, would be inevitably superficial, and wouldn't stick. Many of us know in our own lives the boredom and disinterest generated by such an approach at school, and most of us have seen it too in certain types of adult books. Such an approach locks up knowledge instead of liberating it.

The great educators have been the men and women who have enthralled their listeners or readers by the self-evident passion they themselves have felt for their subjects. Their joy is natural and infectious. We remember what they say and cherish it for ever. The philosophy of *The Joy of Knowledge Library* is one that precisely mirrors that enthusiasm. We aim to seduce you with our pictures, absorb you with our text, entertain you with the multitude of facts we have marshalled for your pleasure—yes, *pleasure*. Why not pleasure?

There are three uses of knowledge: education (things you ought to know because they are important); pleasure (things which are intriguing or entertaining in themselves); application (things we can do with our knowledge for the world at large).

As far as education is concerned there are certain elementary facts we need to learn in our schooldays. The *Library*, with its vast store of information, is primarily designed to have an educational function—to inform, to be a constant companion and to guide everyone through school, college and other forms of higher education.

But most facts, except to the student or specialist (and these books are not only for students and specialists, they are for everyone) aren't vital to know at all. You don't *need* to know them. But discovering them can be a source of endless pleasure and delight, nonetheless, like learning the pleasures of food or wine or love or travel. Who wouldn't give a king's ransom to know when man really became man and stopped being an ape? Who wouldn't have loved to have spent a day at the feet of Leonardo or to have met the historical Jesus or to have been there when Stephenson's *Rocket* first moved? The excitement of discovering new things is like meeting new people—it is one of the great pleasures of life.

There is always the chance, too, that some of the things you find out in these pages may inspire you with a lifelong passion to apply your knowledge in an area which really interests you. My friend Patrick Moore, the astronomer, who first suggested we publish this *Library* and wrote much of the astronomy section in this volume on *Science and The Universe*, once told me that he became an astronomer through the thrill he experienced on first reading an encyclopaedia of astronomy called *The Splendour of the Heavens*, published when he was a boy. Revelation is the reward of encyclopaedists. Our job, my job, is to remind you always that the joy of knowledge knows no boundaries and can work untold miracles.

In an age when we are increasingly creators (and less creatures) of our world, the people who *know*, who have a sense of proportion, a sense of balance, above all perhaps a sense of insight (the inner as well as the outer eye) in the application of their knowledge, are the most valuable people on earth. They, and they alone, will have the capacity to save this earth as a happy and a habitable planet for all its creatures. For the true joy of knowledge lies not only in its acquisition and its enjoyment, but in its wise and loving application in the service of the world.

Thus the Latin tag "Scientiam non dedit natura, semina scientiae nobis dedit" on the first page of this book. It translates as "Nature has given us not knowledge itself, but the seeds thereof."

It is, in the end, up to each of us to make the most of what we find in these pages.

The Structure of this Book

Science and The Universe is a book of general knowledge containing all the information that we think is most interesting and relevant about the physical sciences, astronomy and the exploration of space. In its 304 pages it covers the history of those subjects, the several branches of physical science (mathematics, physics and chemistry), the structure of the universe—from the Moon to the farthest known galaxies—and the story of man's progress in charting and exploring the heavens. It has been our intention to present that mass of facts in such a way that they make sense and tell a logical, comprehensible and coherent story rather than appear in a meaningless jumble.

All too often we dismiss science either because we believe it to be far too complicated for our understanding or because we are uninterested in anything but its practical applications: we applaud the new mousetrap but ignore the scientific discovery that lies behind it. To some degree scientists are to blame for this—only the greatest of them have been able to communicate their discoveries in terms laymen can understand. But it is not altogether their fault; they do have a clear, precise language of their own— mathematics—and they fail only in translating the symbols into words, just as an Englishman might fail in translating "fair words butter no parsnips" into intelligible Japanese. The fact that scientific concepts may be difficult to communicate does not, however, excuse scientific illiteracy. In today's world, not to have some idea of what the theory of relativity is all about, or never to have heard of DNA, marks one down as an ignoramus, just as in an earlier age the man who had little Latin and less Greek was considered illiterate. One of the tasks we set ourselves in constructing *Science and The Universe* was to provide the basic scientific knowledge that everyone needs.

Where to start

Before outlining the plan of the contents of *Science and The Universe* I'm going to assume for a moment that you are coming to those subjects, just as I came to them when planning the book, as a "know-nothing" rather than as a "know-all". Knowing nothing, incidentally, can be a great advantage as a reader—or as an editor, as I discovered in making this book. If you know nothing, but want to find

things out, you ask difficult questions all the time. I spent much of my time as General Editor of this *Library* asking experts difficult questions and refusing to be fobbed off with complicated answers I couldn't understand. *Science and The Universe*, like every other book in this *Library*, has thus been through the sieve of my personal ignorance in its attempt to re-state things simply and understandably.

If you know nothing, my suggestion is that you start with the introductions by Sir Alan Cottrell (pages 16 to 19) and Sir Bernard Lovell (pages 160 to 163). Their views form useful frameworks for the two major sections of the book. If, however, you prefer to plunge straight into the book, but don't have much basic knowledge, I suggest you study eight spreads in the book before anything else (see panel on page 14). These spreads are the "Read First" spreads. They will give you the basic facts about mathematics, energy, the atom, chemistry, astronomical calculations and how the universe evolved. Once you have digested these spreads you can build up a more comprehensive general knowledge by exploring the rest of the book.

Plan of the book

There are broadly seventeen sections, or blocks of spreads, in *Science and The Universe*. The divisions between them are not marked in the text because we thought that would spoil the continuity of the book. They are:

The growth of science

The foundations of modern science are discussed in three articles that begin with man's first attempts to codify the world around him and end with the work of Newton and Lavoisier.

Mathematics

Mathematics is fundamental to science. This section defines mathematics, discusses its history, and explains its branches from arithmetic through algebra and geometry to topology. Many of our basic mathematical ideas derive from past civilizations—almost 6,000 years ago the Sumerians were using a system of numbers based on ten, as well as one based on sixty (this, the sexagesimal system, is still used in our measurement of time and angles), geometry owes its origin to floods in Babylon and Egypt, and the symbol for zero, without which modern mathematics would be impossible, was invented by an ancient Hindu civilization. Modern

Science and The Universe like most volumes in *The Joy of Knowledge Library*, tackles its subject topically on a two-page spread basis. Though the spreads are self-contained, you may find some of them easier to understand if you read certain basic spreads first. Those spreads are illustrated here. They are "scene-setters" that will give you an understanding of the fundamentals of life—its origins and continuity—and of how botanists and zoologists classify plants and animals. With them as background, the rest of the spreads in *Science and The Universe* can be more readily understood. The eight spreads are:

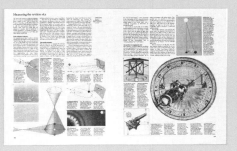

mathematical tools like calculus and logarithms date from discoveries made in Europe in the sixteenth and seventeenth centuries. *Science and The Universe* devotes thirty-four pages to setting out the principles upon which so much of the rest of the book depends.

Atomic theory
The first suggestion that matter might consist of separate particles was made in the fifth century BC, probably by Leucippus of Miletus, but it was not until the end of the nineteenth century that J. J. Thomson made the first discovery that revealed that atoms were themselves made up of even smaller particles. A whole branch of learning—particle physics—is now devoted to finding a single theory that will account for the behaviour of the multitude of particles that make an atom. In this section we examine the intellectual process that lead to the atomic bomb—and to nuclear power stations.

Statics and dynamics
Even when we are sitting still, we are subject to forces—gravity is acting upon our mass in an attempt to pull us down through the seat of the chair, and the chair is pushing upwards with an equal force. This situation is called equilibrium and the study of forces in equilibrium is called statics. The study of the forces that produce movement is called dynamics. Among the most puzzling of these forces are three natural forces—due to gravity, magnetism and electricity—which can act at a distance with no apparent connection between their source and the object they act upon. This section also discusses speed, acceleration and hydraulics—the study of forces and pressures within gases and liquids.

Sound
As well as introducing the physics of sound, and explaining the concepts of loudness, noise and frequency, this section discusses musical sounds—and explains why a clarinet, flute and oboe sound different.

Matter
All matter is either solid, liquid or gaseous. Some substances—such as water—can exist in all three states; some in only one or two, no matter what is done to them. Here you will find explanations of why air escaping from a car tyre is cold, why (unlike most substances) water occupies more space as a solid than as a liquid, why warm treacle pours more easily than cold treacle, and why an alloy is usually harder than the two or more metals of which it is composed.

Heat
Why a vacuum flask retains heat (or cold), how an oven thermometer works, how a car engine conforms to the first law of thermodynamics, why perpetual motion machines won't work—these are some of the topics covered in this section. There is a full spread on the concept of absolute zero—the point at which molecular movement would cease entirely. As absolute zero is approached, certain substances show remarkable properties—for example, some metals and alloys will allow an electric current to flow almost perpetually. The section also contains a spread on what happens at extremely high and extremely low pressures.

Light
Light is one of the most fascinating—and, to most people, most baffling—phenomena. In this section you will find spreads on colour, mirrors and lenses, light waves, the speed of light, the idea of relativity, light energy and energy from lasers.

Electricity and magnetism
In this section you will find an explanation of what electricity is; how magnets and electromagnets work; the principles behind transformers, electric motors and dynamos; the difference between alternating current and direct current; and the basic principles of electronics.

Chemistry
When we strike a match, or start a car engine, we are chemists—or, at least, we are initiating chemical reactions. But chemistry ranges from such simple things as dissolving sugar to the complexities of building a living organism. This section, which concludes the first half of *Science and The Universe*, begins with a definition of chemistry and ends with a chemical definition of life.

Techniques of astronomy
Opening with a history of how man has studied the heavens, this section goes on to look at the world's observatories and at the equipment modern astronomers use—the great optical telescopes and the huge radio installations that probe beyond the reach of the optical telescopes.

The Solar System
In about 5,000 million years from now, the Sun will have exhausted its supply of hydrogen and its structure will

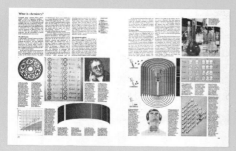

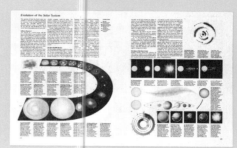
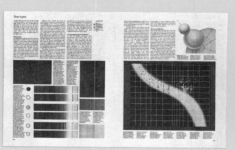
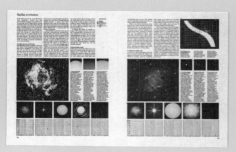

change. As it turns into a red giant star, the inner planets of the Solar System, including the Earth, will be destroyed. After discussing current theories about the birth and death of the Solar System, this section goes on to give a detailed treatment of the members of the Sun's family. There are spreads on missions to the Moon and to Mars, maps of both the Moon and Mars, and coverage of comets, meteors and meteorites.

The Sun

Three spreads examine the Sun in depth, through its spectrum, atmosphere and radiation, and discuss sunspots, eclipses and solar flares.

The stars

Just as man has been cataloguing the treasures of the earth for centuries, so has he been exploring, cataloguing and describing the myriad of pinpoints of light he sees in the night sky. This section begins by explaining how stars are classified by colour, goes on to discuss how they evolve, and describes pulsars, black holes and double stars. There are spreads on pulsating stars, galactic nebulae and stellar clusters.

Galaxies

Our galaxy—the Milky Way—is only one of many, some of them so far away that they can be detected only by their radio pulses. Having examined our galaxy—which contains about 100,000 million suns like our own—this section then discusses the local group of galaxies, the various types of galaxies, radio galaxies and quasars, and concludes by considering the theories that have been advanced about the origin of the universe itself.

Star maps

A spread on the constellations introduces six spreads that provide a detailed observer's guide to the northern and southern skies at various times of the year. These maps contain information that gives both the classification and magnitude of the principal stars.

Man in space

One of the most fascinating stories of our time is the story of the exploration of space, from the first tentative probes to a manned landing on the Moon and the Viking probes to Mars. *Science and The Universe* covers the history of those achievements and then looks into the future to speculate how man might eventually explore all the planets in the Solar System and then go out beyond the Solar System to other systems in our galaxy. The section concludes with a discussion on whether life exists on other planets.

Science and The Universe is a book about discovery, about man's search for those laws of nature that govern the world he lives in. It is an incredible story, and we are only at the beginning. I hope it invokes in you, as it did in me, a burning desire to find out what might happen next.

Science

Sir Alan Cottrell, FRS,

Master of Jesus College, University of Cambridge

Science is an endless voyage of discovery, a continual venture into the unknown, a quest to know and understand the world in which we live. It has taken men to far-off countries and wild places, and even to the Moon. It has discovered fascinating new worlds within the most ordinary things: a wayside flower, a bright pebble, a snowflake. It has taken us in imagination through the telescope, microscope and other scientific instruments to utterly remote places and conditions of existence: to the depth of outer space and the realm of the galaxies; to the infernal underworld a few kilometres beneath our feet; to the perpetual motions of molecules, atoms, nuclei and electrons; to the image of perfection itself, in crystals; to the ice ages and, much farther back in time, to the ages of the dinosaurs, to those of the very first living things and to that of the very beginnings of the earth itself; and it has also taken us to the marvellously self-reliant world within the biological cell.

Our most vital discovery about nature is that it can be studied scientifically. The questions put to it are answered faithfully, provided they are questions of plain fact – questions of "what" or "how" rather than "why" – and that they are put in the right way by means of accurate and impartial observations and well-planned experiments that carefully disentangle the required knowledge from all the confusion of extraneous factors. Knowledge gathered by these strict procedures, after checking and confirming by independent people, becomes the "basic stuff" of science, the "cold facts" of the world as it is seen through scientific eyes.

It is its firmness, reliability and independence of human opinion that gives science its reputation for objectivity and truthfulness. The ionization energy of the hydrogen atom, or the oxygen affinity of the haemoglobin molecule, for example, are exactly the same for Marxist and Liberal, Protestant and Catholic, Black and White, Israeli and Arab. The facts of science are "cold" precisely because they are not warmed by the heat of human passions.

But facts are separate specks of knowledge and scientific facts are often mere single numbers: for example the valency of sodium is +1; light travels through space at nearly 300,000km (186,000 miles) per second; there are 23 chromosome pairs in ordinary human body cells. This bulk of disconnected facts does not make a science. There has to be understanding as well as knowledge in order to gather the facts together into general overall pictures on the basis of common features between them. In a game of billiards there is almost never exactly the same situation twice. The positions and movements of the balls are continually changing through an endless set of possibilities. But there is an underlying regularity there, a uniformity in the way the balls collide and deflect off each other at different angles, which can be recognized instinctively when billiards is played as a game of skill rather than one of chance. The balls clearly move in accordance with one or more "natural" rules. The aim in science is to discover such natural rules, governing large numbers of facts, and to describe them in the most pure and precise way. In the early stages of discovering a rule, while the ideas are still speculative, a tentative rule is known as a hypothesis. Then, if the consequences deduced from it are found to agree with the facts, so that confidence in it begins to build up, it gradually becomes a theory and perhaps eventually even a near-immortal principle or law of nature.

In a healthily growing science there is a rough balance between the increase in the number of facts, through experiment and observation, and the decrease in the number of independent facts, through the unifying effect of more powerful new theories. It is a mistake to suppose that, as science progresses, more and more facts have to be learned. Much of the older scientific knowledge (for example, chemical properties of all the different elements; classification of species of animals and plants) no longer has to be learned in the detail that was once necessary, because it can now be seen "all of a piece" in the light of contemporary understanding of general principles such as atomic theory and the theory of evolution.

The most profound and firmly established generalizations are known as laws of nature, but they are always man-made ideas about the way things work and, as such, are liable to be disproved through the further progress of science, which brings in fresh facts and new insights. Newton's laws of motion, for example, fail for moving bodies approaching the speed of light and the law of conservation of mass fails when atomic nuclei split or fuse together.

A good scientific theory is in principle falsifiable, and the testing of theories to try to prove them wrong is one of the favourite and most fruitful of tactics in modern science. A theory that has stoutly resisted many such attacks nearly always shows itself, when it does finally give way, to be a partial aspect of some larger and more profound theory. Thus, the springboards of Isaac Newton's theory of mechanics and James Clerk Maxwell's theory of electromagnetism helped to launch Albert Einstein towards the profound new insights about space and time in his theory of relativity. Less directly, Thomas Malthus's theory of population helped to stimulate Charles Darwin's theory of the origin of species by means of natural selection.

Although scientific theories are human creations and thus subject to the frailties of our own limited powers of understanding and reasoning, they are nevertheless in general totally unbiased. A work on theoretical physics by a Soviet author will be admired and used, on its scientific merits, in the West just as a work by a Western author will be similarly appreciated by Soviet scientists. The language of science is universal and those who talk of "different sciences" for different social classes are talking of propaganda, not science.

Science is thought to have begun with the ancient Greeks, although the Chinese independently made important early contributions, particularly in astronomy. By 200 BC the Greeks had already measured the circumference of the earth accurately. But science withered under the Romans and then lay moribund so far as Europe was concerned (although the Arabs preserved the knowledge in North Africa) until the Renaissance revived man's spirit of enquiry and reminded him of the treasures that had remained locked up in old Greek and Latin texts. The invention of printing and the discovery of the New World opened new casements in men's minds. But more than anything else, the event that began modern science was the publication of Nicolas Copernicus' book *De Revolutionibus Orbium Coelestium* (1543). This not only set

Flashes of lightning outshine all the street lights of a town; in fifty-millionths of a second they heat the air in their path to more than 30,000°C.

Flashes of lightning outshine all the street lights of a town; in fifty-millionths of a second they heat the air in their path to more than 30,000°C.

the Sun at the centre of the Solar System, it also introduced new principles of the scientific method: the importance of simplicity in explanation; the recognition that only relative positions and motions are significant; and the acceptance that man is not at the centre of the universe, or indeed anywhere special at all. These intellectual tools are even today capable of fashioning new scientific insights.

The real flowering of science came in the seventeenth century. Right at its beginning, William Gilbert explained the north-seeking behaviour of the magnetic compass on the basis that the earth itself is a bar magnet. Soon afterwards William Harvey proved the circulation of the blood and the early microscopists discovered biological cells. In the meantime, Galileo Galilei and Isaac Newton had constructed their epoch-making theories of mechanics and gravitation. Chemistry developed a little later, out of an ancient alchemy, but Antoine Lavoisier and Joseph Priestley carried it forward in the eighteenth century and opened the way for John Dalton's atomic theory of chemistry in the nineteenth century.

The pace accelerated throughout that century. Magnetism and electricity were soon related to each other and in the mid-century Clerk Maxwell's theory showed that light is an electromagnetic wave motion in space. Geology, palaeontology and the biological sciences also advanced rapidly and prepared the ground for Darwin. With the Industrial Revolution came a clarification in the understanding of energy and the thermodynamical laws that govern its conversion from one form to another, a branch of science of great importance for the practical task of converting heat into work in heat engines. At the same time, the atomic hypothesis proved extremely fruitful when used in the "kinetic theory of gases" and it later enabled the laws of thermodynamics to be explained in terms of a new theory of "statistical mechanics". By the end of the nineteenth century the atomic nature of matter was virtually proven, and already the first of the modern "elementary particles" (the electron) had been discovered by J. J. Thomson. In biology, Louis Pasteur proved that micro-organisms cause diseases and Gregor Mendel discovered the genetic basis of heredity.

Physics seemed to have exhausted itself by the end of the nineteenth century, but this was only a short pause before it moved forward again along much more revolutionary lines. The two decisive moves came in Max Planck's quantum theory, which shows that nature acts in steps, not smoothly; and Einstein's relativity theory, which shows that time itself varies its pace for things in relative motion. From these grew still more radical ideas about nature, in particular quantum mechanics (also called wave mechanics), which shows that extremely small things, such as electrons, are neither particles nor waves, but something that can behave equally well as either in different experimental situations, and also general relativity theory, which shows gravitation to be a kind of warping of space and time.

Following these new insights, there have been advances in our knowledge and understanding of the structure of the atom and its elementary particles at one extreme, and of the structure of the universe and its "particles" – galaxies and stars – at the other.

In its simpler aspects the quantum atomic theory has now explained chemistry and in its complex aspects it has gone far into explaining the basic biological processes of reproduction, heredity and genetics in the new science of molecular biology. The atomic theory has also explained innumerable facts about the structure and properties of ordinary matter. Other sciences have also advanced in the twentieth century; notably geology and biology.

At the present time there are three main frontiers of fundamental science. First, the world of the extremely large, the universe itself. Radio telescopes now gather information from the far side of the universe and have detected, as radio waves today, the "firelight" of the big bang which probably

A laser beam punches a hole in the hardened steel of a razor blade; even light, when sufficiently concentrated, can be a potent kind of energy.

started off the expanding universe nearly 20,000 million years ago. Second, the world of the extremely small, that of the elementary particles such as electrons, protons, neutrons, mesons, and others. It is still difficult to understand why nature forms particles with particular masses, electrical charges and other properties, but several types of particles are now seen to be related to one another by means of enigmatic symmetries. There is a great riddle still to be solved before the meaning in these mathematical patterns can be read.

The third frontier is the world of complex matter. It is also the world of "middle-sized" things, from molecules at one extreme to the earth at the other; and it is the world of familiar things, such as raindrops, plants, animals and human beings. Because of its familiarity it contains some of the oldest, most "classical" parts of science. But it also contains the deepest scientific challenges of all, the evolution of life and the nature of thought processes. As regards the evolution of life, remarkable progress has been made in the past few years in proving that the simpler kinds of biological molecules can and are formed spontaneously by conventional chemical processes. This is the beginning of a coherent scientific story that carries evolution right through from simple molecules, such as water, methane and carbon dioxide, to advanced animal life. The problem of the origin and nature of the mind seems much more difficult and it is hard to guess what progress may be made within it in the next few years.

But this is all fundamental science, which is the heart of science. However, it is a small heart that lies at the centre of the much larger body of applied science. The overwhelming majority of all the scientists that have ever been are alive today and most of these are working in technology (applied science) because it is here that they can be of practical value to mankind in developing new agricultural crops and many other important attributes of modern society.

Applied science sets out what can be done and technology shows how to do it; but neither specifies what is to be done, for that is a matter of politics and ethics. The proper use of science sets problems for modern man, but the world will undoubtedly come to depend on science even more than it does at present if the growing population is to be fed, clothed, housed and given a fair standard of living.

Science and technology have provided mankind with many benefits: improved health, longer lives and relief from pain; shorter working days under safer and easier conditions; more and better food, clothes and housing; greater opportunities for education, entertainment, and travel. Most of man's other activities have been profoundly affected by science and technology; this applies even to the arts. The Impressionist movement in the 1800s was made possible by Winsor and Newton's introduction of the collapsible tin tube for paints, which enabled artists to get out of the studio and paint in the open air.

While science and technology have not been able to improve human nature itself or make various parts of society work together harmoniously, they are nevertheless a potent force for change in the outlook of mankind. Cheap radios, newspapers, travel, bring all parts of the world into close touch with each other, into an instant global debate on major topics. This has transformed international politics. Events in, for example, the Middle East, Northern Ireland, South-West Africa, or the Russian-Chinese border immediately reverberate round the world, not only through Foreign Offices, but among the public at large.

The main economic effect of science and technology has been to increase enormously our ability to create wealth, so that the average man in the industrially advanced countries has increased his real earning power ten times over the years since the Industrial Revolution. In Western countries, the science of wealth-creation came before the big populations, so that most of our technology has been directed towards labour-saving, energy-using, highly mechanized, industries. For several reasons, this is likely to change in future, and a lot more applied science will be needed to bring it about. The first reason is that the densely populated poor countries need industries in which the cost of the tools to give a man a job is only a fraction of that in the industrially advanced countries. This calls for an "intermediate technology", to stimulate new job-creating industries. The electronics industry, which requires good eyes and dextrous fingers, is an example. The second reason for the expected change in technology is that unemployment is growing so much in industrially advanced countries that, even in these, new job-creating industries are now needed. The third is that modern man has been gobbling up fuel and raw materials at an alarmingly increasing rate. New kinds of industries, which depend more on human skill and brains than on the mechanical turnover of energy and materials, are needed.

Modern man is an overpopulated species, living beyond his means on a planet poor in available resources but rich in potential resources that are beyond his technological reach today. There never has been a time when man was more dependent on science and technology for his future well-being than the present.

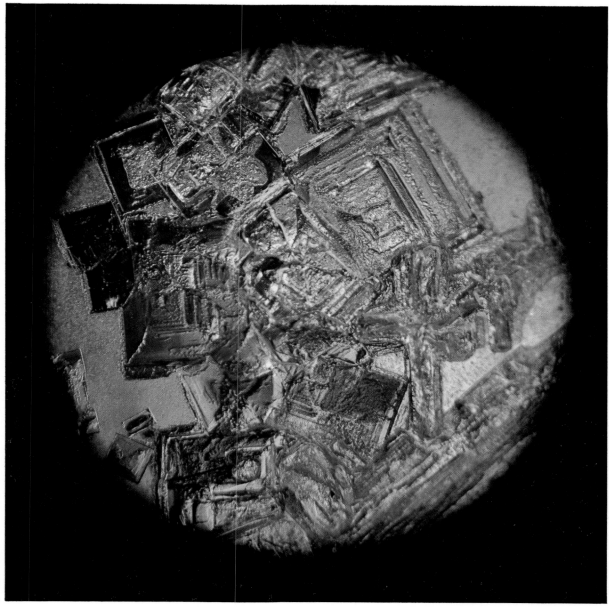

Crystals of salt, seen here through a microscope "growing" from solution, reveal by their shapes the orderly array of the molecules of which they are made.

Prehistoric and ancient science

Nearly two million years ago the ancestors of man used stones as weapons and tools. The need to master the environment and develop tools for the purpose involved a primitive knowledge of science [Key]. Ever since that time the development of science, technology and civilization have been interrelated.

Science of prehistoric man
After perhaps half a million years, the descendants of these first tool-users had become more selective, concentrating on flint, with its useful cutting edges [1, 2]. They discovered how to make fires.

Cave paintings of 15,000 years ago indicate a primitive knowledge of animal anatomy. Some, which show prehistoric elephants with the position of the animal's heart indicated by arrows, may be a record of hunting prowess, or a form of sympathetic magic attempting to influence the hunt.

About 10,000 years ago, men began to adopt a more settled way of life. They invented a repetitive system of food production that involved the domestication of animals and the cultivation of plants. Life in such

settled communities stimulated invention in various building materials, used to provide better shelter and protection. Fires were kept burning continuously for warmth and cooking and frightening off marauders, and it is possible that the women who tended the fires were the first to notice that when clay is left near a fire it hardens into pottery. When certain stones and earths were roasted in a fire, heavy liquids sometimes ran out which, when cool, hardened into useful metals.

Egypt and Mesopotamia
The people who settled in the Nile valley found themselves in an area that was exceptionally safe and fertile. They noticed that the silt brought down by the yearly floods renewed the fertility of the soil so they dug channels and built embankments to divert the fertile silt-laden flood waters onto their fields. These flood-control operations marked the invention of large-scale engineering. The Nile valley dwellers later applied engineering techniques in the construction of the pyramids [5].

It is probable that the science of geometry

arose from the need to fix positions where landmarks had been washed out by annual flooding. Arithmetic was developed to calculate quantities of crops so that they could be shared among the people. Egyptian arithmetic depended on a method of doubling, much as in the operations of a modern computer. Geometrical surveying led the Egyptians to a good method of calculating the area of a circle; they assumed the circumference of a circle to be the square of eight-ninths times its diameter. They also needed to determine the seasons and the times of the Nile floods and so devised a calendar of 365 days.

The people of Mesopotamia, in the double valley of the rivers Tigris and the Euphrates, were developing in a similar way but under rather different conditions. There was little stone in their land, so they recorded information by making marks on soft clay tablets, which were then baked. The Mesopotamians introduced the idea that the value of a digit depends on its position in a number and they even solved algebraic equations. The Egyptians were primarily concerned with simple, practical calculations

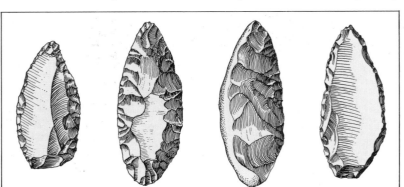

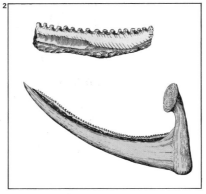

1 **Tools of the middle Palaeolithic** or Mousterian stage in France date from 70,000 to 32,000 years ago. There were various human groups living in Europe in this period. They left many traces of their culture in shelters and entrances to caves. Tools are found dating from the beginning of the last period of glaciation and the scrapers and knives shown here illustrate fine stone tool technology.

2 **Bones, antlers and wood** were used by early men as raw materials for tools and weapons. A wooden tool could be given an efficient cutting edge by adding a row of suitably shaped flint slivers, as in this early Egyptian wooden sickle from about 3000–2500 BC. Pointed wooden implements could also be hardened by charring the point in a fire. Antlers were shaped by carving or heating.

3 **The Neolithic people** who occupied the region of Stoney Littleton, England in about 3000 BC, constructed large barrows for the burial of their dead [A]. A barrow is a long structure with an entrance passage crossed by transept chambers; this barrow has three pairs. The passage [B] contains a vault with corbels or projections from the walls to carry "capstones", in a manner similar to those in vaults found on islands lying off the coast of Scotland.

whereas the Babylonians were much more sophisticated, especially in connection with the science of astronomy.

One of the first Greeks to visit and study other civilizations was Thales of Miletus, who was born in about 630 BC. He returned from Egypt well versed in the techniques of Egyptian geometry. From experience in building, the Egyptians had learned that if a triangle has two sides of equal length, then the angles at its base are also equal. Thales looked for a way of proving this fact. He made two identical triangles, each with two equal sides, and found that when one was picked up and turned over, it could be laid on top of the other and fitted exactly. In this way, a mathematical proof was developed.

Famous Greek scientists

To Pythagoras, born about 60 years after Thales, is attributed a proof of the famous theorem that the square on the longest side of a right-angled triangle is equal to the sum of the squares on the other two sides. Pythagoras sought to explain the properties of matter in terms of numbers.

Another Greek, Euclid (born 330 BC), provided the basic principles for the teaching of classical geometry that have been used ever since. Less than 50 years later, in 287 BC, Archimedes was born in Sicily. He applied the new mathematics with extraordinary power and logic and made many inventions. He established the principle that, when a body is weighed in a liquid, its apparent loss in weight is equal to the weight of the liquid displaced; he is credited with inventing a screw for raising water [7] from one level to another; and he succeeded in launching a large ship using levers.

Astronomy was first placed on an adequate scientific basis by Endoxos, who was born in about 408 BC. He showed that the motions of the Sun and planets could be explained by assuming that they moved with uniform motion in perfect circles, the centres of which are near, but not exactly at, the centre of the Earth. Later Greek astronomers arranged far more complex systems of circular motions, equalling in accuracy the work of Nicolas Copernicus (1473–1543), nearly 2,000 years later.

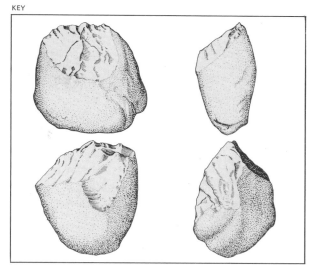

These Oldowan tools were first discovered by Louis Leakey (1903–72) in 1931 in the Olduvai Gorge in northern Tanzania, East Africa. They range from simple broken pebbles to chopping tools, and are 1.2 million to 1.8 million years old.

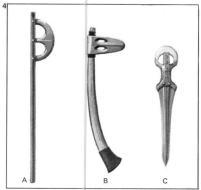

4 The introduction of metals gave early toolmakers much more manageable raw materials – first soft metals such as gold and copper, later bronze and finally iron. Shown here are an Egyptian eye-axe [A] from Megiddo (c. 1900 BC), an Egyptian duck-bill axe [B] from Ugarit (c. 1800 BC) and an Egyptian bronze dagger [C] of the Hyksos period (c. 1650 BC). But iron made the best cutting edge.

5 The Great Pyramid in Egypt was erected by command of the pharaoh Khufu (in Greek, Cheops) of the 4th Dynasty, about 3000 BC. It contains 6.5 million tonnes of limestone. A primary purpose of the pyramids was to provide grand tombs. They may also have had other purposes. The whole group of the major Egyptian pyramids was built within little more than a century. They involved an enormous concentration of labour and it has been suggested that their construction may have provided a convenient means of organizing the whole population of Egypt, creating a centralized state – hence the short building period.

6 The Mesopotamians, lacking the Egyptians' papyrus for writing, made records on clay tablets. Characters were formed by pressing the wedge-shaped end of a stylus into the soft clay and a permanent record could be obtained by baking the clay in a fire until it became hard. From its wedge shape, the writing is known as cuneiform. The associated number system used strokes that, to our eyes, faintly resemble the Roman numerals of 30 centuries later. A numeral's position determined its value.

7 An Archimedean screw is used for raising water. Archimedes (287–212 BC) was a master both of the most refined mathematics and of practical invention. It is said that he invented the screw for raising water to assist irrigation in Egypt. It consists, in principle, of a wedge that can exert sustained pressure in a particular direction by being continuously revolved. In the machine shown in this illustration the water is raised from the sink on the left. When the curved pipe is rotated by pulling the rope, the water is pushed up inside it and delivered into the tank at the top on the right. Screws were not known before Archimedes: those for fastening objects probably arose from his device.

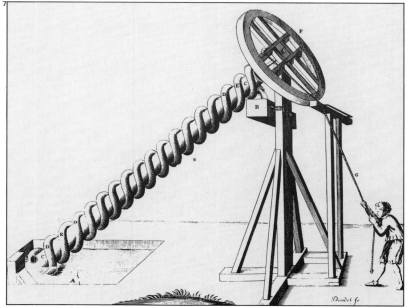

Asian and medieval science

By the fifth century AD the long-ailing Roman Empire lost control of Western Europe. It was overcome by its excessive size, by the "softness" of its citizens, and by a population explosion in Asia that propelled vigorous new peoples against Rome's extended frontiers. At the same time the Eastern Roman Empire, whose capital was at Byzantium (later Constantinople and now Istanbul), flourished until a new attack from the east in the seventh century.

International influences

The Arabs emerged from Arabia as the followers of the new prophet Mohammed (AD c. 570–632), a trader from Mecca. Within about 100 years they captured much of the Middle East and North Africa and invaded Spain and even France [2]. These new conquerors had no culture of their own, but they borrowed learning from Syriac, Greek, Indian and other peoples whom they encountered through their conquests and travels. They became the founders of the internationalism that is one of the most striking features of science. They aimed at all-embracing knowledge and perhaps Avicenna (980–1037) came nearest to attaining it.

The Muslims had a strong trading tradition. Like other peoples, they were interested in the exact calculation of shares in goods and the allocation of family inheritances. They assessed the declining value of female slaves in much the same way as cars are depreciated today. When they invaded India, they discovered Indian mathematics.

The Indians introduced the number system now universally in use, together with the zero symbol and decimals. Their work became known to Al-Khwarizmi (780–c. 850), the greatest of the Arab mathematicians. He was librarian to the caliph Al-Mamum (786–833) in Baghdad and published in 830 his treatise on *Al-jabr wa'l muqabala*, from which the word algebra is derived. He studied various classes of quadratic (second order) equations and called the unknown quantity to be calculated "the root". Knowledge of mathematics in medieval Europe was based mainly on Latin translations of his works. One of his most famous successors was the Persian poet Omar Khayyám (c. 1048–1122), who dealt with special classes of cubic equations.

The Arabs devoted much attention to pharmacy [5] and to astronomy. They calculated elaborate trigonometrical tables, which were used to determine the exact times of prayers and to navigate the Indian Ocean. Córdoba in Spain became the most advanced intellectual centre in Europe. The dependence of medieval Europe on Arab knowledge is illustrated by the example of Adelard of Bath, who went to Córdoba disguised as a Muslim student in 1120. He returned to England with a copy of Euclid's book, which served as a mathematical text for feudal Europe for the next four centuries.

Chinese science and technology

The Arabs also brought knowledge of Chinese inventions and discoveries to Europe, including gunpowder, the magnetic compass, printing with movable type and an efficient horse harness.

The Chinese invented an escapement mechanism for a water clock [1], generally credited to Yi Hoing, in 725. It enabled them

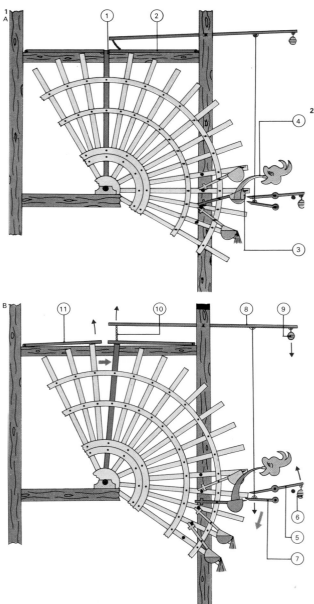

1 This clock escapement mechanism was invented in China in the 8th century. [A] A spoke [1] is arrested by a lock [2] while a scoop [3] fills with water [4] from a tank at a constant rate. The lock is released [B] when the filling scoop trips a checking fork [5], overcomes its counterweight [6] and trips a coupling tongue [7], which pulls down an upper lever [8] with its own counterweight [9]. This jerks a chain [10], freeing the lock and allowing the wheel to swing clockwise until the lock drops again, arresting the following spoke which is also held steady by a ratchet lock [11].

2 The Islamic Empire, by the 8th century, stretched from India to the Pyrenees, making major contributions to mathematics and chemistry and laying the basis for the international spread of learning.

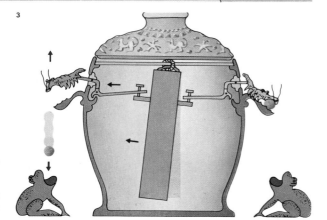

Toledo
Córdoba
Granada
Algeciras
Tunis
Bukhara
Tus
Damascus
Baghdad
Jerusalem
Kufa
Basra
Alexandria
Cairo
Mecca

The Islamic Empire at its greatest extent
0   1,000km

3 A seismograph built in 132 by the Chinese scientist Chang Heng was a vase with a ring of holes round the rim. Metal balls lightly held in each hole fell into receptacles below when there was an earth tremor. It indicated the direction of the tremor according to which balls fell and which did not. Wang Chen-To attempted a reconstruction of the internal mechanism, consisting of a pendulum with arms and cranks governing the motion of the balls.

to build the first accurate mechanical clocks. Chinese science also produced the first seismograph, for detecting earthquakes [3], built by Chang Heng in 132. In 1054 they observed the great new star – a nova, the parent of the Crab Nebula – later to become one of the most important objects in the development of modern radio astronomy.

The Arabs performed a unique service as world informants on ancient and contemporary science. They introduced the ideas of India and China to the Western world. But they were not the founders of modern science, for this arose in Europe.

Medieval science

The Roman Empire in Europe disintegrated into a multitude of individual strongholds of local military chieftains. Slaves became bandits, or peasants tied to the land, but no longer slaves. So the feudal period began. People who settled around the strong-points came to be known as the bourgeoisie, because they lived outside the castle or burg. Many of them were craftsmen. They were dependent on what they could learn from the

Arab encyclopaedists but they looked at the old knowledge in a more individual way. One such man who played a significant role in the advance of medieval science was Leonardo Fibonacci of Pisa (c. 1180–1250). His father was employed on the Barbary coast and there Leonardo learned the Arabic language and arithmetic. On his return to Pisa he introduced Arabic numerals into Europe.

The most eminent of English medieval scientists was Roger Bacon (1214–94). He proposed combinations of lenses for telescopes and microscopes and may have been the first to suggest spectacles [4].

Roger Bacon said that the only man he knew who was to be praised for his experimental science was Petrus Peregrinus of Maricourt, who published a treatise on magnetism in 1269. He explicitly pointed out the importance of manual skill in science. He was one of the forerunners of modern science, in which experiment and theory are equally balanced. This became possible only through the emancipation of the craftsman in medieval Europe, a change that was to lead to the Renaissance.

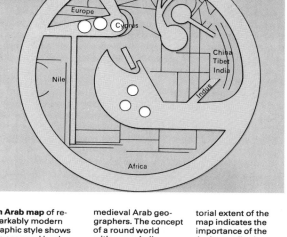

An Arab map of remarkably modern graphic style shows the seas and land masses known to medieval Arab geographers. The concept of a round world with an encircling ocean and the territorial extent of the map indicates the importance of the Arab contribution to man's knowledge.

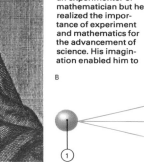

4 Roger Bacon [A] was not personally an experimenter or mathematician but he realized the importance of experiment and mathematics for the advancement of science. His imagination enabled him to make a remarkable collection of scientific suggestions, culled from many sources and ranging from vague hints to clear diagrams. He gave substantially correct optical explanations of [B] why a spherical flask of water [2] acts as a burning glass [3] to concentrate the rays of the sun [1] and [C] how a convex lens [4] produces a magnified image [5] of an object [6] beneath it.

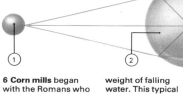

5 A typical pharmacy shop of the Middle Ages was based on Arab traditions. The Arabs brought knowledge of Persian and Indian drugs and spices such as camphor, cloves, cassia, nutmeg and senna. Their influence on European pharmacy was exerted chiefly through Benedictine monks. In the 14th century pharmacy, medicine, chemistry and the grocery business were combined and apothecaries and grocers organized themselves in guilds.

6 Corn mills began with the Romans who spread techniques that helped them to exploit their empire. They built undershot and overshot water mills, one driven by the momentum of flowing water and the other by the weight of falling water. This typical late example has an undershot drive [6] with a hopper [1] for the corn and a chute [2] conveying it to grindstones [3]. The flour produced fell into a chute [4] and then poured into a bag [5].

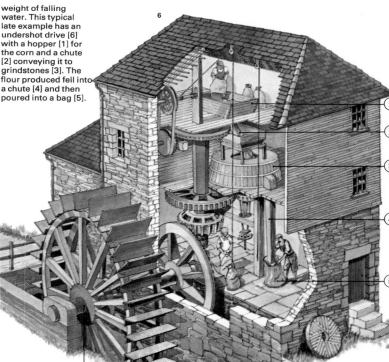

Alchemy and the age of reason

Men gained some practical knowledge of the working of materials in early times but work in the crafts (what is now called technology) was regarded as a lowly pursuit. One reason for this attitude was the disagreeable working conditions associated with them. An Egyptian scribe of 1500 BC noted that the metalworker "stinks like fish-spawn".

Alchemy and the Renaissance

In the second century AD, Diocletian (245–313) ordered that all books on the working of gold, silver and copper should be destroyed to prevent counterfeiting and inflation. The effect was to reduce rational research on practical problems and to increase interest in magic as a method of transmuting base metals into gold. The centre of the development was Alexandria and the Arabs called the new science *alchemy* after Khem, or "black", the name given to Egypt because of its black earth.

The Alexandrians invented apparatus for heating, melting, filtering and distilling substances. They introduced the glass flasks and retorts still typical of chemical laboratories.

The Arabs [1] adopted, extended and transmitted these advances. Their greatest chemist was Jabir ibn Haijan, or Gebir (*c.* AD 721–817), who worked on the transmutation of metals and propounded a theory of their constitution that was not completely superseded until the eighteenth century. Besides being familiar with chemical operations such as crystallization, solution and reduction, he attempted to explain them. His most useful discovery was nitric acid.

Modern science was founded during the Renaissance in the urban society of Italian cities where craftsmen became emancipated and even famous. The supreme example was Leonardo da Vinci (1452–1519) who knew little Latin and no Greek but analysed technical processes scientifically.

Copernicus and Galileo

Nicolas Copernicus (1473–1543) was a Polish–German scholar who studied at Cracow and Bologna in the 1490s. He noted from astronomical references in Latin and Greek literature that Heraclides (388–315 BC) had assigned a motion to the earth "after

the manner of a wheel being carried on its own axis". Copernicus found "by much and long observation" that a consistent account of the movements of the planets could be given on the basis that the earth revolves around the sun. The account of this in Copernicus' treatise *De Revolutionibus Orbium Caelestium* was published in 1543 when he was on his deathbed. The Copernican theory [Key] is perhaps the most important scientific theory in history for it changed man's conception of his place in the universe. Formerly man had believed that the universe revolved around the earth and himself; now he realized that man was but a minute incidental speck in a universe of almost inconceivable vastness.

The Renaissance effort in science was completed by Galileo Galilei [2] who cleared the way for modern science. Copernicus discovered how the Solar System works but Galileo gained the first precise knowledge of how things on the earth move. He was born in Pisa in 1564, the same year as Shakespeare, and died in 1642, the year in which Isaac Newton was born. He went to the local

1 Alchemical processes are represented by the figures on this Arab manuscript reflecting cross-cultural influences. The Arabs gained their main introduction to alchemy through Alexandria and spread it to Western Europe in about the eleventh century. Although they improved on the experimental techniques of the Alexandrians they did not escape the influence of their mystical theorizing which was based on animistic beliefs in objects possessing souls.

2 Galileo [A] was 18 when he discovered the constancy of the swing of a pendulum, which he later adopted into a pendulum clock [C]. His development of the telescope led to original observations on the planets, including the discovery of the satellites of Jupiter, first described in a pamphlet of 1610, a page of which [B] is shown here. His conversion to Copernican theory alienated the Church.

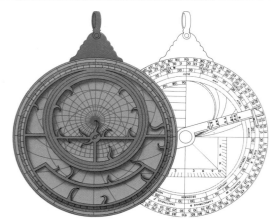

3 The astrolabe enabled an observer to find, at any instant, the position of a known point or object in the sky. It was much used for astrological predictions by the Greeks and Arabs and was introduced to Europe in about the tenth century. Two metal discs bore projections of the celestial and terrestrial spheres. A rotating arm on the back enabled the user to set the inclination of an object from the horizon and to calculate various angles.

university. According to legend, a swinging lamp in the cathedral attracted his attention. He noticed that the time of the swing was independent of the size of the swing (its amplitude). When he arrived home he checked the fact with a bullet and a piece of string. He later used this fundamental property of a pendulum in designing a pendulum clock.

When Galileo was appointed professor he was obliged to teach Aristotelian science. This caused him to make a careful study of Aristotle's ideas, especially those on the motion of objects. Aristotle (384–322 BC) based his theory on the assumption that objects fall with a speed proportional to their weight. Galileo devised experiments to measure exactly how fast objects do fall and found that all that fall freely do so at the same speed. He made many other discoveries. The results of his application of the telescope to astronomical observation were particularly spectacular evidence that Aristotle's picture of the universe was incomplete and mistaken. Galileo did not fully grasp the theological implications of this.

Galileo's demonstration that the move-ment of objects could be exactly determined by a combination of experiment and mathematical reasoning was extended by Isaac Newton (1642–1727). Newton [5] showed that all the then known physical aspects of the universe and nature could be completely described by mathematical theory utilizing laws consistent with experience. Newton's account, in his *Philosophiae Naturalis Principia Mathematica* (1687), is possibly the greatest single intellectual effort yet made.

The Age of Reason

Newton's achievements increased confidence in the power of human reasoning. They had a particularly striking effect in France where Pierre Laplace (1749–1827) and Joseph Lagrange (1736–1813) extended the Newtonian theory and its supporting mathematics. The new confidence in experiment and calculation developed also in other sciences. Antoine Lavoisier (1743–94) revolutionized chemistry, dispatching the magical and mystical remnants from it and in doing so laying the foundations of modern chemistry [6].

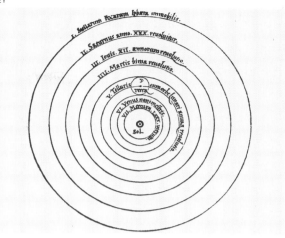

Copernicus pictured the planets as revolving around the sun in a complicated pattern of circular motions, the basis of which is shown here.

His theory did not give more accurate planetary predictions than Ptolemy's of AD 140, but it was a triumph of ideas that showed man in his

true place in nature. Proof that the planets revolve in ellipses came in 1609 from the German astronomer Johannes Kepler (1571–1630).

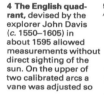
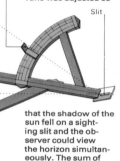

4 The English quadrant, devised by the explorer John Davis (c. 1550–1605) in about 1595 allowed measurements without direct sighting of the sun. On the upper of two calibrated arcs a vane was adjusted so that the shadow of the sun fell on a sighting slit and the observer could view the horizon simultaneously. The sum of the two readings gave the zenith distance, from which the latitude was deduced.

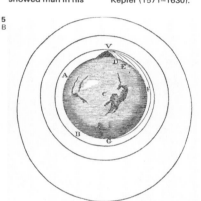

5 Isaac Newton [A] calculated the speed at which a body projected horizontally from the top of a mountain would leave the earth and begin revolving around it. His diagram of the path of an artificial satellite [B] was published in 1728, the year after his death. Newton's genius emerged soon after he left Cambridge University. Within a few years he had laid the basis of differential and integral calculus, elucidated the nature of light and colour and had begun to explore the usefulness of mathematical analysis to physical theories. His main achievements in physical science were expressed in his *Principia* (1687) and *Optiks* (1704).

6 Antoine Lavoisier [B] founded modern chemistry by means of experiments leading to his theory about the nature of combustion. This had previously been ascribed to the transfer of a substance called phlogiston, the main agent of chemical

change sometimes released as fire. Lavoisier heated mercury and air in a flask with a curved neck, which enabled him to measure exactly the decrease in volume of gas and gain in weight of mercury during 12 days' heating. By

means of this apparatus (illustrated here from Lavoisier's own diagram [A]) he showed that the changes could be explained completely in terms of the active constituent of the air discovered by Joseph Priestley (1733–1804), to which

Lavoisier gave the name oxygen. The idea of phlogiston was superfluous and chemistry could develop as a rational science based entirely on quantitative measurements. Experiments gave rise to theories that were then tested by other experiments.

Mathematics and civilization

Mathematics is a continuously expanding system of organized thought. It is employed in science, technology, art, music, architecture, economics, sociology, sport – in fact, in almost every aspect of human activity – and has influenced, and often determined, the direction of philosophical thought concerned with man and his universe. Throughout history mathematics has not only reflected developments in civilization but also made a major contribution to those developments.

Algebra, geometry and calculus

There are three major aspects of mathematics. The assembling and combining of sets of objects led to concepts of number [1], computation and the algebra of number theory. Concern with the measurement of time and space led to geometry, astronomy and chronology. The struggle to understand ideas of continuity and limit led to mathematical analysis and the invention of calculus in the seventeenth century. These three aspects of mathematics overlap considerably. There are now algebras of sets, logic, vector and transformation geometry, probability theory

and a host of specialities that employ the concepts of other fields of study.

Everything natural or man-made has a structure comprising elements that are related in some special way [Key]. A rock crystal, a plant [6], a spaceship and a political system each has a structure, the study of which is mathematical. Mathematics is the result of the thought process known as abstraction, in which activities related to a physical structure can be organized in such a way that the physical structure is replaced by a mental one, an abstract mathematical model. The power of mathematics is further demonstrated when abstract concepts, such as those of number and space, can be represented by concrete symbols, which may be algebraic, geometric or graphical [3].

Mathematics can be described as a form of inquiry made according to defined rules for drawing conclusions from accepted mathematical truths. History shows, however, that mathematics is also a field of creative activity employing great flights of intuition and imagination [7]. The driving force for the creativity is usually the need man has

to solve the problems of his society. But the motivation may also simply be the challenge of intellectual activity for its own sake.

The first mathematicians

All primitive civilizations developed concepts of number and measure as soon as trade progressed beyond the process of barter. Almost 6,000 years ago the Sumerians were using a numeration system based on 10 (denary system) as well as one based on 60 (sexagesimal system). The sexagesimal system still survives in the measure of time and rotation, reflecting the Babylonian preoccupation with the motion of the Sun, Moon and other planets and their influence on man.

The knowledge acquired became not only a religious force but also solved basic problems of agriculture and social organization. The flooding in Babylon and Egypt demanded seasonal surveys of land, the techniques of which led to geometry. Political, commercial and religious pressures to build palaces, ships, temples and tombs stimulated the further development of

1 The concept of number is fundamental to mathematics. It probably developed originally out of the need for farmers to count their animals and produce. Numbers also led to money systems, making buying and selling possible.

2 Stonehenge was built in the Bronze Age as a sort of calendar, which probably also had a religious significance.

The positions of the stone blocks can be used to measure the movements of the Sun and Moon and to "predict" eclipses.

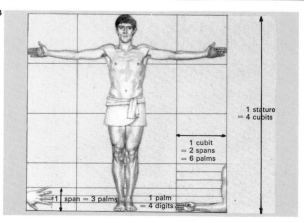

(A) \$131,137

(B) $4 \times \sqrt{27} = 20{\cdot}7846$

(C) $x = \dfrac{-b \pm \sqrt{b^2 - 4ac}}{2a}$

(D)

(E)

3 Mathematics has generated its own "language". Numbers are themselves shorthand forms of words and, linked with units, define exact amounts or measurements [A]. Other symbols stand for operations such as multiplication and square roots [B]. In algebra letters often stand for unknown quantities, as in this formula [C] for finding the solutions to a quadratic equation. A graph [D] can "draw" algebraic functions. Pythagoras created his own geometrical conventions [E].

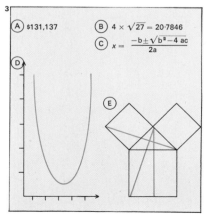

1 stature = 4 cubits

1 cubit = 2 spans = 6 palms

1 span = 3 palms

1 palm = 4 digits

4 Man probably first counted on his fingers and sized objects in terms of his own body. This diagram shows some of the ancient units of length. "Body units" are still used in some countries today. A hand, equal to 4in (about 10cm), is a standard unit for measuring the height of horses and in North America and Britain a foot – 12in (30.5cm) – is still used in measurement as a unit of length. The metric system is now the most widely accepted system of measurement.

geometry. At the same time, astronomy regulated social and religious events and thus served the political ends of ruling priests.

The Greeks established mathematics as a rigid study, placing mathematical argument on a logical basis so that propositions, previously not self-evident, could be deduced from basic assumptions. Euclid's *Elements*, produced in about 300 BC, was a prime example of this approach and dominated geometric thinking for 2,000 years. The Greeks saw beauty in number and shape and their excitement with the Golden Ratio [5] manifested itself in their art and architecture and has been echoed by later civilizations in such places as Notre-Dame in Paris, the architecture of Le Corbusier and the United Nations building in New York.

Every civilization has demanded systems for measuring and each new method has borrowed ideas from previous ones. As civilizations expanded, their influences and trade spread, and the need for standardized units increased. The earlier systems were all based on convenience, so that parts of the body were used for measuring length [4], the

working capacity of oxen for area, stones for weight, skins for volume. Each society learned to standardize; in 1791 the French devised the metric system based on the metre, one ten-millionth of the Earth's quadrant (a quarter of the circumference), a distance calculated from an actual survey. International trade has now forced most of the Western world to adopt the metric system of measurement.

The heritage of numbers

Mathematics resembles a living organism in that its growth is affected by the environment in which it lives. The golden age of Greece produced mathematical beauty that afterwards lay dormant for centuries. The Romans used earlier mathematics but solved no new problems. Not until the sixteenth century was there another great advance. Today the whole world is experiencing change at a pace unequalled in the past. This is mirrored in the development of new mathematics and its applications in solving the problems of science, technology, industry and commerce [8] peculiar to the late twentieth century.

Everything on earth, from the atoms in this crystal to the leaves on a tree, consists of individual components. Mathematics seeks to establish the relationships existing between such components.

5 Greek mathematicians extended their logical thinking into the arts, establishing mathematical relationships in music and art. The Golden Ratio (approximately 1.618) was to the Greeks a pleasing proportion, incorporated here in the Parthenon (built 447–432 BC).

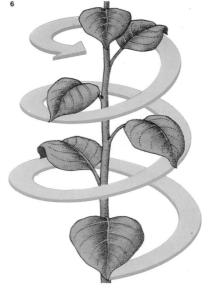

6 Fibonacci ratios are functions in the series 1/1, 2/1, 3/2, 5/3, 8/5, 13/8 and so on. These values approach the Greek Golden Ratio. Both the numerators and the denominators in the series are formed by adding consecutive members of the series. These ratios occur in nature; a spiral following leaves on this stalk has gaps and turns in the ratio of 5/3.

8 An electronic calculator is a modern machine for "doing sums". It and the much more complex digital computer have replaced earlier calculating devices such as the mechanical calculator, the slide rule and, oldest of all, the abacus.

7 The Grand Canal at Venice was a favourite subject of the Venetian painter Canaletto, whose real name was Giovanni Canal (1697–1768). Renaissance painters studied perspective and so laid the foundations of projective geometry in mathematics, map-making and the draughtsmanship used in architecture and engineering, enabling a three-dimensional object to be represented in two dimensions.

27

The grammar of numbers

People use arithmetic so frequently in everyday life that they hardly ever think about it. Yet every time a woman buys something and counts her change [Key] she uses the basic concepts of addition and equality, ideas in use since trading began.

Basic rules of arithmetic
The four main types of calculations are addition, subtraction, multiplication and division. They are carried out following basic laws – most of which are merely statements of common sense. The commutative law holds for both addition and multiplication. It simply states that the sum of seven and two $(7+2)$, for example, is the same as the sum of two and seven $(2+7)$. In other words, the order in which numbers are added does not matter. The same is true of multiplication: $4\times3 = 3\times4$ or, in general terms, $a\times b = b\times a$.

The associative law is an extension of this idea and states that, in adding or multiplying a series of numbers, the order of addition or multiplication does not matter [1]. Using symbols to stand for any numbers, $(a+b)+c = a+(b+c)$, or $(a\times b)\times c = a\times(b\times c)$.

The distributive law states that if two numbers are to be added together and the sum multiplied by a third number, the same result is obtained if each of the first two is first multiplied by the third and the two products added. This law is easier to state using symbols: $(a+b)\times c = (a\times c)+(b\times c)$, and is made clear by an example: $(5+7)\times3 = (5\times3)+(7\times3) = 36$.

Multiplication is equivalent to repeated addition. For instance 7×5, is a shorthand way of writing $7+7+7+7+7$. People learn multiplication tables because it is quicker to apply them than to add columns of figures. Electronic calculators and computers, renowned for their speed and accuracy, cannot multiply; they work by successive addition, but do so extremely quickly.

Just as subtraction is the reverse of addition, so division can be regarded as the reverse of multiplication – a repeated subtraction [3]. This is the method employed in doing "long division" sums. Often it is not possible to subtract successively one number from another an exact number of times – there is generally something "left over",

called the remainder. For example, 380 divided by 70, is 50, and the remainder is 30.

Squares and square roots
When a number is squared, it is multiplied by itself (the area of a square is the length of one side multiplied by itself). Three squared (written 3^2) equals 9. The reverse operation is called taking the square root: what number multiplied by itself makes a given number? Squaring a whole number (integer) gives an integer result, but taking the square root of a whole number often does not. And, as the Greek mathematician Pythagoras and his co-workers discovered, there is not always a rational number (expressible as the ratio of two integers) that when squared will equal a particular integer. The square root of 4 is 2 (both integers), but the square root of 2 is somewhere between 1.4142 and 1.4143. The square of 2 cannot be defined exactly and is called an irrational number.

Fractions, proportions and ratios
Three-sevenths is written as 3/7, meaning 3 divided by 7. It is a fraction (from the Latin

1 Addition is associative – that is, a series of additions can be carried out in any order without affecting the result. This diagram shows the effects of successively weighing [A] 3, 4 and then 5 units of a substance on a spring balance and [B] weighing 4, 5 and then 3 units. In both cases the total weight – the sum or the additions – is 12 units. As in many other mathematical laws, this is applied common sense.

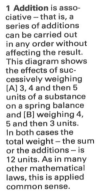

$$3$$
$$+ 4 = 7$$
$$+ 5 = 12$$
i.e. $3 + 4 + 5 = 12$

$$4$$
$$+ 5 = 9$$
$$+ 3 = 12$$
i.e. $4 + 5 + 3 = 12$

2 A series of subtractions can also be carried out in any order. Starting with a piece of wood 10 units long [A], we can cut off first 2 units and then remove a further 3 units from the 8 remaining (so completing the sum $10-2-3=5$). Or [B] we can remove first 3 units and then cut off another 2 units from the original 10. This time the subtraction sum is $10-3-2=5$ but the result is exactly the same.

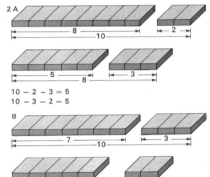

$$10 - 2 - 3 = 5$$
$$10 - 3 - 2 = 5$$

3 Multiplication and division are needed to solve many everyday problems. A man wants to tile the two plain walls of a room [A], which is 5.5m long by 3m wide and 4m tall, using tiles 0.5m square. The walls can be drawn [B] as two areas of 22m² and 12m², giving a total area of 34m². A single tile 0.5m by 0.5m has an area of 0.25m². The number of tiles required [C] can be found by dividing the area of one tile (0.25m²) into the total area to be covered (34m²), giving the result 136 tiles. The same problem can be tackled another way [D]. If the whole area to be tiled is considered, it measures 8.5m by 4m. The long side will accommodate 17 half-metre tiles and the short side only 8 tiles. The total number of tiles required is therefore 17 $\times8 = 136$, the same result as before but without calculating areas.

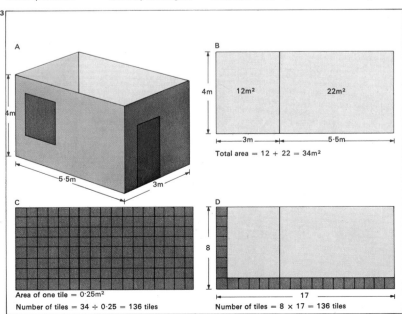

Total area = $12 + 22 = 34m^2$

Area of one tile = $0.25m^2$
Number of tiles = $34 \div 0.25 = 136$ tiles

Number of tiles = $8 \times 17 = 136$ tiles

4 Dividing quantities into equal parts is a method of forming them into fractions. An athlete such as a pole-vaulter intuitively judges his run-up by dividing it into an equal number of paces so that the pole is in exactly the correct place for the jump – he cannot make half a pace. The same is true when we speak of a bottle being half full or say that we have read a third of a book. In a fraction such as three-quarters, written 3/4, 3 is called the numerator and 4 is the denominator, if the numerator is smaller than the denominator, the fraction is termed "proper"; in an improper fraction the numerator is larger than the denominator, although it can be simplified to a whole number and a fraction.

word *fractus* meaning broken). In books, a fraction may be printed as $\frac{1}{3}$ or 1/3. The number below or to the right of the line is called the denominator and is the number of parts that a unit quantity has been "broken" into. The number above or to the left of the line is the numerator and represents the number of such parts being considered. Two pieces of wood 3m and 7m long have lengths in the proportion of 3 to 7, or in the ratio 3 to 7 (often written 3:7). The shorter piece is 3/7 the length of the longer.

There are two types of fractions, called proper and improper. In a proper fraction, the numerator is less than the denominator: 3/7, 7/8 and 29/54 are examples. An improper fraction has a larger numerator than denominator, as in 5/4 and 22/7. Generally these are simplified by dividing out and expressing the remainder as a fraction, as in 1 1/4 and 3 1/7.

The laws of arithmetic also apply to fractions, but special techniques are sometimes needed in manipulating them. Multiplication is simple – the numerators are multiplied together and the denominators multiplied together, and the result expressed as a new fraction. Thus $2/3 \times 7/11 = 14/33$. To divide, invert the second fraction (the divisor) and multiply: $2/3 \div 7/11 = 2/3 \times 11/7 = 22/21$. Here the result is an improper fraction that can be simplified to 1 1/21 (that is, one and one-twenty first).

Addition and subtraction of fractions is more complicated. They must first be written in terms of the same denominator and for simplicity the smallest possible one is chosen (called the lowest common denominator or LCD). Then the numerators can be added or subtracted as necessary, the result expressed in terms of the LCD and simplified if possible [5, 6, 7]. Decimals are a way of writing fractions whose denominators are powers of ten. For example, 1 9/10, is the fraction 19/10 and is written in decimals as 1.9. The decimal point separates the whole number part (the argument) from the fractional part. Every fraction can be expressed as the sum of a series of such fractions (tenths, hundredths, thousandths, and so on) and can be represented in decimal form.

KEY

A cash register at a modern supermarket adds together the cost of each purchase. Some machines will also calculate and even issue the correct change and print a ticket.

5 Before fractions can be added they must all be expressed in terms of the same denominator. To add 1/2, 1/3 and 1/4 they must all be stated in twelfths (in this example, 12 is the lowest common denominator) as 6/12, 4/12 and 3/12. They can then be added to give 13/12, an improper fraction that simplifies to 1 1/12. This sum explains why it is impossible to divide anything into "shares" of 1/2, 1/3 and 1/4 – their sum is larger than 1.

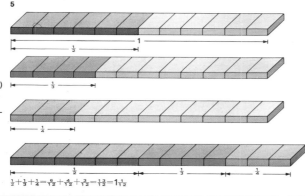

$\frac{1}{2}+\frac{1}{3}+\frac{1}{4}=\frac{6}{12}+\frac{4}{12}+\frac{3}{12}=\frac{13}{12}=1\frac{1}{12}$

6 To multiply fractions merely multiply the numerators and then multiply the denominators. A third of a half is $1/3 \times 1/2 = 1/6$ (the same as a half of a third – order does not matter).

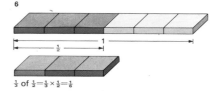

$\frac{1}{3}$ of $\frac{1}{2}=\frac{1}{3}\times\frac{1}{2}=\frac{1}{6}$

7 To divide fractions invert the divisor and multiply. For example, 5/6 divided by 5 is $5/6 \times 1/5 = 1/6$, which is exactly the same as the quantity described as a third of a half, as shown in illustration 6.

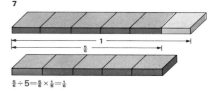

$\frac{5}{6}\div 5=\frac{5}{6}\times\frac{1}{5}=\frac{1}{6}$

8 At a public meeting a vote is often decided by "a show of hands" – those in favour of a motion raise their hands and are counted. But the way in which the results are announced or reported – as fractions or percentages – can convey different shades of meaning. At such a meeting, with 580 people present, 348 voted for the motion and 232 voted against. This basic fact can be expressed in various ways: "Three out of five people voted in favour"; "40 per cent of the voters were against the motion"; and "The motion was carried with a 20 per cent majority" are all true statements based on these figures. Fractions, proportions and ratios (often expressed as percentages) are merely different ways of presenting the same information. But if 200 of the people present did not vote (abstained), a figure of "60 per cent voting in favour" means that of the 580 present only 228 people were in favour of the motion – less than half the people present.

9 Proportions are also used to define slopes – for example, gradients on roads. "A slope of one in nine" means, mathematically, that a slope rises one unit of length for every nine horizontal units. In practice distances are measured along the road's surface and a one-in-nine hill climbs one metre (or one yard) for every nine metres (yards) travelled along the road. Mathematically this hill has a slope of 1 in 8.944m – near enough for a road sign. But proportions can best be compared as percentages. Thus the ratios 7 to 13 and 28 to 53 (corresponding to fractions 7/13 and 28/53) are difficult to compare. But as the percentages 53.85% and 52.83%, the former is obviously larger than the latter.

The language of numbers

The idea of number is a basic concept. The distinction between one and many is probably the easiest for a child to understand. A boy on a beach can pick up one pebble although he can see many more. If he picks up a handful, he obviously has more than one pebble but far less than the total number he can see. To obtain a precise idea of how many he has, he can count the number of pebbles in his hand and find, for example, that there are 12. "Twelve" is the name given to that number of pebbles. It is a property possessed by all collections of 12 objects: 12 cows, 12 seagulls and 12 encyclopaedias.

Positive and negative integers
Whole numbers such as 1, 5 and 212 are called positive integers and have been used ever since men began to count. In the Middle Ages the Hindus developed the concept of negative integers to deal with amounts owing in a trading transaction. A man might own five (+5) sheep and owe three (−3), so that he really owned only 5−3 = 2 sheep.

As long as mathematical operations are limited to counting, integers are sufficient as numbers. But as soon as men started to measure they found that nature is not organized into integer lengths and areas. A farmer could make a measuring stick (a ruler) by marking off a piece of wood into similar lengths equal to, say, the length of his foot. He might find that one of his animals was 5 "feet" long, whereas its offspring was only 2 "feet" long. Then he might find an animal that was 3½ "feet" and another of 2⅓ "feet". He would thus discover a whole new family of numbers, called rational numbers. Any number that can be written in such a form as 8/3 – as a fraction (the ratio of two integers) – is a rational number. Such numbers can be positive or negative and all integers are rational [1A].

In the sixth century BC Greek mathematicians discovered that a square with sides one unit long has a diagonal whose length cannot be measured exactly. No matter what scale of length is used, and no matter how finely it is subdivided into fractions, such a length cannot be measured with precision nor can it be written as a fraction. The system had to be extended again to include this new class of numbers, which are now called irrational numbers [3A].

Today we use zero (0) to denote the absence of a number, but this has not always been so. The Roman numeral system, for instance, had no zero. It was introduced for its present role in about 600 BC by Hindu mathematicians who formulated rules for calculating with it: multiplying by zero always gives a zero result and addition or subtraction of zero leaves a number unaltered. Hindu mathematicians also recognized that dividing by zero does not produce a result that can be defined by the number system.

Infinite and imaginary numbers
The concept that there are infinitely large numbers was first discussed by the Greek mathematician Archimedes. Starting with the largest number in the Greek number system, "a myriad myriads" (a hundred million), he constructed even larger numbers. He then estimated the number of grains of sand in the universe and showed that this was less than his largest number.

Archimedes (c. 287–212 BC) showed

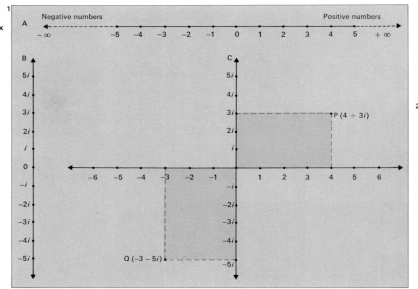

1 Three types of numbers are the real, imaginary and complex numbers. Real numbers [A] can be represented as points along a line extending from minus infinity to plus infinity. They include all negative and positive numbers. Imaginary numbers [B] are based on i, the square root of −1, and can also be positive or negative. Complex numbers [C] each have a real and an imaginary part. They can be pictured as points defined by a distance along the real number axis and a distance along the imaginary number axis. Complex number P, for example is $4+3i$, and Q is $−3−5i$. Complex numbers are much used by scientists.

2 Five oranges, five hens' eggs and five bottles full of wine all possess the identical property of "fiveness". The number 5, fifth along the positive real number scale shown in illustration 1A, can be applied to any such group of five objects. The bottles do not all have the same shape, but this obviously does not affect their number. Only adding or removing some bottles would do that.

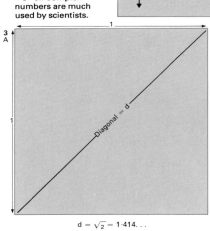

$$d = \sqrt{2} = 1{\cdot}414\ldots$$

$$\frac{C}{D} = \pi = 3{\cdot}1415\ldots$$

3 An irrational number cannot be expressed as a fraction using integers (whole numbers). A square with sides each one unit long [A] has a diagonal equal in length to the square root of 2. This is approximately equal to 1.414 ..., with a never-ending series of numbers after the decimal point. Another irrational number is the ratio of the circumference of a circle to its diameter [B], represented by the Greek letter π. It is equal to 3.1415 ..., again with a never-ending series of numbers after the decimal point. A rough approximation to π is given by the fraction (rational number) 22/7, which is equal to 3.1428 .... Irrational numbers were discovered by the Greek mathematician Pythagoras.

4 Keeping a tally was one of the earliest forms of counting. In this old English game called shove ha'penny the players slide coins along a board into marked-off sections. Ancient farmers probably counted their animals using a tally-stick, a piece of wood carved with a series of notches. In some European beer halls today, the waiter gives a customer a new beer mat with each drink and keeps the old mats as a tally of the total drunk. They keep their scores with chalk tally marks at the edges of the board.

that there is no upper limit to a number system. Infinity, unlike zero, is not a number. No matter how large a number is there is still an indefinite number larger than it. Infinity can never be reached.

With the concepts of zero and infinity, men had a complete number system that could be pictured as every real number along a line stretching from minus infinity to plus infinity. Out with the development of squares (the square of a number is that number multiplied by itself) and square roots (the square root of a number is another number which, when multiplied by itself, equals the original number), mathematicians encountered such problems as: what is the square root of −5? At first such problems were thought to be impossible to solve because there is no real number which, when squared, gives a negative result. Then in the sixteenth century Italian mathematicians introduced the "imaginary" quantity i which, when squared, gives the result −1. Numbers involving i are called imaginary numbers.

Complex numbers consist of a real part and an imaginary part, such as $5+3i$. They can be manipulated in the same way as purely real numbers. Many branches of modern engineering and electronics use them.

The system of numbers commonly used today was adapted from the Arabic numbering system [5] which, in turn, was based on Hindu ideas. In this system the position of a digit (numeral) in a number is significant. Using the basic digits 0 and 1 to 9 it is possible to construct any number. This base-10 or decimal system was introduced into Europe by Adelard of Bath in about 1100 and by 1600 was in almost universal use.

What is the base?

The base, or radix, is the number of digits in a number system. Position is important because in a number such as 333, the first 3 stands for 300 (three hundreds), the second for 30 (three tens) and the third for 3 units. But any convenient base can be used. Modern digital computers, for example, "count" using the base of 2 – the binary system of numbers – because its only digits, 1 and 0, can easily be represented by "on" or "off" pulses of electricity [9B].

An abacus is an ancient type of calculating machine still used in China and Japan. It has a number of beads on wires, generally divided into two sections with two beads (each standing for 5) and five beads (each of which stands for 1). Numbers are added or subtracted by moving the beads.

5 Various numeral systems have been used through the ages. The earliest, such as the Egyptian, used a simple pen stroke or a mark in clay to represent 1; other numbers up to 9 were formed by repeating the 1 symbol. The Romans and Mayans had an additional symbol for 5. Modern Arabic and Chinese have different symbols for each number, although 1 to 3 are formed by adding successive strokes.

	5							
Egyptian	𝍳	∩	𝋍	𝋧	⌐	𝋹		
	1	10	100	1,000	10,000	100,000		
Roman	I	II	III	IIII	V	VI	VII	VIII IX X ↓ CIↃ
	1	2	3	4	5	6	7	8 9 10 50 500
Mayan	•	••	•••	••••	—	⎯•⎯	⎯••⎯	⎯•••⎯ ⎯••••⎯ ═
	1	2	3	4	5	6	7	8 9 10
Modern Arabic	١	٢	٣	٤	٥	٦	٧	٨ ٩ ٠
	1	2	3	4	5	6	7	8 9 0
Chinese	一	二	三	四	五	六	七	八 九
	1	2	3	4	5	6	7	8 9

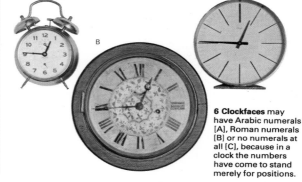

6 Clockfaces may have Arabic numerals [A], Roman numerals [B] or no numerals at all [C], because in a clock the numbers have come to stand merely for positions.

7	English	French	Italian	German	Dutch	Spanish
1	One	Un	Uno	Ein	Een	Uno
2	Two	Deux	Due	Zwei	Twee	Dos
3	Three	Trois	Tre	Drei	Drie	Tres
4	Four	Quatre	Quattro	Vier	Vier	Cuatro
5	Five	Cinq	Cinque	Funf	Vijf	Cinco
6	Six	Six	Sei	Sechs	Zes	Seis
7	Seven	Sept	Sette	Sieben	Zeven	Siete
8	Eight	Huit	Otte	Acht	Acht	Ocho
9	Nine	Neuf	Nove	Neun	Negen	Nueve
10	Ten	Dix	Dieci	Zehn	Tien	Diez

7 The names of numbers in various European languages reveal common word origins. But all these countries use the same number symbols, originally based on early Arabic numerals.

8 Large numbers are awkward to write and have different names even among English-speaking countries. American and British names are different. The scientific use of powers of 10 is unambiguous.

Number	British name	American name	Powers of ten
100	Hundred	Hundred	10^2
1,000	Thousand	Thousand	10^3
1,000,000	Million	Million	10^6
1,000,000,000	(Milliard)	Billion	10^9
1,000,000,000,000	Billion	Trillion	10^{12}
1,000,000,000,000,000	—	Quadrillion	10^{15}
1,000,000,000,000,000,000	Trillion	Quintillion	10^{18}

9 High-speed calculations are needed to deal with a changing situation in which numerical quantities vary continually, as at the tote on a racecourse [A], which computes winning odds for individual runners (horses or dogs) in accordance with the amount of money bet on them. For more complicated varying situations, a computer [B] is needed in order to complete the calculation in "real time" – that is in time for the information to be of use.

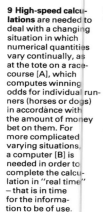

Measurement and dimensions

Four students – a chemist, a physicist, a mathematician and a humanities graduate – were each given a barometer and told to measure the height of a church tower. The chemist knew all about gases. He measured the air pressures at the top and bottom of the tower with his barometer and from the barely perceptible difference produced an answer of "anywhere between 0 and 60m" (0–200ft). The physicist was used to handling expensive equipment casually. He dropped his barometer off the tower and timed its fall, calculating the height as 27–33m (90–110ft). The mathematician compared the length of the tower's shadow with that of the barometer, arriving at a height of 30–30.5m (99–101ft). The humanities graduate sold the barometer, bought the verger a few drinks with the money, and soon found out that the tower was 30.4m (100ft) tall exactly.

Putting numbers on things
This apocryphal story illustrates the variety of ways of "putting numbers on things" and the different results that can be obtained. Life in the modern world depends greatly on man's ability to make accurate measurements, and laboratories throughout the world maintain standards of length, time, mass and voltage to ensure uniformity. Every large factory has a set of reference gauges that have been calibrated against a substandard that, in turn, has been checked against a national copy of the standard metre. As a result, a replacement bearing made in Japan can exactly fit a motor shaft made five years previously in West Germany.

Behind the practice of measurement lies theory. There is the physical theory of the process and also mathematical principles such as dimensional analysis. This derives the "dimensions" of measured quantities in terms of the fundamentals length [L], mass [M] and time [T]. Area, for example (square metres, square yards, acres or hectares), has dimensions $[L^2]$; volume (cubic metres, cubic yards and so on) has dimensions $[L^3]$. If the volume of a paraboloid were stated to be $\pi H^2/8D$, with H as its height and D its base diameter, then without making any calculations at all a student can be sure that the formula is wrong. It involves the product of two lengths divided by a length and so has dimensions $[L^2/L] = [L]$. It must therefore represent a length – it cannot possibly represent a volume. (The correct formula is $\pi H^2 D/8$.)

Similarly, given that the time-of-swing t of a pendulum might depend on its length l, the mass of its bob m and the acceleration imparted by gravity g, any formula for t based on a relationship between l, m and g must yield a number with the dimensions of time. Acceleration is measured in metres per second per second and has dimensions $[L/T^2]$. So
$$t \equiv T = \sqrt{T^2} = \sqrt{L/\tfrac{L}{T^2}} = \sqrt{l/g}$$
($\equiv$ is the sign for "is equivalent to".)

In fact the formula is $t = 2\pi\sqrt{l/g}$ (dimensional analysis can never deduce numerical factors such as 2π). There is no room for any number with the dimensions of mass in the expression, so it has been shown here by pure mathematics alone that the swing of a pendulum is in no way dependent upon the mass of its bob.

The length in metres and acceleration in metres per second per second must give an answer in seconds. "Coherent" systems of

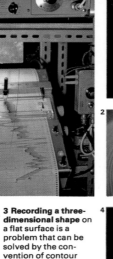

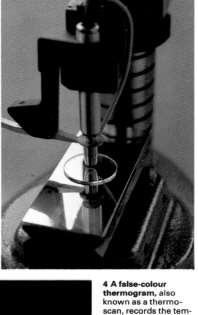

1 Different techniques of measurement have different degrees of precision. The chart-recorder [A, bottom] is precise to about 1 in 100; it is hard to read a chart more accurately than that. The chemical balance [B] can reach 1 in 10^6 and the frequency counter [A, top] 1 in 10^8. Both these have a numerical display; no meter could be calibrated so finely. An air gauge [C] measures extremely small dimensions by sensing the flow of air through a small gap. Its precision of up to 1 in 10^9 is near the limits of current technology (at about 1 in 10^{11} for a laser gauge). The micrometer [Key] is 10 million times less accurate.

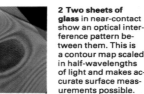

2 Two sheets of glass in near-contact show an optical interference pattern between them. This is a contour map scaled in half-wavelengths of light and makes accurate surface measurements possible.

3 Recording a three-dimensional shape on a flat surface is a problem that can be solved by the convention of contour scaling. This is shown when the cross-sections of a hill at 50-, 100- and 150-metre heights are projected onto a map of the hill. The hill can be envisaged fairly well from such a map, although the "coarseness" of the contour intervals loses some finer detail. The steepness of the sides can be judged by the contour lines on the map. "Newton's rings" and similar interference patterns (as in illustration 2) are fine contour maps. Special techniques make it possible to reveal tiny deformations of stressed surfaces by means of such contours.

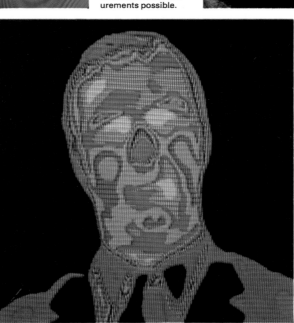

4 A false-colour thermogram, also known as a thermoscan, records the temperatures on the skin of a man's face. The technique allows a doctor to study the extent of skin damage caused by burns and has been adapted to aid in the diagnosis of diseases such as cancer of the breast. An infrared camera has a rotating prism scanner that detects the heat levels in a strip of the picture, and the resulting signal is amplified and displayed on a colour television screen. Blue colours represent low temperatures; the redder the colour, the "hotter" it is. Thermograms can also help architects to design houses for minimum heat loss and aid householders to insulate them.

units such as SI units (the international unit system used throughout science) guarantee an answer in correct units. The speed of a piston multiplied by its area and the pressure it exerts gives its power, for example. With units of feet per second, square inches and atmospheres, the answer would have the dimensions of power $[ML^2/T^3]$ but be in no recognized power units. Using SI units (metres per second, square metres, and newtons per square metre) guarantees an answer in watts. Other common units are the cgs (centimetre-gramme-second) and MKS (metre-kilogramme-second) systems.

Scaling the heights

If a 10-metre (33ft) scale model of a blade of grass were actually made from grass, it would promptly collapse. Similarly a flea the size of an elephant would not be able even to stand, let alone jump. This is because the weight of an object, like its volume, increases as the cube of its height, whereas its strength increases only as its square. Many related properties scale differently so it is quite difficult, for example, to calculate what thrust

will propel an aircraft from the force needed to sustain a scale model in a wind-tunnel. One way out of this difficulty is to think in terms of "dimensionless groups" like Reynold's Number, valuable in many problems of gas and fluid flow. This is LVd/η where L is a length (perhaps of a wing-section), V is a gas velocity, d is its density and η (eta) its viscosity. This combination is a dimensionless ratio – a pure number – having the same value in any units.

Dimensionless ratios

Dimensionless ratios, free from arbitrary units, are fundamental entities. The ratio of the electrical to gravitational force between two protons, for example, is about 10^{39} (that is, 1 followed by 39 zeros). This is also approximately the ratio of the diameter of the knowable universe to that of the proton and of the estimated age of the universe to the time light takes to traverse a proton. The square of 10^{39}, 10^{78}, is about the number of particles in the knowable universe. Cosmologists wonder if this ratio is trying to tell us something.

A micrometer can measure a small object with a pre- cision of up to 1 in 10,000. This is adequate for most high-grade engineering, but far higher precision is possible.

5 With c accurately known, d is found by measuring e and subtracting; error in e is diluted in the longer d. But obtaining e as c−d is inaccurate; it is a small difference between two larger numbers. Obtaining f by Pythagoras as $\sqrt{a^2-d^2}$ is worse; squaring the two similar and large values increases their uncertainty. But e, and f are dissimilar lengths, so that b can be found accurately as $\sqrt{e^2+f^2}$.

6 By the surface volume scaling law, fine structures have relatively more surface to their volume. Each die weighs 2g (0.07oz) and has 9cm² (1.4in²) of surface. The sugar lump is made of 0.5mm (0.02in) grains and has about 200cm² (31in²) of total surface. The 2g of "molecular sieve" on the watchglass is porous to the molecular level and has a remarkable 1,500m² (16,150 sq ft) of total effective surface area.

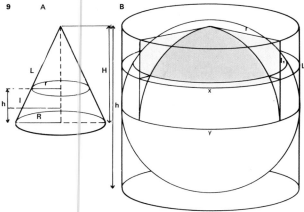

7 If the size of a balloon is doubled, its surface (and therefore weight) goes up four times, but its volume (and therefore lift) goes up eight times. This surface/volume scaling law shows that balloons become more efficient the bigger they are [A]. Conversely, doubling an aircraft's size increases its weight by eight times, but the wing area by only four times [B]. Small aircraft pose fewer design problems.

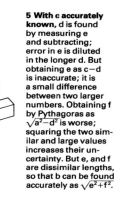

8 Various formulae link the dimensions of plane figures. In a triangle [A] the cosine formula states

that $a^2 = b^2+c^2-2bc$. cosA; and the sine formula that $a/\sin A = b/\sin B = c/\sin C$. If $S = \frac{1}{2}(a+b+c)$, the area of the triangle $= \sqrt{s(s-a)(s-b)(s-c)}$. In a circle [B] angle O = 2B = 2E and the circumference of arc ADC is $\pi rO/180$. In a regular n-sided polygon [C] angle A = 180 $(1-2/n)$; angle O = 360/n; the area of the polygon = $\frac{1}{2}nld$ and the radius r = d.sec(360/n).

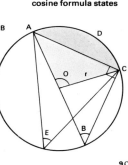

9 The volumes and areas of solid figures can be linked by formulae. In a cone [A] the volume of the full cone = $\frac{1}{3}\pi R^2H$; and the area of the curved surface is given as πLR. In the frustrum of the cone (the lower section) the area of the curved surface = $\pi l(R+r)$; and the volume = $\frac{1}{3}\pi h(R^2+Rr+r^2)$. In a sphere and cylinder [B] the surface area of the sphere =

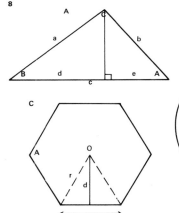

$4\pi r^2$. The curved surface of the cylinder = $2\pi rh$. The volume of the sphere = $4\pi r^3/3$; and of a cylinder it is given as πr^2h. The volume of the section of sphere between planes x and y = $V_s = \pi[L^2(r-L/3)-l_1^2(r-l_1/3)]$. In a regular tetrahedron [C] the distance $r = d/\sqrt{3}$; $H = d(\sqrt{2}/\sqrt{3})$; and $R = \frac{1}{2}d(\sqrt{3}/\sqrt{2})$. The height of the centre of gravity is given as h = H/4.

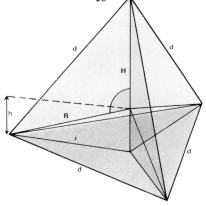

Finding unknown quantities: algebra

In arithmetic various quantities such as lengths, areas and sums of money are represented as numbers (and the appropriate units). But some mathematical problems are concerned with *finding* a number – an unknown quantity. If two numbers add up to 10 and one of them is 6, what is the other? The answer to this simple problem is 4 and yet the method of formalizing it is a basic technique of algebra.

To solve this problem in algebra, let the unknown number be x. Then $6 + x = 10$ (this is an algebraic equation). By subtracting 6 from each side of this equation, it simplifies to $x = 10 - 6 = 4$. By making a letter, x, stand for the unknown quantity, the problem can be solved, and this way of using letters is a basic technique of algebra.

Greek and Arab mathematicians
Greek mathematicians such as Diophantus (*c.* third century AD) used letters in their equations. But the word algebra is derived from the Arabic words *al-jebr*, meaning "bone-setting" (the restoration and reduction of a bone fracture). This formed part of the title of a book by the Arab mathematician Al-Khwarizmi. By the sixteenth century, mathematical problems were fully formulated in algebraic terms, initially in France by Franciscus Vieta (1540–1603). The normal convention of using the last few letters of the alphabet (x, y and z) to denote unknown quantities and the first few letters to stand for known prescribed numbers was introduced by the French mathematician René Descartes (1596–1650).

Algebraic equations and formulae
Common practical applications of algebraic equations are the various formulae used in science, particularly in mathematics and physics. The volume of a cylinder, for example, is given by the formula $V = \pi r^2 h$, where V is the volume, r is the radius of one end and h is the cylinder's height [1]. The formula provides a shorthand way of saying "the volume of a cylinder equals the area of one end multiplied by the height".

Algebraic equations [2] and formulae can be manipulated according to established rules. The subject (V) of the cylinder equation can be changed to find the radius or height of a cylinder of known volume. For instance, $h = V/\pi r^2$. Such formulae are perfectly general – they apply to all cylinders, whether they are tall and thin or short and squat. There are similar formulae for the areas and volumes of all common geometrical figures.

Many problems in algebra involve more than one unknown quantity. Consider the problem of finding two positive numbers whose product is 15 and whose difference is 2. Let the two numbers be represented by the letters x and y. Then the "product" information can be stated as the equation $xy = 15$. There are several possible solutions to this equation: 1 and 15, 3 and 5, 7.5 and 2, and so on. To proceed we must use the "difference" information, which generates the equation $y - x = 2$, rearranging to give $y = x + 2$. Substituting this expression for y in the first equation yields $x(x + 2) = 15$, or $x^2 + 2x - 15 = 0$.

Now this third equation contains only one "unknown" quantity: x. The only positive number that satisfies it is 3 (when the equa-

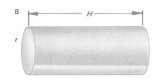
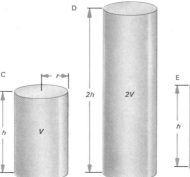

1 The formula for finding the volume of a cylinder is $V = \pi r^2 h$, where r is the radius of one end and h is the height. The two cylinders [A and B] have the same volume, but very different radii and heights. In fact, the diameter of one is almost equal to the height of the other, that is h is about equal to $2r$ and $2R$ is almost the same as H. Another cylinder [C] has volume V. Doubling its height doubles its volume [D] (for C volume $= \pi r^2 h$, for D vol. $= 2\pi r^2 h$). But doubling the radius of the cylinder increases its volume four-fold [E] (for E volume is equal to $\pi(2r)^2 h = 4\pi r^2 h$). These changes are predicted using algebra.

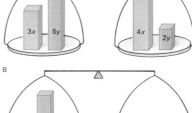

2 An equation in algebra is in a state of balance; the terms on the left-hand side taken together equal those on the right-hand side, just as a collection of objects balances on a pair of scales [A]. In sim- plifying an equation it is essential that the same operation be carried out on each side. For example, $3x$ is subtracted from each pan of the balance [B] (and from each side of the equation). A further simplification is made [C] by subtracting $2y$ from each side. As a result the original equation $3x + 5y = 4x + 2y$ reduces to $3y = x$. It is also possible to multiply or divide each side by a factor.

$3x + 5y = 4x + 2y$

Subtract $3x$ from each side:
$3x - 3x + 5y = 4x - 3x + 2y$
$5y = x + 2y$

Subtract $2y$ from each side:
$5y - 2y = x + 2y - 2y$
$3y = x$

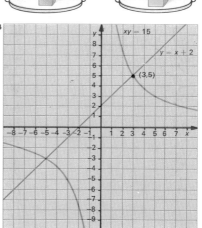

4 Algebraic equations can be plotted as lines on a graph, a technique which is the province of analytic geometry. This graph shows plots of the equations $xy = 15$ and $y = x + 2$. Treated as simultaneous equations, they are both true at the points where the lines cross. When the equation to a straight line is expressed in the form $y = mx + c$, where m and c are numbers, the letter m is a measure of the slope of the line, here equal to 1.

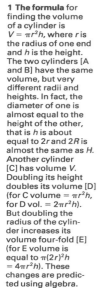

3 Houses in this street are numbered consecutively. A man notices that four times his number is ten more than three times his neighbour's higher up the street. What is his number? Let the house number be x, then the neighbour's is $x + 1$. So $4x = 3(x + 1) + 10 = 3x + 3 + 10$. Subtracting $3x$ from each side of this equation gives $x = 13$. The house number is 13 (and his neighbour's is number 14).

34

tion becomes $9 + 6 - 15 = 0$). Finally, to find y we substitute this value of x in either of the two original equations. According to the first, $y = 15/x = 15/3 = 5$ and for the second $y = x + 2 = 3 + 2 = 5$. The answer to the problem is therefore 3 and 5. In algebraic terms, we have solved two equations that are both true at the same time – called simultaneous equations.

By considering points in space defined by referring to their distances from a line (the x-axis) and another line (the y-axis), the equations of algebra take on a whole new meaning. The equation $xy = 15$, for example, represents a line on which all points have the product of their x-distance and y-distance equal to 15. The equation $y = x + 2$ represents a straight line and all points along it satisfy this equation.

If these two curves are drawn [4] (to a mathematician, even a straight line is a "curve"), they intersect at the point whose x-distance is 3 and whose y-distance is 5 – the point defined as (3, 5). The graphic approach to the problem gives exactly the same solution as the purely algebraic approach. It also

reveals another point at which the curves intersect, corresponding to $x = -3$ and $y = -5$. These solutions are, however, disallowed by the original problem, which called for two *positive* numbers.

The whole procedure of plotting algebraic equations as curves is the province of analytic geometry. It is the branch of mathematics in which algebra and geometry come together.

Algebra also supplies an insight into other puzzles and paradoxes. Any three-digit number whose middle digit is the sum of the other two is divisible by 11. Why? The answer can be supplied by using algebra [5].

Maintaining the balance

The examples already described serve to show the power of algebra in solving problems, particularly by manipulating equations. But there are rules about such manipulation. If there are two unknowns, such as x and y, an equation is simplified by having all the terms in x on one side and all the terms in y on the other. This can be achieved by adding or subtracting equal quantities from each side [2].

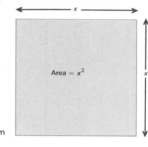

KEY

Arithmetic deals with numerical quantities — for example, the number of people in this crowd. But algebra can tackle problems involving unknown quantites, generally by allocating to them a letter such as x. Then x might stand for the number of men.

5 The three-digit numbers in this table have two properties in common: the middle digit is the sum of the other two and all of them are divisible by 11. If the first digit is x and the third y, the middle one is $(x + y)$. The whole number has the value $100x + 10(x + y) + y$. This expression can be factorized and simplified to $11(10x + y)$; it is a general formula for all the numbers in the table and has 11 as a factor.

110	220	341	473	671
121	231	352	484	682
132	242	363	495	693
143	253	374	550	770
154	264	385	561	781
165	275	396	572	792
176	286	440	583	880
187	297	451	594	891
198	330	462	660	990

6 Squaring the circle – drawing a square with exactly the same area as a given circle – was a problem that defied ancient mathematicians. But it can be solved using algebra. The area of a circle of radius r is πr^2 and of a square of side x is x^2. For equal areas $\pi r^2 = x^2$, or $x = \sqrt{\pi r}$. If $r = 10$, $x = \sqrt{314.2}$, which equals 17.72 approximately. A square of side 17.72cm has an area almost exactly the same as a circle of radius 10cm.

6

Area $= x^2$

If $x = 17.72$
Area $= 314.2$

Area $= \pi r^2$

If $r = 10$
Area $= 314.2$

7
A

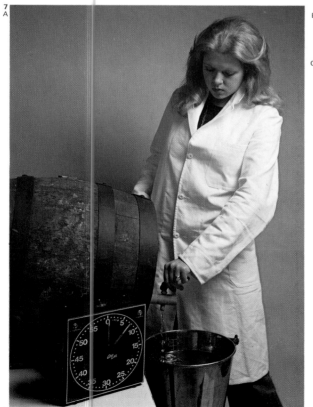

B

Minutes	(x)	1·0	2·5	4·0	6·5	7·5	9·0	11·0
Litres	(y)	0·6	1·7	2·3	4·0	4·5	5·5	6·6

C

$5y = 3x$

Volume (litres) y (vertical axis)

Time (minutes) x (horizontal axis)

7 A woman uses a photographer's clock to time the rate at which a bucket is filling with liquid [A] (the volume of the bucket is calibrated in litres on a scale inside it). The results of the measurements are shown in the table [B]. How much liquid flowed into the bucket after five minutes? To solve this problem she can draw a graph showing the rate of flow [C], which plots volume (litres) against time (minutes). The volume discharge in five minutes can be read off the graph as 3 litres. The graph is a straight line, which shows that during the time in which she made the measurement the rate of flow was constant. The line passes through the origin, at which both axes are zero. It therefore has an equation of the general form $y = mx$, where m is the slope of the line. In this example, the slope is 3/5, so the equation of the line is $y = 3x/5$ or $5y = 3x$. This equation can be used instead of the graph to calculate values of volume or time.

Mathematical curves

Anyone who can catch a ball has an intuitive grasp of mathematical curves and their transformations in space. A thrown ball travels in a mathematical curve which is (very nearly) a parabola: and many sportsmen can, with the ball still rising far away, begin to run at once to where it will fall. This is not just a simple matter of "seeing where it will go" by elementary estimation. A ball on a long piece of elastic, as is used in some tennis-trainers, is almost impossible to catch even if it is close and moving slowly. It travels in some curve other than a parabola and parabolically attuned reflexes are baffled by it.

Curves, equations and laws

Gunnery fire-control and ballistic-tracking systems have to predict curves as does the sportsman. Lacking his intuitive reflexes, they need high-speed computers to represent the trajectories mathematically. This is done by deducing a precise and complete "specification" of the trajectory in the form of an equation. Mathematics is the art of making precise statements and a mathematical curve is merely one that has such a specification. It does not have to be put in equation form and indeed an informal statement is sometimes clearer. Thus the statement that a circle is a curve on which every point is the same distance from a given centre is easier to understand than the curve whose mathematical equation is $x^2 + y^2 = R^2$ [2].

A mathematician can always translate the specification into an appropriate equation that is true for every point on the curve or surface but false for all other points. He can work out all the properties of the curve by manipulating algebraic symbols, which is much easier than pictorial geometry. When Thomas Telford (1757–1834) built the suspension bridge over the Menai Strait, Wales, in 1826, he had to determine the curve of the hanging chains by setting up a large model across a dry valley and measuring it – a sad consequence of mathematical ignorance. Nowadays an engineer can derive the equation of a suspension bridge cable and find out all he needs to know about it without even drawing a diagram.

Because the world is governed by simple mathematical laws, mathematical curves are all around us. A stone falls in a straight line if dropped or a parabola if thrown; the Moon and artificial satellites move in (very nearly) ellipses; the Sun and Earth are nearly spherical; and a static liquid surface is (very nearly) flat – all because of the mathematical form of the law of gravity. A rainbow is a circular arc and the bright cusp you sometimes see in a sun-illuminated cup or pan is an epicycloid [9B] because of the laws of optics. Indeed, much of the process of scientific discovery consists of observing such things, or conducting experiments to reveal them and then deducing "laws" that must hold to give rise to them. But most often the scientist finds mathematical curves in his results only when he draws them (plots them as a graph). Then the equation of the graph tells him the "law" revealed by his experiment.

Spheres occur in nature in objects that adopt this shape as the "line of least resistance" to forces affecting them. Small droplets of water and soap bubbles are spherical to minimize their areas under the effect of surface tension. Lead shot was once made by pouring molten lead down inside a tall

CONNECTIONS

See also
34 Finding unknown quantities: algebra
44 Lines and shapes: geometry
46 Lines and angles: trigonometry

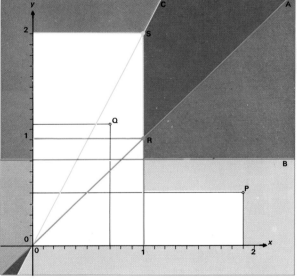

1 To give a mathematical curve an equation, first draw two lines at right-angles: the x axis ([0x] on the diagram) and the y axis [0y]. Then every point on the paper can be defined by its "x distance" and "y distance" along these axes. Thus point Q has $x = 0.7$ and $y = 1.15$; point P has $x = 1.9$ and $y = 0.5$. On the straight line A it is obvious that for every point on it (eg R, with $x = 1$ $y = 1$) the x distance equals the y distance or $x = y$. This then is the equation of line A as a mathematical curve, true for all points on it and false for all others. Line C has equation $y = 2x$ (as point S shows); line B, equation $y = 0.8$.

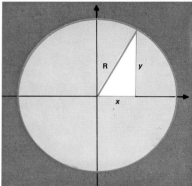

2 To find the equation of a circle, observe that each point on it forms a right-angled triangle whose sides are the x distance and y distance and whose hypotenuse is R. Pythagoras's theorem tells us that $x^2 + y^2 = R^2$. For points outside the circle $x^2 + y^2$ exceeds R^2; inside it $x^2 + y^2$ is less than R^2. Along the intersection of these two regions the equation balances. All mathematical curves divide space in this way.

4 A mathematical surface can be defined like a curve, using three mutually perpendicular axes in space, x, y, z. The equation of the sphere is $x^2 + y^2 + z^2 = R^2$. All points outside it have $x^2 + y^2 + z^2$ greater than R^2; those inside have $x^2 + y^2 + z^2$ less than R^2; the equation balances on the boundary. An equation such as $x^2 + y^2 + z^2 - 2x - 8z + 17 = R^2$ also defines a sphere but its centre is not at the intersection of the three axes.

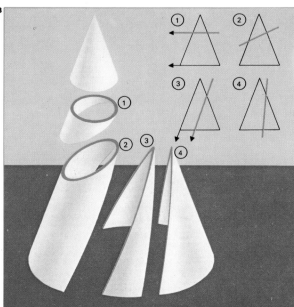

3 An important family of mathematical curves results from sectioning a cone at various angles. A horizontal section [1] gives a circle; an inclined one [2] an ellipse. A section parallel to one side of the cone [3] gives a parabola and still greater inclination [4] a hyperbola. All these have the same general equation: $ax^2 + by^2 + 2hxy + 2gx + 2fy = c$. With h^2 greater than ab, it is a hyperbola; with $h^2 = ab$, a parabola; h^2 less than ab gives an ellipse, of which the circle ($h = 0$, $a = b$) is a special case. The terms a, b, c, h, g and f are chosen constants; with $b = c = h = g = 0$, $a = 1$ and $2f = 1$, the equation becomes $y = -x^2$. As $h^2 = ab$ (both $= 0$) this is the equation of a parabola.

5 A parabola rotated about its axis of symmetry gives a mathematical surface, the paraboloid. A uniformly rotating liquid acquires a paraboloidal surface from the interaction of gravity and centrifugal force, as can be seen by spinning a pan of liquid on a turntable. This surface is perfect for radio- and optical-telescope mirrors; a mirror surface can be made by spinning plastic resin in a dish while it sets and coating it with metal.

building into a tank of cold water. During their fall down the tower the droplets of lead formed spheres and cooled sufficiently to retain their shape when they hit the water.

From theory into practice

Mathematical curves and surfaces are involved in all sorts of human activities. Lenses, for example, have spherical surfaces not because this is the ideal optical shape (although it is quite good) but because it is so easy to make. A spherical surface is the only one that presents the same form and curvature no matter how it is turned around. Therefore a hard surface and a soft one, rubbed together with back-and-forth and mutual twisting movements, will wear together into mating spherical surfaces as these are the only ones that enable them to fit exactly together every time. As a result, simple grinding processes suffice to produce spherical surfaces; where other surfaces are needed (as in telescope corrector-plates and some zoom lenses), manufacture is much more difficult.

Similarly, cylindrical objects (tubes, rods, bolts) and holes are so common because this type of surface is readily created by rotating machinery. Boilers and pressure vessels are cylindrical because this shape resists pressure better than others. Cooling towers are hyperboloidal because the two-way "saddle" curvature resists weight and wind loadings well. A trumpet has an exponentially flared horn because this is the mathematically ideal way of launching the intense sound vibrations in its throat into the open atmosphere.

Beauty and the mathematician

The cooling tower, trumpet, bridge arch and radio-telescope dish all derive their forms from pure mathematical physics. Yet they, and other engineering creations, have an aesthetic appeal that is sometimes sadly missing in those architectural and automobile creations whose "styling" lacks any mathematical necessity. It is an appealing flight of fancy to wonder whether the same intuition that enables the ordinary man to recognize and respond to the unity and inevitability of mathematical curves at work, also enables him to catch a moving ball.

These three bridges spanning the River Tyne, are curves at work. The arches are close to parabolas of different forms. The parabola is ideal for an arch whose weight is negligible compared to the even weight of the roadway. For an actual bridge the arch's own weight must be allowed for.

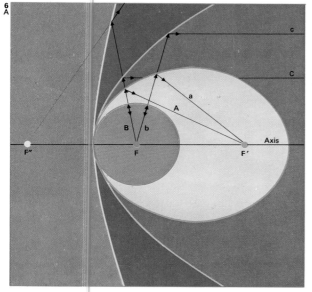

6 The conic sections [A] all have characteristic "reflecting" properties. An ellipse has two foci [F, F'] such that rays emitted from one converge [A, a] on the other. A circle is an ellipse with its foci coincident and an emitted ray is reflected straight back [B, b]. A parabola is like an ellipse with the other focus infinitely far away: rays emitted from the true focus are reflected parallel to its true axis [C, c]. The hyperbola reflects rays from F as if they came from another focus F" behind it. Paris Metro tunnels are almost elliptical and whispering on one platform can be heard on the other by the focusing effect [B].

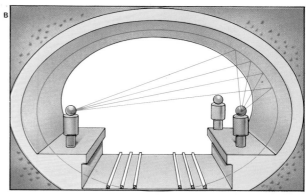

7 Many living creatures grow by compound interest; the bigger they become the faster they grow. The pearly nautilus of the Indian and Pacific oceans extends its shell continuously in spiral fashion as it grows, generating a natural "logarithmic spiral." This mathematical curve is also called the "equi-angular spiral" because a line drawn outwards from the centre intersects the curve at the same angle.

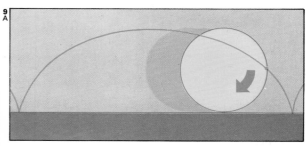

9 A point on the rim of a rolling wheel traces out a cycloid [A]. This famous mathematical curve received attention from Galileo, who suggested it as an arch-form for stone bridges, and Newton, who proved that inverted it was that surface down which a particle slid in the minimum time. A wheel rolling on another wheel generates an epicycloid [B]. To mesh smoothly without "stepping" from tooth to tooth, gear wheels need teeth of defined curves. The rise of an epicycloid from the inner wheel is one ideal curve. The point on the rim of a wheel rolling inside another wheel generates a curve called a hypocycloid, also used in gears.

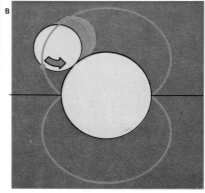

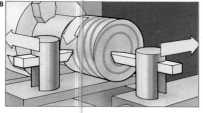

8 An "Archimedean spiral" is traced by a point that travels around a centre, varying in distance from it in proportion to the angle it moves through. Similarly a helix is the curve of a point that moves around a cylinder, travelling along it in proportion to its total angular rotation. Both are created automatically by a lathe whose cutting tool traverses a rotating workpiece. Many cylindrical objects show their machine finishing as a fine helical pattern.

Logarithms and slide rule

As the science of mathematics and its applications progressed, men found themselves having to carry out more and more complicated calculations – especially ones involving multiplication and division. Even a modern computer or electronic calculator takes about ten times as long to multiply two numbers as it does to add them, and this is certainly true of human mathematicians.

Both multiplication and division were simplified in the 1500s by the introduction of decimal notation. Then a Scottish mathematician, John Napier (1550–1617), published his book *Mirifici Logarithmorum Canonis Descriptio* (1614) which announced the discovery of logarithms.

Arithmetic and geometric progressions
A practical form of Napier's ideas about logarithms was a set of numbered rods, or bones, that could be used for carrying out multiplication by merely using the mathematical operation of addition [1]. As with logarithms (and the slide rule) they made use of two types of mathematical series, called arithmetic and geometric.

An arithmetic progression is a series of numbers in which each is obtained by adding a "common difference" to the one before it in the series. The ordinary ordinal sequence of numbers, for example – 1, 2, 3, 4, and so on – is an arithmetic series with a common difference of 1. In a geometric progression, each term is obtained from the previous one by miltuplying it by a "common ratio". In the series 2, 4, 8, 16, and so on, the common ratio is 2.

In the three following series:

1	2	3	4	5 . . .
10	100	1,000	10,000	100,000 . . .
10^1	10^2	10^3	10^4	10^5 . . .

the first is an arithmetic progression and the second a geometric one (with a common ratio of 10). The third row, equivalent to the second, shows how the succeeding powers of 10 in the second (geometric) series are in an arithmetic progression.

The powers (exponents) in the bottom row are called the logarithms of the corresponding terms in the middle row "to the base 10". The logarithm of a given number is the power to which a fixed number (the base) must be raised in order to equal the given number. Thus the logarithm of 100 to the base 10 is 2 (because $10^2 = 100$).

Logarithms, pianos and guitars
Ten is not the only base for logarithms; Napier's original tables were to the base "e" (an irrational number) and are still much used in science, where they are called natural or Naperian logarithms. The pitch of the notes on a piano is in a logarithmic ratio to the base 2, whereas the keys are in a linear sequence of octaves.

The sound wavelength of any note is twice that of the one an octave above it. The pieces of metal called frets across the fingerboard of a guitar also form a logarithmic series (in terms of spacing).

Numbers expressed in terms of powers or exponents are multiplied by adding the exponents. Thus $10^2 \times 10^4 = 10^{2+4} = 10^6$. And since logarithms are also exponents, to multiply two numbers their logarithms are merely added and tables can supply the number whose logarithm is the result. In this way multiplication is reduced to the much easier

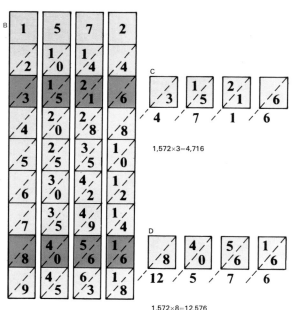

1 Napier's bones consisted of a set of nine square-section rods [A] housed in a tray. They were numbered 1 to 9 in the first segment and the lower segments on each rod were divided diagonally. These segments were numbered down the rods in arithmetic series, the "one" rod numbers increasing by 1 (to give 1, 2, 3, 4, and so on), the "two" rod by 2 (2, 4, 6, etc), the "three" rod by 3 (3, 6, 9, etc) and so on for the whole set ending with the "nine" rod which was calibrated in nines (9, 18, 27, 36, etc). The other faces of each square rod were similarly calibrated, so that each number (1 to 9) had to be represented four times somewhere in the set. To find the multiples of a number – for example 1,572 – the rods numbered 1, 5, 7 and 2 are removed from the tray and laid side by side [B]. To find 3×1,572 the third row of rod segments is used, as at C. The numbers displayed can be added diagonally as shown, to yield 4,716, which is the required product. To find 8×1,572 the segments in the eighth row are used, as at D. When added diagonally, the numbers displayed this time add to 12,576 – again the required product. To multiply by a larger number, say 38, the appropriate products are merely added together (47,160 – a zero is added because we are now multiplying by 30 not 3 – and 12,576) to give 59,736.

1,572×3=4,716

1,572×8=12,576

2 The frequency of a note in music is twice that of the one an octave below it. On a keyboard instrument the frequencies of a note and its successive octaves are in the proportions 1:2:4:8:16, etc. This is a logarithmic scale to the base 2. The spacing of the metal frets across the fingerboard of a guitar are also in a logarithmic sequence and by pressing his fingers against each in turn the guitarist is able to play the notes up a chromatic scale. An unfretted instrument such as a violin works on the same principle but the divisions are not marked on the fingerboard.

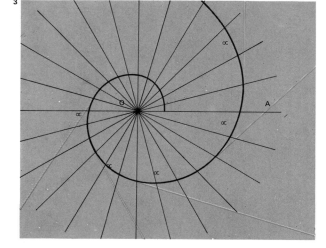

3 In this curve the angle α between a tangent at any point and the radius drawn from the centre is constant. For this reason it is called an equiangular spiral. The lengths of the radii to the curve are proportional to the logarithms of the angles between the radii and the initial horizontal direction OA, so it is also known as a logarithmic spiral.

task of addition. Similarly, logarithms can be used to perform division by actually carrying out a subtraction.

To calculate in decimal numbers, logarithm tables need be compiled only for the numbers between 0 and 9.999 (in four-figure tables; five-figure tables include 9.9999, and so on to as many figures as required). Larger numbers are expressed by adding a whole number (integer) called the characteristic, which represents in base-10 logs the corresponding power. The four-figure logarithm to the base 10 (written $\log_{10}$) of 2, for example, is 0.3010 [4]. The log of 200 is 2.3010 and of 2,000 is 3.3010 (200 is $10^2 \times 10^{0.3010} = 10^{2.3010}$, and 2,000 is $10^3 \times 10^{0.3010} = 10^{3.3010}$).

The slide rule

A slide rule [5] is a mechanical device for multiplying and dividing numbers to limited accuracy. Logarithmic scales are engraved on rods that can be slid in relation to each other, and numbers on them added or subtracted as needed – added for multiplication and subtracted for division. Because of the log scale,

the numbers become closer together along the slide, just like the frets on a guitar's fingerboard. Unlike an ordinary ruler the scale is geometric rather than arithmetic [6].

In its simplest form a slide rule has only two scales – called the X or D scales on a complicated slide rule. To multiply two numbers, the 1 on the upper scale is set opposite one of these numbers on the lower scale and the required product read off it opposite the second number on the upper scale [6A]. For division the two numbers are lined up and the quotient read off against the 1 [6B]. A transparent sliding "saddle" called a cursor can be moved along to line up the graduations and make them easier to read.

The accuracy of a slide rule is limited mainly by its length. A cylindrical slide rule [7] has scales up to a metre long wound round it like a screw thread. Most ordinary slide rules have additional scales to aid various types of calculations: reciprocal (a scale of all the numbers divided into 1), square (numbers multiplied by themselves), square root and even trigonometrical functions such as sine, cosine, and tangent.

The curve shown in illustration 3 is a logarithmic spiral. Such curves occur in nature, generally revealing the effects of accelerating growth as in the spiral shells of snails and various other molluscs and in flowers like this one.

To calculate $1\cdot113 \times 1\cdot456$

	0	1	2	3	4	5	6	7	8	9	1	2	3	4	5	6	7	8	9
10	0000	0043	0086	0128	0170	0212	0253	0294	0334	0374	4	8	12	17	21	25	29	33	37
11	0414	0453	0492	0531	0569	0607	0645	0682	0719	0755	4	8	11	15	19	23	26	30	34
12	0792	0828	0864	0899	0934	0969	1004	1038	1072	1106	3	7	10	14	17	21	24	28	31
13	1139	1173	1206	1239	1271	1303	1335	1367	1399	1430	3	6	10	13	16	19	23	26	29
14	1461	1492	1523	1553	1584	1614	1644	1673	1703	1732	3	6	9	12	15	18	21	24	27
15	1761	1790	1818	1847	1875	1903	1931	1959	1987	2014	3	6	8	11	14	17	20	22	25
16	2041	2068	2095	2122	2148	2175	2201	2227	2253	2279	3	5	8	11	13	16	18	21	24
17	2304	2330	2355	2380	2405	2430	2455	2480	2504	2529	2	5	7	10	12	15	17	20	22
18	2553	2577	2601	2625	2648	2672	2695	2718	2742	2765	2	5	7	9	12	14	16	19	21
19	2788	2810	2833	2856	2878	2900	2923	2945	2967	2989	2	4	7	9	11	13	16	18	20
20	3010	3032	3054	3075	3096	3118	3139	3160	3181	3201	2	4	6	8	11	13	15	17	19

Logarithm of $1\cdot113 =$

$\boxed{0\cdot0453} + \boxed{0\cdot0011} = \boxed{0\cdot0464}$

Logarithm of $1\cdot456 =$

$\boxed{0\cdot1614} + \boxed{0\cdot0018} = \boxed{0\cdot1632}$

To multiply, add the logarithms:
$\boxed{0\cdot0464}$
$+ \boxed{0\cdot1632}$
$= \boxed{0\cdot2096}$ log

i.e. $1\cdot113 \times 1\cdot456 = \boxed{1\cdot620}$

4 Log tables can be used to multiply or divide numbers. In this example 1.113 is found by adding the log under 1.11 to that for 0.003, to give 0.0464. Similarly the log of 1.456 is 0.1632. The two logs are then added to give 0.2096 and the required product is the number that has this logarithm – the number 1.620 in the table. In practice, results are found by consulting tables of antilogarithms.

5 A modern slide rule has various scales such as x, x^2, $1/x$, root x and so on. Some also have trig-onometrical and other functions for calculations by naviga-tors, engineers and others who use them.

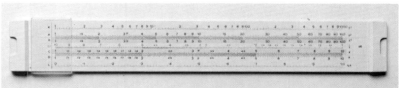

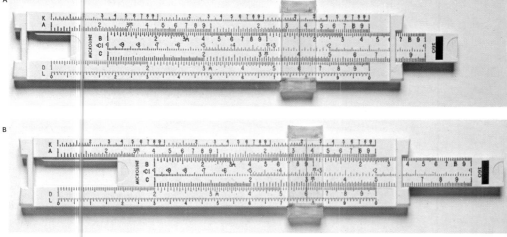

6 To multiply using a slide rule [A], for example to find the product of 1.5 and 4, the 1 on the upper scale is lined up with 1.5 on the lower scale. The required product is read off on the low-er scale opposite the 4 on the upper – in this example to give 6. The example of division [B] is 6÷3. The 3 on the upper scale is lined up with the 6 on the lower scale and the answer read off opp-osite the 1 on the upper scale. In this example the re-quired answer is 2.

7 A cylindrical slide rule can be pictured as a set of long scales wound round the cylinder like a screw thread.

Sets and groups

The mathematical theory of sets was first investigated by Georg Cantor (1845–1918) and later systematized by Ernst Zermelo (1871–1956), but the basic concepts were known earlier. Some adults new to these ideas find them difficult; but children have an intuitive grasp of them early in life. The concepts of number and operations on numbers are abstractions from the experience of sorting and combining sets of objects.

Collecting objects together
The idea of a set is the most fundamental concept in mathematics. A set is a collection of objects with a common description or definition, listed in any order or according to a formal law. The set of oceans, for example, is defined as: oceans = {Pacific, Atlantic, Indian, Arctic, Antarctic} or O = $\{x \mid x$ is an ocean}. The letter O labels the set; x is called a variable; { and } are called braces; and the symbol | means "where" or "such that". This kind of set is a finite set because its cardinality (number of elements) is finite – it has a known value, in this case five. The set of counting numbers is an infinite set because

we cannot say exactly how many elements it has: counting numbers = {1, 2, 3, . . .}, or C = $\{x \mid x$ is a counting number}.

The set of natural numbers is J^+ = {1, 2, 3, . . .}, with the same elements as the set of counting numbers. We say that C and J^+ are equal sets. Sets with the same cardinality are called equivalent sets: the set {blue, green, yellow, orange, red} is equivalent to the set of oceans – they each have five elements.

The language of sets can be understood by studying a particular example. A universal set [1], the set of all elements under consideration, can be partitioned into what are called disjoint subsets – that is, non-overlapping sets. If there are only two such sets, one is the complement of the other [2]. The set of elephants living at the North Pole is an example of the empty or null set, since it has no elements. The null set is written as ∅. In illustration 2, for example, there is no intersection of sets A and B, or P and C, so the intersection equals ∅. The concepts of partition, complement, intersection [3] and union [4] are fundamental to the processes of classification of information.

Networks [5] give rise to the Cartesian product of two sets. This is obtained by finding all possible ordered pairs of elements, taking one from each set. The word Cartesian is derived from René Descartes (1596–1650), who propounded the concept of co-ordinates. If set X is associated with the infinite set of points making one line in a plane and set Y is associated with the infinite set of points making another intersecting line, the Cartesian product of X and Y is associated with the infinite set of points making the plane containing the two lines [6].

Boolean and propositional algebra
The algebra of sets is known as Boolean algebra. It is isomorphic – that is, it has a one-to-one correspondence – with the algebra of propositions or logic. It is named after George Boole (1815–64), who founded the modern study of logic. The two types of algebra use different symbols, with *union* (∪) and *intersection* (∩) corresponding to *or* (∨) and *and* (∧). Propositional algebra analyses the sets of logical possibilities in which various statements and combinations

CONNECTIONS

See also
34 Finding unknown quantities: algebra
54 Maths and mapping
30 The language of numbers

In other volumes
228 Man and Society

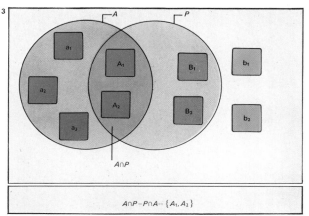

1 Two families – one with three children and one with two – together make up a universal set [A]. It can be represented diagrammatically [B] and letters given to each of the elements in the set. The letters are then sufficient for the mathematical manipulation of the set in what are called Venn diagrams, first introduced by the mathematician John Venn (1834–1923) in 1880. In such diagrams, areas represent sets of things.

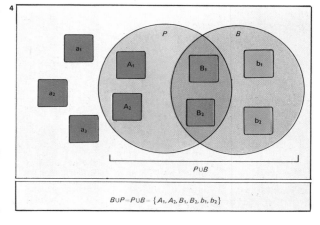

2 These Venn diagrams show how the universal set of illustration [1B] can be split into two non-overlapping subsets. Each family can make up a subset [A] or the parents and children can each form subsets [B]. In each case the subsets are complementary to each other because they include between them all the elements of the first universal set. The complementary relationships in [A], for example, are written as $A' = B$ and $B' = A$.

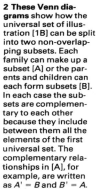

Complementary sets $A' = B$ $B' = A$

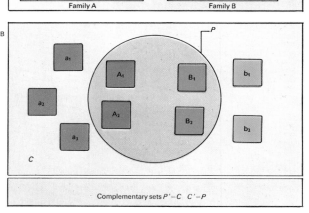

Complementary sets $P' = C$ $C' = P$

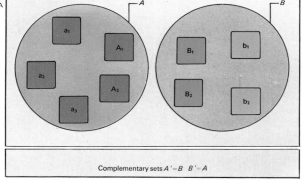

$A \cap P = P \cap A = \{A_1, A_2\}$

3 Intersection of sets generates another subset that contains all the elements common to both. Here the intersection of A and P (written as $A \cap P$) gives a subset containing only A_1 and A_2.

4 Union of sets generates yet another subset that contains all the elements in two original sets, in this example B and P. It is written $B \cup P$ and contains the elements A_1, A_2, B_1, B_2, b_1 and b_2. $B \cup P$ = $P \cup B$ illustrates the commutative law.

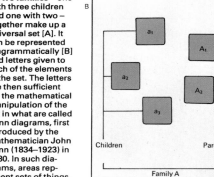

$B \cup P = P \cup B = \{A_1, A_2, B_1, B_2, b_1, b_2\}$

of the statements are either true or false.

A mathematical system is created when one or more binary operations are applied to a set of elements. A binary operation combines two elements into a third of the same set. One of the most valuable systems is the "group", as it occurs in many diverse situations and helps to unify the study of mathematics. The theory was developed by Evariste Galois (1811–32) and later systematized by Arthur Cayley (1821–95). The concept of a group can be illustrated by studying a simple case of formation dancing [8] in which four dancers change their positions (or remain still) to form various patterns. The movements form the set and the operation is a combination of movements called "follows", indicated mathematically by the symbol $\otimes$. Combining any two movements results in one of the four. The identity element is I and each element is its own inverse in this particular example. We have the relationships $(J \otimes K) \otimes L = L \otimes L = I$ and $J \otimes (K \otimes L) = J \otimes J = I$, so that the associative law $(J \otimes K) \otimes L = J \otimes (K \otimes L)$ is valid. In the particular example of dancing there is another law, the commutative law: $A \otimes B = B \otimes A$.

From the four possible choices available in moving a rectangle [9], a set of four transformations arises. These can be paired by the operation "follows" to produce a combination of movements that is in a one-to-one correspondence to those in the dancing example. The two types are said to be isomorphic. The search for isomorphisms is essentially the core of mathematical study.

The usefulness of group theory
Group theory is useful in the study of number systems. The set of integers (whole numbers) with 0 included, $\{\ldots -3, -2, -1, 0, +1, +2, +3, \ldots\}$, is a group under addition with 0 as the identity element. The set of rational numbers is a group under addition if 0 is included. It is a group under multiplication if 0 is excluded. The use of group theory in the study of arithmetic not only enriches it but leads to higher concepts such as those of rings and field, sets of elements subject to two binary operations (addition and multiplication) that satisfy certain axioms.

Any collection of objects constitutes a mathematical set

– a collection of cans of soup, bottles of vinegar, supermarket

trolleys or people in the shop are all definable sets.

5 A map shows the roads connecting two towns A and C. All the roads pass through town B. The two routes from A to B are one set and the three between B and C another set. There are six possible ways of going between A and C. This is known as the Cartesian product of two sets, in this case all possible combinations of paired elements, taking one from each set. The study of networks is one aspect of the subject of topology.

5 Two sets of roads :
$X = \{x_1, x_2\}$
$Y = \{y_1, y_2, y_3\}$

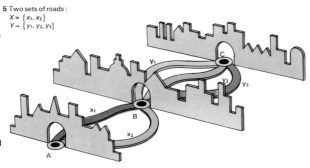

Six possible routes : $X \times Y = \{(x_1, y_1),\ (x_1, y_2),\ (x_1, y_3),\ (x_2, y_1),\ (x_2, y_2),\ (x_2, y_3)\}$

6 A plane defined by two lines is related to the Cartesian product of two sets that represent an infinite number of points on the lines. The point at the corner of the plane is defined by the co-ordinates x and y, written (x,y). These are called Cartesian co-ordinates and are used in co-ordinate, or analytic, geometry in which all lines, whether straight or curved, can be expressed in terms of algebraic equations.

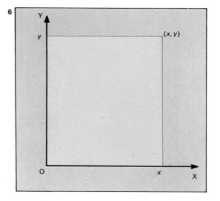

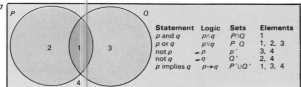

Statement	Logic	Sets	Elements
p and q	$p \wedge q$	$P \cap Q$	1
p or q	$p \vee q$	$P\ Q$	1, 2, 3
not p	$\sim p$	p'	3, 4
not q	$\sim q$	Q'	2, 4
p implies q	$p \rightarrow q$	$P' \cup Q'$	1, 3, 4

7 Union and intersection in set theory correspond to "or" and "and" in logic. This relationship enables particular elements and combinations of elements from sets to be defined by logical statements.

8 Four dancers [A], starting at the corners of a square, can have various positions [B], represented by the symbols I, J, K and L. Carrying out pairs of movements, one after the other, results in new

positions [C] described as J follows I, for example. Sequences of three movements can be analysed as two and the final position predicted. L follows K follows J, for example, reduces to L follows L, equals I.

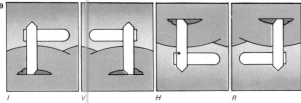

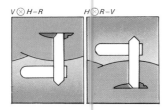

$V \otimes H = R$ $H \otimes R = V$

9 Symmetries of a rectangle involve rotating it in various ways – vertically V, horizontally H or in the plane of the picture R. The letter I represents its original position. Again successive pairs of movements always result in one of the four.

Finding changing quantities: calculus

In a "political vocabulary" compiled for the daily newspaper the *Guardian*, the noun "decrease" was cynically defined as: "reduction in rate of increase – as in unemployment, crime, inflation, taxation, etc". This not only exposes official double-talk [2] but also highlights the universality of a central concept of calculus – rate of change.

Rates of change became important in physics in 1638 when Galileo (1564–1642) concluded that a falling or thrown body had a downward velocity that increases steadily – that is, its rate of increase of downward velocity is constant [1]. What then is its trajectory? It took the genius of Isaac Newton (1642–1727) and Gottfried Leibniz (1646–1716) to solve this problem neatly and completely; the tool they created for the job was calculus.

Velocity, said Newton, is rate of change of position with time – 60km/h, for instance. Similarly, acceleration is rate of change of velocity with time. A car that takes three seconds to reach a speed of 60km/h from a standing start has an average acceleration of 20km/h per second. Galileo's law is that the downward acceleration of a falling body is constant. Calculus provides the methods of obtaining velocity from acceleration and position from velocity, and the whole problem is neatly solved. The calculus operation of deriving velocity from position, for example, is called differentiation. Its inverse is called integration.

The simplicity in symbolism

All mathematics is a sort of symbolic machinery for making subtle conceptual deductions without having to think them out – all the thought has been built into symbolism. Long division, for example, enables 431,613 to be divided by 357 by the unthinking application of a few rules; it is not necessary to know why it works, or what division really means. Calculus is perhaps the supreme example of a symbolism whose economical elegance reduces intractably complex and elusive problems to back-of-an-envelope simplicity.

In mechanics, the branch of physics for which calculus was invented, it is omnipresent in Newton's second law of motion: force equals mass multiplied by acceleration. Given any two of these quantities, the equation defines the third. Consider an internal combustion engine. What is the instantaneous acceleration of the piston as the crank passes top dead centre? Calculus provides the answer so that, knowing the piston's mass, it is possible to find the force on it that must be withstood by the connecting-rod. For what speed will this force become excessive? Again, the pressure on the piston during the power stroke is changing every instant with the burning of the charge and the changing volume of the gases in the cylinder as the piston descends. What then is the total energy imparted to the piston by the whole stroke and for what moment of ignition is it a maximum? This and a myriad of other mechanical problems could hardly be formulated, let alone solved, without calculus.

Electronic applications

Analogous applications occur in electrical engineering [Key]. Take, for example, a resistor (across which the voltage is proportional to the current), a capacitor (in which

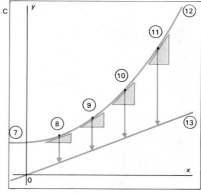

1 A falling hammer is shown [A] in six exposures 0.03sec apart and the hammer position curve [D in diagram B] follows them [1–6]. Its changing slope is important and diagram C explains how this is measured, like a road gradient, by tangent triangles with so many units vertically for each one horizontally. The slopes at points 7–11 of line 12 are 0:2, 1:2, 2:2, 3:2 and 4:2 respectively – or 0, 1/2, 1, 1 1/2 and 2. Line 13 plots the increasing slope of line 12. (Obtaining a slope curve from a curve is called differentiation; the reverse is integration.) In [B], the slope of the distance curve for the first exposure is given by a vertical drop of 2.8cm for a time-interval, measured horizontally, of 0.02 sec. So the velocity here is 2.8cm in 0.02sec = 140 cm/sec. Similar slope measurements give the velocity at other points and yield velocity curve V which initially increases downwards, like the distance. The acceleration (the rate of change of velocity with time) is the slope of the velocity curve – constant while the hammer is falling freely (curve A). At exposure 3, the hammer hits the nail at 200 cm/sec and things start to happen. In a millisecond (thousandth of a second) the velocity is braked back to about 40 cm/sec (curve V); this rapid deceleration drives A off-scale upwards to perhaps 100 times its free-fall value. Newton's law then gives a correspondingly large force on the nail, 100 times the weight of the hammer, and this is how the hammer works – driven by the force generated by rapid deceleration of the head. As the wood seizes the nail the shock kicks the hammer upwards; then in free fall again the hammer resumes steadily increasing downward velocity V¹ and constant acceleration A¹.

the current is proportional to the rate of change of voltage with time), and an inductor (where the voltage is proportional to the rate of change of current with time). Connect them all together and apply an alternating voltage. What happens? Calculus swiftly expresses this seemingly mind-boggling tangle as a differential equation and solves it to show, among other things, that at a certain frequency the whole affair "resonates" and very large currents can flow for very little applied voltage. Resonance is of fundamental value throughout electronics. The tuning control of a radio selects one station out of many by setting a circuit to resonate at the station's transmission frequency.

One powerful application of calculus is in seeking maxima, minima, and optima generally. A vertically thrown ball is momentarily stationary at the top of its flight: when the height is a maximum, the rate of change of height with time is zero. It can be found by differentiating the expression relating height and time and putting this equal to zero. This is a general rule of great value – for all technology is governed by the search for optima.

What is the optimum speed for a journey, for example? Too slow wastes time, too fast wastes fuel; both of these have an assignable cost. Calculus enables the rate of change of overall cost with speed to be found and the speed for which it is zero. This must then be the speed for minimum cost. Such calculations are essential in making the best use of ships and aircraft and the same principles apply in all searches for the best design or flow-rate or working temperature of almost every industrial system [6].

Universal principle
Many physical laws embody the same principle [3, 4, 5]. Thus light traverses an optical system by a path that takes less time than any other possible path – a principle from which the whole of classical optics can be derived. Indeed Leibniz, possibly carried away by the power of his creation, proposed that the whole universe had been designed in some such mighty self-optimization process and that this was the best of all possible worlds. He seems not to have considered the possibility that it might be the worst.

Faraday's "anchor ring" was the first-ever transformer. It revealed the law that the voltage on the output coil depends on the rate of change of voltage on the input coil.

2 Calculus can be used to analyse a headline such as: "Government acts to hold prices; rate of increase of inflation cut back". Suppose curve 1 is the rate of increase of inflation. Inflation itself will be the integral of 1 – curve 2, whose rate of increase at each point is proportional to the height of curve 1. Thus the slopes shown (black) are equal. Inflation is the rate of increase of prices, so prices (curve 3) are the integral of curve 2. It is then obvious that prices are not being "held".

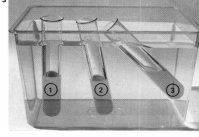

3 The stability of ships, buoys and other floating objects depends on whether a small tilt raises a greater weight than it allows to fall. In calculus terms, stability requires a positive rate of change of height energy with tilt. The picture shows that with more than a critical volume of ballast-liquid, a cylindrical tube floats vertically [1]. Tubes 2 and 3 have too little.

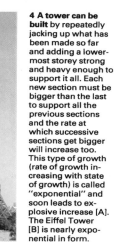

4 A tower can be built by repeatedly jacking up what has been made so far and adding a lowermost storey strong and heavy enough to support it all. Each new section must be bigger than the last to support all the previous sections and the rate at which successive sections get bigger will increase too. This type of growth (rate of growth increasing with state of growth) is called "exponential" and soon leads to explosive increase [A]. The Eiffel Tower [B] is nearly exponential in form.

5 The curvature of a beam at any point depends on its loading. A projecting strip is bent at each point by the weight of the length beyond it and calculus-based beam-theory adds up all these changing curvatures to arrive at the final shape of self-loaded beams. These plastic strips have different thicknesses and their degrees of deflection vary roughly as the inverse square of their thickness. Engineering beams do not sag as much but the same design principles apply.

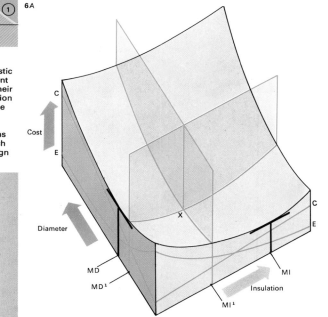

6 The cost benefit of increased insulation on a pipe [A] depends on reduction of heat loss (E on the pale blue) offsetting higher capital costs C. Their sum is least at MI. Increasing diameter reduces pumping costs (E on the dark blue) while raising capital cost (least at MD). The overall minimum cost is (MI¹, MD¹) at X. A chemical plant [B] must minimize cost over hundreds of variables.

Lines and shapes: geometry

Imagine the United Nations decided to encircle the world at the equator with a steel band symbolizing international unity. If the contractor made it too long by one part in ten million – 4m in 40,075km (13ft in 24,900 miles), how high would it stand above the surface all round the globe? The answer is 63.7cm (approximately 25ins).

Lines and shapes working for a living

The foregoing is an example of simple geometry, the mathematics of size and shape. Since all solid objects have size and shape geometry is one of the most practical mathematical studies. If someone wants to know how thick to make a rotating shaft to transmit a certain amount of power, or what contour to give a ship's propeller, or even how much paint is needed to cover a room or concrete to lay a path he uses geometry to provide the solution [5]. Indeed, geometry arose from the surveying needs of the early Egyptians, who had to share out fairly the featureless hectares of fertile mud left by the annual flooding of the River Nile.

The Greeks took geometry over [3] and

built an amazing intellectual edifice out of it. Euclid's *Elements of Geometry*, which was written in about 300 BC, develops a complete "axiomatic system" – a web of interlocking proofs all derived from a few basic axioms. "If you can't prove it, you don't know it!" challenged the *Elements* and ever since the admitted business of mathematicians has been the clarifying of basic axioms and the proving or disproving of statements derived from them.

A practical engineer seldom bothers with proofs; he generally accepts the mathematician's formulae and uses them. And almost instinctively, because geometry makes it simple, he designs objects from rigid parts linked at pivot joints. Many mechanisms around us embody the truths of plane geometry. The motions of a typewriter, the pantograph of an electric locomotive [2], the suspension of a car, the linkages in a sewing-machine or an autochange gramophone can all be described as working "models" of a set of geometrical theorems.

Some machines – printing presses and knitting machines for example – appear

almost magical in the motions they generate by ingenious geometrical linkages. And most rigid structures use the geometrical fact that a triangle is the only rigid figure. A triangle of rods joined at pivots cannot deform whereas a square, say, can distort to a diamond shape. As a result, girderwork is generally made up of triangles (a big girder bridge is a good example, as is a geodesic dome [Key]).

Pi in the sky, and elsewhere

The circle is a simple geometrical shape but one that is mathematically rich. The Greeks succeeded in proving its circumference to be $2\pi r$ and its area πr^2, where r is its radius and π some number between $3^{1}/_{7}$ and $3^{10}/_{71}$. In fact π cannot be expressed as any whole-number fraction. Expressed in decimals it begins 3.1415926535 . . . and goes on for ever, with a never-ending series of numbers after the decimal point with no numbers repeating. It is a fundamental constant in trigonometry, a numerical branch of geometry invented for mapping the stars and now fundamental to astronomy, navigation, surveying and all kinds of practical measure-

1 Pythagoras' theorem is the famous one children learn at school. The square on the hypotenuse (the longest side of a right-angled triangle) is the sum of the squares on the other two sides. The big bottom square (the hypotenuse square) divides into four corner sections, which can be reassembled into the top left square, and the one in the centre, which is the size of the top right square.

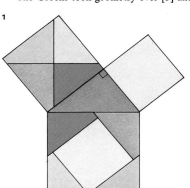

3 The ancient Greek Eratosthenes measured the earth's circumference by geometry. He found that when the sun was overhead at Syene it was 7° from the vertical at Alexandria. He knew the distance between them, about 800km (500 miles), and he reasoned that it represented 7° at the earth's centre. The full 360° of circle representing the earth's circumference must be 360 ÷ 7 × 800 (500) = about 41,140km (25,700 miles).

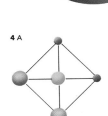

East
Vertical
West
To sun
Vertical
Longitude angle of sun from vertical

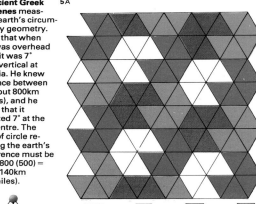

5A

2 An electric locomotive must always maintain contact with the overhead wire, which itself is not perfectly level. The geometry of the pantograph achieves this. A spring system urges points P and

5 Tiling a floor with identical tiles can be done in various ways. Obviously it can be done with equilateral triangles [A], hexagons [B], squares, or with tiles made by fusing shapes together [C]. But it can-

Q together, pivoting arms A and B to shorten distance ST. This distortion of triangle STC keeps C against the overhead wire. As arm A pivots linkage RQ makes B pivot too, preserving symmetry.

not be done with pentagons or any tiles with pentagonal symmetry. Geometry proves that there are just 17 basically different tiling patterns – including the most ornate – using identical tiles.

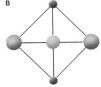

B

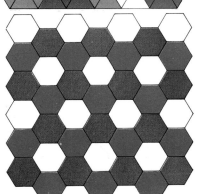

C

4 Molecules are too small to be visible so chemists use geometry to deduce molecular structures. Dichloromethane (a solvent used in paint-strippers) has one carbon atom, two hydrogen atoms and two chlorine atoms in its molecule. If they

were arranged as a square with the carbon in the middle, two forms of dichloromethane should exist, one with the chlorines adjacent [A] and one with them opposite [B]. But if the atoms are arranged tetrahedrally only one form is poss-

ible [C]. Only one form has ever been obtained so the square structure is wrong. By such reasoning chemists deduced the spatial arrangements of thousands of molecules long before methods such as crystallography provided direct evidence.

ment. In fact π has "escaped" from geometry and pervades all numerical measurements.

Some of the most elaborate geometry based on the circle is used in lens design. Almost all lenses – for cameras, spectacles, telescopes and so on – have circular cross-sections. Tracing the light path through a multi-component lens system is a complex geometrical task now carried out by computers. The computer programs calculate the characteristics of many possible lens designs and select the one with the fewest aberrations (for no lens system can be absolutely perfect). The result is a compromise, but the best that can be reached bearing in mind the practical difficulty of actually grinding the lenses.

Geometries beyond intuition

Euclidean geometry takes a number of intuitive notions for granted – the idea of a straight line, for example. Euclid thought of it as a line of zero curvature, the shortest line that could be drawn between two points. In practical matters, such as sighting and surveying, we assume that light travels in straight lines. But the physicist feels free to question

these suppositions. He considers it possible that light flashed out from the earth might go all round the universe and return to its starting-point, just as a person would who travelled in what he regarded as a "straight line" on the spherical earth. Indeed cosmology, the study of the universe as a whole, currently favours a "closed curved" universe with a finite volume but no boundaries just as the earth's surface has a finite area but is without edges.

Mathematicians see Euclidean geometry as just one of many imaginable geometries, each true of space of a particular curvature [10]. Their theorems may be strange, but provided they can be rigorously derived from the stated axioms (assumed facts), mathematical protocol is satisfied. And which of them is true of our real space is a matter of scientific experiment, not of axiomatic assertion. Fortunately, any curvature must be very small, so that Euclidean geometry works well in the small volumes we can deal with, just as in mapping a small area of the earth it can be assumed to be flat without significant error.

A geodesic dome is a rigid structure

made from many triangles and designed

for both lightness and strength.

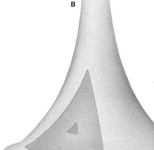

6

6 An air compressor uses the subtle geometry of interlinked cycloids. The end-lobes of each "paddle" have the curve traced by a point on a small circle [1] rolling outside the pitch circle. Its waist has the curves

from a similar circle [2] rolling on the inside of the pitch circle. As the paddles mesh like 2-toothed cogwheels, they always touch each other, trapping successive volumes of air and compressing them.

7 This model of Felton and Murray's early steam engine is a geometrical theorem in action. The inner cogwheel rolls around a fixed outer gear of twice its diameter. Geometrically, this implies that one

point on it reciprocates in an exact straight line. A piston rod attached at this point is driven from a cylinder; a crank takes the drive from the centre of the rolling wheel, to drive other machines.

7

8

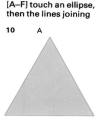

8 The principle of duality in geometry states that any two lines define a point (their intersection) and any two points define a line (the one joining them). If six points [A–F] touch an ellipse, then the lines joining

them form three opposite pairs whose points of intersection meet at a single line [GHI]. The dual of the theorem is that if six lines touch an ellipse then the points at which they intersect [J–O] form three op-

posite pairs whose lines of connection meet at a single point [P]. So points and lines are "duals" of each other and if these words are interchanged in a theorem, a new theorem results.

9

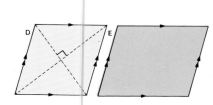

9 Quadrilaterals include a square [A], with right-angles and all its sides equal and parallel; a rectangle [B] with only opposite sides equal; a trapezium [C] with only two opposite sides parallel; a rhombus [D] and a parallelogram [E], both with opposite sides parallel and no right-angles.

10

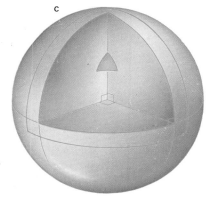

10 Euclidean geometry [A] is not inevitable and may not be true of real space. Mathematicians accept any geometry that is not self-contradictory and recognize many different kinds. In Lobachevskian geometry [B] the angles

never reach 180°, like geometry on a trumpet surface. In Riemann geometry [C] angles of a triangle always exceed the Euclidean 180°, like geometry on the surface of a sphere. Extended to three dimensions this is geometry of "curved space".

Lines and angles: trigonometry

The Simplon tunnel, between Italy and Switzerland, is 20km (about 12 miles) long and was bored from both ends through the Alps. When the headings met in the middle, in 1906, they were in exact horizontal alignment and only 10cm (4in) out vertically. The engineers managed to smooth out the discontinuity. Using trigonometry they had set up their machines to cut along the 10km (6 mile) sides of two huge triangles in the mountain.

Sines, cosines and tangents

Trigonometry is the art of calculating the dimensions of triangles. The basic idea [1] is that the ratios between the sides of a right-angled triangle depend on its base angle [A]. The ratios have been named the sine of A (sin A), the cosine of A (cos A), the tangent of A (tan A) and others. They have been tabulated for many values of the angle A. Sin A is the length of the triangle side opposite the angle A divided by the longest side; cos A is the length of the side adjacent to the angle A divided by the longest side; and tan A is the ratio of the length of the opposite and adjacent sides of the triangle.

Armed with trigonometrical tables anyone can determine the dimensions of any triangle with great accuracy. Since nearly any shape can be broken up into a series of triangles this is a powerful method of solving even complex spatial problems. To use it in tunnelling engineers set up a station from which both the ends are visible or (as this may be difficult with mountains all around) a station from which other stations are visible, from which in turn the ends can be seen. They measure the angles between all the stations by optical sighting and thus relate the two ends. Trigonometry then tells them the tunnelling angles that will align the two headings. The required accuracy of a thousandth of a degree implies a certain expertise; but the mathematical principle involved is nevertheless extremely simple.

Trigonometry in everyday life

Trigonometrical ratios have, however, "escaped" from their simple geometrical interpretation and uses in surveying and measuring, and now crop up in all sorts of mathematical problems that do not seem to be at all "angular". Some of their most fruitful applications are in circuit theory, radiation physics and information-handling, in which the angles are not real but introduced merely for convenience.

The sine of 0° is 0 and it increases with increasing angles up to 90°, whose sine is 1. Between 90° and 180° the sine reduces again to 0. From 180° to 270° the sine is negative, decreasing to −1. And from 270° to 360° the sine increases again from −1 to 0. Thus if a trigonometrical angle is regarded as winding up continuously [5], its sine swings between +1 and −1 and back at each revolution of 360°. This periodic behaviour gives mathematicians a framework for handling waves, vibrations, oscillating radiation such as light and radio waves, and alternating current (AC) electricity. In most European countries a power station generator spins at 50 revolutions a second. As a result its output voltage (which depends on the sine of the angle of rotation) swings back and forth between positive and negative at 50 cycles per second (50Hz) to generate mains-frequency AC. Any other source of oscillation, even

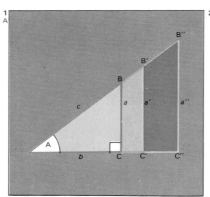
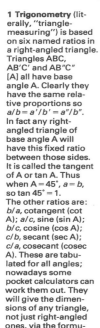

1 Trigonometry (literally, "triangle-measuring") is based on six named ratios in a right-angled triangle. Triangles ABC, AB'C' and AB''C'' [A] all have base angle A. Clearly they have the same relative proportions so $a/b = a'/b' = a''/b''$. In fact any right-angled triangle of base angle A will have this fixed ratio between those sides. It is called the tangent of A or tan A. Thus when $A = 45°$, $a = b$, so $\tan 45° = 1$. The other ratios are: b/a, cotangent (cot A); a/c, sine (sin A); b/c, cosine (cos A); c/b, secant (sec A); c/a, cosecant (cosec A). These are tabulated for all angles; nowadays some pocket calculators can work them out. They will give the dimensions of any triangle, not just right-angled ones, via the formulae [B]: $a/\sin A = b/\sin B = c/\sin C$ and $a^2 = b^2 + c^2 - 2bc \cos A$. (Side a is always opposite angle A, side b opposite angle B, etc.)

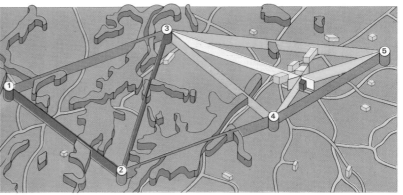

2 Surveying by "triangulation" uses the formulae that fix any triangle if one side and two angles are known. Distance [1–2] is carefully measured as the fundamental base line. Reference point [3] is selected and the angles of triangle 123 determined by optical sighting. This fixes point 3 and enables distance 2–3 to be calculated. Sighting from these established reference points will then locate any others [4, 5].

3 On a sphere such as the earth any distance can be represented by the angle it makes at the centre. Thus distance PB may be represented by angle POB. Accordingly positions are defined by angles of latitude (with the equator) and of longitude (north/south line). Point A has latitude x° west, longitude y° south; B has x'° east, y'° north. The "spherical trigonometry" of "spherical triangles" such as PAB tells a navigator the distance AB and the compass bearing (angle A) of the journey. Similarly in mapping the heavens astronomers locate stars on spherical celestial triangles like παβ.

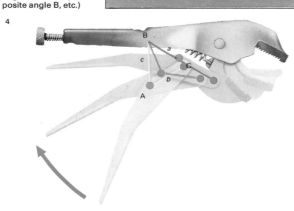
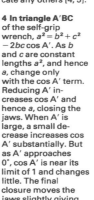

4 In triangle A'BC of the self-grip wrench, $a^2 = b^2 + c^2 - 2bc \cos A'$. As b and c are constant lengths a^2, and hence a, change only with the cos A' term. Reducing A' increases cos A' and hence a, closing the jaws. When A' is large, a small decrease increases cos A' substantially. But as A' approaches 0°, cos A' is near its limit of 1 and changes little. The final closure moves the jaws slightly giving a big leverage ratio with gripping force.

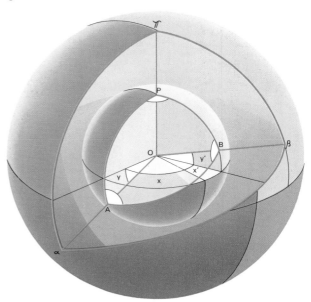

light with a frequency of 600 million million Hz, can be similarly assigned a notional "phase angle" winding up at the appropriate rate of time.

Any vibration, however complicated, can be made up of a set of sine-wave components (or cosine-wave ones which are similar), each with its own frequency. Each frequency is quite independent of the rest. (Two stones thrown into water together generate two sets of spreading ripples which intersect and go right through each other, emerging quite unaffected.) Similarly, the human ear can pick out the notes in a chord although they make a single vibrational pattern in the air or a single groove on a gramophone record.

Angles in a radio beam
Many electronic techniques process these frequency components of vibrations in ways governed by trigonometry. An AM (amplitude modulated) radio transmitter, for example, has to take a sine-wave audio frequency A (say the musical note A, 440Hz) and attach it somehow to a radio sine-wave "carrier" C, being broadcast at perhaps one

million Hz (1MHz, in the medium-wave band). It does this in effect by multiplying the audio voltage at each instant by the carrier voltage at that instant and transmitting the result. Now one of the many trigonometrical formulae for simple angles asserts that sin $A \times \sin C = \frac{1}{2}\cos(A-C) - \frac{1}{2}\cos(A+C)$. Since A and C are phase-angles of audio and carrier frequencies the result of the multiplication is two cosine-waves (just like sine-waves), one at $(1,000,000 - 440)$ Hz and the other at $(1,000,000 + 440)$ Hz, each of half the intensity of the original carrier.

The splitting of the carrier into two closely spaced "sidebands" is called amplitude modulation, or AM. A transmission generally has many such pairs of sidebands continuously changing in their spacing and intensity with the changing frequency-components of the audio signal. At the receiver the audio signal is recovered by the reverse process of demodulation. It may seem incredible that a mathematical formula first proved for static triangles on paper can be impudently applied to the imaginary rotating angles of an electronic signal.

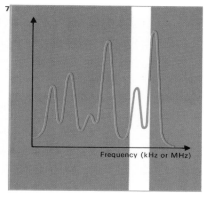

The quadrant was an early instrument used by astronomers to find the altitude of the heavenly bodies. The surveyor's quadrant developed as a portable version for surveying and artillery

ranging. This example was made by Jacob Lusuerg of Rome in 1674. Its most interesting feature is the Vernier scale – invented by Pierre Vernier (c. 1580–1637) in 1631 – for measur-

ing to $1/60$°. This is the lower arc-scale joining the legs of the pivoting V-shaped unit which slides over the static quadrant base plate. Another scale shows the tangent of the measured angle.

5 As a rotating radius sweeps out an ever increasing angle, the angle's sine varies cyclically, repeating itself for every additional 360° of rotation. For a circle of unit radius the sine is the

height of the end of the radius above the horizontal. Such sinusoidal waveforms occur in vibrations. The frequency is the number of radius-rotations per second. Two simultaneous sine-waves of dif-

ferent frequency will add together to a complex waveform: thus sine-waves [1] and [2] add to give the waveform [3], which might represent the variation in sound-pressure of two notes sounding together.

6 A waveform can be made by combining sine-waves; it can also be broken down into them. This diagram shows the amplitude (intensity) spectra of the waveforms of illustration 5. Waveform 1 has

only one component in its spectrum, at the frequency f. Waveform 2 has a single component of frequency 2f but of lower intensity. Their combination [3] has both these lines in its spectrum.

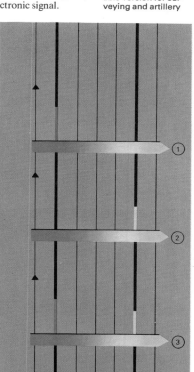

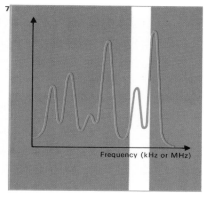

7 The complex waveform of all the signals entering the radio receiver's aerial will have many components in its frequency spectrum. Each peak is a broadcast on a specific frequency. Some

stations are weak, some strong; tuning the radio moves a narrow frequency-acceptance band along the frequency scale to select just one of them. The small modulation is then decoded to give sound.

Frequency (kHz or MHz)

f   2f   Frequency

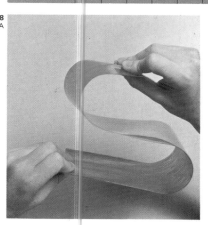

8 A bent steel strip [A] does not adopt a sine-wave form but a related "sine-generated" curve. The direction of the strip from point to point varies sinusoidally with distance along the strip. This minimizes the energy of bending stored in the steel. This curve is created on a grand scale [B] when a slow-moving river winds to the sea. The water has little energy and so seeks the line of least resistance as it cuts its channel.

Surface and volumes: solid geometry

In 1826 the German astronomer Heinrich Olbers (1758–1840) asked what may seem to be a silly question: Why is it dark at night? Silly questions are sometimes the most profound and Olbers tackled this one using straightforward solid geometry. He imagined the universe divided into a series of concentric shells around the Earth, like the layers of an onion spreading out to infinity. He supposed that the stars were more or less uniformly distributed. Then, through solid geometry, Olbers calculated that a shell twice as far away is bigger and contains four times as many stars. But, in theory, only a quarter as much of their light should reach the Earth. Each shell therefore contributes the same radiance to the night sky no matter how far away it is. Because there are an infinite number of shells the night sky should be infinitely bright – or at least as bright as the face of the Sun.

So why is the night dark? Even today, astronomers do not agree on the structure of the heavens. The universe might be finite in space, with only a limited number of shells; or finite in time, so that light from the most dis-

tant shells has yet to reach here; or it may be expanding, weakening the light from distant shells. Olbers's paradox remains an outstanding example of how simple mathematics can provoke the most surprising conclusions from uncontroversial assumptions.

Sizing things up

Mathematicians and engineers alike have to be able to calculate the areas and volumes of various solid objects. For an object with flat faces the surface area equals the sum of the areas of the faces. Thus for a cube, the surface area is merely six times the area of one face. For a sphere the area is four times π times the square of the radius. The volume of a cube is the length of one side multiplied by itself three times (the length cubed), and the volume of a sphere is 4/3 times π times the cube of the radius.

Pyramids, prisms, cylinders, cones and ellipsoids present more complex problems, but all can be calculated using solid geometry, that is the geometry of shapes in three dimensions. Mathematicians use solid geometry to find the surface areas of such

shapes and whether they can be made by forming a flat paper shape (a cylinder, prism, pyramid and cone can be made this way, but not a sphere or ellipsoid). The path that a grinding-wheel of known dimensions must traverse to cut a given shape from a metal blank, or how much earth must be shifted to make a railway embankment of a given height, or what size cylinders can be bored in an engine-block for a given safe spacing between them (and the resulting swept volume) – to determine all these quantities, engineers make use of solid geometry.

Networks of force

The subject-matter of solid geometry includes not just the shapes of objects and assemblies, but the invisible strains and forces that traverse them. The centre of gravity of a cylinder is half-way up it; stood on end and tilted it will not fall over provided that any part of the top surface is still vertically above any part of the bottom surface. But the centre of gravity of a cone is a quarter of the way up it. It can be tilted until its tip is one-and-a-half times as far to one side as the

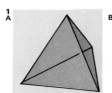 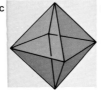 

1 A regular polygon has all its sides and angles equal, as in the equilateral triangle, square and pentagon. Euclid proved that there can be only five regular solids whose faces are all identical regular polygons: the tetrahedron [A], the cube [B], the octahedron [C], the dodecahedron [D] (with 12 faces), and the icosahedron [E] (with 20 faces). Only cubes pack together to fill space completely.

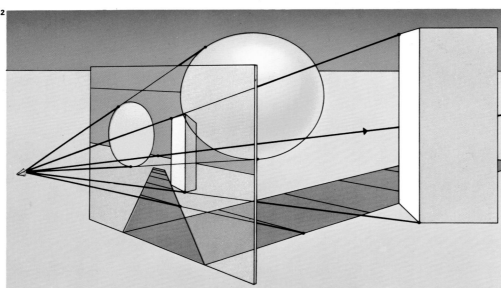

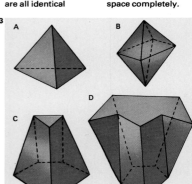

3 All solids that do not have holes through them obey Euler's theorem: $V + F = E + 2$ where V is the number of vertices (corners), F the number of faces and E the number of edges. For the tetrahedron [A] $4 + 4 = 6 + 2$; for the octahedron [B], $6 + 8 = 12 + 2$. The shapes C, D also obey the rules. The theorem is intriguing because the shape and size of the solid does not matter at all.

2 Solid geometry controls the perspective appearance of the world because light travels in straight lines. The laws of perspective envisage a picture-plane between the eye and the scene to be represented. Connect each point in the scene to the eye by a straight line: the place where this penetrates the picture-plane is its position in the perspective representation of the scene. From the eye's viewpoint, a perspective picture is seen.

4 A perspective picture has a central "vanishing point" to which parallel lines perpendicular to the picture-plane all converge. Other horizontal lines (such as edges of the cuboid box) converge to other points on the picture's horizon. A circular disc is distorted by perspective representation unless it is directly in front of the eye: the effect is small for most deviations from this position. This picture is the perspective view of illustration 2.

5 Uniform polyhedra can have several different regular polygons contributing to their faces. There are 13 "Archimedean solids" (not counting the infinity of simple prisms allowed by this definition) each of which has a regular polygon top and bottom, joined by square faces round the middle. If faces are allowed to intersect, 53 additional uniform polyhedra result. This one is composed of star-shaped dodecagons and equilateral triangles.

edge of the base extends on the other side.

Such simple results, elaborated for far more complex shapes, determine for example what form a dam must have in order that the water pressure should not push it over; how high in the water a boat of given shape will float, and how far over it will heel if loaded lop-sided; and what overloading of a tower-crane will just topple it.

For forces more complex than gravity still more intricate questions arise. What pressure can a round-ended gas cylinder withstand, and where will it fail if overpressurized? (Answer: on the inner surface, at a point midway between the ends.) What structure must an aircraft wing be given in order that, when loaded by lift and thrust and weight and drag, it will deflect into the desired shape without overstressing any of its parts? Problems such as these can be solved by modelling, or by computing and translating the solid geometry of the model into its numerical equivalents.

There are systems whose geometry reveals the active forces directly. A magnetic-liquid labyrinth [Key] reveals the opposing magnetic forces that mould it. And in nature, a bone or a tree that grows against the forces on it reveals those forces to the intelligent eye by the shape it grows into – the ideal shape for the loads it has to bear.

Molecular architecture

The solid geometry of molecules is surprisingly important in modern chemistry. It determines not only how they pack into crystals, but how they react. It is particularly significant for understanding enzymes, the powerful biological catalysts that bring about reactions which the chemist is often helpless to imitate. An enzyme is a huge molecule with a complicated active surface on which only the right reacting molecules can fit. And having fitted they are then held in the right positions to react. In doing so the reactants' geometry alters and they spring from the surface, leaving it ready to accept more reagents. The double helix of the DNA molecule consists of two interlinked twisting strands. The whole marvellous mechanism of the human body depends on the sub-microscopic solid geometry of the fundamental catalysts of life.

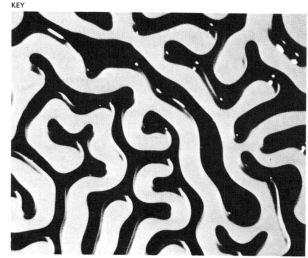

This mixture of a magnetizable liquid and an immiscible transparent one is in a magnetic field. Every part of the magnetic liquid then repels every other part, so it seeks to divide into many small sections. But every division uses energy so the liquid compromises to the shape shown here.

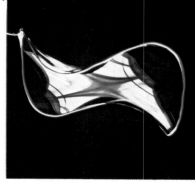

6 The geometrical shapes of engineering objects are often beautiful. This sludge-pump impeller is a stack of discs whose centres are helically disposed around the central axis of the impeller.

7 There is no complete mathematical construction to the general problem (given a closed line) of finding a minimum-area surface that has that line as a boundary. A soap film solves it automatically for any closed line. The film is always in tension and shapes itself to minimize its area. Here is one outlining the smooth and elegant minimum-area surface for a three-lobed loop made of copper wire.

8 Optical components have surfaces governed by the laws of optics and the solid geometry of ray paths. The big rectangular outline mirror of this infra-red spectrometer has an ellipsoidal sur-face. It intercepts a diverging beam of infra-red radiation coming from the left and reflects it so that it all converges on the radiation detector, that is the stalk mounted in front of the mirror. The mirror [bottom left] has a spherical surface and processes the invisible beam at an earlier stage of its journey through the instrument. The complete instrument has about 20 reflecting surfaces.

9 For maximum volume from a given area of tin-plate, the height should equal the diameter. A standard 425g (15oz) can [D] has a height 1.4 times the diameter because stamping out and forming the ends wastes some tin-plate. This is thus the practical optimum. The polish tins [A, B] are wider than ideal for easy access to the contents; the aerosols [C, E] are narrower to resist internal pressure.

10 The absolute minimum surface area for a given volume container is a sphere. (A soap bubble proves this by minimizing its surface area round an enclosed volume of gas.) This also gives the greatest resistance to internal pressure. Spherical tanks are used to store liquids under pressure. Such tanks are also used for liquids held at low temperatures where the absolute minimum wall area minimizes leakage of heat from outside into the liquid.

Shape and symmetry

If in the same boarding houses there are two rooms of equal aspect and furnishings, then their rents will be equal. For suppose the rents are different: then one tenant will be paying less than he might, which is absurd. This elegant theorem formulated by the Canadian humorist Stephen Leacock (1869–1944) illustrates the mathematical idea of symmetry.

Symmetry is a powerful concept and its workings can be seen in many aspects of the world. The two halves of a bridge span, the wings of a bird or of an aircraft, the blades of a propeller, all have symmetry – for otherwise one of them is worse than it need be, which is also absurd. Mathematicians recognize many different types of symmetry all described by the group of real or imaginable "symmetry operations" which leave the symmetrical entity apparently unchanged. A square, a cube or a four-bladed propeller can all be turned through 90 degrees without apparent change: they are said to have a "fourfold axis of symmetry". An irregular object has the lowest symmetry because any twist or turn is detectable. A sphere has the highest possible symmetry; no twist or turn is detectable. This made it the "perfect" figure to the ancient Greeks and makes it highly useful today. A ballrace is so simple because the balls need no aligning; no matter how they roll they cannot jam the bearing. A roller-bearing of lower symmetry needs guides to keep rollers parallel to the bearing axis; a tapered roller-bearing of lower symmetry has even more geometrical constraints.

Symmetry of nature

A snowflake [Key] shows how the laws of nature give symmetry to their products. It has 120-degree angles between many faces because in the water molecules of which it is comprised two hydrogen atoms form a 120-degree angle with an oxygen atom. The crystal lattice in ice is formed by the regular interpacking of the molecules and reflects this symmetry.

But this does not explain why the whole elaborate structure has a sixfold axis of symmetry. How does one branch of the flake know how its fellows are growing, so as to imitate them exactly? The physicist Samuel Tolansky (1907–73) made the suggestion that a snowflake, as it falls and takes up water vapour from the cold air, is vibrating with the symmetry of its crystal structure. All the branches move and twist together in a complex and changing pattern; the fastest points on each branch intercept the most water vapour and so grow together.

Such symmetries of process are common throughout nature. The radial shatter-pattern of a broken window betrays the symmetrical stresses that radiated outwards from the impact point.

Symmetry in the abstract

Mathematics manages symmetry by "group theory", a fascinating topic which, from a few apparently trivial axioms, develops rapidly into a structure of amazing subtlety and elegance. The oddest thing about it is that, unlike number theory, it allows $a \times b$ not to equal $b \times a$. This lack of symmetry in the mathematics of symmetry may seem like complete nonsense, but in practice the order of events can also be important. Sanding down a door, then painting it, for example,

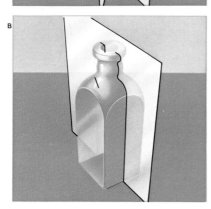
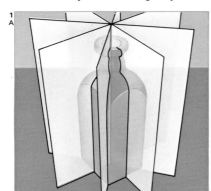

1 An object or mathematical entity has symmetry if some defined "symmetry operation" on it leaves it unchanged. If the bottle [A] is rotated through 90° about its vertical axis it presents its original appearance again. Because this symmetry operation would recur four times in a complete revolution the axis is a "fourfold axis of rotational symmetry". The bottle's other "symmetry elements" are four mirror planes. Reflecting every point on the bottle through such a plane to the corresponding position on the other side is a symmetry operation. A bottle with a blank label [B] has no rotation axis and only one mirror plane as symmetry element. The pseudo-bottle [C] has new symmetry elements: a horizontal mirror-plane H and four twofold axes in it, as well as a centre of inversion [I], about which the bottle can be rotated and remain unchanged.

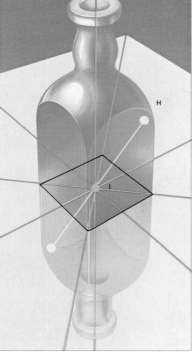
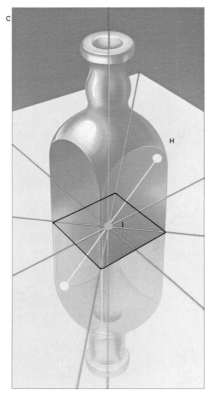

2 Crystals of chemical substances show symmetry that derives from the lattice of molecules composing them. The urea crystal has a vertical "improper axis of symmetry" for a symmetry element. This means that rotating the crystal through 90° and then reflecting it in a bisecting horizontal plane leaves it apparently unchanged. It also has two mirror planes and two 2-fold axes.

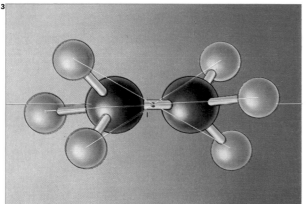
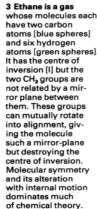

3 Ethane is a gas whose molecules each have two carbon atoms [blue spheres] and six hydrogen atoms [green spheres]. It has the centre of inversion [I] but the two CH_3 groups are not related by a mirror plane between them. These groups can mutually rotate into alignment, giving the molecule such a mirror-plane but destroying the centre of inversion. Molecular symmetry and its alteration with internal motion dominates much of chemical theory.

4 The common starfish (*Asterias rubens*) has five appropriate planes of symmetry and a 5-fold rotation axis. Among animals only a few specialized sea creatures (radiata) have such high symmetry. They probably evolved from ancestors of lower symmetry, as inferred from their larvae, which have the approximate mirror-symmetry of most creatures, including man. The starfish has no horizontal mirror plane; it has a true "top" and a true "bottom".

gives a different result from painting it and then sanding it down.

For an object with symmetry its group consists of the "operations" that can be carried out on it: turning it through 90 degrees, reflecting it in a plane and so on. Take a squat square-shouldered pill bottle without its label, hold it upright and pivot it through a right-angle about its top-left and bottom-right corners. It will then be horizontal with the neck on the left. Turn it clockwise through 90 degrees and it will be upright again. That is a symmetry operation. But if the latter is performed first and then the former the bottle will finish upside down: $a \times b$ does not equal $b \times a$.

The uses of group theory

Group theory is one of the many inventions of nineteenth-century mathematics that later found scientific use. Indeed, the rapid spread of its strange but potent "arithmetic" in twentieth-century physics earned it the title, among an older generation of physicists, of *die Gruppenpest* ("group nuisance"). But its incorporation into modern physics and

chemistry, with their need to understand the subtle symmetries of molecules and crystals and their energy states, has made possible the theories which give us such modern marvels as semiconductor electronics.

So much symmetry exists in scientific theories and mathematics that researchers acquire a feeling for it and any "lopsided" features of a theory or experiment make them uneasy. In electromagnetism, for example, the fact that electric charges (positive and negative) can be isolated, whereas magnetic poles (north and south) cannot, seems somehow to be "wrong". Many physicists have sought magnetic monopoles to complete the symmetry of the situation but, so far, attempts to discover such particles have been unsuccessful. But the most daring of all such insights was that of Albert Einstein (1879–1955) when he reasoned that the speed of light (and indeed every phenomenon of physics) must be the same for all observers, no matter how fast they themselves were travelling. Implicit in that mighty assertion of symmetry was nuclear power and the atomic bomb.

The symmetry of a snowflake echoes the symmetry of its molecules but owes its elaborate perfection to the subtle process of crystal growth by vapour-deposition on a vibrating surface.

5 Just as an object can have symmetry, so can an infinite repeating lattice. The symmetry operations for objects also apply to lattices. But there are other operations which, applied to an infinite lattice of appropriate symmetry, will leave it apparently unchanged. One is "translation", that is, shifting the lattice sideways. Every lattice can be divided into repeating "unit-cells" and displacement by one unit-cell spacing is a symmetry operation. Another is "gliding", reflection in a line followed by translation along it. This painting, "Angels and Devils" by M. C. Escher (1898–1970), has symmetry elements decreasing in size.

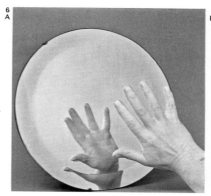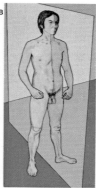

6 Why does a mirror turn an image right-to-left and not upside down? The answer to this confusing question is that the mirror's transformation is neither right-to-left nor up-to-down, but back-to-front. The left side stays on the left and the top stays on top, but the back becomes the front. Because hands [A], like other pairs of body parts related by mirror symmetry, are called "right" and "left", a right hand becomes a left one, starting the confusion. Some lack of symmetry is not obvious: in man [B] the heart and other organs are on the right if he considers his mirror image.

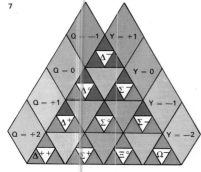

7 Symmetry considerations are basic to nuclear-particle physics. Many fundamental particles are now known, but the laws governing their occurrence and properties are poorly understood. One attractive theory has them composed of "quarks". The diagram shows how the d-quark [blue], the u-quark [red] and the s-quark [green] might combine in threes to form each of 10 particles called hyperons ($-$, 0, or $+$). The charge [Q] and "hypercharge" [Y] of each hyperon is correctly predicted by this type of symmetry classification, which also predicted the Ω^- particle before its discovery.

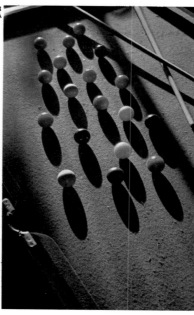

8 One of the most paradoxical of physical laws is that of time symmetry: any process can go backwards. This may seem absurd but in a film of two or more billiard balls colliding [A], it might not be possible to tell if it was run backwards. The reversed film would still show a possible physical event. But it might not be a probable one and most processes (eg throwing a stone into a pond) are unlikely to reverse. Nonetheless, dispersed molecular motions could converge on a stone and eject it spontaneously through a calm surface. Similarly, light rays can always retrace their paths exactly, so a camera could be used backwards [B] as a projector, exchanging object and image, and remain in focus.

The language of space: topology

When the comic film character Monsieur Hulot traces along a tangled hosepipe from a tap and finds it leads back to the tap again, why does the audience laugh? And what is so odd about a household hint from *The Times* of London: "Mending a hole in a tablecloth: lay the cloth on a table with the hole uppermost . . ."? Both items offend our instincts about topology – a branch of mathematics that deals not with shape or size but with much more fundamental properties of objects and of space.

Spheres, nets and knots

It is a topological truth that, regardless of its length or curvature, a hosepipe has two ends. Similarly, we feel sure that no matter what the size of a tablecloth, or the outline of a hole in it, it would be hard to spread the cloth out with the hole underneath. Topology takes such intuitive matters and formalizes them into mathematical logic. It is concerned with all those properties of objects that are unaffected by any change of form, however extreme. For instance, any simple solid object without holes is a "ball" to a topologist, for if made of deformable clay it could be rounded into a ball without being torn. Thus topology is sometimes called "rubber sheet geometry".

A button with four holes in it is not a topological ball. It is a "quadruply connected solid" because you would have to make four cuts in it, opening out the holes to the edge of the button, to make a shape that is a topological sphere. Small creatures living on the button would find it a different kind of space to the surface of a sphere. Any closed curve on a sphere [1], for example, must fence in and enclose a definite area. On a button this is not so. A closed curve round one of the holes does no such thing. The pure topologist's main concern is to decide whether particular abstract entities (objects or spaces of many forms and dimensions) are or are not topologically equivalent. Human intuition in comprehending the basic topology of even simple figures is relatively limited and sometimes leads to wrong conclusions.

This strange branch of mathematics has links with the real world [7]. An electrical circuit is a topological entity, for example; its exact layout does not matter because only the pattern of interconnections is electrically significant. Graph theory [3, 4], the branch of topology that handles networks, is fundamental in advanced circuit design. And the age-old crafts of knitting and weaving are really exercises in applied topology. A loop with a knot in it retains that knot however it is deformed and cannot be "undone"; it is topologically different from an unknotted loop. Textile manufacturers practise topology in their efforts to produce garments with specific topological properties: ones that can be knotted in one piece or that will not unravel if a fibre breaks [Key].

A mathematical playground

Most serious topology has, as yet, little to do with the practical world. No branch of it is as closely tied to human affairs as, say, arithmetic is to banking. It is therefore a subject full of potential – for time and time again in the history of mathematics such theorists' playgrounds have become the workshops of a new science or discipline. At present, however, theorems in topology, although proved

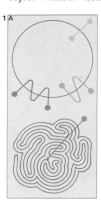

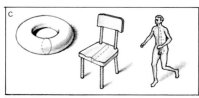

1 "Any closed curve divides a surface into an inside and an outside, and a line connecting these crosses the boundary an odd number of times." This theorem may seem obvious [A] but is not true of all surfaces. A table and statue [B] are topological balls on which the theorem holds. But a chair and man (because of his alimentary canal) have holes through them, like a doughnut [C], which need not obey the theorem.

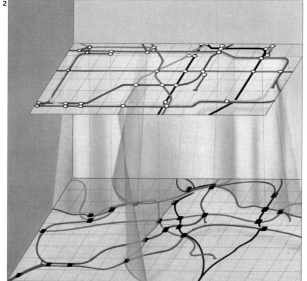

2 The London Underground map is a very distorted plan of the lines. But they correspond point-by-point, and two points joined on the map are connected in reality. This fundamental test makes them topologically identical.

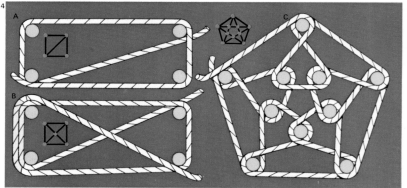

3 Königsberg, a Prussian town, posed a teaser that led to topological "graph theory". Could you take a stroll crossing each of its seven bridges only once? In 1734 the Swiss mathematician Leonard Euler analysed the problem to form a theory of the traversability of a network or "graph" (shown superimposed on the city). One bridge has to be crossed twice. The problem depends on connections, not distances.

4 A graph traversable in one pass must have an even number of lines meeting at each junction. The pentagram C is traversable but the rope makes a double pass on rectangle B and begins and ends at different points on rectangle A.

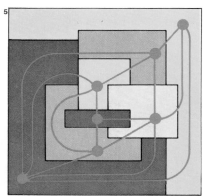

5 A flat map needs no more than four colours to prevent adjacent areas sharing the same colour. This unproved theorem is part of graph theory, for every map can be drawn as a graph with areas as junctions and boundaries as lines.

with full rigour, are less directly useful than those of a subject like geometry.

A typical topological theorem says that in colouring a flat map no more than five colours are ever needed to ensure that adjacent areas need not share the same colour. The theorem does not state how this can be accomplished for any given case but merely asserts that it can always be done somehow. In fact, four colours may be sufficient [5], although this has not been mathematically proved. Similarly, it is topologically certain that however briskly a cup of tea is stirred, at any instant at least one point in the liquid is not moving. The topologist is not concerned to identify this point; he just proves that it must exist. In different types of space different theorems hold good. There need be no fixed points in a stirred inner-tube full of water; on a doughnut, up to seven colours may be needed for a map without adjacent colours [6]; and on a Möbius strip, up to six colours may be needed.

A Möbius strip is some sort of vindication of Monsieur Hulot for it is a contradiction of the strong human intuition that a piece of paper must have two sides. It is named after the German astronomer and mathematician August Möbius (1790–1868). His strip can be made simply by cutting out a ribbon of paper, making a half turn in the middle of it and sticking the ends together to form a twisted loop. This loop now has only one side, as you can prove by drawing along it with a pen, never going over the edge until you meet your starting-point again. Cutting along the line creates another surprise.

Twisted space

Topologists study such "twisted spaces" in more dimensions than two, hard though they are to imagine. Indeed it is topologically entirely possible that the universe itself has a Möbius twist in it. One result of this might be that a traveller who went far enough out into space would return reversed in mirror-image fashion with his heart in the right side of his chest. Glove manufacturers might then be able to make only left-handed gloves and ship half their output around the universe from where they would return as matching right-handed ones.

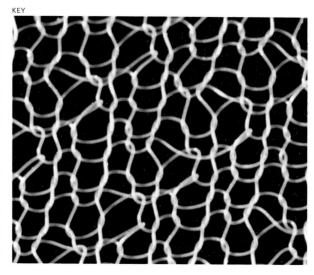

The significance of topology in textile manufacture is shown in the structure of a pair of tights photographed through a microscope. The complex system of knots is designed to avoid a "run" in the garment if a fibre is broken.

6 On a doughnut or torus a map can need up to seven colours to prevent adjacent areas sharing the same colour. The map shown (with its mirrored reflection for completeness) needs all seven because each area touches the other six. The sections form a continuous helix winding round twice before closing on itself.

7 Any structure has many "modes of failure". Engineering disasters such as the collapse of the River Yarra bridge in Melbourne can occur because of modes which the designers have not recognized. As a novel design such as the box girder bridge is refined and made lighter and cheaper, unsuspected modes of failure may be discovered the hard way. Classification of abstract entities by their "spatial" properties can help here. Topological "catastrophe theory" is concerned with the ways shapes can change; its principal application is to morphogenesis, the biological study of the ways in which organs and tissues develop and form.

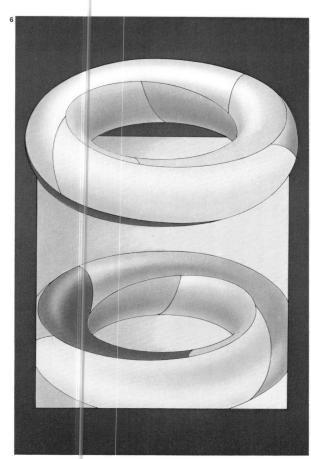

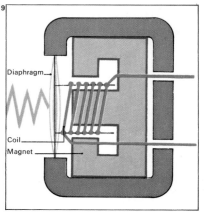

8 A ball covered with fur cannot be smoothed down all over; at least two crowns must remain where the fur radiates from a point or piles up at one, or a set of partings. This "hairy ball" theory governs the way directions, like hairs, can be aligned on a sphere. If they are lines of magnetic flux, it shows that every magnet must have two poles. If they are wind directions on the globe, the theory proves that somewhere in the world the wind is not blowing.

Diaphragm
Coil
Magnet

9 In electromagnetic machines such as motors, loudspeakers or meters, motion occurs and electric and magnetic fields intersect. Each field is a closed loop; topologically, two loops must intersect at an even number of points. In a moving-coil microphone the reverse effect takes place, again involving two intersecting fields. Sound waves vibrate a diaphragm which moves a coil in a magnetic field. The relative movement generates a current in the coil.

Maths and mapping

Can you read a map? A blueprint? A circuit diagram? Morse code? Then you are a mathematician because these are all examples of mathematical mappings. The idea is simple: a map is any way of relating one set of objects to another set [1]. In a geographical map [3], every one of the infinite number of points on the earth's surface corresponds to (or is "sent" to) just one of the infinite number on the map. Similarly blueprints and circuit diagrams map certain features of a physical object onto a pattern on paper.

Maps and their meaning
One interesting thing about such maps is what they can and cannot do. It is impossible, for example, to map the whole globe onto flat paper without sacrificing some features to preserve others. True directions on a map are impossible to achieve without some distortions. But even distorted mapping is still mathematically acceptable – like the hidden painting of a skull [Key]. The apparently meaningless set of distorted smears is an anamorphic painting (one that appears in proportion when viewed from a particular angle, generally using a lens or mirror), designed to map to a recognizable image when viewed correctly.

Mathematical maps embrace much more than these simple correspondences of points in space. They deal with anything: points, numbers, sets and abstract entities with no meaning beyond themselves. They even handle the mapping of a set of objects onto itself. This seeming paradox is commonplace, for example, in secret coding. A code is a rule for replacing each letter of a message by another from the same alphabet; it is a complex mapping of the alphabet onto itself. Similarly the two-times table relates a number to its double and (if we include fractions in it too) is a mapping of the set of all real numbers onto itself.

One-to-one maps are not the only kind. The "zero-times table" that takes all numbers to zero is a good mathematical map. But the mapping must specify a definite image or images for every element in its "domain" of operation. Therefore there is the old puzzle of the village whose barber shaves everyone who does not shave himself. This purports to describe a mapping that sends all the "shavers" to themselves and all the "non-shavers" to the barber. It is not a well-formed mapping because it leaves the barber himself in a paradoxical position. Does he map to himself? (That is, does he shave himself?) If so, he shouldn't, and if not, he should.

By contrast the marriage map between n men and n women (where n is any whole number) is a proper map and defines a possible set of marriages. Assuming that each individual has some order of preferences for his or her n possible partners, then it is a wry outcome of mapping theory that of all possible such maps – ways of pairing off the men and women – one and one only is stable. In all others cases will inevitably occur in which a couple, not married to each other, will prefer each other to their own spouses.

Maps between one thing and another
The above mappings are examples of "discrete" instead of "continuous" maps. Unlike those of points on a surface or a set of all real numbers they handle a finite set of elements only. The mapping of telephone subscribers

CONNECTIONS

See also
52 The language of space: topology
40 Sets and groups

In other volumes
34 The Physical Earth

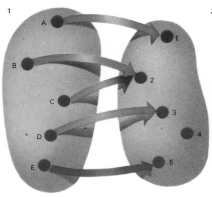

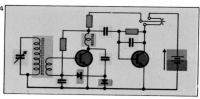

1 A mathematical map relates one set of objects (eg A, B, C, D) to an "image" set (eg 1, 2, 3, 4), symbolized by the arrows. One-to-one correspondence is not necessary; both B and C are sent to 2, D is sent to 3 and nothing is sent to 4. But the map must act on every object in its domain. Without the purple arrow E would be unimaged and the mapping therefore improper.

4 An electronic circuit diagram is a map unconcerned with scale or shape. It shows the connections between components using stylistic conventions – wires, for example, become straight-line segments.

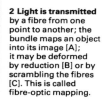

2 Light is transmitted by a fibre from one point to another; the bundle maps an object into its image [A]; it may be deformed by reduction [B] or by scrambling the fibres [C]. This is called fibre-optic mapping.

3 Each point on the globe is sent to one on the map in this diagram. The "zenithal projection" is mapping the Southern Hemisphere; each point is projected along a line from the North Pole onto a plane touching the South Pole. The Equator becomes a circle, as do other lines of latitude. Lines of longitude become radii. The scale is not constant; it increases dramatically towards the edge of the map.

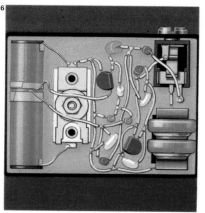

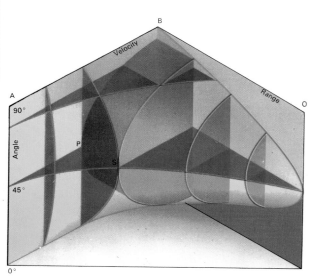

6 A reflex AM radio-receiver circuit has the circuit diagram shown in illustration 4. They do not look alike but are related by a mapping which ensures that the connections to each component are the same in both the physical and schematic layouts. The manufacturer is unconstrained by circuit-diagram conventions and routes the wiring for tight packing of components, for example. But adjacent components may interact, by their electric or magnetic fields, through space.

5 The muzzle velocity and elevation angle of a gun are mapped onto the range of the shell. The plane OAB is a "map" of angle and velocity; the height of a point [P] represents elevation angle and its horizontal distance from OA represents muzzle velocity. The perpendicular distance of the surfaces from P represents the range for those settings. The surface contour shows that range increases with velocity and is greatest at 45° elevation.

onto their telephone numbers is like this too. Each subscriber has his own number, but not all numbers are represented. For example, there is no number 000 0000. A reservoir of possible but unused numbers is held by the telephone companies. This map illustrates a new point too. Our previous examples sent points to points, numbers to numbers, people to people; this one sends people to numbers.

The mapping of one set of entities onto another apparently quite different set is a powerful mathematical technique. For instance, analytical geometry maps geometry onto algebra [11]. Each geometrical curve or line is sent to its corresponding equation; and for each geometrical theorem there is a corresponding algebraic identity. The mapping preserves the relational features of geometry so that geometrical problems of great difficulty (for example, those in many dimensions) can be mapped by easier-to-handle algebra, solved and then mapped back to give the required geometrical truth.

Morse code is another example of such mapping. Letters and numbers are mapped onto combinations of dots and dashes. These in turn can be transmitted as short and long flashes of light or pulses of electric current. At the receiving end the dots and dashes are mapped back to letters and numbers.

A vast range of scientific and technical enterprise depends on mapping from the real world into symbolic systems that preserve the important features. An astronomer maps the positions of the heavenly bodies into terms of a set of equations. From the ensuing calculations, he can recover terms that map back into future positions of the bodies and perhaps predict an eclipse.

Mapping: theory into practice

It is the business of mathematical science to make sure that mappings work. All scientific theories are maps in this sense and so are the calculations and designs of an engineer who decides on paper that an aircraft as yet unbuilt can fly. Like geographical maps of the globe, these maps sacrifice some features to preserve others and so they are incomplete. They fail totally in some areas of experience and distort so badly in others that they are useless for any practical purposes.

Anamorphic art is a technique in which an artist draws or paints a familiar shape in a grossly distorted form. It is an example of mapping, just as maps of the earth's curved surface can be drawn on a flat sheet of paper by a suitable choice of projection. In this painting part of "The Ambassadors" by Hans Holbein (1533), the stretched shape at the bottom is a "map" of a skull; it can be seen [below] by viewing from the lower left.

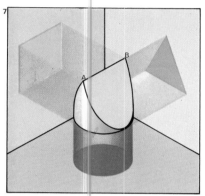

7 A draughtsman's projection is a mathematical map in which many points on the object go to the same point on the image. In the blue triangular projection all the points on the line AB are sent to the top vertex of the triangle. They retain their identity in the yellow square projection, but lose it in the circular red one. All projections "compress" information in this way, so draughting makes use of several.

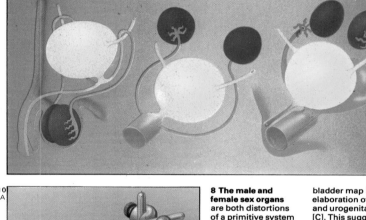

9 The psychologist Plutchik mapped the emotions onto this "emotional solid". The most intense ones map to the top. The next layer shows diluted versions of these (by their initial letters): delight, anger, expectancy, disgust, sadness etc, and the next layer happiness, annoyance, interest, boredom, pensiveness etc. The lower marks neutrality. Any sequence of emotions is a "worm-track" through this solid.

10 A weight suspended by a spring in water [A] and an electrical circuit [B] are mathematical maps of each other. If disturbed, the weight will vibrate with decreasing vigour and an electrical pulse will cause oscillations in the circuit to fade away gradually. Both examples have an energy-storing element (the spring or capacitor), an inertial element (the weight or inductor) and an energy-dissipating element (the water or resistor).

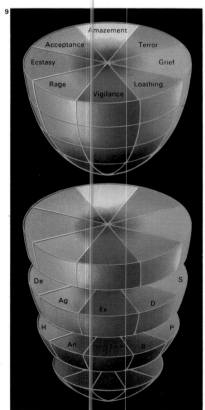
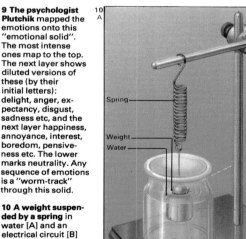
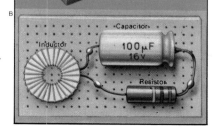

8 The male and female sex organs are both distortions of a primitive system and are mappings of each other in the mathematical sense. The testicles [A] map into the ovaries and the penis and bladder map into the elaboration of uterus and urogenital tract [C]. This suggests that both systems may have evolved from a common one [B] and that, in nature, different structures can have the same form.

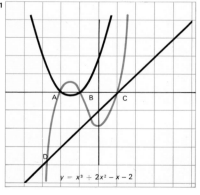

$$y = x^3 + 2x^2 - x - 2$$

11 "Mathematics is the art of saying the same thing in different words" (Bertrand Russell). The curves are geometry, their equations algebra. The mapping between these is analytical geometry. The two curves intersect at A and B; their equations will yield two solutions whose values give the co-ordinates of A and B. Similarly the straight line intersects one of the curves at C and D whose co-ordinates are found by solving the equations simultaneously.

55

Facts and statistics

In the next minute at least 60 and not more than 310 babies will be born. This statistical claim requires no knowledge of any individual women. It just assumes the average world birth-rate of three per second – and it has only one chance in a thousand of turning out to be wrong.

Making reliable statements about chance events is the business of statistics. Think of a tossed penny – the classic uncertainty. An unbiased penny tossed a million times gives, with 99 per cent certainty, between 498,700 and 502,300 heads and the rest tails. Conversely, if bias is suspected, one toss will not confirm it. But a million tosses giving 500,000 heads indicates, again with 99 per cent certainty, that the bias of the coin used is between 0.4987 and 0.5013 (a perfect coin has a bias of 0.5000).

Chance and certainty

Phrases such as "99 per cent certainty" are common to all reliable statistical statements. Certainty is never 100 per cent and for this reason a reputable statistician always states his error and confidence limits. Ninety-nine

times out of 100 he would be right in bracketing the penny's bias between the given limits. Only once in 100 times would a coin of greater bias give, by chance, equal numbers of heads and tails. If it is really necessary to bracket the penny's bias with greater confidence, it would have to be tossed more times. In statistics there is always this trade-off between the information necessary and the reliability of the knowledge it yields, with complete certainty forever unattainable. The art of practical statistics lies in knowing the probability that suffices for the task in hand and knowing what is the sufficient amount of data to collect in order to derive it.

Making good and bad guesses

Insurance companies depend on statistics. Will a client aged 20 die in 40 years time? Nobody knows. But an insurance company, with its records of thousands of men and women, estimates how many clients are likely to die and be the subject of claims and therefore how much it must charge in premiums to keep in business. From its widely amassed survival statistics it can deduce how

dangerous overweight, smoking and so on, are to health. This is achieved by seeking "bias" in the death records of various groups, just as one might seek it in the tossing records of various pennies.

Medical science gains from this too. It was statistical analysis that correlated the taking of the drug thalidomide during pregnancy with deformed babies and cigarette smoking with lung cancer. But such correlations need careful interpretation. Statistics cannot say why smokers are more likely to contract cancer than non-smokers. Perhaps people predisposed to lung cancer also tend to have a taste for smoking – an odd hypothesis, perhaps, but one that is statistically feasible. Similarly cancer of the cervix in women shows a slight but definite correlation with the number of children they have borne. Does this mean that childbirth causes cancer? Further studies show that the correlation fails for Jewish women. This clue leads to the conclusion that the correlation is with sexual activity rather than with its natural outcome. In fact the correlation arises because of the irritant substances that can form under the

CONNECTIONS

See also
58 Ods and probability
34 Finding unknown quantities: algebra
42 Finding changing quantities: calculus

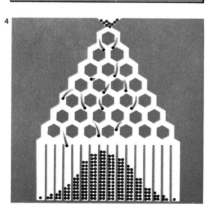

1 Conception depends on many small chances, even when intercourse occurs during ovulation. A fertile couple's chance of conceiving might be as curve A, reaching 0.9 (90%) in 3 months. An average couple [B] might have a 60% chance of conceiving in this time and one of low fertility [C] of only 25%, with only 60% chance of conceiving in a year.

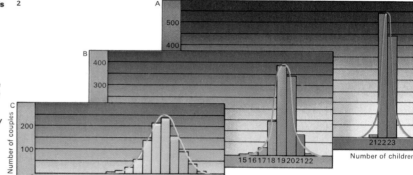

2 Computer calculations show the range of family size 1,000 couples of each type might expect after 25 years. A will probably have 21–23 children; B 15–21 and C probably 10–17.

3 Even in a primitive community very large families are rare. Limiting factors include the death of the mother or her becoming infertile. If, for such reasons, the chances of a family being complete after the birth of the first child is assessed as 4%, at 8% after the birth of the second child and so on, then families of high, average and low fertility each have an average of five or six children and produce the same curve, shown here.

4 A rain of balls through this Galton board is distributed in the bell-shaped "normal error" curve. This and similar curves are commonplace in statistics. It shows the outcome of events under many individual chances – most stay near the average; a few stray farther away.

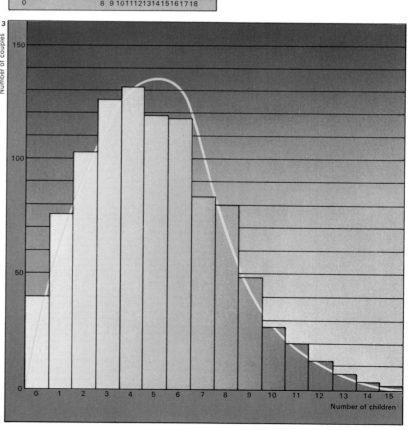

foreskin of an uncircumcised man if he is careless about personal hygiene. Jewish men, being circumcised, do not expose their wives to this slight hazard. So mathematicians must not jump to hasty conclusions. Correlations are not causes (merely clues), and statistical data are dangerously easy to misuse.

Molecules and magnetic tape
In a sense the whole world is ruled by statistics for its individual atoms and molecules are, by the uncertainty principle, not completely predictable. Only when considered in countless millions is their behaviour reliable. It is most unlikely, for example, although theoretically possible, that all the air molecules around someone would chance to rush away spontaneously and leave him to suffocate. On a smaller scale chance molecular fluctuations are inevitable and modern technology, in its quest for sensitivity, occasionally encounters them. A good audio amplifier with the volume turned up, for example, produces a slight hissing which is the amplified random motion of the electrons in the input circuit. It is as if the

amplifier handles information by "tossing electrons" and residual uncertainties cannot be avoided.

In a similar way magnetic recording maps an audio signal onto millions of metal-oxide particles on the tape [Key]. Each can have one of just two magnetic states, equivalent to heads and tails. The faster the tape runs, and the wider the track processed by the recording head, the more particles are used to record a given sound by the changes in their distribution and magnetic states. For this reason the best quality machines use high speeds (38cm/sec [15in/sec]) and wide tracks (up to 1.26cm [0.5in]) to reduce tape-hiss to the minimum. Domestic recorders use lower speeds down to only 3.75cm/sec (1.5in/sec) and track widths down to 0.05cm (0.02in). They suffer accordingly from the smaller sample of magnetic particles from which they must reconstruct the signal. The same statistical principles underlie extraction of information from the tossing of pennies, the fate of smokers and the reading of magnetic tape – and make it highly likely that you will be able to draw your next breath.

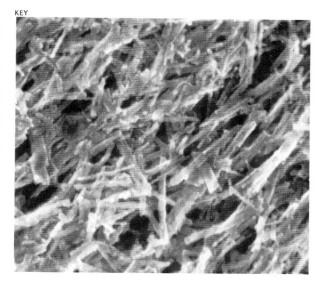

Metallic particles on recording tape, here highly magnified, have their magnetic state changed by the recording process. The quality of the recording depends on how many are affected.

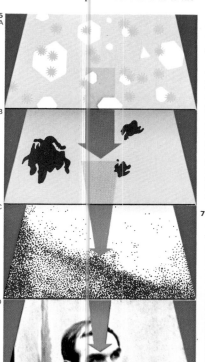

5 In photographic film light-sensitive grains are distributed in gelatine. Two photons (light particles) must hit a grain [A] to render it developable [B]. In a random hail of photons this is pure chance for any one grain. But there are so many [C] that statistically the number of developed grains follows the illumination closely [D]. There is a remote chance that the picture might look like something completely different.

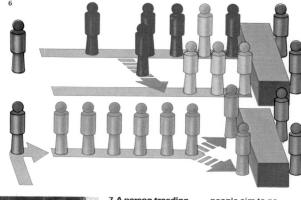

6 People arrive at a counter quite unpredictably. How many clerks are needed for an efficient service? This is a question for queuing theory, which predicts that one queue served by two clerks is more efficient that two queues served by one each. It also decides how much switchgear a telephone exchange needs to handle randomly arriving calls, how many machinists a machine-shop needs to cope with irregular repairs, and so on. It proves that if some queues are not to grow infinitely long there should be one queue that splits at the head, although a few machinists, clerks or telephonists may be idle for a certain fraction of the time.

7 A person treading on a step removes a certain amount of stone from his path. Over the years the stone wears away according to the average distribution of paths down the steps. Since most people aim to go down the middle, but deviate randomly to either side, the steps tend to wear into the bell-shaped "normal error" curve. This is a statistical curve that, in time, draws itself.

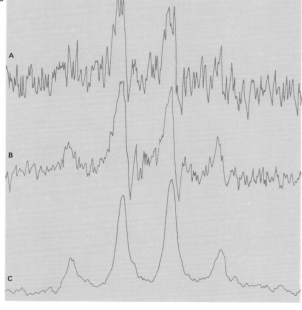

8 Many scientific instruments must register a weak signal against a background of random interference. One strategy is to keep repeating the measurement. The signal is always there, the interference is positive or negative and statistically tends to cancel out. The top trace [A], from a nuclear magnetic resonance spectrometer, shows a spectrum heavily degraded by random interference. In B 16 scans have been added and in C 256 scans.

9 A liquid of one colour is decanted into another and left. Soon the boundary becomes fuzzy and in due course the mixture is uniform. Any one molecule wanders at random, but the statistical effect of all their travels is a perfect mix, the most disordered arrangement possible. The gradual but inevitable increase of disorder is a basic law of physics.

Odds and probability

A businessman worried by the prevalence of aircraft sabotage consulted a mathematician. "Don't worry," he was told, "there is only one chance in a thousand of a plane having a bomb on board." "But I do such a lot of flying," said the businessman. "Then always carry a bomb yourself," came the reply, "because there's only one chance in a million of a plane having two bombs on it!"

How to find the probabilities

This is an elementary but popular fallacy about probability theory. If two independent events each have a known probability such as one-thousandth, the chance of their both occurring together is indeed obtained by multiplying the two probabilities giving in this example one-millionth. But they must be independent: the chance of one cannot be altered by tampering with that of the other – such as ensuring its certainty.

This multiplication rule is one of the two great pillars of probability theory. The other, the addition rule, says that given two mutually exclusive events (such as rolling a one or a two with a die – both cannot be rolled), then the chance of either occurring is the sum of their probabilities. In this case each has a 1/6 probability; so if either one or two win, the chance of success is 1/6 + 1/6 = 1/3.

These two rules, carefully used, can solve most problems of probability. They rest on a subtle sort of probablistic "atomic theory" that takes any chance event as being compounded from a set of basic "equiprobable events". By calculating what combination of these will result in the desired chance coming up, its probability is obtained. But the notion requires subtle handling. Many misleading arguments depend on a deceptive choice of basic equiprobabilities. What is the chance of there being monkeys on Mars, for example? Either there are or there are not – and it could be argued that, since nobody has yet been to Mars, these mutually exclusive situations are equally probable. Then each has half a chance of truth and there is a 50 per cent chance that there are monkeys on Mars.

More subtly, what is the chance of getting one head and one tail on two tosses of a coin? It might be reasoned that there are only three basic possibilities: two heads, head and tail, and two tails. Only one of these is favourable, so the chance is 1/3. But this is not so. There are actually four "atomic" equiprobabilities: HH, HT, TH and TT (where H stands for heads and T stands for tails), of which two are favourable. The chance is 2/4, or one half.

Calculating the chances of success

In mathematical notation, chances vary from 0 (impossible) to 1 (certain). If there are 7 equiprobable possibilities, and 2 of them result in success, the chance of success is 2 in 7, or 2/7, or 0.2857. This can also be expressed as 28.57 per cent, or in betting parlance 2 to 5 on, or 5 to 2 against. Such figures make most intuitive sense when applied to situations that can occur many times. In a run of 7,000 trials each with a 2/7 chance of success, about 2,000 successes would be expected. A gambler would break even in the long run by accepting odds of 7 to 2 (that is £7 return for a £2 stake). Where the basic equiprobable events are clear and knowable (as in the fall of coins, dice or cards), probability theory can give unambiguous chances of success for any outcome. All casinos and

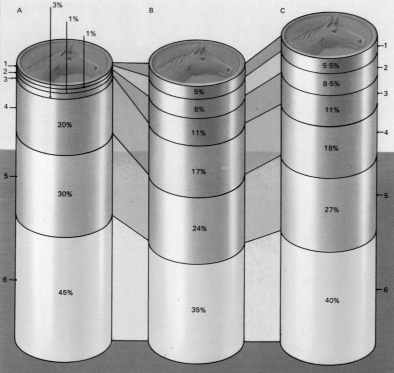

1 The bookmaker aims to offer odds that give him the same predictable profits whichever horse wins. Thus if he received £3 on one horse and £5 on another, he might offer odds of 7:3 on the first (4 to 3 on) and 7:5 (2 to 5 on) on the second. Whichever wins, he pays out £7 and makes £1 profit. So his odds reflect the money bet. "Outsiders" attract little money, so he offers long odds on them. The chances of six horses are shown in A; 1, 2, 3 are outsiders with very long odds against them. But novice betters find them reasonably seductive, so the money placed distributes itself as in B. The bookmaker changes the total odds upwards in his own favour, as in C. He is sure of a profit in the ratio of C to B. But even so, some winning odds – on the favourite [6] – are undervalued in C compared to the "reality" of A: 40% (ie offering a return of 100 to 40) compared to its actual chance of winning, 45%. Hence 6 favours the bookmaker and a series of such bets should clear an average of 10% profit to him. But the gullible backers of outsiders, in the long run, also lose. The same mathematical calculation of odds – probabilities – occurs throughout science. In atomic theory, for example, the location of an electron within an atom is defined in terms of probabilities.

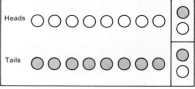

2 A tossed coin can land either "heads" or "tails". On each toss the probability of a head (or tail) is 1/2 (0.5) – the chances are evens. If a coin lands heads (or tails) eight times in succession, a gambler might be tempted to expect that a tail (or head) is more likely to occur on the ninth toss. But the mathematical probability of either outcome is still exactly 1/2 – an even chance.

Heads

Tails

3 Crown and anchor uses three dice inscribed with the six symbols of the matrix below, which shows all outcomes for the first die "diamond" (five other matrices are similar). Players bet on their symbols against a banker, who returns twice the stake for one symbol displayed, three times for double and four times for triple. Assume each symbol is backed, giving six stakes "input" per throw. 20 out of 36 times (by the matrix), three different symbols come up and the banker makes no gain, returning three 2-fold stakes to the winners. On doubles (15 in 36) he pays out two on the singlet and three on the doublet and keeps one. On the only triple, he pays four and keeps two. So in 36 rounds he has gained (with one unit staked on each symbol) 15+2 = 17 of the 216 stakes: 7.9% return.

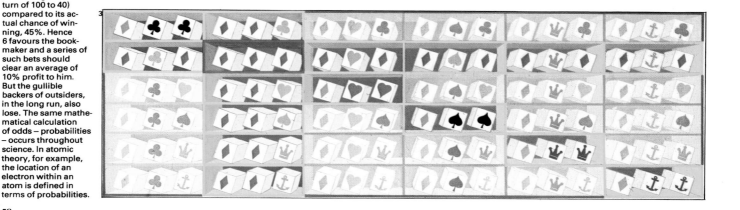

gambling houses use this principle to set fixed odds that give them a small advantage.

In sports and business assessments, odds are subjective and different people guess them differently. By betting on the favourite in a horse-race with a number of unproven "outsiders", however, the gambler's chances of winning are demonstrably better [1]. If one of the horses is known to be doped, or a rival's business strategy is known, it is possible to place investments with better-than-average insight. This is the province of "game theory" – the theory of competing for gains against opponents who possess assumed aims and knowledge.

In the child's game of button-button a button is hidden in one hand and the opponent has to guess which. He wins a penny if he is correct and loses one if he is wrong. What is the best strategy for the holder? If the same hand is always played, or hands are switched regularly, the opponent will soon outguess the holder. Game theory proves that the best strategy is to decide the switch at random, for example by tossing a coin before each round. This is entirely foolproof; even if the oppo-

nent discovers the strategy he cannot win more than he loses in the long run. But if two pennies are lost for a right-hand disclosure and only one penny for a left-hand one, the opponent could then win steadily by always choosing the right hand, and making on average bigger gains than losses. For this modification, game theory prescribes for the holder "weighted random switch" of 2:1 towards the left – say by tossing a die and playing to the right on 1 and 2, but to the left on 3, 4, 5 and 6.

Uses in real-life conflicts

In real-life conflicts such as war and business, game theory is often used for clarifying options, but seldom slavishly followed. If two people make an agreement, for example, game theory recommends to each that he double-crosses the other, for he will gain more if the other is honest. And in a world of unique events that either happen or do not, the whole concept of probability needs careful handling. Be warned by Peter Sellers's parody of a politician, who "does not consider present conditions likely"!

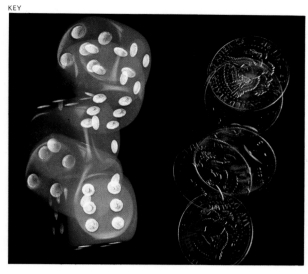

Probability theory cannot predict the outcome of a chance event such as the rolling of a die or the tossing of a coin. But in the long term (thousands of rolls) any one number on a die will occur with a probability of 1/6 (0.16666).

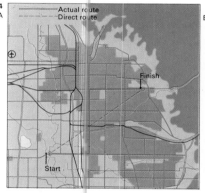

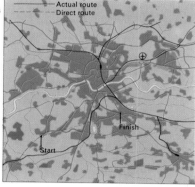

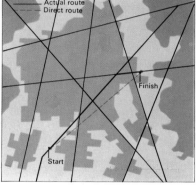

4 Is the rational grid layout of Salt Lake City [A] more efficient than the rambling European city of Cracow [B]? A diagonal journey on a grid forces you to traverse the equivalent of two sides of a triangle, even if you zigzag. Probability theory shows that to facilitate many unpredictable point-to-point journeys, a random distribution of straight lines is best [C], a style close to Cracow's.

5 In the "buffon needle problem", a match is thrown at random on a striped cloth. If the stripes are n match-lengths wide, the chance of it coming to rest across a line is $2/n\pi$. It is surprising to find π in this answer: it enters because the match can lie at any angle; like a spoke in a wheel thrown onto the stripes. Mathematicians have evaluated π experimentally by repeated throwing.

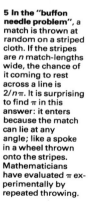

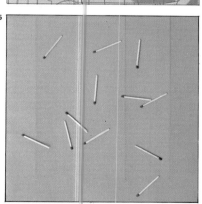

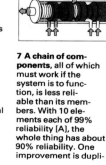

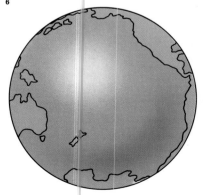

6 Are the shape and distribution of the continents random? The fact that they are crowded into half the globe, leaving the Pacific Basin occupying a whole hemisphere, suggests they are not. But modern plate tectonics proves otherwise. The continents are "floating" on the crustal surface and for most of past geological time converging subcrustal rock currents held them all in one smallish area, although they are now spreading out.

7 A chain of components, all of which must work if the system is to function, is less reliable than its members. With 10 elements each of 99% reliability [A], the whole thing has about 90% reliability. One improvement is duplication [B]; with two such chains in parallel, the chance of at least one of them working is 95%. But it is better to parallel each element separately, so that a paralleled pair will still function if either of its members is not working [C]. Then each pair has a reliability of 99.9% and the whole chain of them has 99.9%. The principle is built into a car's dual braking system [D]. Pressing the pedal [1] moves pistons in the master cylinder [2]. Three brakes work even with a leak [3].

8 A fly has an instinctive system of evading predators using an aerobatic pattern. As game theory recommends, it keeps making random alterations of course at random times. Its course is then safely unpredictable, even to itself.

59

The scale of the universe

Every object around us – indeed, all matter – is made up of countless tiny fragments called atoms. And the Earth is but a tiny speck in the vastness of the universe [Key]. But how large are these fragments? How big is an atom? And how large is the universe?

According to current thinking, the universe is about a billion billion billion (10^{36}) times as large as a single atom. But this statement gives no clue to the absolute size of either of them. To define the sizes of atoms, galaxies and the universe – as for a table or a garden – scientists use a series of units. An understanding of these is essential to a proper understanding of modern science – and to helping the imagination to grasp the range [1] between the immensity of the universe and the smallness of an atom.

Units of scale
Small objects can be measured in millimetres (about 0.04 inch) and longer distances are quoted in kilometres (about 0.621 mile). It is difficult to imagine the number of millimetres in a kilometre. But 10mm=1cm; 100cm=1m; 1,000m=1km. Or, writing the numbers as powers of ten, 10^1mm=1cm; 10^2cm=1m; and 10^3m=1km. Therefore one million, or 10^6mm=1km. To denote something smaller a negative index is used: 10^{-1}cm (a tenth of a centimetre)=1mm.

Today an atom is visualized as being almost all empty space with a few tiny subatomic particles near the centre. Very roughly, a sub-atomic particle [2] may be thought of as having a diameter of 10^{-13}cm. Ten billion (10^{13}) of them stretched out in a row might extend through a centimetre. The nucleus of an atom is made up of such particles – protons and neutrons – and may be 10^{-12}cm in diameter. An atom is the next jump in size; measured by pioneers of X-ray crystallography in ångström units, Å=10^{-8}cm, an atom is about 100 thousand times as large as a proton. Atoms can be bound together to form molecules that can be grouped to make a volume of any size: molecules of gas; a crystal; a droplet of liquid; or all the water in the oceans. The paper of this page is about a few million atoms thick.

The wavelength of visible light is 4×10^{-5}cm to 7.2×10^{-5}cm. As a result, particles with a larger diameter than this can be seen using an ordinary microscope. To make smaller objects visible scientists use electron microscopes, because fast electrons have much shorter wavelengths. The smallest living organisms, such as bacteria, are microscopic. Smaller bodies such as viruses [3], which are submicroscopic, cannot live and develop alone but are parasitic on the cells of living organisms. All visible living things are made up of many millions of atoms.

Distances – from men to the stars
The tallest men are about 2m (6.5ft) in height and the Earth is more than 12,000km in diameter. The diameter of the Sun is more than a million kilometres. The nearest heavenly body to Earth is the Moon, about 384,000km away. Since man landed on the Moon and looked back to the Earth [7] this distance has acquired a more tangible reality.

The Sun is about 1.5×10^8 (150 million) km away from the Earth and the planet Pluto nearly 6×10^9km. These numbers are already becoming difficult to visualize and the whole

CONNECTIONS

See also
62 What is an atom?
66 Beyond the atom
252 The expanding universe

1 Within the known universe the dimensions of tiny, sub-atomic particles and the distance attainable by astronomers' telescopes stand in a ratio of about 1:10⁴⁰. The objects shown spanning this staggering range are a proton [1]; an atomic nucleus [2]; an atom [3]; a giant molecule [4]; a virus [5]; a small cell, an amoeba [6]; a large cell, a diatom [7]; a flea [8]; a hen's egg [9]; a man [10] 2m (6.5ft) high and one of his buildings [11]; the Earth [12]; a giant star [13]; an interstellar gas cloud or nebula [14]; our Galaxy [15] and the limits of the theoretically observable universe [16]. The 10m symbol is 10m (32.5ft) tall – five times as tall as a man - and the skyscraper [11] is more than 10²=100m (325ft) tall, enough to dwarf the men and the bus.

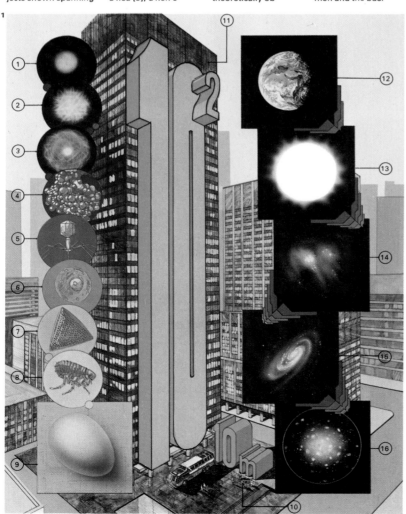

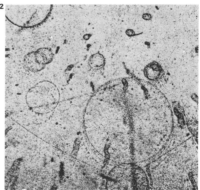

2 Particles are so remote from our experience that indirect methods must be used to make them visible, such as the cloud chamber invented by Charles Wilson in the 1920s and developed by Patrick Blackett. It uses water vapour condensing on ions to reveal particle tracks and therefore identify them. Particles can also be tracked in bubble or spark chambers, or by using stacks of special photographic plates.

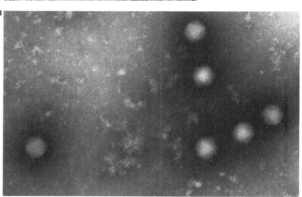

3 Viruses are too small to be seen under an ordinary microscope but can be seen by using an electron microscope. They are non-living matter but affect the properties of living cells.

4 The unique pattern of a fingerprint exemplifies the enormous number of individual cells that make up even the smallest piece of living matter visible to the naked eye. There are about 10 million cells in each cm² of skin.

Solar System, the Sun with its attendant planetoids [8], is the merest speck in space. Therefore to describe the geometry of the stars a different unit of distance is used: the light-year, or the distance light travels in a year. In just one second light travels 3×10^5 (300,000) km. In a year light travels about 10^{13} km and the nearest star is more than four light-years from the Sun.

It is only in relation to these kinds of figures that the granulation of the universe can be sensed. The atomic nucleus is formed of densely packed particles, but the atom is almost all empty space. Similarly, in the universe the atoms on the planets and stars are closely packed to form solids, liquids or gases. But the stars are separated by huge distances compared with their diameters and so the universe around us, like the atoms of which we are made, is nearly all empty space.

Towards infinity

The stars themselves, at their immense distances, are grouped into great clusters called galaxies or nebulae. So huge are their numbers that through telescopes they look like great white clouds. In our Galaxy alone, the Milky Way, there are more than 10^{11} stars. The Galaxy is about 10^4 light-years thick in the middle and 10^5 light-years across.

Galaxies have greatly differing appearances according to the wavelengths of the light used to observe them. One of the most rapidly developing branches of astronomy is radio astronomy in which observation is made by radio waves. The largest known object in the universe is a galaxy designated 3C236, observed by radio telescope. A curious object in which the principal radio sources are two bulges at each end, it measures about 2×10^7 (20 million) light-years from end to end. The nearest large galaxy to our own is 2×10^6 light-years away. The farthest objects probably lie at a distance of about 10^{10} light-years.

The scale of the known universe from the atomic nucleus to the farthest star is about 10^{40}. But what is beyond the limits of present-day astronomy? Does the universe stretch to infinity? Or could it be curved in a curious way so that the nebulae that seem farthest away are in reality near to our own?

The spiral galaxy in Andromeda has a measurable size, but it is difficult to grasp: 120,000 light-years or 10^{18} km across.

5 "Man is the measure of all things" said Protagoras in the 5th century BC – a humanist view that established man at the centre of the universe and related objects to a human scale. The personification of nature was a theme of such Renaissance painters as Botticelli (1444–1510), who painted "The Birth of Venus". Modern science uses other units and man is no longer the standard of length.

6 Man's most visible artefact, the Great Wall of China, runs for more than 2,400 km (1,500 miles), about five per cent of the circumference of the Earth. It can be seen from well out in space.

7 The Earth, as a body floating in space, has acquired a new reality since man left his own planet and looked back at his terrestrial home. For the first time, it has been seen as merely one, minute, heavenly body in immeasurable space. The idea of other planets having other life forms is no longer regarded as improbable and current estimates put the number of possibly inhabited planets at many millions.

8 The Solar System has been brought within man's reach by interplanetary rockets. But beyond it the galaxies stretch endlessly – the nearest to Earth is 20,000 million million km away.

What is an atom?

The first recorded suggestion that matter might consist of separate particles was made in the fifth century BC probably by Leucippus of Miletus [1] and the idea was developed by his pupil Democritus, who adopted the word *atomos* (from the Greek word meaning indivisible). John Dalton (1766–1844) revived the word at the beginning of the nineteenth century when he provided a scientific basis for the simple Greek idea. To Dalton an atom was a tiny indivisible particle, the basic unit of matter that takes part in chemical reactions.

The atom and electricity
The simple Daltonian view of the atom was overturned in 1897 when J. J. Thomson (1856–1940) discovered that atoms could emit even smaller particles of negative electricity (later called electrons) [5]. Clearly the atom itself must have some form of internal structure. Thomson's discovery also implied that an atom must also contain positive electricity. He suggested that electrons were like currants dispersed throughout a positively charged bun. This model failed to explain a

number of the properties of atoms, but a better one had made use of the discovery of radioactivity by Antoine Becquerel (1852–1908). He found that certain heavy atoms spontaneously emit radiation. Three forms of this are now known: beta rays (negatively charged electrons), alpha particles (positively charged helium nuclei consisting of two protons and two neutrons) and gamma rays (short-wave X-rays).

The Rutherford model
In 1911 Ernest Rutherford (1871–1937) produced an entirely new model of the atom based on the results of his own experiments and those of Hans Geiger (1882–1945) and his co-workers (who measured the scatter of alpha particles when shot at gold foil). Rutherford's suggestion was that the positive charge and most of the mass of the atom were concentrated in a central nucleus and that the electrons revolve around it. We now know that the atom is mostly empty space with a minute central nucleus some tens of thousands of times smaller than the atom. The atoms themselves are extremely small –

ten million of them side by side would form a line measuring only about 1mm (0.039in).

Rutherford later discovered that the positive charge of the nucleus is carried by particles 1,846 times heavier than electrons; he christened them protons. The charge of the proton is equal, but opposite, to that of the electron. A hydrogen atom consists of a single positively charged proton (the nucleus) with one electron travelling in an orbit around it [Key].

Heavier atoms have increasing numbers of protons in their nuclei, but the number of protons in the nucleus (called the atomic number) is always balanced by an equal number of orbiting electrons. It was later discovered that all atoms except hydrogen have another type of particle in their nuclei. These are uncharged particles (and are therefore called neutrons) and they have almost the same mass as the proton.

Quantum theory and spectroscopy
Two other fields of investigation helped the Danish physicist Niels Bohr (1885–1962) to construct the next important atomic model

1 The city of Miletus was the first known home of natural philosophy. Thales (*c.* 630 BC) was born there. He was a member of the Ionian School, the earliest known in Greek philosophy. He discovered the electrical properties of amber. Anaximander also dwelt there, as did Leucippus (*c.* 400 BC), credited by Aristotle with atomic theory that formed a central part in the evolution of Western scientific thought.

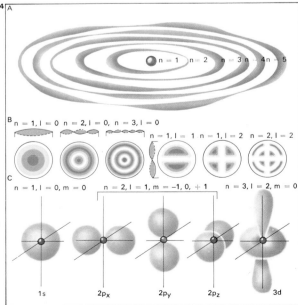

2 Modern science pictures matter as having a dual existence as waves and particles. Waves are seen on the sea and when a pebble drops into a pond, and sound and electromagnetic radiation such as light and X-rays are known to travel as waves. The wave theory of atomic particles such as electrons, protons and neutrons has led to an improved understanding of atoms and nuclei.

3 Lines in the spectrum can be observed in light given out by incandescent elements. These are emission lines, which result from the emission of light by atoms. One of the tests of the Bohr theory was its ability to explain the wavelengths of the lines in the spectrum of hydrogen in terms of electron energy level changes.

4 Possible orbits of an electron around an atomic nucleus can be pictured [A] as circles that exactly accommodate a whole number of wavelengths, denoted by the principal quantum number n. A two-dimensional analogy (a vibrating drum skin) is described by two quantum numbers n and l [B], and the shape of a real atom [C] in terms of three (n, l and m).

[Key]. The first was the quantum theory, the other was the science of spectroscopy. Quantum theory was proposed by Max Planck (1898–1947) [7] in 1900 as a way of explaining the emission of heat (and light) by a hot body. He realized that energy can be emitted and absorbed only in discontinuous amounts, discrete "packets" of energy that he called quanta.

Spectroscopy began when Isaac Newton (1642–1727) passed a ray of sunlight through a glass prism, breaking the ray into all the colours of the visible spectrum. In 1814 Joseph von Fraunhofer (1787–1826) had discovered that the spectrum of sunlight contains a number of black lines, which were later found to coincide with the position of coloured lines in the spectrum formed by electric discharge in hydrogen gas [3]. Bohr postulated that the circulating electron in an atom of hydrogen can exist only in fixed orbits [4A] and that the spectral lines correspond to the absorption (black lines) or emission (coloured lines) of a quantum of energy when this electron jumps from one fixed orbit to another. This theory, later modified by Arnold Sommerfeld

(1868–1951), has been extremely successful in explaining the hydrogen spectrum.

Modern developments of the quantum theory suggest that the fixed orbits of Bohr should be visualized less precisely and that the position of an atomic electron should be treated as a probability that it will be in a certain place at a certain time. This treatment, known as quantum mechanics [2], was largely the work of Louis de Broglie (1892–) and Erwin Schrödinger (1887–1961). The substitution of a probability for a fixed orbit is a reflection of the uncertainty principle of Werner Heisenberg (1901–76). This says that if the energy of a particle is known precisely there must be an uncertainty as to its position. In de Broglie's treatment matter in the form of atomic particles is like light in that some of its properties are best explained in terms of particles, others in terms of waves. A stream of electrons, for example, behave like particles in cathode rays and like waves in an electron microscope. But for the purposes of chemistry the concept of the atom as the smallest unit of matter that can take part in reactions remains supreme.

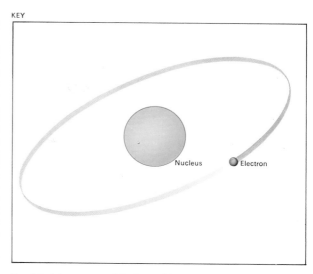

The pictorial representation of the model of the atom proposed by Niels Bohr is an established part of the iconography of modern physics, even though Bohr's ideas have largely been superseded by various forms of quantum mechanics. This example is a hydrogen atom.

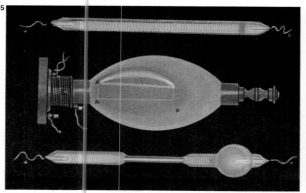

5 Geissler tubes, Victorian toys for adults, depended on electronic rays in a near-vacuum long before the principles of cathode rays were understood. The study of the rays, principally by J. J. Thomson, a British physicist, was the crucial one in the elucidation of the structure of the atom, by establishing the mass and charge of the electron, in conjunction with other experiments.

6 The new ideas about atomic physics were brought together at a series of conferences, such as this Solvay meeting at Brussels in 1911, attended by Bohr, Rutherford, Curie and others.

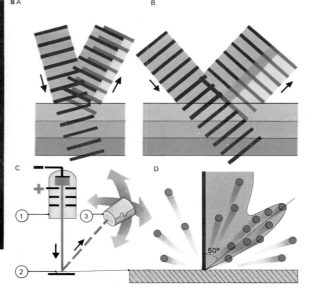

8 When waves are reflected from parallel surfaces they are out of step [A] or in step [B]. An electron beam [C] from a gun [1] can be reflected from nickel [2] into a detector [3] and the angles plotted [D].

9 Erwin Schrödinger played a principal part in the mathematical development of the modern model of the atom. He developed wave mechanics from de Broglie's picture of wave-particle duality.

7 Max Planck suggested in 1900 that light was absorbed and emitted in packets or "quanta", with energies proportional to the frequency of light. This is known as the quantum theory.

Nuclear physics

Nuclear energy plays a decisive part in shaping the modern world: nuclear weapons not only haunt the statesman but cast a threatening shadow on every living person [Key]. And while the mirage of limitless nuclear power attracts a civilization hungry for energy, the disposal of radioactive waste threatens lasting pollution of the world. In fact, life has always depended on nuclear energy: nuclear fusion heats the sun [1], and radioactivity in the earth [2] heats the liquid core and contributes to the mobility of the continental plates. Nuclear energy is derived from splitting the atomic nucleus by radioactivity or fission, and secondly from fusion, which is the joining of a pair of light nuclei.

Radioactivity: its discovery and source

Radioactivity was discovered by Antoine Becquerel (1852–1908). With the isolation of radium it became clear that enormous amounts of energy were involved. Radium decays over many decades and in fact contains 2×10^5 times as much energy as an equal mass of coal. A nucleus, a few times 10^{-12}cm in diameter, is made of protons (positively charged particles) and neutrons (neutral particles of nearly equal mass to the proton). Hydrogen is unique in having a single proton (and no neutrons) in its nucleus. Most elements consist of a mixture of isotopes, whose nuclei differ in their numbers of neutrons. The total number of constituents (protons and neutrons) in a particular isotope is indicated by a superscript, for example He^4. It is on the properties of individual isotopes that nuclear power depends.

Deliberate transformation of one nucleus into another was achieved by Ernest Rutherford (1871–1937) [4] in 1919: $He^4 + N^{14} \rightarrow O^{17} + H^1$. In words, an alpha-particle (the nucleus of helium) and a nitrogen nucleus momentarily combine and then split into the oxygen isotope O^{17} and a proton.

As mass spectrometers [6] – instruments that measure the individual masses of ions and thus of the nuclei – became more accurate, it was found that the masses of the nuclei of the various isotopes were not equal to the sum of the masses of the constituent protons and neutrons. This discrepancy, according to Einstein's relativity formula, $E = mc^2$, is the source of nuclear energy. Modern theory views the nucleus to be rather like a "liquid" droplet of neutrons and protons. Any such system tends to decay – that is to say transfers itself – into a state of lower energy. If it does so by breaking into two nearly equal parts, that is called fission; if a nuclei gives off one or more particles, that is radioactivity; if two nuclei join together, that is fusion. Two nuclei are both positively charged so one of them must be accelerated to high speed to achieve fusion, or both must be moving fast due to high temperature.

Generating nuclear energy

The generation of nuclear energy in large quantities by fission requires a chain reaction, first achieved with uranium [6]. When a neutron is absorbed by the isotope uranium-235 it induces fission into two major fragments and two or three neutrons. If on average two neutrons are absorbed by other U-235 nuclei the process will spread explosively through the fissile material.

In order to generate electricity the process must be slowed down and controlled,

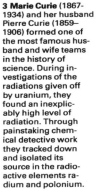

1 **The sun** is powered by nuclear fusion which needs temperatures of the order of hundreds of millions of degrees centigrade. On earth the necessary conditions and temperatures for fusion reactions have so far been achieved only in bombs. Nuclear energy by controlled fusion remains a dream.

2 **Volcanic energy** is provided by radiation – a source of energy so powerful that a scattering of radioactive atoms with slow decay rates is sufficient to heat the centre of the earth. The liquid core that results underlies the continents and powers volcanic eruptions.

3 **Marie Curie** (1867-1934) and her husband Pierre Curie (1859–1906) formed one of the most famous husband and wife teams in the history of science. During investigations of the radiations given off by uranium, they found an inexplicably high level of radiation. Through painstaking chemical detective work they tracked down and isolated its source in the radioactive elements radium and polonium.

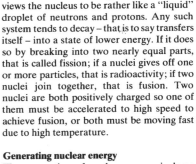

4 **Ernest Rutherford** [A], a New Zealand-born physicist who first worked at Cambridge University in 1903, established the nuclear theory of the atom in 1911 and later achieved the first splitting of the atom when he produced protons from the nuclei of nitrogen atoms. He and his team used remarkably simple (including "homemade") apparatus at the Cavendish Laboratory [B] and were able to change our picture of the structure of atoms.

and means must be provided for removing the heat. An atomic pile to produce electric power is merely a special kind of furnace. There is a higher probability of fission occurring in U-235 if the neutron absorbed is moving relatively slowly, about 2km (1.2 miles) per second. For this reason special materials, called moderators, are incorporated into the atomic pile in order to slow the neutrons down.

Atomic piles may be classified according to the material used as moderator, including graphite, water and heavy water. Heavy water is water in which the hydrogen is a heavy isotope, deuterium, and it absorbs far fewer neutrons than ordinary water.

The amount of fissile material brought together is crucial for a sustained chain reaction. If more neutrons are lost by absorption or escape than are produced, the reaction will not be self-sustaining. If more neutrons are produced than are lost on average, a self-sustaining and expanding reaction occurs. The smallest amount in which fission is self-sustaining is called the critical mass. In an atomic pile it is necessary to keep the flux of

neutrons nearly in balance and constant. To control a pile, rods of neutron-absorbing material can be moved in and out [8].

Fast reactors and their make-up

Structural supports in atomic piles are made of materials that absorb as few neutrons as possible. Fast reactors have a small core of fissile material and no moderator to slow the neutrons. There is little absorbent material and few neutrons are wasted. Natural uranium is 99.3 per cent U-238 and 0.7 per cent U-235. While U-235 is fissionable, U-238 is not, but after absorbing a neutron can decay radioactively to plutonium Pu-239, which is. Atomic bombs and fast reactors require fairly pure fissionable material so U-235 must be separated out – for example in a giant diffusion plant – or U-238 turned into Pu-239 in a reactor and separated chemically. In a fast reactor, the fissile core is surrounded by a blanket of natural uranium so that neutrons escaping from the core can turn U-238 into Pu-239. If more fissionable material is made than consumed the reactor, or pile, is called a breeder reactor.

The mushroom cloud of an atomic explosion haunts our civilization. Although the spread of nuclear weapons has been banned by treaties, and some nations have accepted technical limitations on testing, the number of nations with access to nuclear weaponry continues to grow.

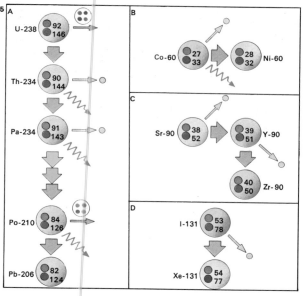

5 The nucleus of an atom, with protons (red) and neutrons (brown), may change, giving radioactivity: gamma rays (electromagnetic radiation, violet), electrons or beta rays (grey), positrons, or alpha particles (helium ions, orange). Naturally radioactive uranium-238 [A] decays as shown to form lead. [B] shows the decay of cobalt-60, [C] strontium-90 and [D] iodine-131.

6 When uranium-235 is hit by a slow neutron [1] it may split by stages and release energy and more neutrons [2]. One of these may strike more uranium-235 [3] and lead to a chain reaction, or be absorbed by other atoms [4] or U-238 [5].

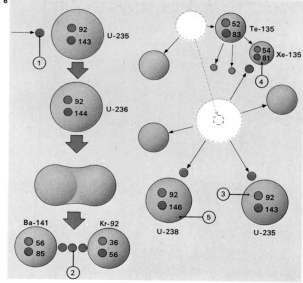

7 The mass spectrograph of Francis Aston (1877–1945) showed that elements are formed of separate isotopes, each nearly an integral multiple of the mass of a proton. Later spectrographs gave exact measurements of nuclear masses and are used to distinguish isotopes.

8 Nuclear reactors are the powerhouses of the future and to some extent of the present. But the formidable problems they create in disposal of radioactive wastes have not yet been satisfactorily solved. Here engineers are carefully stacking the rods that go to make up the central core of an atomic pile. Control rods are also inserted to regulate the reaction rate.

Beyond the atom

One of the characteristic features of science is the way in which it attempts to explain a collection of different phenomena in terms of a few basic concepts. A striking example is the atomic theory of John Dalton (1766–1844) in which many different substances are considered to be made up of a few different types of atom. According to this view atoms are the fundamental "building blocks" of all matter.

In the late nineteenth and early twentieth centuries evidence accumulated to show that atoms themselves have an internal structure. By 1932 it had been realized that atoms are combinations of sub-atomic particles: protons and neutrons (together forming a small positively charged nucleus) with orbiting negatively charged electrons.

Interactions between particles
To give a full description of matter it is necessary to describe not only the particles but also the way in which they are held together – that is, the way in which they interact with one another. Four types of interaction are recognized; two of these are fairly well known

because they are observed in matter in bulk as well as on the atomic scale. The gravitational interaction [1] produces an attraction between objects that depends on their masses. It is an extremely weak effect and plays no part in the binding within atoms, but it is responsible for the forces between heavenly bodies. The electromagnetic interaction [2] occurs between particles that have an electric charge. This force is many millions of times stronger than the gravitational effect and is responsible for the force of attraction between the nuclei of atoms and the orbiting electrons.

Within the nucleus itself a quite different effect must occur. Here neutrons and protons are held together strongly in spite of the electromagnetic repulsion between them. This strong interaction is independent of charge, for it acts between neutrons as well as protons, and is about 7,000 times stronger than the electromagnetic interaction. Moreover, it falls off sharply with distance – its influence extends only over distances comparable with the dimensions of the atomic nucleus, generally less than 10^{-12}cm.

The fourth type, known as the weak interaction, is about one-thousandth of the strength of the electromagnetic interaction. It is observed in certain processes in which transformations of particles occur, as in beta decay, where a neutron changes into a proton, an electron, and an antineutrino.

Fields of force
The four types of interaction take place through free space. One way of explaining this action-at-a-distance uses the idea of a field of force. A charged particle, for instance, is thought of as affecting the surrounding space in such a way that another charged particle placed in this region experiences a force. The region of influence is called an electromagnetic field. Similarly a mass has an associated gravitational field in the space surrounding it.

A different model, based on quantum mechanics, uses the idea of an exchange of virtual particles. Two charged particles interact by emitting and adsorbing photons (particles of light). Gravitational interaction is similarly explained by exchange of

1 The force of gravity is encountered in all its power and immediacy by a weightlifter. The gravitational force was the first to be studied quantitatively and the first to receive, in Isaac Newton's *Principia* (1687), detailed discussion of its theoretical principles.

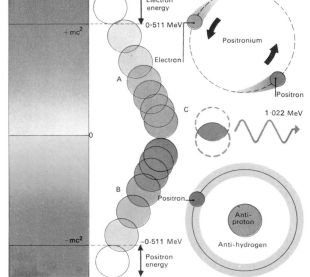

2 Electromagnetism (used in an electric bell) was the second interaction of which man became aware. Magnetism and static electricity were known earlier but the combination of electricity and magnetism in a single theory was achieved only in the nineteenth century by the work of Oersted, Faraday and Maxwell.

3 Hideki Yukawa (1907–) predicted a particle, later identified as the pi meson, as the quantum of nuclear force. It was found in cosmic ray photographs by Cecil Powell (1903–69) at Bristol University.

4 Enrico Fermi (1901–54) developed the theory of beta decay, which depends on weak interaction. He was largely responsible for the development of the first atomic pile (nuclear reactor).

5 The existence of a particle equal in mass to an electron [A] but having negative energy was predicted by Paul Dirac (1902–). The discovery of the positron [B] confirmed this. If the two meet, they annihilate each other, releasing their combined mass-energy [C]. Positronium – an electron and positron in orbit – is known to exist briefly and as other anti-particles are known it is possible to imagine anti-atoms such as anti-hydrogen in some other world.

hypothetical particles called gravitons. In 1935 Hideki Yukawa [3] suggested that the strong interaction holding the nucleus together was due to the exchange of a particle with a mass between that of the electron and the proton. This particle is now known as the pi meson (or pion). Another particle, the intermediate vector boson, has been suggested as being responsible for weak interactions, but so far scientists have been unable to prove its existence.

Other fundamental particles
In 1932 only three particles were necessary to explain atomic structure. Since then the situation has been complicated by the discovery of many more particles through work on cosmic rays and experiments using particle accelerators [6, 8]. It is found that high-energy collisions between particles lead to the production of new ones. Now more than 200 are known, most of them very unstable [7]. They are characterized by their mass and charge. They also have other characteristic properties, such as average lifetime, that describe the ways in which they interact.

The numerous sub-atomic particles are classified into groups: particles that partake in strong interactions are called hadrons (including nucleons, hyperons, and mesons). Particles that do not take part in strong interactions are called leptons (including electrons and neutrinos).

The problem of high-energy physics is to produce a single theory explaining the existence and behaviour of this multitude of particles. One suggestion is that the particles themselves are made up of even more basic particles. It is possible, for instance, to describe all hadrons as combinations of three particles called quarks. These have charges that are one-third or two-thirds the size of the electron charge. According to this theory protons and neutrons, for example, are not completely indivisible.

Another goal is the creation of a single theory to account for all the types of interaction. So far some success has been achieved in unifying the electromagnetic and weak interactions, but a single mathematical theory encompassing all four types of interactions is still far off.

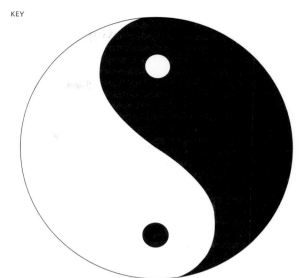

The essence of matter summed up in the Chinese symbol of yin and yang was a symmetry of complementary principles – aptly representing the modern theory of particle-wave duality.

6 Particle accelerators use the principle that an electric field [A] accelerates or deflects positive [red] or negative [blue] particles parallel with the direction of the field, whereas a magnetic field [B] makes them curve at right-angles to the field. In a "drift-tube" accelerator [C, E] oscillating electric fields speed up at the same rate as the particles. A synchrotron [D] is used for particles brought close to the speed of light by a linear accelerator [1]. A magnetic field [2], increasing to balance the growth of centrifugal force, causes particles to circulate through an accelerating field [3], seen through a viewer [4].

Type of particle	Symbol	Mass in units of electron mass	Electric charge
Photon	γ	0	0
Leptons			
Neutrino (electronic)	υ	0	0
Antineutrino (electronic)	$\bar{\upsilon}$	0	0
Neutrino (mesonic)	υ_μ	0	0
Antineutrino (mesonic)	$\bar{\upsilon}_\mu$	0	0
Electron	e^-	1	-1
Positron	e^+	1	1
Mu (muon) plus	μ^+	207	$+1$
Mu (muon)	μ^-	207	-1
Mesons			
Pi plus	π^+	273	$+1$
Pi minus	π^-	273	-1
Pi zero	π^0	264	0
K plus	κ^+	967	$+1$
K minus	κ^-	967	-1
K zero	κ^0	974	0
Anti-K zero	$\bar{\kappa}^0$	974	0
Nucleons			
Proton	p^+	1,836	$+1$
Antiproton	p^-	1,836	-1
Neutron	n	1,839	0
Antineutron	$\bar{n}$	1,839	0
Hyperons			
Lambda zero	Λ^0	2,183	0
Antilambda zero	$\bar{\Lambda}^0$	2,183	0
Sigma zero	Σ^0	2,332	0
Antisigma zero	$\bar{\Sigma}^0$	2,332	0
Sigma minus	Σ^-	2,328	-1
Antisigma minus	$\bar{\Sigma}^-$	2,341	$+1$
Sigma plus	Σ^+	2,328	$+1$
Antisigma plus	$\bar{\Sigma}^+$	2,328	-1
Xi plus	Ξ^+	2,566	0
Antixi zero	$\bar{\Xi}^0$	2,580	0
Xi minus	Ξ^-	2,580	-1
Antixi minus	$\bar{\Xi}^-$	2,582	$+1$

7 Fundamental particles are now so numerous that the term fundamental seems inappropriate, but there is insufficient information for a unified theory and some vital clues may be missing. Perhaps one day order will be brought to the apparent chaos of information, just as Niels Bohr clarified atomic theory.

8 The ring cyclotron was invented by Professor Ernest Lawrence (1901–58) at the University of California in 1930. Its vacuum chamber, in which charged particles were accelerated by being circulated past two D-shaped electrodes with a high-frequency voltage, measured only 9.8 cm (4in) across. The giant machine now at CERN in Geneva has a diameter of 4.8km (3 miles). Accelerated to high speeds in cyclotrons, particles can be used to bombard nuclei – with the creation of other particles.

The nature of energy

Energy is required for work to be done – work being the operation of a force over a distance. Thus energy is expended when a golf ball is struck, when a dumbbell is lifted, when a spring is compressed or stretched, when a bomb explodes and when electrons flow in a wire as an electric current.

Additionally, energy is needed to raise the temperature of any substance and living organisms need energy for movement and growth. Green plants get their energy as light from the sun [Key], which they utilize by photosynthesis. Animals use the chemical energy of food – the energy of plants or other animals that they eat. From these examples it can be seen that energy exists in many forms. But to a scientist, energy exists in only two distinct ways: as potential energy (the capacity for doing work) and as energy-doing-work, or released energy.

The essence of potential energy
Potential energy can be considered as stored energy. The potential energy of food and fuels such as coal and oil, for example, is the chemical energy stored in these materials.

The potential energy in the water of a dam is equivalent to stored gravitational energy; the earth tries to "pull down" the raised mass of water with a force proportional to its mass [1] – the force of gravity.

In a coiled spring, the potential energy is proportional to the amount of compression or extension. A copper sphere insulated from electrical leakage can be charged with static electricity (unmoving electrons) and the electrical potential energy of the sphere is determined by the amount of static electric charge and the associated voltage.

What is internal energy?
The total energy of any system is called its internal energy. This quantity is not usually measurable and not all of it can be used to do work. A hot object does work in cooling, for example, but even if it is cooled to near absolute zero, namely −273°C (−460°F), its molecules still possess most of their internal energy. The potential energy of a sytem is measured not as the total internal energy but as the part of it available to do work.

When matter is in motion it is said to pos-sess kinetic energy. For this reason, molecules of a gas always have kinetic energy because they are always moving. The temperature of a gas is a measure of the average kinetic energy of its moving molecules – the faster they move, the higher the temperature.

The pressure of a gas is also a measure of its kinetic energy, because the pressure is a measure of the number and energy of the collisions made by gas molecules on the walls of its container. Gas pressure is often used, as in pneumatic drills and lifts, to do useful work. Finally, the kinetic energy of the gas molecules, or of any other moving object, can be expressed in absolute mathematical terms as $\frac{1}{2}mv^2$, where m is the mass of the object and v is its velocity.

Conservation of energy
Heat energy can be regarded as energy of movement. To take the example already used, a gas has kinetic energy proportional to its absolute temperature (its temperature above absolute zero). If the same gas heats an object, the molecules of the object gain

1 **A dam holds water at a height** and when the water is released its potential energy changes to kinetic energy that can be converted into useful electrical energy by water turbine generators. Other hydro-electric processes rely on the kinetic energy of water moving in the form of waves and tides.

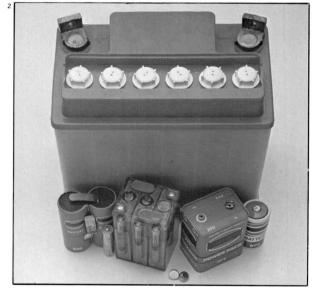

2 **In batteries** the energy of chemical reactions is converted into electrical energy, which is used for many familiar purposes. Batteries run down as their chemical activity declines. Accumulator batteries (secondary cells) can be recharged by electricity, which is converted back to stored chemical energy.

3 **A nuclear power station** uses the great energy released when the nucleus of an atom disintegrates. Under controlled conditions the nuclear energy of uranium or plutonium is carefully released, mostly as heat, which is used to raise the temperature of water in boilers and produce steam. This heat energy is converted in steam turbines into mechanical energy, then to electrical energy in generators.

4 **In a microphone-loudspeaker system** a double energy conversion takes place. Sound energy is converted by a microphone into electrical energy. The loudspeaker reverses the conversion process.

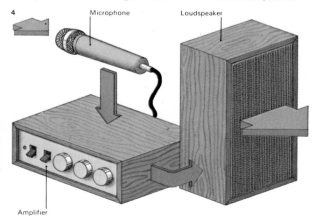

Microphone   Loudspeaker

Amplifier

kinetic energy. This transfer of kinetic energy takes the form of a flow of heat from the hot gas to the cooler body.

Useful energy is usually thought of in such terms as the heat of burning coal, the electricity flowing in electric fires and the mechanical energy produced by burning petrol in a car engine. But in each of these familiar examples, energy has had to be converted from one form to another to be of practical use. The chemical potential energy of coal is released by burning the coal and is thereby converted into useful hot gases and radiant energy; in a power station these are further converted into useful electrical energy by a system of water boilers, steam turbines and electricity generators. The chemical potential energy of motor fuel is released by rapid burning as kinetic energy of gas, which is translated into the useful mechanical energy that propels the car.

Energy conversions always involve a loss: no conversion is 100 per cent efficient. A coal fire, for example, releases only about 20 per cent of the chemical energy of the coal as useful heat. An electric motor converts about 80 per cent of the electrical energy supplied to it into mechanical energy [7].

The principle of conservation of energy is usually stated as "energy can neither be created nor destroyed". In energy conversions the amount of energy output to do useful work is always less than the input energy. The total energy of the system, however, always remains the same, the "missing" energy being wasted energy. Not all the electricity flowing through a lamp filament, for instance, is converted into light, most being wastefully converted to heat. The heat and light together are equivalent to the input electricity and so energy is conserved. But after the conversion, less energy is available to do work; as a whole, the energy is at a lower level.

What happens in this particular instance is true for the sum of all energy reactions in the universe at any one time; the overall result of these reactions being a degradation in energy level. At some remote time in the future, all energy will have degraded to a level where no work can be done: the universe will then have "run down".

The sun is the ultimate source of energy for all life on earth. Here it is reappearing from behind the moon after an eclipse.

5 Solar cells, as fitted on this satellite, convert radiant energy from the sun directly into electrical energy. A solar cell contains a wafer of semiconductor material, usually silicon. This is made in such a way that when light falls upon it, electrons move in one direction within it and "holes" (positively charged regions) in the other direction. In a circuit, each solar cell produces about half a volt.

6 In the filament of an electric light bulb some electrical energy is converted to light but most is wasted as heat. A fluorescent lamp contains a vapour and fluorescent coating that together convert electricity to light with less heat and greater efficiency. The element of an electric fire converts electricity mainly to heat.

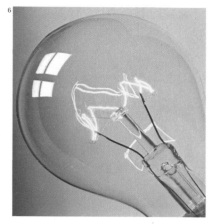

7 An electric motor converts electrical energy into mechanical or kinetic energy. Most familiar is the rotary type in which a rotor revolves within a fixed stator.

8 The hydraulic ram raises water by converting its kinetic energy into gravitational potential energy. A lake or reservoir [A] supplies water [1] through a pipe [2] to the ram chamber, which is at a lower level, ensuring that the water has adequate kinetic energy on reaching the chamber [3]. This fills with water [B] which briefly escapes through a spring valve [4] before closing it. The water passes on round a one-way valve [5] into a second chamber [6] where air is first compressed by the water [C] then re-expands to force water up the delivery pipe [7]. A back-surge allows the spring valve to reopen and the process to be repeated.

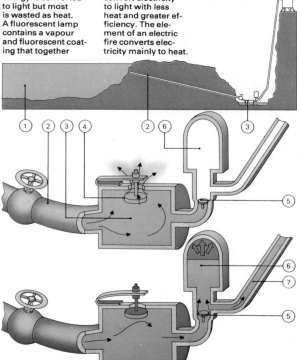

Statics and forces

The sudden movement of an object with seemingly no cause – for example, the unexpected movement of a table in the middle of a room – would obviously cause consternation to the observer. Most people would probably think a trick was responsible for such a happening because they expect a cause for this effect of motion. The scientific name for the cause is "force" – it is anything that causes an object to start to move when it has been at rest, or vice versa. A force is also required to change a condition of motion that already exists – to change the direction of an already existing motion or to alter the velocity of the motion. Once made to move, an object continues to move without stopping or changing direction until it is acted on by another force. This idea has now become almost self-evident following its original expression and generalization by Isaac Newton in the late 1600s, and is generally known in physics as Newton's first law of motion.

In many cases the cause that stops, or modifies, already existing motion is provided by the force of friction. This acts in the direction opposite to the movement of an object.

It is produced by the rubbing together of the surface of the moving object and the surface it is moving on or, in a gas or a liquid, the medium it is moving through.

States of equilibrium

When several forces act at the same time on the same object, each tries to move the object along a line pointing in its own direction at a rate that depends on the size of the applied force. If it happens that the object does not move as a result of all these forces, then it is said to be in equilibrium [5].

The magnitude and direction of any one of the forces is balanced by the total effect of the other forces and there is no resultant movement. The study of forces applied to objects in a state of equilibrium is called statics (as opposed to dynamics, the study of forces acting on moving objects).

Someone sitting still on a chair is an everyday example of an object in equilibrium – the upward force of the chair on the person balances the downward force of the earth's gravitational attraction that is trying to pull the person through the chair to the floor. A

tree standing upright in the ground is a similar example – its downward gravitational force, commonly known as its weight, is balanced by the upward force of the ground in which it is rooted. As a consequence of this equilibrium, neither the person nor the tree is in motion up or down. This state of affairs can be modified only if another external force is brought into the arrangement – for example, by the person moving about on the chair and possibly making it topple or, in a most extreme case, by chopping down the tree [7]. Thousands of everyday objects stay where they are and do not move unaided because of the existence of this equilibrium of forces both in magnitude and direction.

Moments and levers

So far, only those forces that try to move objects along straight-line paths have been considered. There are many other forces, however, that can act on objects and try to rotate them around a central point. These forces have an effectiveness that depends on how far from the central point they are acting. Everyone knows that a much greater

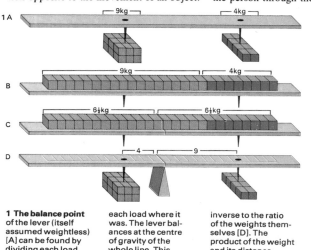

1 The balance point of the lever (itself assumed weightless) [A] can be found by dividing each load into equal small weights and distributing them along the lever [B], keeping the centre of gravity of each load where it was. The lever balances at the centre of gravity of the whole line. This point is halfway along the line [C] of equal weights at the distance from the weights that is inverse to the ratio of the weights themselves [D]. The product of the weight and its distance from the fulcrum (pivot) is the same on each side – the moments are equal and opposite.

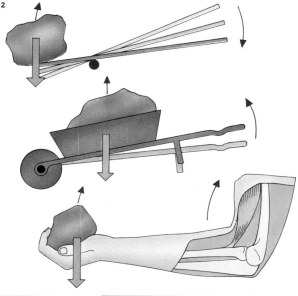

2 The see-saw principle is used in the first order of levers and has the fulcrum, or balance-point, between the load and the applied effort. The wheelbarrow is a familiar example of the second order of levers, with the load between the fulcrum and the effort. Finally, the third order of levers, with the effort between the fulcrum and the load, is used when lifting a weight with the forearm – the elbow is the fulcrum. If both lever and load remain the same in all three cases, the principle of moments shows the required force is the greatest for the third order and least for the second order, which is the best form of leverage.

3 Force applied by an animal tethered to the arm of a grinding mill produces a moment or turning effect about the central axis. This turns the mill and grinds the corn poured between heavy stones at the centre of the mill.

4 A steelyard is a balance often used by butchers for weighing carcasses that are too heavy for ordinary scales. The small sliding jockey weight is moved along the long arm of the steelyard until the whole beam is horizontal. The

Object being weighed, suspended from point close to the pivot

Small counterweight distant from the pivot

sliding weight, at a greater distance from the fulcrum than the heavy load, can be much lighter. The principle involved here is the same as the first order of levers – like the see-saw. Its first recorded use was in 315 BC.

turning effect is achieved by pulling on a wrench or spanner that has a long, rather than a short, handle. The combined effect of the magnitude of the force multiplied by the perpendicular distance from the turning point (called the axis of rotation) is called the moment of the force. The greater the applied force or the distance of action, the greater is the resulting moment or turning effect. Simple machines that employ the phenomenon of moment of a force belong to one of the various classes of levers [2].

Of course several forces may simultaneously act on an object through more than one point within its boundary. In this case, each has its own moment about a particular axis, through which the object can still be in equilibrium if these moments produce no collective rotational effect. That is, if the total magnitude of the clockwise moments about the axis exactly balances that of the anticlockwise moments, there is no movement. This result is called the principle of moments and it may apply to an object at the same time as the equilibrium, mentioned previously, that arises from forces acting through a par-

ticular point trying to move the object in a straight line.

For the example of a person sitting still on a chair and being in equilibrium for vertical motion, it is clear that if someone tries to tilt the chair backwards, the seated person will topple over unless someone or something pulls the chair in the opposite direction with an equivalent force. The vertical equilibrium through the point of contact of the chair with the floor acts at the same time as the equilibrium for the moments of the forces trying to twist the chair around at the same point.

How couples operate

One further type of force that a study of statics includes is called a couple [6]. Actually this is the application of two equal forces arranged so that they both tend to rotate the object in the same direction. The couple produces only rotation with an *equal* moment about any point between the forces, and does not produce any straight-line motion. Consequently it can be balanced only by another equal or opposite couple if an equilibrium state is to be achieved.

The principle of moments is used in most of the activities in a children's playground. The see-saw obviously demonstrates the lever, as shown here; in addition, the umbrella, roundabout and swings all use the moment of a force about their axis or fulcrum to achieve movement.

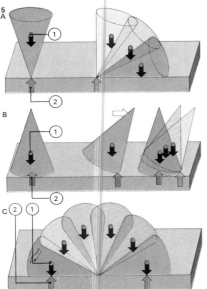

5 If a cone standing on its point [A] is slightly displaced, its weight produces a moment that continues to topple the cone about the point of contact. This is an unstable equilibrium position. With the cone on its base [B], displacement produces a moment that restores the original position – a stable equilibrium. If the cone lies on its side [C], displacement produces no moment since the weight and its reaction [1, 2] still act along the same line – the cone remains in its new position and is in neutral equilibrium. The very low centre of gravity of the bus [D] keeps it in stable equilibrium even for large displacements.

6 The moment of a couple is equal to the product of one of the forces and the distance between the two. This hand is applying an equal force to each end of the top of a tap, producing a rotating effect about the central axis to turn it on and off. Both these forces are trying to turn the tap in the same direction, and together form a couple; the larger the tap-top, the larger the couple.

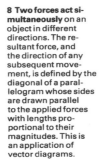

7 A tree stands in equilibrium (its centre of gravity acts downwards in a straight line through its base) until a wedge is cut into its trunk. This starts to destroy the equilibrium by allowing the tree's weight to develop a toppling moment that is not balanced by an equal and opposite reaction.

8 Two forces act simultaneously on an object in different directions. The resultant force, and the direction of any subsequent movement, is defined by the diagonal of a parallelogram whose sides are drawn parallel to the applied forces with lengths proportional to their magnitudes. This is an application of vector diagrams.

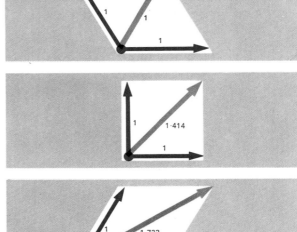

Attraction and repulsion

To many people, the most mysterious natural forces are those that produce an effect on objects at great distances, reaching across even empty space without any material contact between the body producing the force and that being affected by it. This phenomenon is often called "action at a distance", and there are several fundamental forces of nature that act in this way. They are gravitational, magnetic and electric. Other fundamental forces act within the limits of the atomic nucleus, over short ranges only.

The inverse-square law

The three forces acting at a distance obey a common law that describes how the magnitude of the force depends on the object producing it, the object affected by it and the distance separating the two. It is called the inverse-square law [Key]. If the distance between two bodies is doubled, then the force between them falls by one-quarter (the inverse square), and so on.

Isaac Newton (1642–1727) was first to realize this type of law applied to the gravitational force of attraction when performing calculations on the speed of the moon in orbit round the earth. His law of gravity [1] states that "the attractive force between two bodies is proportional to the product of their masses divided by the square of the distance between them" and it should be noted that this force is a mutual one – there are equal and opposite forces on the two objects.

About 100 years later, in 1789, Henry Cavendish (1731–1810) used this theory of gravitation to produce the first estimate of the mass of the earth [2]. He was also responsible for one of the best experimental verifications (in 1785) of the use of the inverse-square law to describe the force between electrostatic charges (that is, electric charges at rest). Its use was further verified in 1870 by James Clerk Maxwell (1831–79) who showed that the inverse-square law was true to one part in 20,000. More modern methods have taken this limit to one in 1,000 million.

The volume of space within which the force exerted by a body produces a detectable effect (usually by causing movement of a second body) is normally called the field of force. The directions along which movement can occur are known as the lines of force. These are merely imaginary "lines" in space that are used to help describe the possible directions of any motion within the field – they spread out from the force-producing object, filling the complete field.

Gravitational and magnetic force

Any object with mass, however small, can produce a gravitational force field. The force is always positive because no object can have a negative mass. This indicates that the force of gravity can only be one of attraction. (A negative value for a force indicates repulsion, not attraction.)

It is gravitational attraction that gives all objects their weight, trying to pull them towards the centre of the earth [3], and which keeps the planets in orbit around the sun. In some circumstances, gravitational attraction can be balanced by an equal and opposite force to achieve a weightless condition. This is most commonly accomplished by the centrifugal force caused by the rotation of an orbiting satellite. As a consequence, objects

1 The earth's mass W can be measured by means of a tall balance with two pans on each side. Initially, m1 balances m2 [A] but when m1 is moved to the lower pan it weighs more, being closer to the earth [B]. Mass c is used to restore the balance [C]. Next, the large mass M is used to make m1 weigh more, by gravitational attraction [D]. Balance is restored [E] with weight n. If R is the earth's radius, d is the distance between m1 and M, then Newton's law states:
$$\frac{m1 \times M}{d^2} = \frac{n \times W}{R^2}$$
The distance between the pans on each side must be great enough to prevent M from having significant gravitational pull on the other masses.

2 In 1798, Henry Cavendish determined the earth's mass using a quartz fibre torsion balance. Two small lead spheres were attracted by two larger lead spheres and from the size of the deflection, the force of attraction between the spheres was calculated. To work out the mass of the earth, Cavendish compared this force with the gravitational pull of the earth on the spheres (that is, their weight).

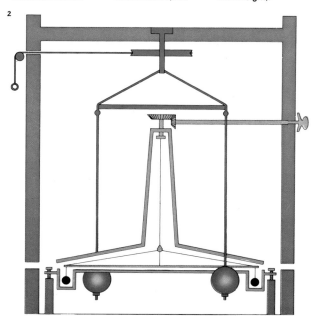

3 On a roller-coaster, power is needed to pull the car to the top of the highest incline against the gravitational force exerted by the earth. The resulting potential energy of the car is then transformed into kinetic energy of motion as it is allowed to free-wheel down the track. According to the principle of energy conservation, the total kinetic energy at the bottom of an incline should equal the potential energy at the top; the car would be able to travel to the top of another equally high incline with no further application of external power. Some of the potential energy, however, is lost as a result of its conversion to heat energy by friction between the speeding car and the track. Therefore the remaining kinetic energy will allow the car to climb only a smaller vertical height each time it travels downwards. To raise the height of the car again, it would be necessary to apply additional power against the force of gravity.

within the satellite are able to float freely within its confines.

Magnetic force [4] is familiar in connection with the working of a normal compass. This force also acts over large distances, although in this case both attractive and repulsive forces may occur, a fact that is easily tested with two simple bar magnets. The ends, or "poles", of the magnets are distinguished by being called north and south and it is always found that two north or two south poles will not remain in contact, whereas a north and south combination will. This fundamental effect is summarized by the rule: "like magnetic poles repel each other, unlike poles attract".

The different poles are given the descriptive labels north and south for historical reasons – the north pole of a magnet is the one that is always attracted to the North Pole of the earth, though in fact a magnetic south pole must be sited there to achieve this magnetic effect [5]. The inverse-square law determines the magnitude of the magnetic force, although this is now proportional to the product of two magnetic pole strengths

that can be positive or negative (instead of two masses that can only be positive, as in the case of the gravitational force).

Electric force [6] can also be either attractive or repulsive since its source, an electric charge, can be either positive or negative. According to the type of charge either a positive (attractive) or a negative (repulsive) force is obtained by the inverse-square law.

What is an electric charge?

Every electric charge is in fact a multiple of a unit charge that is equal in magnitude to the irreducible charge associated with a single electron, a fact first noted and evaluated in 1909 by the American physicist Robert Millikan (1868–1953). However the dual nature of the electric charge had been known from Greek times and is best demonstrated, as then, by electrostatic effects. If a glass rod is rubbed with silk the two objects attract each other, while two glass rods rubbed in the same way repel each other. Each rod acquires a net positive electric charge, whereas the silk collects excess electrons by the action of friction and becomes negatively charged.

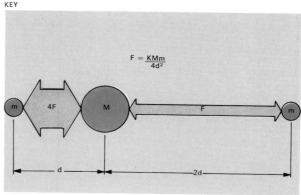

$$F = \frac{KMm}{4d^2}$$

The inverse-square law governs gravitational, electric and magnetic forces. M and m represent the masses of two bodies for the gravitational force, their charge values for the electric force and their pole strengths for the magnetic force.

In all cases, d represents the distance between the bodies. K is a constant quantity that has a different value for each of the three forces. Gravitational force is much the weakest, the gravitational attraction for two electrons

being about 10^{42} times less than the electric force of repulsion. (Consequently one is aware of gravity only when great masses such as the earth are involved.) The inverse-square law equation thus varies for the three forces according to the value of K.

4 A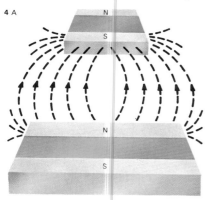

B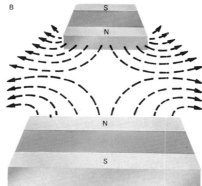

4 A field of force exists all round a magnetic material and the associated lines of force indicate paths along which a unit magnetic north pole would move. Iron filings sprinkled on paper laid over a pair of magnets will display negative [A] or positive [B] lines of force. In a similar way, a small compass needle will line up along lines of force with its north pole pointing out the field direction [C].

C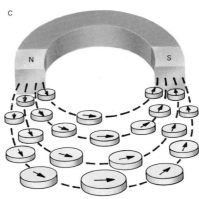

5 The earth's magnetic field is created by phenomena below the crust as if it had been formed by an immense bar magnet with its south pole approximately aligned with the north geographic pole sited at one end of the earth's rotational axis [A]. The needle of a ship's magnetic compass [B] swings to a position where its ends point to north and south, along a line of force of the earth's magnetic field.

5 A North pole
South pole

B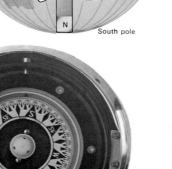

6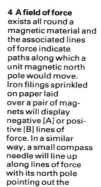

○ Molecule in bulk liquid
● Molecule at surface

6 Electric forces that exist on a molecular and atomic scale within a substance are responsible for the cohesive forces that hold it together and give it shape and strength. Both attractive and repulsive forces exist in a state of equilibrium, the attractive holding the substance together and the repulsive effectively preventing the atoms collapsing in on each other. A striking result of these cohesive forces between atoms is seen in liquids. Within the body of the liquid, any one atom has equal and opposite forces on either side of it in any direction; there is therefore no resultant force. But at the surface there is a net force pulling atoms into the body of the liquid and as a result the surface appears to behave like an invisible elastic "skin" pulled across the liquid. This surface tension effect is used by insects such as water boatmen and mosquito larvae to keep themselves floating on a pond's surface.

Speed and acceleration

Cars, rockets, falling weights and footballs all move under the action of forces. The branch of physics that studies movement, and the forces that produce and influence it, is called dynamics. It was given a firm scientific basis as a result of the work of Isaac Newton (1642–1727), who formulated the fundamental three laws of motion.

Principles of motion

The first law [1] summarizes the principle of inertia – the basic tendency for anything moving to continue moving and for an object at rest to remain at rest. It states that "an object will remain at rest or in motion at constant velocity unless acted on by a force". Once a car is in motion, both it and its passengers carry on moving unless acted on by a force – such as a braking force. A head-on collision may stop the car, but the inertia of the passengers will cause them to fly forwards from their seats. They may be thrown against the windscreen unless held in place by safety belts which exert a restraining force.

As a result of this law, it is apparent that the greater the applied force on an object the

greater its change of velocity. Velocity is merely speed in a certain direction and change of velocity in a given time is called acceleration. So the greater the applied force, the greater the acceleration. The second law of motion states that, in addition, acceleration is inversely proportional to the mass of the object being moved.

The third law considers the way forces act against each other. If an object rests on a table, the table exerts an upward force equal and opposite to the downward force of the object's weight. The third law generalizes this by saying that "for every applied force there is an equal and opposite reaction". Two spring balances hooked together and pulled in opposite directions register the same force. Another much more spectacular application of this law is the rocket. The force of the expanding gases in a rocket's combustion chamber acts equally in all directions. The forward thrust is produced by the reaction to the displacement of mass that occurs when the burning gases escape through the nozzle of the rocket.

Newton's laws of motion also connect

with several other concepts. Thus the second law describes how acceleration (a) is dependent on mass (m) and applied force (F) in the equation $F=ma$. This can be used to calculate the weight of an object, because weight is the force with which a body is attracted towards the centre of the earth. This force equals the product of the mass and the acceleration with which a falling object drops to the ground (called the acceleration due to gravity). Consequently mass and weight are completely different quantities, having different units to describe their magnitudes. Mass is a property of an object due to the quantity of matter it contains; weight, a force which acts on it due to gravity. Unfortunately in the past the same units (pounds, kilogrammes and so on) have been used for both mass and weight; the modern scientific units are the kilogramme (or pound) for mass and the newton (or poundal) for weight.

Definition of momentum

The equation of the second law also shows that, because acceleration is the rate of change of velocity, force can be expressed as

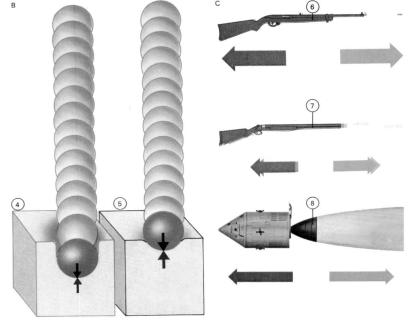

1 Newton's first law describes inertial effects. [A] An object resists being moved from rest by toppling backwards [1], although moving steadily it is undisturbed, as if at rest [2]. When stopped it resists slowing and tends to continue moving [3]. The second law explains that acceleration or deceleration is proportional to the force producing it. [B] A ball falling on to a

soft material [4] sinks deeper than into a harder one [5] because the deceleration force is smaller. The third law states there is an equal and opposite reaction to every force. [C] A rifle recoils when fired [6], although the bullet's velocity is much greater. Successive firings cause successive recoils [7]. A rocket ejects gas and moves forwards [8] because of reaction.

2 A heavy gun recoils when firing and an equal and opposite reaction propels the shell forwards. The principle of conservation of momentum, states that the total momentum before and after is zero but the shell is given its forward velocity because it is lighter than the gun. The chemical energy stored in the propellant explosive charge is transformed into the kinetic energies of gun and shell.

3 The collision of two balls of different masses travelling in the same direction occurs as the smaller ball is moving faster

[A]. The total momentum before and after impact [B] is unchanged. Before contact both balls contribute to the

total momentum, whereas only the heavier is still moving afterwards. It moves faster than it did originally, having

experienced the impulse of the lighter ball, but slower than the original velocity of the small ball [C].

the rate of change of the product of mass and velocity. This product is called momentum and can be thought of as a quantity of motion that, for a definite velocity, increases with the object's mass. In effect the momentum indicates the effort needed to move an object or to stop or change the direction of its motion. For instance, a brick carefully placed on someone's foot does not hurt. But the pain caused when the brick is dropped from a height of a metre or so testifies to the effect of momentum.

Probably the most important reason for calculating momentum is that it is conserved throughout events involving changes of motion – for example, sudden collisions or explosions. This means that the total momentum before and after such an event stays absolutely constant [3]. Following the event, momentum may be lost through the action of friction reducing the velocities, but during its occurrence this law of conservation of momentum holds exactly.

Another phenomenon occurring during this kind of event is the transfer of energy. It can exist in many forms such as heat, light, sound, chemical and electrical energy. All of these can be transformed into one another – petrol's stored chemical energy is transformed in the combustion engine to mechanical energy for moving a car. If the energy transfers occurring in any process are considered collectively, very careful measurements have shown that energy is never created or destroyed. This is the law of conservation of energy and it implies that no machine can produce a net gain of energy.

Two forms of energy

For the study of dynamics two forms of energy are fundamental – kinetic energy, possessed by objects that are in motion, and potential energy, possessed by objects that are at rest and able to do work by virtue of their position [5]. The pile driver has work done on it to lift it against the earth's gravitational force and then can expend the stored energy when again allowed to fall. In fact, the potential energy is transformed to kinetic energy as the hammer falls and it is the kinetic energy that finally does the work of driving the pile into the ground.

Galileo's legendary experiment, dropping a cannon ball and a pebble from the tower at Pisa, showed that objects of different masses fall to the ground together, proving that the acceleration due to gravity is the same for all objects. With objects of different cross-sectional area, air resistance to downward motion may also be different and prevent the objects hitting the ground together. If resistance equals the gravitational attraction a limiting velocity is reached. Also there will not be a total transformation of potential to kinetic energy as the objects fall because some heat energy is lost by the action of friction with the air.

4 A motorway pile-up can occur after one or two cars stop. Other cars that are unable to stop in time collide with the stationary vehicles and transfer their forward momentum to them, causing a "knock-on" effect which ripples along an ever-extending chain. Conservation of energy controls the transformation of a car's mechanical kinetic energy to wasted heat and the energy imparted to the stationary cars.

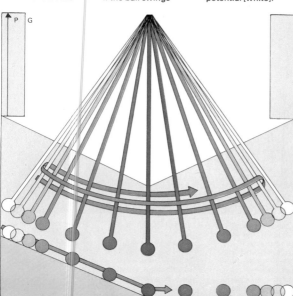

5 The total energy in the world remains constant even though it may be converted from one form to another. Any object in a gravitational field [G] has potential energy [P] proportional to its height. If a ball is allowed to roll down a slope, its potential energy lessens as an equal amount of kinetic energy is gained. Some energy is wasted by friction, so that if the ball rolls horizontally it eventually comes to rest. If the ball swings at the end of a string, kinetic energy is reconverted to potential energy during the upward motion. The pendulum's energy can, in this situation, oscillate between kinetic [red] and potential [white].

6 A steam catapult, driven by the pressure of gas built up in the rams, is effectively storing a great quantity of potential energy that, when released, can be transformed to the kinetic energy of movement. As a result, even the large mass of an aircraft can be accelerated to a speed that will allow it to take off from rest in the very short distance available on an aircraft carrier's deck.

7 In 1798 Count Rumford (1753–1814) noticed that, in the boring of cannon barrels, the barrels, borer and metal chips all became hot despite there being no apparent heat source, except for friction. He realized that the mechanical energy needed to turn the barrel against this friction was being converted to heat energy. Subsequently other experimenters, notably James Joule (1818–89), demonstrated the transformations between other forms of energy.

Circular and vibrating motion

If a driver presses the accelerator pedal of his car, the car speeds up – in scientific terms its velocity changes. But even when he goes round a bend at constant speed, the car's velocity also changes. This is because velocity is speed in a certain direction; if either its magnitude or direction alter, velocity changes.

Motion in a circle
The rate of change in velocity is called acceleration. Thus an object which changes direction while travelling at constant speed experiences an acceleration. When a stone is tied to the end of a string and whirled round in a circle at constant speed, the velocity's magnitude is unchanged but the direction alters continuously. If the string were cut at any instant, the stone would fly off along a tangent to the circle (as seen in the sparks flying off a rotating catherine-wheel).

A force must be acting on the rotating stone to produce its acceleration. Here the force is due to the tension in the string and one can feel this force on the hand as the stone is whirled round. It is called the "centripetal force" because it acts towards the centre of the circle. The acceleration it produces is therefore similarly directed towards the centre. For a car moving round a circular track, frictional forces at the ground acting on the wheels provide the centripetal force.

Also, by Newton's third law of motion, there must be an equal and opposite reaction force to the centripetal force that acts on a rotating object – it is called the "centrifugal force" because it acts outwards from the orbit centre. The effect of centrifugal force is used in the centrifuge machines that subject pilots and astronauts to the high acceleration forces they will encounter in their flights and, on a smaller scale, in domestic spin driers.

Periodic motion
Both centripetal and centrifugal forces depend on the mass (m) of the object and its velocity (v) in the circular motion (circle of radius r). A heavy object needs a greater centripetal force to hold it in orbit and a greater force is also required for high speeds of rotation. Experiment shows that, in addition, the required force (F) is inversely proportional to the radius of motion, and $F = mv^2/r$ where

v^2/r gives the magnitude of the centripetal acceleration.

Uniform circular motion is periodic – that is, the events recur over and over again. The time taken for a complete revolution of the object remains constant. This periodic character is further demonstrated by considering how the object's distance from any fixed diameter of the circle varies with time. If a graph of these distances is plotted, the resulting curve is one of a uniformly oscillating amplitude of motion [1], resembling a sine wave.

For a swinging pendulum, the force which pulls the bob back through the central vertical position is its weight. It undergoes oscillatory motion with an acceleration that is proportional to its distance from the point of suspension and it is also directed towards that point. Movement of this type is called "simple harmonic motion" and for the pendulum the time taken for a complete cycle of forwards and backwards swing (the oscillation "period" t) is proportional to the square root of its length (l). The number of cycles completed per second is called the "fre-

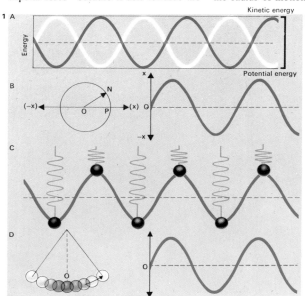

1 All periodic motion involves the continuous interchange of kinetic and potential energy, as shown in the graph [A]. Simple harmonic motion (SHM) is one form of periodic motion that is characterized by the shape of a sine wave [B]. The point [P] on the diameter of the circle around which N is moving is an example of SHM. The mass on a spring [C] performs linear SHM and a pendulum [D] performs angular SHM.

2 Circular motion of the "aeroplanes" produces an outward centrifugal force that lifts the planes off the ground. The equal and opposite centripetal force is provided by tension in the arms.

3 The constant periodic time of oscillation of a pendulum is used to control clocks, especially case clocks [A]. The periodic time needed to drive the escapement correctly can be exactly matched by choosing a pendulum [B] of correct length. It can be calculated from the equation $t = 2\pi\sqrt{l/g}$ where t is the oscillation period, l is the length and g is the acceleration due to the earth's attraction – gravity.

4 The moons of Jupiter, as well as all other natural and artificial satellites, move in orbits around their mother planet at great speeds. The force of gravity between them provides the centripetal force towards the centre of motion that keeps a moon in orbit and produces its centripetal acceleration. As the speed of rotation remains almost constant, it is the continuously changing direction of the moon which implies acceleration. If the inward gravitational force were removed the equal centrifugal force would cause the moon to hurtle out into space. The orbits are not exactly circular, so that the speed of rotation does not remain completely constant; the associated forces are analogous to those of a system performing circular motion.

quency" of oscillation (measured in hertz).

The amplitude in all types of simple harmonic motion rises and falls like a sine wave. This also characterizes many other wave motions. If a long rope is clamped at one end and its free end is whipped, a wave-like disturbance travels along the rope and the amplitude of the rope's displacement at any point from the fixed end is described by the waveform shown in illustration 6.

Moving and standing waves

The characteristics of this wave picture describe amplitude changes for the plane waves that move through the sea, the circular waves (ripples) that spread from the point at which a stone is dropped in a pond, the air pressure waves of sound and the electromagnetic waves of radio and light (which are distinguished by their wavelengths). For all these waveforms, energy is transmitted in the direction of the wave motion. Actual vibrations of the medium in which they travel may occur in the same direction, producing "longitudinal waves" such as those of sound, or in a direction at right-angles to the motion,

producing the transverse waves of all the other examples. In water, a floating object merely bobs up and down as the waves pass, and it does not move in the direction of the wave motion.

If a string is clamped at both ends and then plucked, it still vibrates but in this case the wave appears to stay in the same place – a stationary, or "standing", wave has been produced as the ends are fixed and allow no forward motion. The string vibrates at its natural frequency, which is inversely proportional to its length and directly proportional to $\sqrt{T/m}$ where T is its tension and m is its mass per unit length.

This state of vibration is called "resonance" because it occurs with the natural frequency of the string. Blowing into the end of an organ pipe makes the air in it resonate at a natural frequency that depends on the pipe's length. Similarly, when struck, a tuning fork vibrates with one particular frequency. There are many useful applications of resonance, as in tuning circuits on radio receivers [8], and in musical instruments – perhaps the most satisfying application of this effect.

Many fairground rides depend on circular motion and the resulting centrifugal forces for the excitement they can provide. They all use different variations of this type of motion, ranging from the slow revolutions of the merry-go-round to the much faster and more complex movements of the aeroplane "spinners". The magnitude of the centrifugal force depends on the mass of the circling object. When this fairground "chairoplane" gets up to speed, the heavier children will "fly" higher than the lighter ones. The same force squeezes water out of wet laundry in a domestic spin drier and causes precipitation in a centrifuge.

5 Sea waves move with a transverse wave motion that transmits energy in their direction of motion, while floating objects move only vertically up and down as the waves pass them.

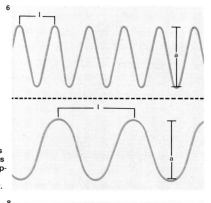

6 The forward velocity (v) of a wave of frequency (f) and wavelength (l) is given by v = fl. Such waves can be drawn as sine waves. Two waves can have the same amplitude (a) even though the wavelengths differ.

7 The air inside a pipe will vibrate at a natural frequency that depends on the pipe's length. If the external impulse has this frequency, resonance occurs. Powder in the lower tube shows the wave.

8 This circuit, with the AC source from a radio's aerial, can be "tuned" so that its natural frequency is matched to that of the incoming radio signals. Resonance excites it to oscillate and work the radio.

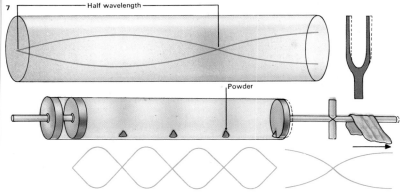

Half wavelength

Powder

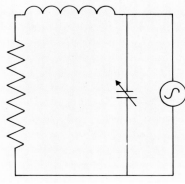

9 Freak winds can begin to produce oscillations in very large and heavy structures. If they continue, the oscillations may gradually increase in amplitude and if their frequency equals the natural vibration frequency of the structure resonance may occur, leading to catastrophic break-up of the structure. This has happened with bridges (the most renowned probably being the Tacoma Narrows Suspension Bridge, USA, shown here) and tall structures such as skyscrapers.

Pressure and flow

The branch of physics that deals with the forces and pressures that act within liquids and gases is called hydrostatics. It also examines how these forces affect any surface they touch and what uses they can be put to. It includes, for example, problems ranging from deep-sea diving to aeroplane altimeters and from floating and sinking to the design of hydraulic lifts and other machinery.

Archimedes' principle
Hydrostatics explains what happens to an object immersed in a fluid. Both liquids and gases are fluids and each is able to exert or transmit a force. If a cork is pushed below a liquid surface and then released, it immediately bobs to the surface. The cork has experienced an upward force, called the "upthrust", due to the liquid, and it is this that keeps the cork floating on the surface. In exactly the same way, an upthrust must act on a floating balloon, although here it is produced by a gas rather than a liquid.

It was the Greek scientist Archimedes (287–212 BC) who first quantified this fact by stating that "when an object is totally or partially immersed in a fluid the upthrust on it is equal to the weight of fluid displaced". Using Archimedes' principle [1] the magnitude of the upthrust can always be found and it makes itself apparent as a loss of weight of the object.

For this reason, a floating object has its weight exactly balanced by the upthrust. But if the object is too dense, the upthrust may not be sufficient to counterbalance its weight and the object sinks. This principle of flotation is used directly by a hydrometer, an instrument for measuring liquid densities. A hydrometer floats at a level that depends on the weight of a liquid it displaces and, since the submerged volume is known, the density of the liquid can be calculated.

Internal forces in fluids
As well as being able to exert a force such as an upthrust, a fluid can produce an internal force at any depth due to the weight of the fluid above it. It is normal to measure this effect as the force per unit area, or pressure, developed by the fluid's weight; this increases with depth (in both liquids and gases). Thus the greater water pressure at the bottom of a dam requires a form of construction in which strength also increases with depth – a triangular cross-section getting thicker with depth is generally used. For a similar reason, the suit of a deep-sea diver must contain a jacket of compressed air whose pressure counteracts the external water pressure, so that he can breathe without his muscles having to expand his chest against this pressure.

Another example of a fluid is the earth's atmosphere. The weight of air produces a pressure at the surface, commonly called atmospheric pressure, which is about 14.7 pounds per square inch at sea-level. In other unit systems it is equal to 760mm of mercury, 101,325 newtons/m² or 1,013 millibars. Variations in this pressure affect the weather and are caused by atmospheric disturbances. They are measured by the common barometer. The simplest form of barometer measures the height of a column of liquid, often mercury, supported by the atmospheric pressure. An aneroid barometer transforms the effect of pressure on a thin-walled metal

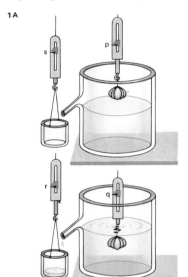

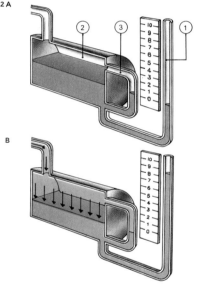

1 Archimedes' principle states that for an object immersed in a fluid the upthrust on it equals the weight of fluid that its volume displaces [A]. The round object suspended on the balance has weight p in air and lesser weight q when immersed in a liquid. The difference in its weight is equal to the upthrust on it in the liquid; that is, to the weight of displaced water in the beaker, which equals r minus s and which in turn equals p minus q. An application of Archimedes' principle is the hydrometer [B], which is used to test the condition of a car battery by checking the acid density. The tube is immersed in the battery acid and the bulb is squeezed to expel air; it is then released to suck acid into the main stem. The hydrometer then floats in the acid, at a depth that depends on the liquid's density.

2 The manometer is a U-shaped liquid column gauge used to measure differences in fluid pressure. The "well-type" has one column [1] of relatively small diameter whereas the second that is wider acts as a reservoir [2]. The difference in columns ensures that the level in the reservoir does not change much with pressure, but that in the small diameter column does, so making for accuracy in reading these variations. Small positional adjustments of the scale with the help of the level indicator [3] compensate for the small reservoir changes. At first [A] both columns are at equal levels. With the reservoir pressurized [B] the new level shows pressure.

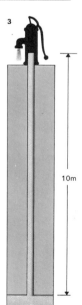

3 In the lift-pump, the piston [1] is moved upwards by the downwards stroke of the lever, producing a vacuum in the cylinder [2]. The piston valve [4] is kept closed by the water already filling the pump chamber. Water is then forced into the cylinder by atmospheric pressure acting on the surface of the water outside the pump. This water passes through the open valve [3] filling the upper chamber. On the upward lever stroke the piston moves down, valve 3 closes and valve 4 opens to allow the piston to move through the trapped water. On the next downward stroke this water is lifted out of the pump and the cycle begins again. In theory, atmospheric pressure "lifts" up to 10m (34ft).

4 A water cannon has water pumped into it under great pressure and forced through an exit nozzle of relatively small diameter. As a result a great force propels the water forwards (pressure is the force per unit area) and accelerates it to a high velocity. This means that the water acquires considerable momentum and kinetic energy which can be used to cut relatively soft china clay from a quarry wall. The great power of the water cannon is also often used to clean the outside walls of buildings. In this case, an abrasive in the form of a powder can be put into the water to increase the corrosive effect of the water jet as it strikes the surface to be cleaned, removing accumulated dirt.

cylinder into the mechanical movement of a needle moving across a calibrated dial. This form is used as an altimeter in many aircraft.

The possibility of supporting a column of liquid by gas pressure is also used in a manometer [2]. This generally consists of a U-shaped glass tube containing a liquid that moves round the U-bend by an amount depending on the difference between two pressures applied at each end. The same principle is employed in the common pump that lifts water from a well. It "taps" a column of water supported by atmospheric pressure acting on the surface of the water source. The height of this column can theoretically be 10.36m (34ft), although in practice the so-called lift-pump [3] can raise water from a depth of only about 8.5m (28ft).

External pressure can be used to move a fluid, but its own internal pressure can also be most effectively employed. A liquid is virtually incompressible, so that pressure developed at one point is transmitted equally in all directions. This fact can be utilized in a hydraulic press or jack to exert very large forces. A force acting on a very small area

produces an enormous pressure (force per unit area) which can be transformed into a much greater force acting over the large area of the hydraulic jack ram [5], since the pressure is constant throughout a liquid.

The study of hydrodynamics

All these effects use the static properties of fluids. But by definition a fluid is something that flows and the properties resulting from motion are described by hydrodynamics.

Motion changes the pressure within a fluid, and this can be difficult to predict accurately. The flow can be either smooth (streamlined) or turbulent, when the fluid is broken up into eddies; it is then harder to calculate the pressure at different points within the fluid. The Swiss scientist Daniel Bernoulli (1700–82) first noticed that pressure decreases as fluid velocity increases (although this holds accurately only for streamlined flow) and this principle creates the necessary lift on an aircraft's wing [Key]. The wing shape is arranged to produce a greater flow velocity above than below and so there is a net upward pressure or lift.

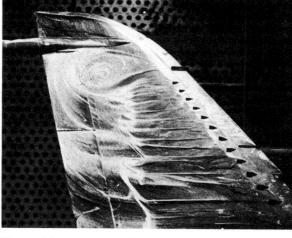

The way air flows around objects can be studied using a wind tunnel. Thus the aerodynamic properties of aircraft wing shapes, or of complete planes and cars, for example, can be investigated. Tiny plastic spheres are injected into the airstream so that the flow patterns can be photographed and perfectly streamlined designs created.

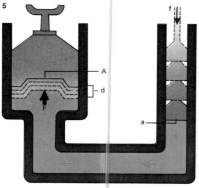

5 An hydraulic lift works on the principle that an incompressible liquid transmits pressure equally in all directions. The small force f acting on area a produces pressure f/a which is converted at the much larger area A to the correspondingly larger force F so that F/A = f/a. The larger force can be used to move great weights. But the weight moves through only the small distance d.

6 The principle of hydraulic lift has many applications, the most familiar of which is probably the car lift used by motor mechanics in garages. Other types of hydraulic machinery have many other applications in agriculture and industry, as in this hydraulic hoist used by an engineer to rig or maintain overhead telephone and electric power lines. If such machines worked exactly as described in illustration 5, the pressurizing force would have to move several metres to produce a small movement of the actuating piston. The hydraulic fluid is therefore pressurized by a pump, which transfers the necessary pressure.

7 A shock absorber is used to reduce or dampen oscillating motions. Several damping systems can be employed – oil damping is shown in A, air damping in B and friction damping in C. The most common type is the oil damper D, which is used especially in road vehicles. It consists of a plunger that on its downward stroke allows oil to pass through a small valve in the piston. The viscosity of the oil and the size of the valve determine the damping characteristics; that is, the manner and speed with which the oscillation is reduced. For rapid loading conditions, where speed of application makes damping solid, a second valve channels oil into a reserve.

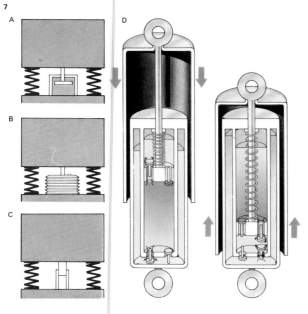

8 A stream of gas tends to adhere to any adjacent solid surface, a phenomenon called the Coanda wall-attachment effect. A gas stream that is free to enter two identical channels [A] chooses one of them by adhering only to its walls. A small side-stream [B] can then deflect it to other output channels [C] where it remains even when the disturbance ceases. This is the fluidics analogue of an electric switch.

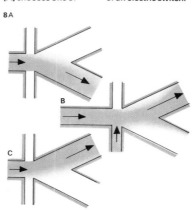

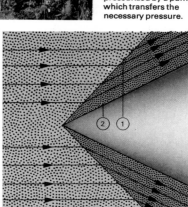

9 When the velocity of a gas is low compared to that of sound its flow can be described without considering the compressibility. Above 0.3 of the speed of sound (Mach 1) a gas is compressed when it meets a solid and there is a consequent temperature change. Above Mach 1 compression occurs abruptly and a shock wave forms. The diagram shows a Mach 4 shock wave [1] at an aircraft nose-cone [2]. At the shock-front the direction of flow changes and the gas becomes more dense.

What is sound?

Sound is energy and, like other forms of energy, can be useful to man. The vast range of expression that characterizes both speech and music makes sound a highly efficient medium of communication and ultrasound – sound above the hearing range of man – has many practical uses. Even loud sounds do not represent a great deal of energy. The noise of a symphony orchestra playing as loudly as possible involves, for example, sound energy equivalent to the light and heat energy from only a low-powered electric lamp. Our hearing sense is more easily saturated (in energy terms) than our visual sense.

How sound is produced

Sound is a particular form of kinetic energy (energy of motion) produced when any object vibrates. Vibration is the cause of all sounds, although generally it is not visible. The sound of a car crash booms out as the surfaces of the two colliding vehicles vibrate with the force of the collision; music comes from a radio as the cone of a loudspeaker vibrates; and talking and singing result from vibrations of the vocal cords in the larynx.

As an object vibrates it sets the air molecules around it vibrating. The vibrations move out through the air, forming a sound wave, but the air does not move along with the wave. Where the air molecules gather together a region of higher pressure (compression) forms. Where they move apart a region of lower pressure (rarefaction) occurs. A succession of compressions and rarefactions move through the air as the sound wave passes. At the ear, they set the eardrum vibrating and we hear sound.

If a surface vibrates more strongly the pressure difference between the compressions and rarefactions is greater and the sound is loud [1]. The frequency of vibrations affects the pitch, or note, of the sound. If it is fast, the compressions and rarefactions are close together and the pitch is high. A slower speed of vibration causes the compressions and rarefactions to be farther apart and the sound is lower in pitch.

A sound wave moves out from its source in all directions, travelling at a speed of 331m (1,087ft) per second or 1,194km/h (741mph) in air at sea-level. The speed is slower at high altitudes as air is less dense there and faster in water and metal because these substances are more elastic than air and transmit vibrations more rapidly. Sound cannot move through a vacuum because there are no gas molecules to vibrate and transmit the sound.

Like other waves of energy, sound normally travels in straight lines, but sound can turn corners. It is reflected whenever it strikes a surface such as a wall [2] or floor and is diffracted or spreads out as it passes through an opening such as a window [3].

Dynamics, pitch and frequency

The loudness of a sound can be measured with a decibel meter and the result given as a number of decibels (dB). The scale is logarithmic – a sound that is twice as loud as one at the threshold of hearing is 10dB greater, not twice as great. Strictly, the meter measures the intensity of the sound, which is related to the pressure differences in the sound wave. (Loudness is the strength of the sensation received in the eardrum and transmitted to the brain.) The human ear does not

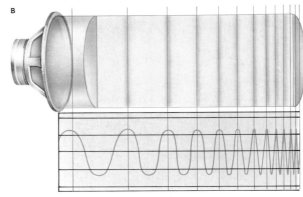

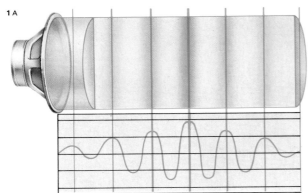

1 A sound wave consists of pressure differences, shown as dark and light bands [A]. The curve shows how pressure changes with time. This wave has a constant frequency (a single note) but decreases and increases in intensity. It would have a "wah" sound. A falling note [B] would be heard as this sound wave passes. The frequency decreases as the note becomes lower, but the intensity remains the same.

2 The Whispering Gallery in the dome of St Paul's Cathedral in London is renowned for its acoustics. A sound whispered against the wall on one side of the gallery can be heard clearly on the other side. Being circular in shape and made of stone, the walls reflect the sound of the whisper all round the gallery and concentrate the sound at the opposite side, 32.6m (107ft) away. Normally, a whisper would be inaudible at such a distance.

Source of sound

Window

3 Both reflection and diffraction enable a sound to be heard even though the person or object producing the sound is hidden from view. Sound is reflected from surfaces such as walls, floors and ceilings and, moreover, undergoes diffraction at an opening such as a door or window. As they pass the edges of the opening the sound waves spread out and hence the opening appears to be the source of the sound.

hear all frequencies of sound in the same way and a low sound is perceived as being less loud than a high sound of the same intensity.

The number of compressions that pass in every second is called the frequency of the sound wave and is measured in hertz (Hz), equal to cycles per second. This scale is not logarithmic and a note of 440Hz (the A above middle C in music) sounds twice as high as, or an octave above, one of 220Hz (the A below middle C). In other words, the higher the frequency, the higher the pitch.

Noise and acoustics

Noise does not have any particular pitch and covers a wide frequency range [4]. Very loud noise is dangerous as well as a nuisance, because continuous exposure to sound of more than 100dB – the levels produced by jet aircraft [5] and machines in many factories – soon results in a permanent reduction in hearing ability. Low-frequency noises are particularly hazardous because they do not seem to be as loud as higher sounds and tests have shown that very high levels of low-frequency sound and infrasound (sound below the hearing range of the ear) quickly result in vertigo, nausea and other physical effects; military scientists have even conducted experiments with infrasound as a potential weapon.

Acoustic engineers work to reduce noise and improve sound in many ways. A consideration of acoustics in the design of a machine such as a jet engine can reduce the amount of noise it makes. Buildings can also be designed to prevent the transmission of sound through them. A steel framework tends to distribute sound throughout a building, but the use of soft sound-absorbing materials in and on floors, walls and ceilings prevents sound from getting into and out of rooms. In concert halls the reflection of sound inside the hall is rigorously controlled to provide an exact amount of echo and give the best quality sound [6]. This may be assisted by electronic amplification, although very loud music loses clarity in a concert hall. Some recording studios have completely absorbent walls to remove all echo and ensure total clarity whatever the type of music being played.

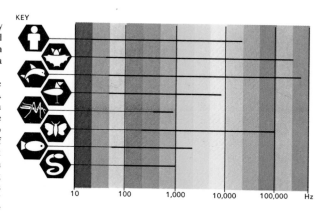
KEY

10   100   1,000   10,000   100,000   Hz

The range of hearing varies widely in man and other animals. Birds and man have fairly similar hearing ranges and both use sound to communicate. Bats and dolphins are sensitive to ultrasound (beyond human hearing), which they use to avoid obstacles and to find their prey by echo-location. There is good evidence that dolphins and other whales communicate by means of ultra-sound. Night moths make use of ultra-sound to avoid predators. Mosquitoes hear a narrow range of sound, corresponding to their own buzzing. The range heard by fish is also extremely small.

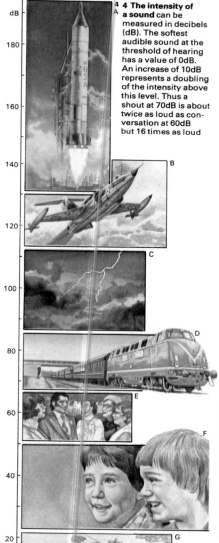

4 The intensity of a sound can be measured in decibels (dB). The softest audible sound at the threshold of hearing has a value of 0dB. An increase of 10dB represents a doubling of the intensity above this level. Thus a shout at 70dB is about twice as loud as conversation at 60dB but 16 times as loud as a whisper at 30 dB. The chart shows the loudness of some sounds near their sources. At 140dB sound causes pain. These fairly common sounds illustrate the range that can be heard by human beings: [A] space rocket at lift-off, 140–190dB; [B] a jet aircraft on take-off, 110–140dB; [C] thunder, 90–110dB; [D] a train, 65–90dB; [E] loud conversation, 50–65dB; [F] quiet conversation, 20–50 dB; and [G] a rustling of dry autumn leaves, 0–10dB.

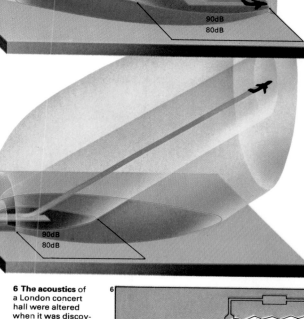

5

90dB
80dB

90dB
80dB

5 Noise is a hazard, particularly near an aircraft, and airport personnel who work on the tarmac may wear earmuffs for protection. Farther away, noise is a severe nuisance but not physically dangerous. Near an airport noise levels are about 90dB – sufficient to drown all conversation – and for miles around noise levels may be 80dB – roughly that of heavy traffic. The problem is worse when aircraft are landing than at take-off. A landing aircraft flies nearer the ground for a longer time and its noise consists of disturbing high-pitched whines. After take-off an aircraft climbs rapidly and leaves a smaller area affected; also its sound is more of a rumble. The new generation of jet aircraft – supersonic airliners excepted – have quieter engines and may affect an area only a tenth the size of that disturbed by older jet aircraft.

6 The acoustics of a London concert hall were altered when it was discovered that the reverberation time – the time taken for the sounds made on the stage to die away in the hall – was too short for the lower frequencies. Electronically amplified resonators were placed in the ceiling to add echo to the hall. The graph shows the reverberation time before [1] and after [2]. The result was a more balanced and pleasing sound throughout the hall.

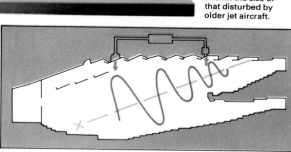

6

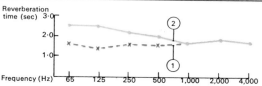

Reverberation time (sec)

2

1

Frequency (Hz)   65   125   250   500   1,000   2,000   4,000

Musical sounds

Why should one musical instrument sound so different from another? Instruments are played in various ways; some are struck, some are blown, while others are bowed or plucked to produce many kinds of sounds. But what is different about the sound itself?

Frequency and pitch

Every instrument produces a sound by making something vibrate and the frequency of the vibration is related to the pitch of the note produced. If the vibration is more rapid, the number of vibrations in the sound wave that reaches the ear (the frequency of the wave) is greater and the pitch is higher or more to the treble. If the frequency is less, the pitch will be lower or more to the bass.

The frequency of a sound wave (number of vibrations per second) is measured in hertz (Hz). The audible range of frequencies for most human beings lies between 20 and 20,000Hz. But some animals, bats and dogs for example, hear over a far wider range.

Every instrument produces a certain set of notes within a particular range of pitch. But each note is in fact a combination of

many more notes. The pitch of the main note heard by the ear is called the fundamental, and above it every instrument also produces a group of higher-pitched notes called harmonics. The harmonics are produced because the vibrating object making the sound vibrates at several frequencies at once and the extra frequencies are simple multiples of the fundamental frequency.

These higher notes can sometimes be produced deliberately on certain instruments – on brass instruments by blowing harder and on string instruments by a particular method of fingering – but normally they are not heard individually. If they were, each note on an instrument would sound like a vast chord. Instead, all the harmonics combine with the fundamental note to produce a complex waveform. Each instrument produces its own particular waveform because the relative intensity of the harmonics is different. The modern music synthesizer works by producing several waveforms of basic shapes – a sine wave, a saw-toothed wave and a square wave – and then combining them to make all kinds of sounds.

Not all instruments produce a note of definite pitch. Several, such as drums and cymbals, produce noise, which consists of a wide range of frequencies without any particular dominant frequency.

The effect of volume

Volume, or the degree of loudness, is another quality of musical sound. Music employs contrasts of volume on a large time scale for dramatic effect, but on a small time scale the change of volume at the beginning of a note is essential to the quality of a sound. The starting characteristics [3], called transients, determine whether a note begins quickly or takes some time to build up; transients are complex and involve changes in the waveform as well as in volume as the instrument begins to sound. Transients are vital to recognition; if the transients are removed from a recording of an oboe, for example, the character of its sound changes until it sounds more like a mouth organ.

Two other qualities often present in a musical sound are echo and vibrato. Echo is often believed to improve music, giving it a

CONNECTIONS

See also
80 What is sound?

In other volumes
104 History and Culture 2
272 History and Culture 2

1 Sound waves combine and the waveforms show that the first tuning fork [A] has twice the frequency of the second one [B]. The combination [C] produces a sound equal to the differences in fre- quencies and the altered shape of the waveform [green curve] shows that it has changed in tone. Two waves that are only slightly different in frequency combine to give a slow beating (pulsing) sound.

2 The waveforms of a flute [A], oboe [B] and clarinet [C] show the differences in tone among them. The flute's rounded waveform betrays the instrument's gentle, fluid sound. The clarinet wave has a similar shape with "jinks" indicating a reedier sound. The oboe's jagged waveform shows that its sound is very reedy. The lowest note of the violin [D] has a fundamental frequency of 196Hz

(G below middle C) but it also vibrates at frequencies that are simple multiples of this, giving a range of harmonics above the fundamental note. The second harmonic has a frequency twice that of the fundamental (the first harmonic). These harmonics collectively colour the basic note by combining with it to give a complex waveform. The relative intensity of the harmonics [E] shows that the fundamental [red] is less intense than some of the harmonics. But it is reinforced because the harmonics interact to give notes with frequencies equal to their differences.

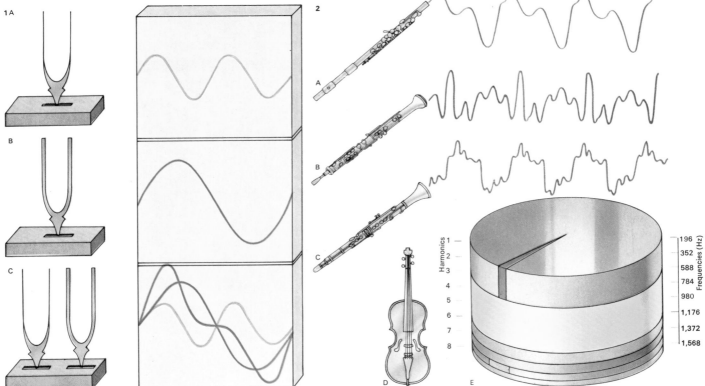

3 The start of a sound must be heard if it is to be recognized. The graph shows that a piano note, for example, reaches its peak volume very soon after it has been struck [A]. It then begins to fall away with great rapidity at first, before tailing away slowly for several seconds. This is what gives the instrument its "attack". The same note produced on a gong takes a comparatively long time to build up and, although of the same pitch, sounds completely different. The same piano note played backwards on a tape recording [B] slowly increases in volume and then suddenly stops. Shorn of its start, the sound is totally unlike a piano and rather like an organ.

more rounded sound, and it is produced by the reflections of sound from the walls of a concert hall or added artificially to recordings. Vibrato is a slight wobble in pitch that many musicians like to use; a violinist moves his left wrist to and fro to produce vibrato.

The nature of the vibrating object is the basis of family grouping of instruments. In string instruments – the violin [2D], viola, cello, double bass, guitar, piano, harpsichord and harp – a taut string is vibrated by stroking it with a bow, plucking it with the fingers or a plectrum, or striking it with a soft hammer. A longer string produces a lower note, and the pitch of notes from an instrument is altered either by pressing the string against a fingerboard to change its length, or by playing a string of a different length. The tension and thickness of the string also affect the note, a tauter or thinner string giving a higher note.

Types of musical instruments

Wind instruments work by making a column of air vibrate. In brass instruments – the trumpet [5], trombone and horn – the player's lips vibrate in the mouthpiece. In some woodwind instruments such as the bassoon, oboe [2B] and clarinet [2C], the mouthpiece contains one or two vibrating reeds and in the flute [2A] the player blows across a hole to set the air column in the instrument vibrating. When a player presses down keys or valves, he alters the length of the air column and produces notes of different pitch [6]. Also, he can obtain some harmonics instead of the fundamental.

Some percussion instruments are played by striking either a taut skin, as in a drum, or a solid object of some kind – a disc of metal in a cymbal, for example. Tuned percussion instruments give definite pitches. They include the vibraphone [7] and xylophone, in which metal or wooden bars of different lengths are struck to sound various notes.

Electric instruments pick up the vibration of a string, as in an electric guitar, or a rod in an electric piano, and convert the vibration into an electric signal that passes to an amplifier and loudspeaker to produce the sound. Electronic instruments include the electronic organ and synthesizer, in which oscillator circuits produce electric signals.

KEY

Musical instruments, from the deepest to the highest members of each family, cover almost the entire range of human hearing. The woodwind family has a particularly wide compass, the lowest note of the contrabassoon and the harmonics [dotted line] of the piccolo nearing the limits of audibility.

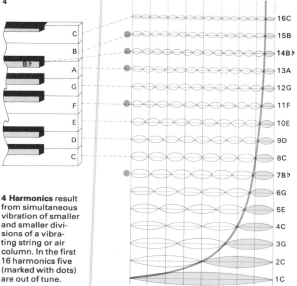

4 Harmonics result from simultaneous vibration of smaller and smaller divisions of a vibrating string or air column. In the first 16 harmonics five (marked with dots) are out of tune.

16C
15B
14Bb
13A
12G
11F
10E
9D
8C
7Bb
6G
5E
4C
3G
2C
1C

5 A modern jazz-rock band contains all families of musical instruments, mingling the acoustic sounds of trumpet, saxophone and drums with the electric sounds of synthesizer, electric piano, bass guitar and electric guitar.

6 Resonance is an important part of musical sounds. It can be demonstrated with a milk bottle and a tuning fork. The fork is struck and held over the neck of the bottle. The bottle can also sound, but its frequency depends on the length of the air column inside. It is possible to "tune" the bottle by adding a liquid until the remaining air space resonates at the same frequency as the tuning fork. The sound of the fork is then much louder. Many musical instruments make use of resonance. The low-volume sound produced by the vibrating string of a violin or guitar, for example, is made much louder by the air resonating inside the accurately shaped body of the instrument.

7 Sound production in musical instruments often involves resonance. Amplification of sound is achieved in the vibraphone in the same way as in the tuning fork and milk bottle. Beneath each bar is a tube of sufficient length to resonate at the frequency produced by the bar when it is struck. Small motor-driven fans over the top of each tube blow air into it and at slow speeds produce a "wavy" vibrato quality in the notes produced.

States of matter: gases

Everything in the world – all matter – exists in just one of three basic states. It is a gas, a liquid or a solid. Some substances can exist in all three states, depending on the temperature. Water, for example, is a liquid at ordinary temperatures. But above 100°C (212°F) it changes to a gas (steam) and below 0°C (32°F) it becomes a solid (ice).

All matter is composed of atoms or molecules and these are held together in liquids and solids by what are called intermolecular forces. The molecules are in continuous motion whose vigour depends on temperature. This "thermal motion" is restrained by the intermolecular forces of attraction, which hold the molecules together. Scientists call this view of matter the kinetic theory ("kinetic" means relating to movement or motion).

Kinetic theory of gases
In gases, thermal motion predominates and the molecules move rapidly in space, constantly colliding with each other and with the walls of their containing vessel. Collisions account for the pressure exerted by a gas [1].

Scientists have made measurements that confirm the kinetic theory of gases. A litre (1.75 pints) of oxygen, for instance, is known to contain about 3×10^{22} (30,000 million million million) molecules. At 0°C (32°F) and a pressure of 760mm (29.9in) of mercury, known as standard temperature and pressure (STP), the molecules move with a speed of about 430m/sec (1,411ft/sec). Molecules are extremely small and measured in ångström units ($1\text{Å} = 10^{-10}\text{m}$). Each oxygen molecule is 3.5Å across. The molecules are, on average, 70Å apart, and they travel about 905Å between collisions; this distance is called the mean free path.

Boyle's law and Avogadro's principle
As temperature rises the kinetic energy of the molecules (their energy of motion) increases by an amount that is proportional to the change in absolute temperature. In a mixture of gases the average kinetic energy becomes the same for each kind of molecule.

When a gas is compressed – that is, if its volume is made to change at a constant temperature – the volume is inversely proportional to the pressure. This relationship is known as Boyle's law [3] after its discoverer, the British scientist Robert Boyle (1627–91). It exists because, when the volume is reduced, collisions with the container walls become more frequent and the pressure rises. If the temperature of a gas rises but it is not allowed to expand (its volume is held constant), the pressure again increases because molecular collisions with the walls become more forceful as well as more frequent.

Another basic gas law is called Avogadro's principle after the Italian physicist Amadeo Avogadro (1776–1856). It states that, at the same temperature and pressure, equal volumes of all gases contain the same number of molecules. A litre of a dense gas such as carbon dioxide contains the same number of molecules as a litre of a light gas such as hydrogen.

Gases slowly diffuse through the walls of a porous vessel because their molecules are smaller than the minute holes in the walls of the container. The rate at which they do so is inversely proportional to the square root of

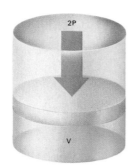

1 The molecules of a gas move continuously at various speeds and in various directions. Collisions with the walls of the container cause pressure and there are so many molecules in even the smallest volume that the pressure is the same everywhere in it. The actual pressure is proportional to the number of molecules in a unit volume and to the average kinetic energy (energy of motion) of the molecules.

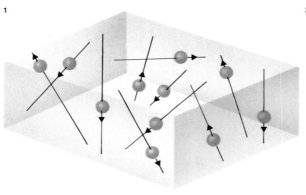

2 A compressed gas is a store of potential energy and it can be made to do useful work when it expands to atmospheric pressure, as in pneumatic drills, hammers and "guns" for spraying paint.

3 When the pressure exerted by the piston is doubled, the gas volume is halved (provided temperature does not change). This is an example of Boyle's law: pressure is inversely proportional to volume.

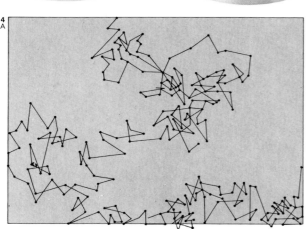

4 In this diagram of Brownian movement [A] – which is evidence for kinetic theory – the dots represent the position of a particle recorded after equal small intervals of time and the lines joining them indicate the paths taken by the random motion of the particle. A beam of sunlight passing through smoke [B] is made visible by reflection from the smoke particles. The same principle is used to view Brownian movement with a microscope.

their density. Discovered by the British physicist Thomas Graham (1805–69), this relationship is called Graham's law [7]. It is explained by the kinetic theory: if different gas molecules have the same average kinetic energy, the lighter ones must move more quickly. Thus light gases diffuse through pores more quickly than dense ones.

Brownian movement
More direct evidence for the kinetic theory is provided by the phenomenon known as Brownian movement [4]. Smoke can be seen in a sunbeam crossing a room. This effect can be produced in the laboratory by looking through a microscope at smoke particles in a box. Specks of light can be seen moving in a haphazard manner, first a short distance in one direction, then in another, and so on. The cause is unequal bombardment of the smoke particles, from different sides, by air molecules. The movement is less with larger smoke particles because they and the air molecules have the same kinetic energy, so the larger particles move more slowly.

When a gas expands it has to do work

against the external pressure. As a result it becomes cooler because the necessary energy must come from the kinetic energy of the gas. This phenomenon accounts for the coldness of the air escaping from a car tyre, for example. When such a change in pressure or volume occurs without heat entering or leaving the gas it is called an adiabatic change. The pressure changes in air as a sound wave passes are adiabatic.

The amount of heat needed to raise the temperature of unit mass of a substance through 1°C is called its specific heat capacity. A gas has two principal ones: that measured at constant pressure (c_p), and one at constant volume (c_v).

Subjected to sufficient pressure most gases turn into a liquid. But above a certain "critical temperature" it is impossible to liquefy a gas using pressure alone [5]. This is because, above this temperature, the kinetic energy of the molecules is sufficient to overcome the intermolecular attractions of their neighbours. Scientists therefore apply the cooling effects of adiabatic expansion in order to liquefy gases [6].

Gasometers store coal gas or natural gas and do not, as their name suggests, merely measure quan- tity. They enable the pressure of the supply to consumers to be kept almost constant, even with fluctuating demand. A gasometer consists basically of a large movable cylinder with a water seal at the base.

5 A

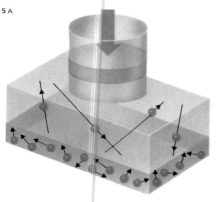

B

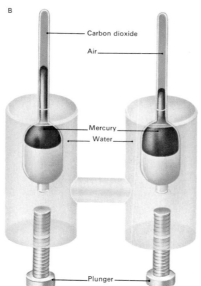

C

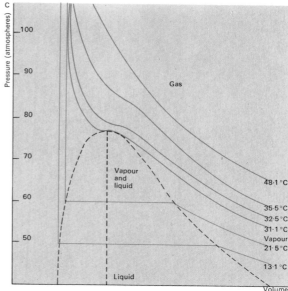

5 In early attempts to liquefy gases [A] they were subjected to high pressure. Some gases liquefied under these conditions and a dynamic equilibrium was established between the liquid and vapour states – molecules left and entered the surface at the same rate. The experiments of Thomas Andrews (1813–85) [B] resulted in techniques that allowed other gases to be liquefied. Pressure is increased by screwing in the plun-

gers and transmitted through the water to the gas and air in the upper tubes. The air is assumed to obey Boyle's law so that its change in volume is a measure of the pressure. Graphs of volume against pressure, called iso- thermals [C], reflect the various states of the gas. The hori- zontal blue line shows liquefaction and does not appear until temp- erature falls below 31.1°C (86°F), the critical temperature of carbon dioxide.

6 In an air liquefier, air free from water vapour and carbon dioxide is com- pressed and cooled by a refrigerator to −25°C (−13°F). It moves the piston part of an adiabatic engine and is further cooled

to −160°C (−256°F). This air cools the other part of the high- pressure air flowing down the central pipes. Final cooling occurs by the Joule- Thomson effect as the air expands through the valve.

7

6

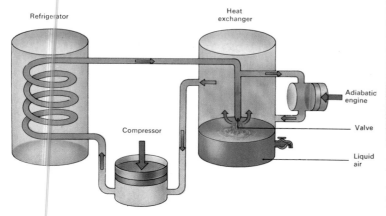

Refrigerator
Heat exchanger
Compressor
Adiabatic engine
Valve
Liquid air

7 The rate of dif- fusion of a gas is inversely proportional to its density. This principle (called Graham's law) was used to separate the isotopes of uranium during World War II to make the first atomic bombs and nuclear reactors in this plant at Oak Ridge in the USA. The uranium had to be converted into its fluoride, which is a volatile solid.

States of matter: liquids

A liquid occupies a definite volume and yet it can flow. The first property is evidence that a liquid's molecules are attracted to each other, whereas the second shows that they have greater freedom than those locked in the lattice of a solid. In a liquid, the molecules vibrate continually (at a rate of a million million times a second) and they change places with each other at nearly the same rate.

A stationary liquid cannot support any stress trying to shear it (as can a solid), because the pressure at any point is the same in all directions. The actual value of the pressure is the product of the depth, the density of the liquid and the acceleration due to gravity. For this reason, solid objects can float in a liquid and even a submerged one is acted on by an upthrust equal to the weight of the liquid displaced. This fact is known as Archimedes' principle.

Structure of liquids

Scientific methods used to study the structure of solids (such as X-ray diffraction) reveal that there are sometimes small volumes in a liquid with molecules in an ordered array.

But there is no overall order as in a solid. In a hexagonal close-packed solid, for example, each molecule has 12 nearest neighbours. In a liquid the number varies between four and 11 and is continually changing.

The average distance between a liquid's molecules is greater than between those of a solid, which explains why most solids take up more room (expand) when they melt. But in a liquid the molecules cannot be squeezed close together (a liquid is almost incompressible). As a result, a liquid can transmit pressure along a pipe [5].

Evaporation and boiling

When a liquid is heated, its molecules move more and more until, at the boiling-point, the liquid turns into a gas or vapour. The heat energy needed to achieve this is called the latent heat of vaporization. Similarly, when a liquid is cooled, its molecules move less quickly until they take up fixed positions and the liquid freezes into a solid. The heat needed to melt a frozen solid back to a liquid at the same temperature is called the latent heat of fusion.

Even at ordinary temperatures (below boiling-point) some molecules "jump" out of the surface of a liquid to form vapour – they evaporate. In a closed vessel there is an equilibrium between a liquid and its saturated vapour; the rate at which molecules leave the liquid is the same as that at which they re-enter it.

When a liquid boils, some work has to be done by the escaping vapour to overcome atmospheric pressure. If the pressure on a liquid is reduced it boils at a lower temperature. If the pressure is increased the boiling-point rises. But if there are no tiny particles in the liquid on which vapour can form bubbles, boiling is suppressed. This effect causes the "bumping" that takes place when pure water is boiled and is the principle of the liquid hydrogen bubble chamber [6]. Water does not follow many of the general rules that apply to liquids [3]. Most substances expand by between five and 15 per cent on melting, for instance, but water contracts by about ten per cent and expands on freezing. These properties arise from the highly directional nature of intermolecular forces in water (due to hyd-

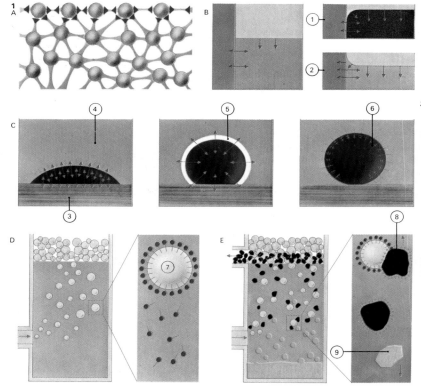

1 At the surface [A], the force between a liquid's molecules causes the surface to behave like a stretched membrane. The surface in a glass container [B] curves depending on whether the glass attracts liquid weakly (eg mercury [1]) or strongly (eg water [2]). An oily liquid [C] adhering to cloth [3] is weakly attracted by water [4]; a detergent forms a new surface layer [5] attracted to both (detergent [6] is partly oil-like and partly salt-like). Some detergents [D] prevent gas bubbles [7] collapsing, making it possible [E] to separate ore [8] clinging to them from earth and sand [9] by flotation.

2 Oil is a prime source of energy and chemicals, and huge quantities of crude oil are carried in tankers from the producing countries to highly industrialized ones, such as the USA, Japan and Western Europe. By Archimedes' principle a tanker displaces its own weight of seawater in order to float and the cargo (oil) is less dense than water. The tanker sinks slightly deeper in warmer water and on the return journey water ballast is carried in the tanks to keep the vessel stable. If a mishap occurs, escaping oil floats on the water: oil slicks are a form of pollution.

3 In all materials the atoms or molecules are in continuous motion. The energy of this motion depends on the temperature. In a liquid, the motion prevents any permanent intermolecular structure from forming, but forces of attraction govern the overall volume. In water there are many temporary linkages (shown blue) between the molecules; very small cavities [1] form and vanish, giving water an ever-changing structure.

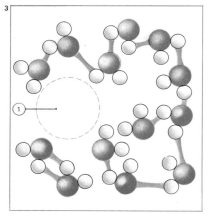

4 All the molecules in a liquid exert attractive forces on their immediate neighbours. Within the main body of the liquid the effects of these attractions cancel each other out. But on the surface the attraction can take place only inwards (there are no molecules outside the surface to counteract it). As a result the surface is in tension. This surface tension behaves like a "skin" which pulls a droplet of water into a spherical shape.

rogen bonding). In the solid state (ice) it produces a very open structure, which disappears on melting. Water has its maximum density at about 4°C (39°F), probably due to a crystal-like ordering of small groups of molecules which disappears on warming.

Surface tension and viscosity

In the centre of a liquid each molecule is attracted by all those surrounding it and their net effect is zero. But at the surface there can be no upward forces to balance the attractive downward forces. As a result a surface molecule tends to be pulled into the body of the liquid [4]. The number of molecules at the surface becomes the smallest possible, and the surface behaves as if it were in tension and had a "skin" on it. The membranous effect of surface tension allows small, dense objects, such as needles or insects, to "float" on the surface of water. If the cohesive forces of attraction between a liquid's molecules are large, it has a high surface tension and a large viscosity (stickiness).

Water wets glass because the cohesive force between a water molecule and a glass molecule is greater than that between two water molecules [1]. The opposite is true of glass and a liquid such as mercury. A liquid such as water rises in a fine capillary tube dipped into it and the meniscus (shape of the surface) curves inwards, or is concave. Mercury, on the other hand, is depressed in a capillary tube and possesses a convex (outward curving) meniscus.

Water pours more easily from a jar than does treacle and treacle is said to be more viscous than water. A simple model of what happens shows one layer of liquid molecules under a shearing stress sliding over another layer. For this to happen any molecule in the faster-moving layer must overcome the attraction of the nearest one in the adjoining layer. And having moved one place along the line it must repeat the process. To do this it must use some energy and this slows it down. The relative velocity between the layers is reduced and the result is viscosity. Heating a liquid provides it with more energy and, as expected, viscosity falls with increasing temperature. For this reason, warm treacle pours more easily than cold treacle.

Various liquids differ in their physical properties such as boiling-point and viscosity (stickiness). Water and wine have roughly the same boiling point and low viscosity. Oil is more viscous and boils at a higher temperature whereas exceptionally viscous liquids such as honey and tomato ketchup have extremely high boiling temperatures.

5 Liquids are almost incompressible and as a result can transmit pressure. This important principle finds many applications in the branch of engineering called hydraulics. Many trucks have an hydraulic jack in which pressure transmitted by means of oil is used to tilt the load. A pump is used to provide the pressure and provision has to be made for the oil to run back when the pressure is released.

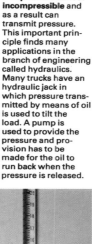

6 In a bubble chamber, a dust-free liquid in a perfectly clean vessel is heated to a temperature above its boiling-point and extra pressure is applied to stabilize it. If charged particles are then directed into the chamber, bubbles form on the charged "nuclei" left by the particle along its track, making it visible. Liquid hydrogen is generally used as it is a good source of protons on which bubbles can form.

7 A water drop at the end of a glass tube takes its shape because of surface tension. The attraction between its molecules leads to a spherical shape as the "skin" effect caused by inward-acting forces in the surface holds the bulk of the liquid back. But a water droplet becomes a distorted sphere because gravitational as well as surface tension forces are acting on it.

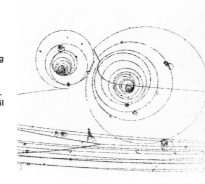

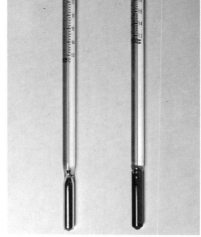

8 The most common method of measuring temperature makes use of the expansion of a liquid on heating. A mercury thermometer (left) has a wide range (−39°C to 360°C), but needs a large bulb (reservoir) and a narrow stem if small temperature changes are to be detected. Alcohol can be used in thermometers for measuring lower temperatures; it expands more, but boils at 78°C. The great advantage of liquid-in-glass thermometers is that they can be read directly and carried easily.

9 Motor oil is more viscous than water, as can be seen when each is poured; the water flows much more easily than oil because layers of water molecules slide over each other more easily than do layers of oil molecules. Normally, the viscosity of a liquid decreases with rise in temperature. Oil for a particular application must have the right lubricating properties and much research has been done to produce the correct oils for car engines and gearboxes, including oils whose viscosities change only a little when they get hot. When a liquid flows in a pipe in streamlined motion, the region in contact with the pipe is still and that near the axis has the greatest velocity.

States of matter: solids

The crystalline shapes of many solids indicate that the atoms in them take up some kind of regular arrangement. In the amorphous or non-crystalline substance, there is no regular order. There are seven main crystalline structures, of which the cubic system is the simplest. Sodium chloride (common salt) is composed of sodium ions and chloride ions. In the solid salt these ions take up what is called a face-centred cubic structure. This and other arrangements can be confirmed by making the crystals diffract an X-ray beam, then experts can use such X-ray photographs to work out the structures of complex crystals.

Sodium chloride is an example of an ionic crystalline substance [1]. Other crystalline substances, such as diamond, consist of a regular array of atoms linked to each other by covalent chemical bonds, in which one or more electrons are shared between neighbouring atoms. In waxes and similar substances, molecules are held together only weakly by what are called Van der Waals' forces. And a metal has a lattice of positive ions in which free electrons occupy the spaces. Applying a voltage across the metal makes electrons drift between the ions, which is the reason why metals and their alloys are good conductors of heat and electricity.

Vibrating and slipping atoms

All intermolecular forces can be thought of as electrical in origin, causing attraction between molecules at relatively large distances and causing repulsion at close quarters. The elastic properties of solids can be explained in terms of such forces. When a material is stretched, the distances between its atoms increase slightly and the resulting strain is found to be proportional to the stress producing it (the relationship known as Hooke's law). Compression moves the atoms closer together, while shearing makes layers of atoms slide over each other.

The atoms in a solid – even a crystalline one – vibrate about their average position in the lattice. Heating a pure solid makes its atoms vibrate more vigorously. Sufficient heat energy overcomes the forces holding the atoms together, the crystalline structure "falls apart" and the solid melts.

A single crystal of a pure metal is much weaker than might be expected. This may be due to imperfections in the lattice which cause dislocations [7]. Under stress, the layers of atoms move in such a way that the dislocation shifts towards the edge. An ordinary metal is polycrystalline, consisting of an irregular arrangement of many small crystals. Stress makes layers of atoms in individual crystals slip over each other. But atoms of an impurity within a crystal can "anchor" dislocations and prevent slipping. Consequently an alloy is usually harder than the two or more metals of which it is composed.

Metal fatigue: cause and cure

Strain deforms some solids over a period of time in the phenomenon called creep. This can be due to the movement of dislocations in crystalline grains, slip between grain boundaries, or slip along well-defined glide planes. "Fatigue" is the name given to a change in metallic properties that may result in sudden breakage. It resembles work-hardening (caused by hammering) in that dislocations in crystallites interlock, causing brittleness.

Scientists can make thin fibres of sub-

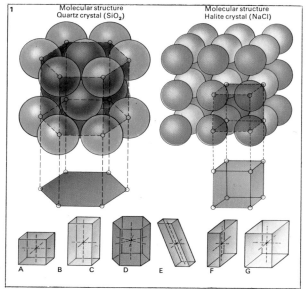

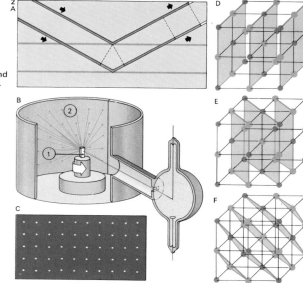

1 **The main systems of crystal structure** are cubic [A], tetragonal [B], hexagonal [C], monoclinic [D], triclinic [E], orthorhombic [F] and trigonal [G]. Scientists studied the shapes and properties of crystals. Their research established the seven basic systems. The molecular structure of a quartz crystal is hexagonal and relates to the three-dimensional arrangement of its atoms. This crystal has three equal axes [C] inclined at 120° to each other with a fourth unequal and perpendicular to the other three. The crystal of halite (sodium chloride) has three axes equal and mutually perpendicular to each other [A].

Molecular structure Quartz crystal (SiO₂) — Molecular structure Halite crystal (NaCl)

A B C D E F G

2 **Regular layers of atoms** in a crystal [A] diffract a beam of X-rays passing through it. A goniometer [B] uses this effect to reveal crystal structures. The crystal [1] rotates on a pillar and scatters X-rays onto a photographic plate [2], to produce an X-ray diffraction photograph [C]. From the pattern of spots, a scientist can work out the structure. If halite is used, the X-rays reveal sheets of atoms arranged in planes [D], [E] and [F]. It is these planes that reflect the X-rays when the crystal rotates around an axis that coincides with one of its main axes.

3 **Tensile strength** is tested by stretching. A metal extends overall at first, but later expansion concentrates around the point of fracture. Curves A and B show typical extensions. Curve A, for mild steel, remains linear to its elastic limit. If the load is released early, the metal returns to its original length. Curve B is typical of softer metals. A range of relative strengths of metals is also shown (right).

Load — Yield strength — Elastic limit — A — B — Elongation

Alloy steels — High carbon steels — Nickel alloys — Carbon steels — Aluminium alloys — Brasses — Copper — Silver

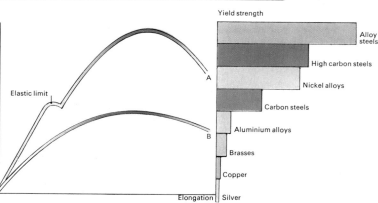

4 **Engineers** use high-grade steel girders in their work because good steel is usually highly resistant to cracking. If a crack does appear, the metal is generally ductile enough for the edges of the crack to flow together, which will diminish the danger of the crack extending far. In poor steels cracks may develop rapidly. This is a particular danger in bridges where the steel has to resist changes in temperature.

stances free from any dislocations. Called whiskers, they are very strong and when incorporated into a matrix of another substance produce a strong composite material.

Structural effects in solids

Molecules in natural and man-made polymers have a complex arrangement. X-ray photographs reveal that, when rubber is stretched, its coiled long-chain molecules line up. When tension is released, the molecules snap back into their former shapes, and as a result giving rubber its elasticity. Similarly a stretching process during the manufacture of nylon lines up long chains of molecules.

Semiconductors – the key to modern solid-state devices such as transistors – make use of subtle variations in an otherwise normal crystal lattice. The basic element, such as silicon or germanium, has incorporated in it minute traces of a deliberately introduced impurity element. The impurity has either one more or one fewer electron in its atoms than the basic semiconductor element. As a result, there is a slight excess of

electrons or a slight deficiency (an absence of an electron is in this context called a "hole"), and it is the movement of these electrons or holes that gives the materials their special electrical properties. When extra electrons are present the semiconductor has negatively charged current carriers and so is called an *n*-type semiconductor. The holes in the second type are considered to be positive carriers and it is called *p*-type.

X-ray diffraction enables scientists to study the microstructure of solids. Other methods reveal the overall or macrostructure. The grain structure of a metal, for example, can be revealed by etching the surface and viewing it by reflected light through an optical microscope.

Higher magnifications are possible using an electron microscope. Scientists make a copy of the etched surface by depositing on it a layer of carbon or plastic and stripping it off as a thin film to be viewed with the microscope. They may use a very thin foil of the metal to be studied. The fine structure of the surface is revealed in three-dimensional detail by a scanning electron microscope.

Seen under a microscope, crystals reveal their regular shapes. Perfect – that is, unbroken – crystals have precise shapes, such as the feathery needles of ammonium chloride There are six other basic crystal shapes, or systems.

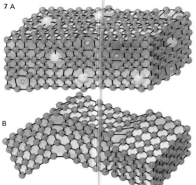

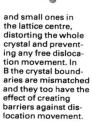

5 Metalworkers hammer sheet metals to harden them without making them brittle. The hammering moves dislocations along intersecting slip planes until they meet and stop. These meeting places act as barriers to the movement of any other dislocations, making the metal stronger.

6 Foundryworkers separate metals from ores by smelting them. Metals are heated to break down the lattice of atoms so that the metal flows. They are poured into moulds or cooled and rolled into sheets. Melting-points of metals range from mercury which melts at −38.8°C (−38°F) to tungsten which melts at 3,410°C (6,170°F).

7 A

B

C

D

7 Strong metals restrict the free movement of dislocations. A metallurgist may achieve a strong metal by making an alloy, or he may make the metal's crystals as small as possible. In A, large atoms are at the crystal corners and small ones in the lattice centre, distorting the whole crystal and preventing any free dislocation movement. In B the crystal boundaries are mismatched and they too have the effect of creating barriers against dislocation movement.

Where dislocation of the atomic lattice occurs, or where there is slipping between grain boundaries and along glide planes, the metal will fail. Frequently repeated strains and fluctuating loads may eventually cause metal fatigue [C]. The edges of the fractures may show signs of metallurgical recrystallization. Strain over a long period produces a similar effect, called creep failure [D]. The example of failures C and D occurred in a nickel alloy turbine blade.

8 Increased pressure lowers the melting-point in substances such as water, which expand when they solidify. Ice melting under the pressure of a skate acts as a lubricant that makes the skater's motion both smooth and easy.

89

Heat and temperature

A bicycle pump becomes warm when pumping up a tyre, an effect that can be explained by the kinetic theory of heat, first put forward by Isaac Newton (1642–1727) and Henry Cavendish (1731–1810) in the eighteenth century. The theory explains what heat is – the kinetic energy (energy of motion) of the vibrating atoms or molecules that make up every substance. In the bicycle pump, air molecules are speeded up by collisions with the pump's piston. The increase in their kinetic energy takes the form of heat.

Thermal agitation and molecular motion
According to the third law of thermodynamics, absolute zero – the temperature (−273°C) at which molecules cease all motion – is unobtainable. As a result, molecules perform a continuous motion known as "thermal agitation", which increases in vigour as heat is transferred to an object. Indirect evidence of this incessant motion was first obtained by the botanist Robert Brown (1773–1858) in 1827. He discovered that tiny pollen grains suspended in water were continually making jerky

movements. It is the continual unequal bombardment of each tiny speck by the molecules of the liquid that produces this "Brownian motion". The smaller the particle, the more violent is the motion.

The kinetic theory also explains why, when a hot gas is mixed with a cooler one, a common temperature is eventually reached. Molecular kinetic energy of the hot gas molecules is transferred by thousands of collisions to the cold gas molecules, until the average kinetic energies of both gases are the same. The molecules all travel with different velocities, changing after each collision. For this reason the temperature of the gas (or other substance) is a measure of the average molecular kinetic energy.

The changes of state that may occur when a substance is heated can be described by the kinetic theory [1]. In a solid, the atoms or molecules are tightly bound together and vibrate only about an average position. As the solid is heated its internal kinetic energy increases and the particles vibrate more vigorously. They move farther and farther apart until eventually the attraction between

them is insufficient to keep them in a fixed position. They can then slide about and exchange partners – the solid melts and a liquid is formed. The heat needed to achieve this change of state from tightly bound solid to loosely bound liquid at the same temperature is called the latent heat of fusion. If yet more heat is applied, the atoms or molecules gain more kinetic energy, move with greater velocity within the liquid and the proportion of these escaping from the surface increases (vapour pressure increases). Eventually, at the boiling-point, so many atoms have enough energy to escape from the liquid that the vapour pressure equals atmospheric pressure. In the gaseous state the atoms or molecules move almost independently; the conversion of a liquid at its boiling-point into a gas requires latent heat of vaporization.

Changes of temperature
Instead of changing the state of a substance (for which latent heat is needed), applied heat energy may merely raise its temperature. The temperature change depends directly on the quantity of heat transferred

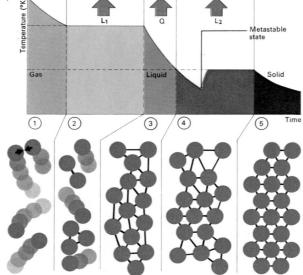

1 Vibration of atoms and molecules in a substance governs its temperature. In a gas [1] atoms move independently and their average velocity and mass determine the internal energy and temperature. After cooling, loss of latent heat of condensation [L₁] converts the gas at boiling-point to a liquid state [2] when its atoms become locked in a weakly bonded arrangement. Further cooling [3] to freezing-point loses a quantity of heat [Q] and a solid then forms through [4] release of latent heat of fusion [L₂]; the atoms then become rigidly bonded together [5]. A colder "metastable" phase can precede actual freezing.

2 The three ways in which heat moves all take place when a pan is heated [A] – conduction through the metal walls of the pan [1], convection by fluid motion [2] and radiation from the heat source to the pan [3]. In theory an insulated good conductor with ice at one end and boiling water at the other varies in temperature linearly with distance along the bar [B], as in the straight-line graph. Without insulation the dotted line results. A vacuum flask [C] has a vacuum [4] to prevent conduction and convection, and silvered walls [5], like mirrors, to minimize heat loss by radiation.

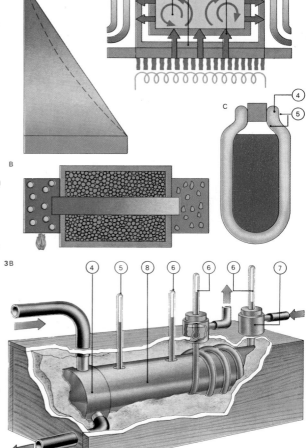

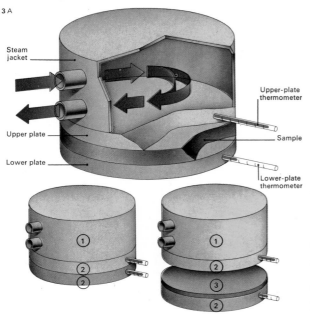

3 Conductivity is the amount of heat passing in unit time across unit cross-section per unit of temperature gradient. It can be measured [A] by noting the time taken for a known quantity of heat to pass through a sample. Two plates of equal area [2] are put against the material [3]. Thermometers on each side measure the temperatures as the upper plate is heated by a steam jacket [1]. Materials of high conductivity [B] are tested in a cylinder [8] heated at one end by steam [4] to 100°C [5]. Other thermometers [6] measure the temperature of the sample and the rise in temperature of water circulating through a jacket attached to the other end [7], and give a way of calculating conductivity.

and this is measured in units called calories or joules (4.2 joules=1 calorie). The calorie is defined as the quantity of heat that raises the temperature of 1 gramme of water by 1°C. So the quantity of heat needed to raise the temperature of 1 gramme of any substance by 1°C can be measured in these units – and this is called the specific heat of the substance. The quantity of heat that raises the whole bulk of substance by 1°C is called the thermal capacity of the given mass.

It is possible for heat to be transferred from place to place and there are three ways in which this can occur: by conduction, convection and radiation [2]. All methods rely on the fact that atoms that have received kinetic energy from a heat source can transmit this to their neighbours by collisions with them. In a tightly bound solid only nearest-neighbour collisions occur and any heat transfer through the solid is called conduction. In a fluid (liquid or gas), the medium itself can move and transport atoms of high kinetic energy to the cooler parts of the fluid where they then transfer their heat – this is convection. Even when no physical

contact exists between atoms, heat can still be transferred. For instance, heat from the Sun reaches the Earth through the near vacuum of space. This method is radiation. Different substances conduct heat at different rates. Their ability to transmit heat is known as their thermal conductivity [3].

Measuring temperature

All methods of measuring temperature changes are based on the ways in which materials change physically when heated. The most commonly used characteristic is the expansion of solids or liquids when heated. Usually the physical change caused by a temperature change is made visible against a calibrated scale on the measuring instrument or thermometer. In making any type of thermometer [5, 6, 7], two constant temperatures or "fixed points" must first be marked. The range between them can then be subdivided as finely as desired. The numbers assigned to the fixed points and the numbers of degrees between them define the temperature scale, such as Celsius (or centigrade), Fahrenheit and Kelvin (or absolute) [4].

A furnace converts the chemical energy in coal to heat energy used to produce the steam that drives an engine. The heat of the furnace brings the water to its boiling-point and provides latent heat of vaporization to turn it into steam.

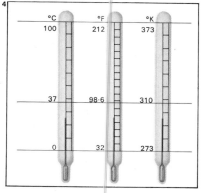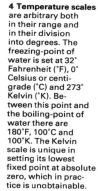

4 Temperature scales are arbitrary both in their range and in their division into degrees. The freezing-point of water is set at 32° Fahrenheit (°F), 0° Celsius or centigrade (°C) and 273° Kelvin (°K). Between this point and the boiling-point of water there are 180°F, 100°C and 100°K. The Kelvin scale is unique in setting its lowest fixed point at absolute zero, which in practice is unobtainable.

5 The indicators in a maximum and minimum thermometer [A] are pushed along by mercury in a U-tube and stay put at the farthest point of travel. Liquid in a cylindrical reservoir senses the temperature and its contraction [B] or expansion [C] displaces the mercury along scales at each side of the U. The steel indicators can be reset simply by moving a small magnet along the outside of the tube.

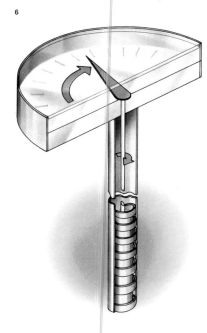

6 A bimetallic strip thermometer uses a helical metal strip that unwinds when heated and rotates a pointer over a calibrated scale. When warmed, the inner metal (usually copper) expands more than the outer one, so causing the bimetallic strip to unwind. The material used for the outer strip is usually Invar, an alloy of iron and nickel, which has a low coefficient of thermal expansion.

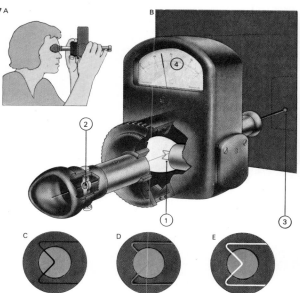

7 An optical pyrometer allows very high temperatures to be measured at a distance from the temperature source [A]. It exploits the fact that two solids at the same high temperature radiate light with the same spectrum and show the same colour. To measure the heat of a furnace [B], an electrically heated filament [1] fixed in the tube of a telescope with a special lens [2] is heated until it glows with the same colour as that emitted by the furnace [3]. Comparison with the background image shows if the furnace is hotter [C], as hot [D] or colder [E] than the wire. A meter [4] indicates the current passing through the wire and this can be calibrated directly in temperature degrees.

8 Liquid air can be produced by cooling air, which is normally gaseous at room temperature. During the change to a liquid state, latent heat of condensation is released. To cut down heat input the liquid is kept in a Dewar vacuum flask. But if it is poured out of this flask, heat from the surroundings makes the liquid boil rapidly. Large amounts of gas are generated during this process and cause a fog of condensed water vapour to form above the liquid.

Order and disorder: thermodynamics

Thermodynamics (meaning "the movement of heat") deals with the ways in which heat energy travels from one place to another and how heat is converted into other forms of energy. In a heat-transfer process temperature, pressure and volume may each or all undergo various changes. Much of thermodynamics consists of ways of mathematically manipulating these and other parameters to be able to make predictions about the ways in which they will and do change.

The four laws of thermodynamics
Historically scientists first derived three laws called the first, second and third laws of thermodynamics. Then an even more fundamental law was recognized. It has been labelled the "zeroth" law of thermodynamics.

If a hot and a cold object are brought into contact, they finally reach the same temperature [1]. The hot object emits more heat energy than it receives and the cold object has a net absorption of heat. Both objects absorb and emit energy continually, although in unequal quantities, and the exchange process continues until the temperatures

equalize. Each object is then absorbing and emitting equal amounts of heat and the objects are said to be in "thermal equilibrium". The zeroth law states that, if two objects are each in thermal equilibrium with a third object, then they are in thermal equilibrium with each other.

The first law really has two parts – the first is the law of conservation of energy and the second effectively defines "heat energy" and how types of energy can be converted into one another [2]. If heat energy is supplied to a system, then the first law states that this equals the change of internal energy of the system together with the mechanical energy that allows the system to do external work. Thus in a petrol engine an air/petrol vapour mixture is ignited after being compressed. The burning of the mixture releases heat energy from the chemical reaction, thereby causing the gases to expand and do work against the piston by moving it. The burned gases are finally hotter than the system was before the explosion, so there is a change in the internal energy of the petrol engine system. The sum of this energy change and

the external work done equals the released heat energy.

Following the zeroth law, which defines temperature, and the first law, which describes energy conversion, the second law of thermodynamics governs the direction of flow of heat energy between objects at different temperatures. It says that, of its own accord, heat can flow only from a hot to a cold object. The heat transfer increases the motion of the molecules of which the colder object is composed and so effectively increases its internal "disorder"

Cooling substances
The third law of thermodynamics states that it is impossible to cool any substance to absolute zero. This zero of temperature would occur for example in a gas whose pressure was zero. All its molecules would have stopped moving and possess zero energy, so that extracting further energy and achieving corresponding cooling would be impossible. As the temperature of a substance approaches absolute zero, it becomes progressively harder to cool further.

1 The thermic lance produces enormously high temperatures. If it is directed onto a substance a thermal equilibrium condition is achieved, raising the temperature of the substance near the lance up to its melting-point, allowing it to be carved at will. Here US scientists are using a lance to cut the Arctic ice cap.

2 Friction occurs whenever two surfaces rub together, and this in turn generates heat. The sudden surge of power at the beginning of a dragster race spins a car's rear wheels and the heat generated burns the rubber of the tyres. Similar heating occurs in a car's brakes when they are used.

3 Carbon dioxide, released as a solid from a fire extinguisher, turns into a gaseous form and prevents oxygen reaching the fire.

4 Heat can be converted to mechanical work by allowing a heated gas to expand. Maximum efficiency comes from the Carnot cycle of alternate adiabatic [BC, DA] and isothermal [AB,CD] processes. The former take place without gaining or losing heat, the latter without changing temperature. From A to C the gas raises the piston and performs work equal to the area below the curve, that is, $V_A ABCV_C$. From C to A it uses work equal to area $V_A ADCV_C$. The net work obtained equals area ABCD.

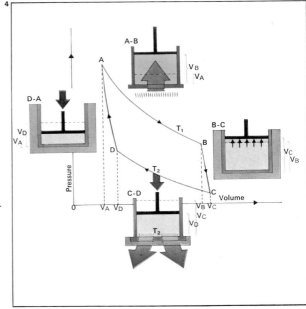

From the statement of the second law, a heat transfer process naturally proceeds "downhill" – from a hotter to a cooler object. There must be some property or parameter of the system that is a measure of its internal state (its order or disorder), and which has different values at the start and the end of a possible process (one allowed by the first law). This parameter is termed the "entropy" of the system, and the second law maintains that entropy cannot be destroyed but can only be created.

Careful observation of all types of machines shows that they consume more energy than they convert to useful work. Some energy is wasted in wear and friction or lost by necessity, as in a shock-absorber or a radiator. In all these ways the energy is degraded. The entropy of the system is a reflection of its inaccessible energy, and the second law says that it is always increasing. Heat is a random motion of atoms and when the energy is degraded towards the inaccessible energy pool, these atoms assume a more disorderly state – and entropy is a measure of this disorder.

Under the constraints imposed by the laws of thermodynamics it is possible for a system to undergo a series of changes of its state (in terms of its pressure, volume and temperature). In some cases the series ends with a return to the initial state, useful work having been done during the series.

Heat cycles and efficiency

The sequence of changes of the system is called a heat cycle and the theoretical maximum efficiency for such a "heat engine" would be obtained from following the so-called Carnot cycle [4] which is named after the Frenchman Nicholas Carnot (1796–1832). If it were possible to construct a machine operating in cycles which, in any number of complete cycles, would generate more energy in the form of work than was supplied to it in the form of heat then the dream of the perpetual motion machine would be possible [5]. The first law states the impossibility of achieving this result and the second law denies the possibility of even merely converting all the heat to an exactly equivalent amount of mechanical work.

KEY
A flame applied to one end of a metal bar transfers heat energy to the atoms of the metal. This raises their kinetic energy so that the atoms begin to vibrate much more vigorously about their fixed mean positions within the lattice network of the metal. As the extent of vibration increases, collisions with neighbouring atoms occur so that energy is transferred to these atoms causing them to vibrate. Heat energy is eventually transmitted to the other end of the bar and if the flame is kept in position for some time the temperature at each end of the bar may equalize.

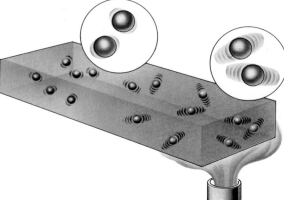

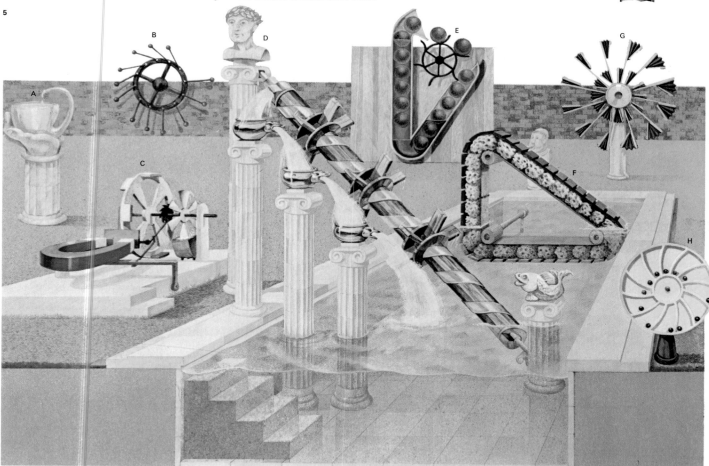

5 Perpetual motion machines have been the aim of many inventors. Usually these have been mechanical arrangements designed to continue moving for all time once set in motion. The failure of all such attempts provides one of the best verifications of the laws of thermodynamics. Machine A contains a quantity of liquid that it is assumed will drop back into the large vessel from the bent spout and so maintain circular movement of the liquid. But once the height of liquid in the spout equals that in the vessel, the pressures within the liquid on either side equalize and the fluid motion cannot continue. Machine B has a set of equal weights attached to a rotating wheel. Although there are few weights on the downward side, their longer distance from the axis provides the extra movement of force needed to raise the greater number of weights on the upward side. Unfortunately energy is wasted in overcoming the friction opposing the turning motion of the wheel, this upsets energy balance and the rotation stops. Friction prevents most of the machines from working; machines C, G and H all employ almost the same thinking as B and therefore do not function as perpetual motion machines. Machine D does not work because energy is lost against friction as the screw-like construction is turned and as heat is generated and lost when the water cascades down against the paddle wheels and lower reservoir levels. Machine E tries to keep the balls moving round continually by having the resultant upthrust force acting in one direction through the liquid. But the water levels in each tube would equalize under the force of gravity so that the balls would not get buoyed up one tube further than the other. In machine F the sponges raising and lowering water on the conveyor belt would not retain all the water they first absorb so that the lighter sponges on the downward belt would not raise the heavier ones just leaving the water. Energy loss by friction would occur and prevent perpetual motion of even ideal sponges.

5

Towards absolute zero

Every substance contains a certain amount of heat, even a relatively cold substance such as ice. The heat is the result of the continual motion of the substance's molecules which, by that motion, possess kinetic energy. Temperature is a measurement of the average kinetic energy of the molecules. The cooler a substance becomes the less its molecules move. Thus it should be possible to continue cooling to the point at which molecular movement ceases completely. This point, "absolute zero", is of great interest to scientists but in practice is unattainable. At temperatures close to absolute zero some materials exhibit remarkable properties, such as superconductivity [6] and superfluidity [Key].

Calculation of absolute zero
On the centigrade temperature scale, absolute zero is 273.15 degrees below the freezing-point of water. Its value can be predicted as a result of the behaviour of gases when they are heated or cooled. When heated, a "perfect" gas expands in volume (V) proportionally to its absolute temperature (T) if its pressure (P) is kept constant. Its pressure increases in the same proportion if its volume is kept constant. The reverse occurs on cooling, according to the equation $PV = RT$ where R is known as the universal gas constant. The pressure actually falls by a factor of 1/273.15 for every 1°C temperature decrease. Thus at −273.15°C zero pressure would be reached and this must be the absolute zero of temperature.

Absolute zero is usually denoted as 0° on the Kelvin scale of temperature, named after the British scientist William Thomson, Baron Kelvin of Largs (1824–1907). Its temperature increments equal those on the centigrade or Celsius scale [1]. Thus 0°K is the same as −273°C (absolute zero temperature is usually rounded to −273°C, or −460° on the Fahrenheit scale) and 273°K equals 0°C − the freezing-point of pure water.

Aiming at absolute zero
Gas temperatures can be lowered by first compressing the gas in a fixed-volume enclosure and then removing the resultant heat with, for example, a surrounding water jacket. If the gas is then allowed to escape into a larger volume it becomes even cooler because its molecules lose kinetic energy during the expansion. This cycle is used in a refrigerator and can liquefy and even freeze many gases.

The gas most useful in experiments at very low temperatures has been helium, the gas with the lowest boiling-point, 4.2°K (−269°C). The temperature of liquid helium can be further reduced to 1°K by vacuum pumping the gas above the level of the liquid to reduce its pressure and thereby force down the boiling-point. Liquid helium is generally produced in an air liquefaction plant as one of the rare gases left after oxygen and nitrogen have been liquefied [2].

Below 1°K it is much more difficult to achieve further cooling and a low-temperature effect that occurs in some solids is used [3]. Some salts act as magnets when close to a strong magnetic field but stop being magnetic when the field is removed, a phenomenon known as paramagnetism. When the salt is magnetized its molecules line up in the field but are disarranged when the

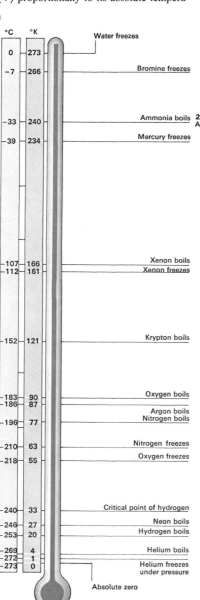

1
°C	°K	
0	273	Water freezes
−7	266	Bromine freezes
−33	240	Ammonia boils
−39	234	Mercury freezes
−107	166	Xenon boils
−112	161	Xenon freezes
−152	121	Krypton boils
−183	90	Oxygen boils
−186	87	Argon boils
−196	77	Nitrogen boils
−210	63	Nitrogen freezes
−218	55	Oxygen freezes
−240	33	Critical point of hydrogen
−246	27	Neon boils
−253	20	Hydrogen boils
−269	4	Helium boils
−272	1	Helium freezes under pressure
−273	0	

Absolute zero

1 The temperatures on this diagram are stated in both the Kelvin (absolute) and Celsius (centigrade) scales, below a temperature equivalent to the melting-point of ice in equilibrium with water.

2 In a helium liquefier, which can be carried by road [A], a mixture of helium gas and air is first compressed and the heat generated removed. Air contains, in addition to oxygen and nitrogen, other "inert" gases such as argon, neon, krypton and xenon. At about 20°K all the gases of air except helium can be liquefied in the separator [B]. The helium can be expanded through a nozzle and liquefied.

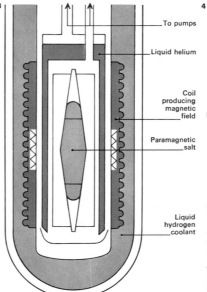

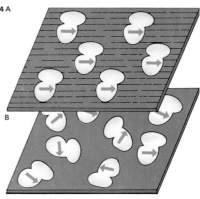

3 To approach absolute zero – below 1°K (−272°C) – a phenomenon called adiabatic demagnetization of paramagnetic salts is used. A paramagnetic salt is one that acts as a magnet only near a strong magnetic field. The field forces the molecules into an orderly array. If there is no exchange of heat energy during this process then it is said to be adiabatic. But when they are ordered the molecules have less energy and the balance is given up to the salt, raising its temperature above 1°K. Liquid helium is then used to re-cool the salt. If it is now demagnetized, its molecules become disordered. The energy for this can come only from the salt and its temperature falls below 1°K.

To pumps
Liquid helium
Coil producing magnetic field
Paramagnetic salt
Liquid hydrogen coolant

4 The molecules of a paramagnetic salt are normally in continual disordered motion, even if the temperature is as low as 1°K (−272°C). As long as the molecules behave in this way there is no part of the salt that appears to be like either pole of a magnet. As soon as the salt is placed in a magnetic field [A], however, the molecular magnets line up along the field and endow the bulk substance with north and south magnetic poles. Molecular disorder returns when the field is removed [B].

field is removed [4]. If a paramagnetic solid is cooled to 1°K by liquid helium that is allowed to evaporate, heat energy is removed from the solid. When a strong magnetic field is switched on, the molecules align themselves and create heat by their motion. This is removed by the surrounding helium gas, which is pumped away. When the field is switched off, the molecules become disordered and cause a further lowering of the solid's internal energy. The cold salt can then absorb heat from a second helium container. A cycle of magnetization and demagnetization can produce temperatures of a few thousandths of a degree Kelvin.

Superfluidity and superconductivity

Liquid helium at very low temperatures is not only difficult to produce but behaves in a most unusual way. Fast boiling occurs as the vapour pressure falls but at 2.18°K the internal bubbling of helium gas suddenly ceases, although boiling continues. Below this so-called "lambda point" liquid helium exhibits "superfluidic" properties [Key].

Near absolute zero, certain substances show remarkable properties; for example, a kind of perpetual motion in electric current becomes possible – that is, some metals and alloys exhibit superconductivity [6]. As their temperature is lowered (for example, to 7.2°K for lead) the electrical resistance of the material disappears completely. If an electric current can be made to flow in a ring of such metal it continues to flow. Current has been kept flowing unattenuated in this way for up to several years.

Superconductivity was discovered by the Dutch physicist Heike Kamerlingh Onnes (1853–1926). It can theoretically be used as the basis for some computer memories, for once stored in a superconductor, information remains unaltered. A magnetic field of sufficient strength can destroy the superconducting state and this effect can be used to achieve a high-speed current-switching facility. As a superconducting material has zero electrical resistance, very high currents can pass through it. As a result, superconducting windings for electromagnets can be used to generate extremely powerful magnetic fields.

KEY
Liquid helium cooled below its boiling-point behaves strangely. If a tube is dipped into liquid helium [A] at a temperature of 2.18°K (−270.97°C) an invisible film of liquid creeps up the outside of the tube and then down the inside. The helium fills the tube until the liquid levels are the same inside and out. If the tube is raised a little [B] the "superfluid" helium flows the other way to equalize the levels. Drops of the liquid drip off the bottom of the tube. The thickness of the liquid film can be measured (by light polarization) as about 3 millionths of a centimetre at a height of 1cm above the liquid.

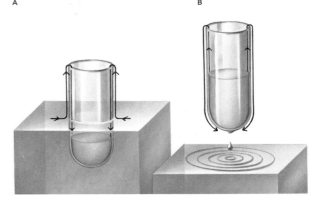

A   B

6
A

5 Liquid air has a temperature of only 83°K (−190°C) and thus a flower dipped in it will solidify completely as all the fluid in its cells freezes [A]. When this happens the flower becomes so brittle that it can be broken into small pieces with a blow from a hammer [B]. Liquid air is used industrially for freezing other substances and for the commercial production of oxygen and nitrogen.

6 The superconducting magnets [A] of a particle accelerator, such as those used in the giant proton synchrotron at CERN in Geneva, Switzerland [B], are products of low-temperature physics. Normally superconductivity is destroyed by a high magnetic field. But materials such as niobium-zirconium alloy, with distorted crystal structures, remain superconductive in fields of up to 100 thousand gauss.

5
A

B

6B

Proton synchrotron

Main offices

Tunnel: 4·8m (15·6ft) in diameter and 7km (4·35 miles) long

Intersecting storage rings

Proton synchrotron 200m (650ft) across

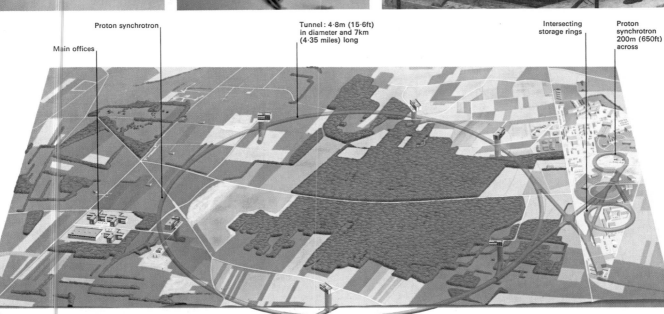

Extremes of pressure

The extremes of pressure – ultra-high vacuum (low pressure) and very high pressure – have varied and sometimes remarkable effects on different materials.

Matter exists as either solid, liquid or gas, and all are compressible to various degrees. Perfect gases are compressible to almost any extent, following at low pressures Boyle's law (which states that the volume of a gas varies inversely with its pressure). But liquids are much less compressible and the changes in their volume brought about by pressure follow no simple law.

Solids are the least compressible. Their rigid structure, in which atoms have their mean separation distance fixed by very strong forces, is the most resistant to externally applied pressure. Their structure can be distorted or destroyed by sufficiently high pressures, but the way in which they actually behave is governed by their internal atomic or molecular arrangement.

The compressibility of a gas can be calculated by its equation of state, but that of a liquid or a solid has to be determined experimentally. For the liquid metal mercury at 0°C (32°F), for example, it has been shown that the volume changes by less than one-millionth part over a pressure range of 0–7,340 kg/cm² (0–7,000 atmospheres).

Effects of pressure

Pressure applied to any substance can, under certain conditions, cause a change of state. Thus below a certain critical temperature pressure can turn a gas to liquid.

Extremely high pressures find many industrial applications. Hydraulically generated and transmitted pressures are employed for lifting extremely heavy loads. In the motor industry they are used to press completely shaped car body panels [3] from flat metal sheet. The behaviour of metals under compression is also the basis for processes involving rolling and forging. Again, if sufficiently high pressures are brought to bear, metal enters what is called its "plastic" range. That means that the metal continues to yield (ie, to extend its dimensions), even though the load applied remains constant. In the normal elastic range, the dimensions change in direct proportion to the applied load and return to normal when the load is removed. The plastic property is used in processes such as extrusion [2], but enormous pressures must be applied to achieve it.

At the other end of the pressure range is the vaccum, which can be of varying degree. To obtain a vacuum, gas atoms or molecules are removed from an enclosed vessel. The number of intermolecular collisions is correspondingly reduced and thus there is a reduction in the internal energy and pressure of the gas.

The creation of vacuums is not, however, limited to vessels that previously contained nothing but gas. A partial vaccum can also be achieved above the surface of a liquid – but the space is filled by the vapour of the liquid. As the gas pressure above the liquid is reduced by pumping, the liquid boils at a lower temperature than it does at atmospheric (that is, normal) pressure.

Vacuums in industry

As with high pressures, high vacuum states also have many industrial uses. The process of vacuum deposition by evaporation allows

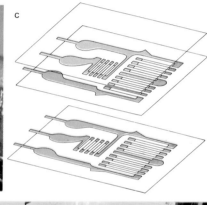

1 The pressures operating on large structures, such as the bridge spanning the Severn [A], are measured by an electrical strain guage [B]. This device uses the phenomenom (first noted by Lord Kelvin in 1856) that if a wire is strained, its electrical resistance is changed. The principle was first employed practically in the United States in 1938 and now the strain gauge is the most common instrument used for analysis of stress. Gauges can consist of a grid of fine metal wire that is then bonded to a thin backing but, more commonly, a grid of wire filaments is obtained by printing onto a metal foil [C]. In either case, the gauge is cemented onto the structural surface and changes in its resistance are measured; the readings are then recorded automatically.

2 Metal forms of complicated shapes can be made by the process of extrusion. Cold metal is forced through a hole of the required shape and size and the great pressures exerted on the metal cause it to assume a "plastic" condition, so that it is able to "flow" smoothly through the extrusion die, as in this machine.

3 Hydraulic presses can make shaped body panels, in one stage, from single flat sheets of metal. A piston carries a former of the shape required and then a hydraulic ram presses this against the metal sheet with tremendous force. The high pressures generally needed call for costly presses, but their speed justifies the expense.

solid objects to be thinly coated with metal [4]. Here, the object to be coated is placed in an enclosed vessel that can maintain a high vacuum. When the coating metal is vaporized within that vessel, it forms a thin mirror-like film over the object.

There are other manufacturing processes in which the controlled deposition of impurities on substances or objects is performed by means of high vacuums. Sophisticated electronic circuitry is based upon this use of an ultra-high vacuum [5].

But technological processes are not the only applications of extremes of pressure. Given that all substances are compressible to some degree, consideration of the effects of increasing pressure on materials to be used in construction work is obviously important. Compression and tension tests of various building materials provide the necessary information and increase our knowledge of the extremes of stresses and strain sustainable by different substances.

The behaviour of materials under the two extremes of pressure is well demonstrated in the varying conditions of space. Extremely high vacuum conditions exist in interstellar space, with probably only a few atoms per cubic centimetre. Separate chemical radicals have been found in interplanetary space.

When atoms are smashed
In the denser clouds of gas and dust (nebulae) that are found in galaxies, the pressures become greater due to gravitational attraction, generating higher temperatures so that eventually the electrons are forced to change orbit within the atom. As they jump back to their original orbit, energy is given up in the form of light [7].

Within the stars, the radiation temperature and pressures are extremely high. At the centre of the Sun the temperature is probably about 10 million °C (18 million °F) and the density fifty times that of water; the pressure amounts to about 400,000 million kg/cm² (6 million million lb per sq in).

In other stars, the pressures can be so great that the normal arrangement of an atom is completely broken. Protons and electrons are squashed together, creating immense densities to several tonnes per cm³.

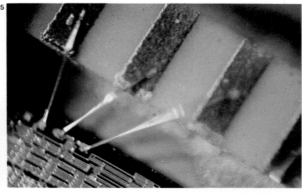

Artificial diamonds are produced when graphite is subjected to very high pressures and temperatures. Most of the stones are inferior to those that have been produced by nature.

4 Vacuum deposition is widely used to form metal coatings on plastic objects. The object to be so treated is enclosed in a chamber [A], along with a wire filament carrying beads of the metal that is to be deposited. Then the chamber is pumped down to a very low pressure and an electric current is passed through the wire. The beads melt, vaporize and deposit a metal film on the objects [B].

5 Silicon "chips" can be modified by the use of high-vacuum methods so that whole electronic circuits can be integrated into a single piece. To achieve this, certain "impurities" are diffused into or layered onto the base material; these impurities affect the conductivity of the base and perform equivalent functions to electronic components such as transistors, diodes, capacitors and resistors. The pure silicon base is held in a vacuum during the implantation or layering process. Connections to the circuit are made by fine wires welded to it from gold-plated contacts.

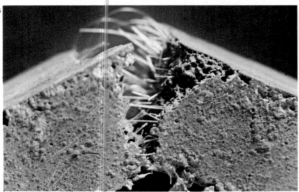

6 Testing materials to destruction is often necessary to discover their real strengths and their weaknesses. These tests are carried out in a machine that is capable of exerting both compressive and tensile (pulling) forces and it records automatically the applied pressure on the material at the very instant of its failure. Such tests are not confined to ductile metals (those that can be drawn out as wire), but also applied to brittle materials such as this fibreglass.

7 The evolution of stars illustrates the dramatic effects of extremes of pressure on matter. Beginning as gas at very low pressure it gradually condenses, becomes hotter and forms galaxies and stars, and finally dense neutron stars and pulsars.

Light and colour

We are surrounded by various forms of energy, namely light, heat, chemical and mechanical energy. Of these, light is as necessary as heat energy and chemical energy – and all are essential to life; few sightless people can survive without the aid of people who can see. By virtue of its basic nature light enables people to sense the world around them in great detail. This is because light consists of a wave motion of extremely high frequency. If human beings were sensitive instead to radio waves, which have a much lower frequency than light, they would detect no more detail in their surroundings than the blurred outlines seen on a radar screen.

How light travels

Light travels in waves, as sound does. In light the vibrations of the wave consist of vibrating electric and magnetic fields, whereas in sound they are vibrations of a medium such as air or water. Both kinds of waves may vary in intensity, producing stronger vibrations. In sound an increase in intensity causes an increase in loudness and in light it produces an increase in brightness.

Light waves also possess a range of frequencies – the number of vibrations passing every second. In sound, people hear different frequencies as sounds of different pitch; in light they see different colours. Blue light, for example, has a higher frequency than red light. Light may also be considered in terms of wavelength – the distance between successive vibrations in a wave; blue light has a shorter wavelength than red light. Light frequencies are very high and the wavelengths are very short (about 55 millionths of a centimetre).

Paint manufacturers include white and black among their ranges of colours, but these are not strictly colours at all. Black is simply an absence of light and therefore of colour and white light is made up of a mixture of several basic colours. This can be shown by passing sunlight through a prism. The white light of the sun is split up into a band of colours called a spectrum. The spectrum looks exactly like a rainbow [5], which is no coincidence because raindrops act as prisms and split up sunlight to produce a rainbow. The beam of colours from a prism can be made to

recombine and produce white light once more. This is a proof that white light is a mixture of all the colours.

The science of spectroscopy

Splitting up light to form a spectrum is important in science [Key]. Different elements glow with different colours when they are heated sufficiently or subjected to an electric discharge – examples are gases in sodium street lights or neon advertising signs. By passing the light from a glowing substance through a prism and examining the resulting spectrum, which has a different pattern for different elements, the glowing substance can be identified. This is useful in all kinds of analysis, but particularly in finding out which elements are present in the Sun and the stars. This branch of science is called spectroscopy.

Most of the colour that reaches our eyes comes from objects that are naturally coloured, painted or dyed. When white light strikes the surface of a red object, red light is reflected from it, but all the other colours in the white light are absorbed by the surface. Colour is also produced in other ways. A sub-

1 **Colour** is used as a means of imparting simple information quickly and unambiguously. On the road, for example, red tail-lights on cars indicate "stop" or "danger".

2 **Mixing colours** depends on whether coloured lights or pigments are being used. Lights combine by additive mixing [A] in which three basic or primary colours, red, green and blue, combine to give white. Yellow, cyan and magenta are secondary colours formed by mixing equally two primaries. Pigment colours combine by subtractive mixing [B] in which some colour is absorbed before mixing of the remaining colours occurs.

3 **The Munsell colour tree** is a system of grading any colour. The hue (basic colour), chroma (amount of colour) and value (degree of lightness or darkness) are measured and the colour's position found among all those in the tree. Hue is denoted by its place on the circumference of the tree; chroma by its distance from the trunk; value by its place up the trunk.

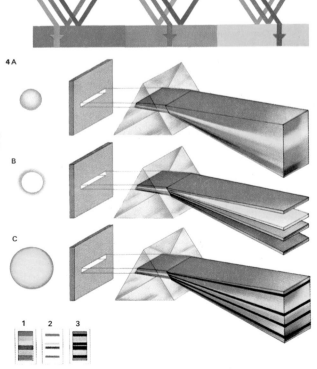

4 **A solid or plasma under pressure** heated to incandescence emits a continuous spectrum [A]. At low pressure a gas produces an emission spectrum [B]. In the Sun [C], light from the inside [1] is partly absorbed as it passes through the outer regions [2] to form an absorption spectrum [3].

stance can be heated so much that it glows with colour and luminous compounds such as the phosphors in a colour television screen light up with colour when they are struck by invisible cathode rays (beams of electrons) or ultra-violet rays.

Humans (and many other animals) can perceive colour because the retina in the eye contains three kinds of light sensors. These detect different ranges of light frequencies, roughly corresponding to red, green and blue. All other colours can be produced by combining light of these three basic colours in various amounts. Red and green combine to produce yellow; green and blue to give cyan; and blue and red to make magenta. All three basic colours combine to give white light.

Additive and subtractive mixing

It may seem strange, to anyone used to mixing red and green on a paint brush and obtaining brown, to read that red and green make yellow. This is because coloured lights and coloured paints combine in different ways. A colour television set produces coloured light and close examination of a lit screen reveals that it contains patterns of red, green and blue dots or stripes. At a distance the dots or stripes merge into a colour picture. But close up, the yellow light can be seen to be made up of red light and green light. This kind of colour mixing, in which light combines directly, is called additive mixing [2]. Any three coloured lights, such as red, green and blue, that combine to form any other desired colour or in the right proportions to form white light are known as the primary colours.

In producing colours by mixing paints, dyes and inks, subtractive mixing occurs [2]. The colours form not by mixing the three basic colours directly, but by absorbing some of them from the light that illuminates the surface of a painted object. Thus yellow paint absorbs blue from the illuminating white light but reflects red and green, which combine to reach the eye as yellow. Cyan paint absorbs red light, leaving blue and green to mix and make cyan. Mixing yellow and cyan subtracts red and blue from the white light, but leaves green to be reflected: so cyan and yellow paints mix to give green.

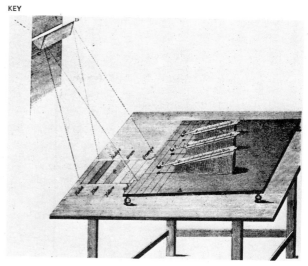

Dispersion of light by a prism produces a spectrum of colours. In 1800 the British astronomer William Herschel placed thermometers just beyond the red end of the spectrum and observed a rise in temperature. He deduced that the prism was dispersing invisible heat rays (now called infra-red rays).

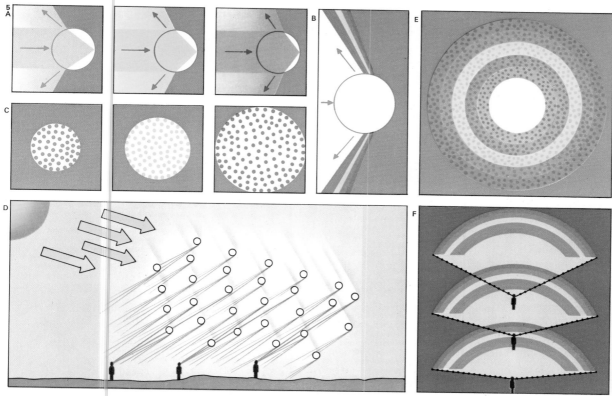

5 A rainbow occurs when raindrops reflect sunlight back towards the observer, dispersing the light into its component colours [A]. The colours are reflected back at various angles from each drop, blue being bent more than red [B]. Each raindrop lit up by the Sun produces a circle of colour [C], but only the lower part of each circle is directed towards the ground [D]. The observer sees the colours coming from many raindrops as a rainbow. Of the reflected rays that reach him the red are at a greater angle than the blue [E] so that, to the observer, the red side appears higher than the blue. To other observers the rainbow may look nearer or farther away, but it is always in the same direction and has the same apparent size [F]. Outside the first bow a secondary rainbow in which the colours are reversed may also be seen.

6 The spectrum of visible light is only a small part of the much greater spectrum of all electromagnetic radiation. Beyond the blue end lie the invisible ultra-violet rays, X-rays and gamma-rays, while infra-red rays (heat rays), microwaves and radio waves lie beyond the red end. All electromagnetic radiation has the power to penetrate matter to a certain extent. High frequency radiation – that is, X-rays and gamma-rays – penetrates most.

X-rays					Ultra-violet	Infra-red		Microwaves		Radio waves
Gamma-rays										

10^{24} 10^{22} 10^{20} 10^{18} 10^{16} 10^{14} 10^{12} 10^{14} 10^{8} Frequency in hertz

Mirrors and lenses

Nearly everyone looks in a mirror at least once a day and people with less than perfect vision spend most of their lives looking through spectacle or contact lenses. Telescopes, binoculars, microscopes, cameras and projectors help us to examine the world about us in far more detail than can be perceived by the unaided human eye. All these visual aids and optical instruments use mirrors or lenses. They work using the simple laws of optics, but before these laws and their application can be understood it is necessary to appreciate how an image is formed.

Light rays and images

Any illuminated or luminous object sends out light rays that spread in all directions in straight lines. An image forms if any of the light rays coming from the same point on the object happen to meet. Normally images do not form because there is nothing to bend the light rays to make them meet, but a lens will do this. The image produced can be seen on a card or a screen placed at the point where the rays meet. If the rays meet exactly a sharp image is formed and the image is said to be in

focus. It is known as a real image and is the kind produced in a camera or by a projector. If the incoming light rays are parallel, the image is produced at a distance called the focal length of the lens.

Plane (flat) mirrors, on the other hand, produce images that cannot be shown on a screen. In this case the light rays are bent but they continue to become farther apart (diverge) rather than closer together (converge). But the human brain always assumes that light rays reach the eyes in a straight line, and we therefore see an image at the point that the object would be occupying if the rays were not bent [6]. This kind of image is called a virtual image and it is always sharp.

Except when illuminating the deepest black objects, light is reflected from every surface it strikes. A dull or matt surface scatters the rays at all angles. But a very smooth surface acts as a mirror and reflects all rays so that the angle of incidence equals the angle of reflection. A plane mirror, being flat, bends all the rays that strike it by the same angle and so a virtual image, unchanged in size and shape, is seen in it. Left and right are inter-

changed because a light ray leaving one side of the object is reflected by the mirror to the opposite side of the eye from the one it would otherwise strike [1].

Curved mirrors [2] produce images changed in size and shape. A convex mirror, one that bends outwards towards the observer, gives a smaller virtual image. It makes the rays diverge more than if they came directly from the object. This produces the same effect as if the object were farther away and it therefore appears smaller, as in a car's rearview mirror. A concave (inward-bending) mirror has the opposite effect. It makes the rays converge more than they otherwise would, making a close object appear nearer and therefore larger.

Bending of light

If light rays meet a transparent object most of the rays enter the object and emerge from the other side. The rays are bent as they pass through the surfaces, being deflected away from the surface if they are entering a denser medium and towards the surface if they are leaving a denser medium. This effect is called

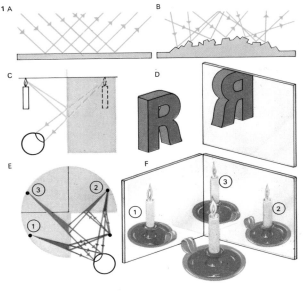

1 **A plane mirror** reflects all light rays at the same angle [A] whereas a matt surface [B] scatters light. The brain imagines that light rays reaching the eye from a mirror come to it in straight lines and it therefore sees an image at that point where the rays would originate if their paths were not bent by the mirror [C]. The image is seen laterally inverted [D], because the reflected rays reach opposite sides of the eye. An image can be seen right way round in two plane mirrors at right-angles [E]. Although an image reflected once appears laterally inverted [1, 2], one reflected twice [3] is seen correctly [F].

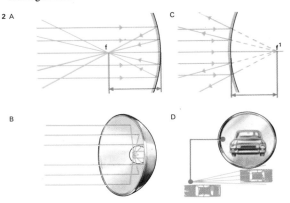

2 **Curved mirrors** form real and virtual images. A concave mirror [A] reflects the rays of a parallel beam of light so that they converge and meet at the focus of the mirror [f]. A real image of a distant object will be formed on a screen placed at the focus. If the eye is placed closer to the mirror than the focus, a magnified virtual image will be seen behind the mirror. A lamp reflector works in this way [B]. A convex mirror [C] makes the rays of a parallel beam diverge as if they were coming from the focus [f¹], which produces a diminished virtual image, as in a driving mirror [D].

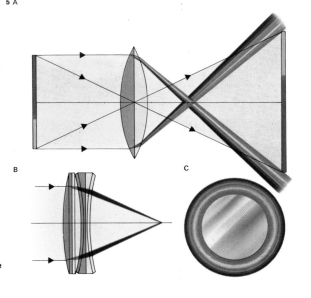

3 **Lenses form images** because light rays are bent by refraction as they pass through a lens. The rays are made to converge on passing through a convex lens [1] or diverge through a concave lens [2], regardless of the direction in which the rays are moving. Thus convex lenses can make light rays meet and produce real images, whereas concave lenses give only virtual images.

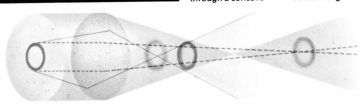

4 **Spherical aberration** produces a blurred image. Rays passing through the centre of the lens are brought to a focus at a different point from rays passing through the edge of the lens and there is no place at which all the rays come to a single focus and give a sharp image. Spherical aberration may be reduced by narrowing the lens so that rays do not pass through the edges and by combining lenses to cancel out the defects in each kind of lens.

5 **Chromatic aberration** produces coloured fringes around the lens edges, and parts of the image may not be sharp [C]. This aberration occurs with single lenses because they behave like prisms and bend blue light more than red light [A]. Combining the lens with a weaker concave lens [B] made of a different glass cancels out this dispersion effect, and both red and blue rays are brought to the same focus to produce a sharper, more distinct image.

refraction. Some of the light is reflected instead of being refracted. If, on leaving the denser medium, the light rays strike a surface at or below a certain angle, called the critical angle, they are all reflected back (none is refracted) and the light stays inside the denser medium. The amount of bending depends on the refractive index of the medium; the greater the refractive index the greater the amount of bending.

Refraction and lenses
Refraction explains why objects seen in water appear to be less deep than they really are. Light rays from the submerged object bend as they leave the water but the eye, as always, imagines that the rays have come in straight lines from the object.

Lenses work by refraction [3] and are shaped to bend the light rays passing through them by different amounts. In a convex lens – one in which the surfaces curve outwards – light rays passing through converge and meet to form a real image. The lenses in a camera, a projector and in the eye work in this way. When viewed close up, the converging rays

from the lens enter the eye of the observer before they can form a real image, giving a large virtual image of objects on the other side of the lens. This is how a magnifying glass works. Concave lenses – which curve inwards – make light rays diverge and produce small virtual images. These lenses are used in spectacles to correct short sight, converging lenses being necessary for long sight.

Single lenses produce several kinds of aberrations [4, 5], or distortions, of the image. Coloured fringes may be seen around the edges, and parts of the image may not be sharp. When several lenses are combined the aberration is reduced by cancelling out defects in each kind of lens, and high-quality lenses consist of several different elements grouped together to give a perfect image. Each element has to be carefully shaped and positioned with great accuracy.

Most optical instruments use lenses to produce images [8, 9, 10]. But the largest telescopes used in observatories all have a concave mirror to produce a real image of a distant planet, and this image is then viewed with a magnifying lens to enlarge it.

Mirrors and lenses alter the paths of light rays in accordance with simple laws. A mirror reflects light rays so they leave the mirror at the same angle as they strike the mirror. The brain assumes that the light rays reaching the eye have travelled from an object in straight lines, and it therefore sees an image behind the mirror. A transparent material refracts light passing through it so that the path of the rays is bent at the surface by a certain angle, which depends on the refractive index of the material. The brain assumes that light moves only in straight lines and the image it sees of an object is displaced.

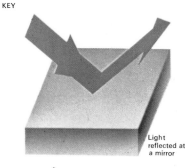

Light reflected at a mirror

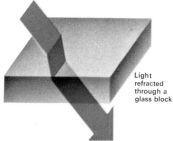

Light refracted through a glass block

6

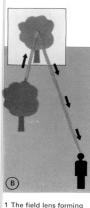

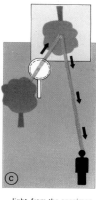

6 Images form when the eye receives light rays coming from the same point on an object and bends them to meet at the retina. The tree reflects light rays in all directions. The eye may see it directly [A], see a virtual image of the tree "behind" a mirror [B] or view a real image of it formed on a screen by a lens [C]. The images are located at the points from which the light rays appear to originate.

7 A periscope, in its most simple form, consists of two mirrors angled at 45° one above the other so that an image is reflected from the top of the instrument down to the observer at its base. A submarine periscope works on the same principle, but has prisms instead of mirrors and a system of lenses to produce a magnified image or a wide field of view. Optical adjustments can be made with the handles.

7

8 In a microscope an image of the specimen is formed by the objective lens and this image is then viewed by the eyepiece lens so that it is magnified and an extremely close view obtained.

1 The field lens forming part of the objective
2 The eye lens of the eyepiece
3 Specimen placed on the glass slide
4 The objective of the eyepiece
5 The microscope effectively increases the angle at which

light, from the specimen, enters the eye, and the final virtual image appears to lie in this plane
6 Mirror reflects light on to specimen

9

9 Opera glasses have a pair of telescopes known as Galilean telescopes. The concave eyepiece lens [1] is placed inside the focus of the convex objective lens [2] to obtain a magnified upright image [3] of the object. The lenses have low-power magnification. Galileo discovered the moons of Jupiter using a similar instrument.

10 An astronomical telescope has a similar optical system to a microscope. It consists of an objective lens [1] (of focal length f_o) and an eyepiece lens [2] (of focal length f_e). Parallel light from a distant object converges to form an inverted image [3]. Light rays then appear to come from the large upside-down image [4].

8

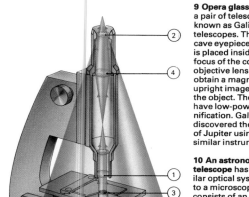

10

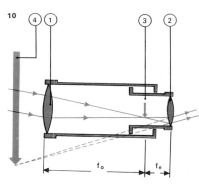

f_o f_e

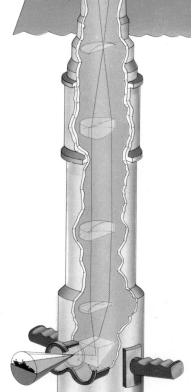

Light waves

In the seventeenth century scientists were divided in their opinions about the nature of light. Some believed it to be made up of streams of particles of some kind, while others argued that it consisted of waves. Reflection was easy to explain by the particle theory; it could be pictured as a bouncing of particles off surfaces rather as balls bounce from the sides of a billiard table. (The modern form of this theory envisages light travelling as "packets" of energy called photons.) Refraction was more difficult to explain – why should some particles bounce off and others pass through a surface? Other effects were impossible to account for by the particle theory and the wave nature of light became accepted as more convincing.

The wave motion of light
In those early days no one knew exactly what vibrated in light to make it behave as a wave motion, nor how the waves could be produced. These problems began to be solved in the 1870s. It was discovered that a light wave consists of vibrating electric and magnetic fields travelling through space; the two fields vibrate at right-angles to each other and to the direction of motion. In fact light waves are part of a whole group of electromagnetic waves that include X-rays, ultra-violet rays, infra-red rays and radio waves. Light waves can be produced by changing the orbits of electrons inside atoms. If an atom receives energy in some way – perhaps as heat, light or electrical energy for example – the electrons move away from the nucleus to orbits of higher energy. They then jump to a lower energy orbit and give out energy in the form of electromagnetic waves as they do so. In this way objects produce light [1].

Light spreads out from any point producing it or reflecting it in ever-expanding spheres, rather as ripples spread out in circles over a pond. Each ray of light can be thought of as moving in a straight line, producing a continuous series of ever-expanding vibrational movements through space. In all the rays leaving a point the vibrations add up to give a set of spherical wavefronts consisting of alternate peaks and troughs of energy. Each peak and trough are maxima of vibration but in opposite directions.

The shadow of an object is rarely seen to have sharp edges, but this is because a source of light always has a certain size [Key]. If the source were infinitely small we would expect it to give shadows that are totally sharp because light rays are considered as straight lines, but this is not so. All waves spread round the edge of an object – an effect called diffraction [2]. In the case of light the edge is illuminated and points close to it can act as sources of light waves that spread out in all directions so that the rays are effectively bent by the edge. The wavelength of light is so short that this effect is hard to detect at edges, but it becomes clearly apparent when light passes through very small openings about the same size as the wavelength. This happens in a diffraction grating [3] in which light passes through or is reflected from extremely narrow slits.

The effects of interference
The wavefronts that spread out from the two edges of the opening cross each other. Where two peaks of the waves meet, an increase in brightness occurs, but where a peak meets a

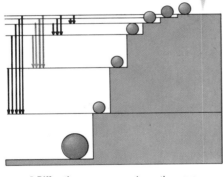

1 Light production occurs when atoms lose energy. Normally an electron circles the nucleus of an atom in a certain orbit [A]. If the atom takes up energy such as heat, light or electrical energy the electron moves to a higher orbit. It may then fall to a lower one [B], giving out energy in the form of visible light (shown as coloured arrows) or as invisible radiation (shown in black).

2 Diffraction occurs when a wave passes an edge, causing it to spread around that edge. Often this effect is too slight to be noticed, but it is marked when waves pass through an opening of the same size as the wavelength (with light, about 55 millionths of a centimetre). The light waves spread out from the edges of the opening and patterns of light and dark are formed where they cross.

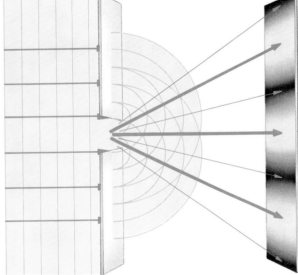

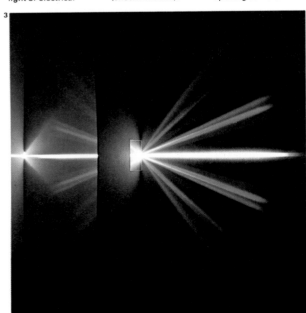

3 A diffraction grating has a fine mesh. When white light passes through it is bent in many directions and split up into a spectrum of colours: each wavelength is bent a different amount.

4 Interference occurs when two waves of the same wavelength [1, 2] travel over the same path. The waves interact to give a new light wave [3]. If the waves are in phase [A] the new wave is brighter than either of the original waves. If they are slightly out of phase [B] the new wave has about the same brightness as the original waves. If they are totally out of phase [C] the peaks and troughs cancel each other out and so no new wave is produced.

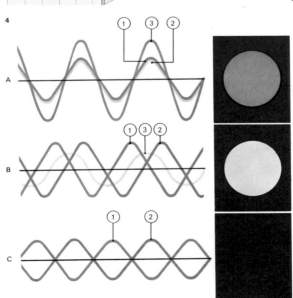

trough they cancel each other out so that no vibration occurs and there is no light. As a result a series of light and dark fringes is produced instead of a single image of the opening [2]. This effect, in which waves reinforce each other or cancel each other out, is called interference [4].

If a ray of light is divided into two rays that later recombine then interference effects are seen if one of the divided rays travels a longer path than the other before the recombination. The peaks and troughs may be out of phase (not exactly together) and the light is affected. This happens between two surfaces that are very close together, as in a thin film or two pieces of glass pressed together, and it produces colourful fringed patterns [5]. The iridescent colours seen in the plumage of some birds and some butterflies' wings are produced by the phenomenon of interference; the fine structure of the feather or wing resembles either a diffraction grating or a thin film.

Because interference can be produced by a path difference of only a wavelength or so, interference effects can be used to detect very small changes in length. Interferometers are used for this purpose. They produce interference by dividing a ray of light into two or more beams and then recombining them.

Polarization of light waves

Another effect to be seen with light waves is polarization [7]. In an ordinary light wave the electric and magnetic fields vibrate in many randomly orientated planes about the direction of wave motion; in polarized light they vibrate in only one plane. Light is polarized by passing it through a filter that cuts out all vibrations except those in one particular plane. The polarized beam will then pass through a second filter only if it is set at the correct angle to allow the vibrations through. Otherwise the beam is stopped. Light reflected from surfaces at certain angles is polarized, and polarizing sunglasses [8] cut out glare by stopping reflected beams in this way. Solutions of some chemical substances, such as various sugars, rotate the plane of polarization of light passing through them. The effect is used in chemistry for analysing such solutions.

Shadows form with sharp edges when they are cast on a nearby object, but the outlines become less distinct the farther away they are cast. This can easily be explained because light travels in straight rays and every light source has a certain size. The ray paths show that a region exists at the edge of the shadow that is partially illuminated; this region, the penumbra, makes the outlines of the shadow fuzzy. The dark part of the shadow, the umbra, is completely shielded from the light source. The penumbra is less broad the closer the shadow is to the object casting it, and so nearby shadows look sharper.

5

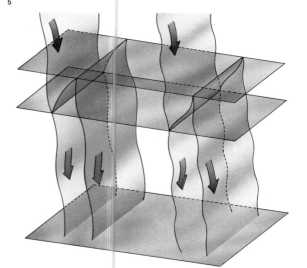

5 A thin film, such as a soap bubble or oil film, glistens with colour. Part of the light passing through the film is reflected between the inner surfaces of the film and emerges to interfere with the rest of the light that passed straight through. Travelling paths of different lengths, some of the waves are in phase and reinforce each other [red] while others [blue] cancel each other out and are not seen.

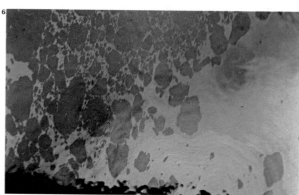

6 Interference is responsible for the coloured reflections from bubbles and oil films on water. The light reflected from the top of the film interferes with light reflected from the lower surface.

7 A

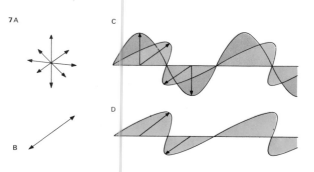

7 Unpolarized light consists of vibrations in all planes at right-angles to the direction of the light wave; the arrows show the wave approaching head on [A]. Polarized light consists of vibrations in one plane only [B]. Light rays consist of vibrating electric and magnetic fields at right-angles [C]; only the electric vibration denotes the plane of polarization [D].

8 Light reflected from glass or water is partly polarized. Here [B] a reflection makes it difficult to see through a shop window. A similar photograph taken with a polarizing filter over the camera lens gives a reflection-free view [A]. Polarizing sunglasses reduce glare this way.

9 Stresses and strains in transparent materials such as glass and plastic become visible when viewed with polarized light. Here the regions of strain in a heat-treated car windscreen become visible as spectral colours.

9

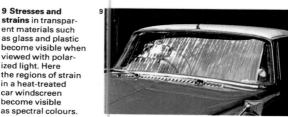

The speed of light

Every time we press a light switch, light floods the room instantaneously – or almost instantaneously. It does take a fraction of a second for the light to pass from the light bulb to our eyes but the time taken is far too brief for us to be aware of it. To early scientists light seemed to take no time to propagate; many claimed its velocity to be infinite.

Determining the velocity of light

Others, notably the Italian astronomer Galileo (1564–1642), challenged this view. He attempted to measure the velocity of light by trying to find out how long light took to travel between two hills a known distance apart. His experiment was inconclusive but it did show that if light has a particular velocity then it is very great. Confirmation of this view came with observations of the moons of Jupiter by the Danish astronomer Olaus Roemer (1644–1710) in 1675 [1]. The moons, which had been discovered by Galileo in 1610, are often eclipsed by Jupiter but Roemer found that predictions of the eclipse times were as much as 22 minutes out. Roemer reasoned that the variations occurred because the dis-

tance between Earth and Jupiter varies depending on their positions in their orbits around the Sun, and light therefore takes different times to reach the Earth from Jupiter. Knowing the distances Roemer made a good estimate of the velocity of light, obtaining a value of 227,000km (141,000 miles) a second. The true velocity is almost 300,000km or 186,000 miles a second.

Another astronomical determination of the velocity was made by the English astronomer James Bradley (1693–1762) in 1728. He observed that the stars are seen in slightly different directions depending on the position of the Earth in its orbit. This phenomenon, called stellar aberration, is caused by the Earth's motion and the differences in direction are simply related to the difference between this motion and the velocity of light. Bradley was therefore able to obtain a value for the velocity of light, and it was of the same order as Roemer's figure.

Later determinations of the velocity were completely terrestrial and sensitive instruments were used to measure very precisely the time light took to travel a known distance.

The instruments contain mirrors to reflect a ray of light along a particular path and time its passage by a variety of shutter mechanisms. Modern methods use an electronic shutter capable of very rapid action.

The accepted value for the velocity of light is now 299,792.58km (186,181 miles) a second. This is the velocity in a vacuum, for light slows when it enters a medium such as air, water or glass. The change in velocity causes the light to bend on entering a different medium and refraction occurs. The refractive index of a medium is the ratio of the velocity of light in a vacuum to its velocity in the medium. For example, the refractive index of water is 1.333 or $^4/_3$, and thus the velocity of light in water is only three-quarters of its velocity in a vacuum.

The mystery of the ether

Having determined that light has a certain velocity scientists began to wonder how light waves could travel through space. Other wave motions need a medium in which to travel – sound, for example, moves through air – and light had to have a medium too.

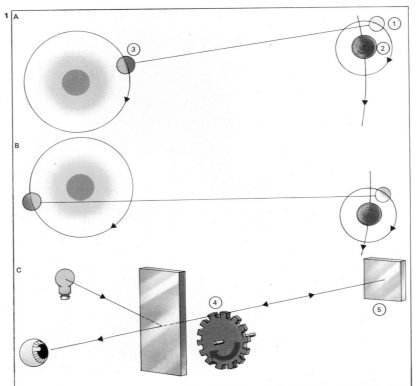

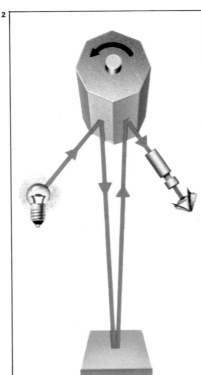

1 **The velocity of light** was first determined by Olaus Roemer in 1675 [A]. He saw the eclipses of Jupiter's moons [1] by Jupiter [2]. Light from the moons takes less time to reach the Earth [3] when its orbit nears Jupiter than when it is far away towards the other side of the Sun [B]. Knowing the distances and times involved he could calculate the velocity of light. Another determination was made by Armand Fizeau (1819–96) in 1849 [C]. Light was reflected through the teeth of a rotating wheel [4] to a mirror [5] and back through the teeth to the observer. The light was seen only when the wheel spun so fast that no teeth blocked its return journey. From the spacing of the teeth, the speed of rotation of the wheel and the distance of the mirror (8km [5 miles]), the velocity of light could be accurately calculated.

3 **A mirage** is seen in a desert [A] because the hot air acts like a mirror. The heat of the sand causes layers of air at different temperatures to lie above the ground and rays of light from the ground bend as they move through the layers [B]. Each layer has a different refractive index. In extreme conditions the light rays bend so much that they are deflected back towards the ground and an image of an object over the horizon is seen.

2 **A rotating mirror** was used by Michelson to measure the velocity of light in 1927. Light travelled from one face of the mirror to a plane mirror 35km (22 miles) away and then back to another

face and an eyepiece. An image of the light source was obtained with the mirror first stationary and then rotated at sufficient speed for the image to be seen in the same position. At such a speed the

mirror turned so that the next face moved into position as the light made its 70km (44 mile) journey to and from the plane mirror. Velocity of light was calculated from the speed of rotation.

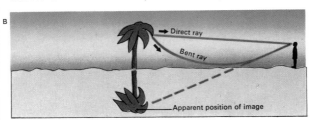

As the medium through which light moved could not be seen to exist, one was invented; it was called the ether and it was supposed to pervade the whole universe. Thorny problems surrounded the ether. Known wave motions move more rapidly in denser, more elastic substances and a wave motion as fast as light should theoretically need a medium denser than steel. Yet the planets continue to sail through space, unimpeded by the ether. There were many other contradictions and so an experiment was made to detect the motion of the Earth through the ether.

In the 1880s two American physicists, Albert Michelson (1852–1931) and Edward Morley (1838–1923) [5], made a simple instrument to detect the ether. In it a beam of light was split into two beams at right-angles and the two beams reflected from mirrors before recombining. Combined beams show interference effects if one travels a slightly longer path than the other. Michelson and Morley observed the combined beams in one direction and then turned the instrument at right-angles and observed the beams again. If the light were travelling in an ether it would have to move over a different path in the direction of the Earth's motion from at right-angles to it. Turning the instrument at right-angles should show a difference in the interference effects if the ether existed. None was observed and none has been observed in many repeats of this classic experiment.

The basis of relativity
The conclusion of the Michelson–Morley experiment was that ether does not exist and light does not need a medium for its propagation, or that the ether can never be detected. Without a stationary ether there is no basis in the universe against which the absolute motion of everything can be measured, except for light. The Michelson–Morley experiment showed that the velocity of light is the same in the direction of the Earth's motion as at right-angles to it and is always the same whatever the observer's motion. These conclusions had profound implications but to realize them it took a genius – Albert Einstein (1879–1955) – who used them as a basis for the theory of relativity.

Light that reaches us from heavenly bodies does not travel instantaneously. It takes 1.25 seconds to get to Earth from the Moon, 8 minutes from the Sun, over an hour from Saturn and the outer planets, and over four years from the nearest star. We see the galaxies as they were millions of years ago.

4 The blue glow coming from the water surrounding this nuclear reactor is known as Cerenkov radiation. It is produced because nuclear particles emitted by the reactor are moving faster than light does itself in water, which slows light by about a quarter. The particles cause a shock wave to be produced in the water, just as a supersonic plane produces a shock wave in the air. We hear the sound shock wave as a supersonic boom and see the light shock wave as blue light. The production of Cerenkov radiation is used as a method of detecting fast-moving particles in nuclear physics. The radiation was first observed by the Russian physicist Paul Cerenkov (sometimes known as Cherenkov) in 1934.

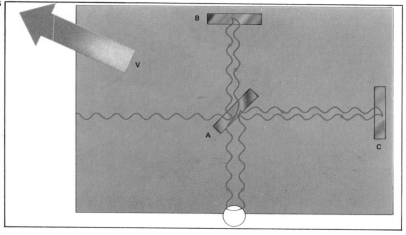

5 The Michelson–Morley experiment, first made in 1881, used an interferometer to produce a pattern of interference fringes from two beams at right-angles. The Earth's motion [V] was expected to make the light move faster along one path [AB] than along the other [AC], so that a change in pattern would be seen on turning the interferometer. No change was detected from the experiment.

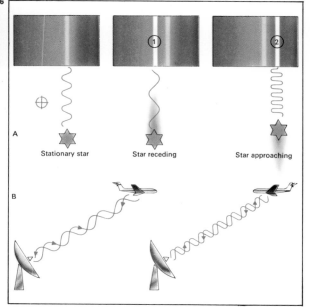

Stationary star

Star receding

Star approaching

6 The motion of an observer does not affect the velocity of light but it does change the frequency or colour of light known as the Doppler effect. However, only stars [A] move fast enough to show the effect. If the star and the observer are moving apart the frequency decreases because the individual waves are encountered less frequently. The light is redder than if the star were stationary and this red shift shows as a shift in the lines of the spectrum of the star [1]. If a star and observer are approaching, frequency increases and the light appears bluer [2]. A similar frequency shift is used in radar to detect the motion of aircraft or cars [B].

The idea of relativity

Relativity sought to eliminate from physics the idea of absolute values for space and time. Such values were held to be fixed and quite independent of the person measuring them or of the instruments used. To Isaac Newton (1642–1727) they existed as a backdrop against which he could formulate general "laws" about such quantities as acceleration and force. It was the genius of Albert Einstein (1879–1955) that, through the special and general theories of relativity, showed that such absolutes did not exist and that Newton's laws were not universally true.

The special theory of relativity

Einstein's special theory of relativity (1905) was based on the idea that all uniform motion is relative – that is, an object can be seen to move only in relation to some stationary frame of reference. The classic experiment made by Albert Michelson (1852–1931) and Edward Morley (1838–1923) determined that the speed of light is always the same in a given vacuum regardless of the speed of the source of light, of anyone observing it or of its wavelength [1, 2]. From these results Einstein deduced an astonishing set of conclusions. They showed that the mass, length and time interval of an object will appear to change when the object begins to move relative to an observer.

If, say, an astronomer were to observe an extremely fast-moving spaceship, then his instruments would indicate that the mass of the spaceship had increased, that all lengths in the direction of the spaceship's motion had decreased, and time aboard was slower. Yet in the spaceship itself nothing would appear to have changed, although if the pilot looked back at the astronomer – who would be in the same motion relative to him – he would observe that mass, length and time there had changed in exactly the same way.

The light clock [4A] shows why time varies with motion and by how much. Normally the effects of special relativity are undetectable in an object until it is travelling at nearly the speed of light (300,000km [186,000 miles] a second) [3], although very sensitive atomic clocks have been used to detect clocks "going slow" on aircraft in flight. The effects do become large for sub-atomic particles moving at close to the speed of light. Thus, because of their high speed, very fast unstable particles in cosmic rays live longer in the Earth's atmosphere than would otherwise be expected [4B]. Sub-atomic particles can be so speeded artificially that their masses are increased many thousand-fold; particle accelerators have to be specially designed to allow for this effect.

It is Einstein's famous equation "$E = mc^2$", relating the energy E and mass m of a moving particle with c the velocity of light that shows why in a special relativity a particle given ever greater energy will increase its mass. Because c^2 is so large, only a small amount of mass is equivalent to a vast amount of energy. The conversion of mass into energy takes place in nuclear reactors, in atomic power stations, in nuclear weapons and in the Sun and other stars.

As the speed of a particle approaches that of light its energy increases indefinitely. But there is a limit to the amount of energy available to any particle and so it can never travel faster than light. The light barrier cannot be crossed, but there may exist parti-

1 Relativity hinges on the simple idea that all motion is relative. A sailor in a yacht hauls a pennant up the mast [A]. To him, it appears to move vertically up [1]. To a man on the shore, the pennant appears to move forwards and up [2], because it is being carried past him as it is raised. A passenger in a passing aircraft sees the pennant disappearing rapidly behind him as it is raised [3]. Each observer records the same motion differently [B]; none is any more "correct" than the rest, for the planet on which all this happens is also moving. Their views confirm the relativity of all motion.

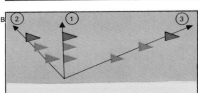
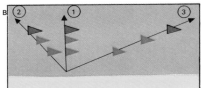

2 The special theory of relativity states that all motion is relative and that the velocity of light is always constant. When two spacecraft pass each other in orbit, each travelling at 8km (5 miles) a second as measured by radar at the tracking station below, the pilots detect that they are travelling relative to each other at 16km (10 miles) a second. If the spacecraft and the tracking station then measure the velocity of light from the Sun, they all get the same result. The spacecraft moving towards the Sun does not get a value that reflects its motion relative to any other body.

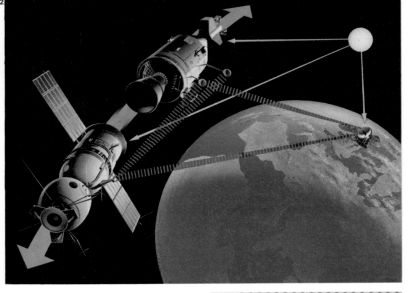

3 Einstein's first theory states that the measurement of mass, length and time depends totally on the relative motion of the measuring instrument and the object being measured. Compared with measurements made at rest, the mass will be increased, length decreased in the direction of motion, and time will be slowed. The effects are apparent only at extremely high speeds. At 90% of the velocity of light, mass more than doubles, length reduces by over a half and a clock takes an hour to record 26 min. At the velocity of light, mass would become infinite, length zero and time would slow to a complete stop – an impossible situation – which means that nothing can overtake the speed of light.

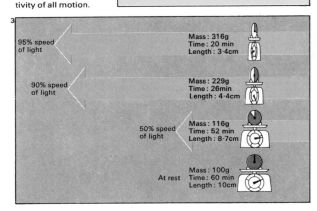

	Mass	Time	Length
95% speed of light	316g	20 min	3·4cm
90% speed of light	229g	26min	4·4cm
50% speed of light	116g	52 min	8·7cm
At rest	100g	60 min	10cm

4 A light clock [A], in which light moves to and fro between mirrors, shows in theory how time is slowed by motion. When the clock is in motion, its mirrors travel along and the light travels farther between reflections. It thus takes longer to reach the mirrors, although its own velocity is constant and observed time is slowed down. Time slowing was actually proved when particles from space first observed at high altitudes were found on the ground in greater numbers than predicted [B]. Because of their speed, they had lived longer than expected.

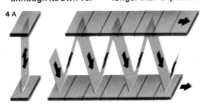

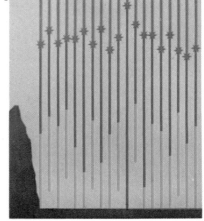

cles that are always travelling faster than light. These particles, called tachyons, have been looked for but not yet found.

The general theory of relativity

To take account of acceleration [8] and of the force of gravity, Einstein's general theory of relativity (1915) incorporated the fact that all bodies fall equally fast at the surface of the Earth. In other words, the effect of the Earth's gravitational field is an intrinsic feature of the space around the Earth. Einstein described this feature in terms of the curvature of space: the greater the distortion the greater the gravitational force. If time is included with space in this distortion it is possible to incorporate the idea that all motion is relative. The amount of space-time distortion caused by massive bodies can be quantified and it was Einstein's genius that showed how the amount of this curvature depends on nearby massive bodies.

Experimental observations, for example of small deviations in the motions of planets from those predicted by Newton [7], make the general theory of relativity the most

satisfactory of a whole range of similar theories. Confirmation also comes from the bending of the path of a ray of light near a massive body. Light has energy – and hence mass – and therefore moves in a curved path in the distorted space around the body [9]. Such bending of light by the Sun was confirmed at an eclipse [10].

Holes in the heavens

All these effects involve weak gravitational fields and cannot put general relativity through the most searching test. When stars have used up their nuclear fuel they may evolve into extremely condensed objects in which strong gravitational fields occur and so they are good testing grounds for general relativity. It is postulated that very heavy stars collapse in on themselves so completely that the escape velocity on their surface is greater than the speed of light. As a result, nothing can ever escape from them again – not even light – and so they are known as "black holes". Good candidates for black holes in our own Galaxy are the variable X-ray stars such as Cygnus X-1.

ALBERT EINSTEIN 1879-1955

HUGUENIN
R

Albert Einstein, working without the aid of a laboratory or university post, thought out the revolutionary concepts of the first theory of relativity from simple and seemingly unconnected ideas. Einstein was 26 and working as a patent officer when he published his special theory of relativity in 1905. Ten years later he announced his general theory of relativity. Einstein's fame was worldwide, but this did not stop the Nazi rulers of his native Germany from persecuting him for his Jewish blood. From 1933 he lived in the USA.

5 Nuclear weapons are one consequence of Einstein's discovery that mass can be converted into energy. But so too are atomic power stations and our understanding of the Sun's energy.

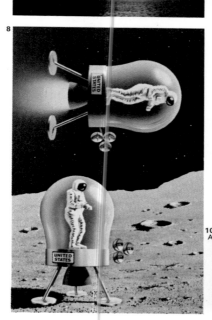

6 Low velocities accumulate by simple arithmetic. If a tank moving at velocity V fires a shell that leaves the gun at velocity v, then the shell will be travelling at $V + v$ [A]. Addition of velocities near that of light (c) is different. If a hypothetical body moving relative to Earth had a supergun that fired a shell at 0.5c, the shell would appear from Earth to move at only 0.8c [B].

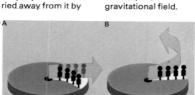

6 A

B

7 The orbit of Mercury puzzled astronomers because its perihelion (point of nearest approach to the Sun) continually shifted more than could be accounted for by the influence of the other planets. Einstein's general theory of relativity accounted for this movement. He explained that gravity distorts space so the orbits of the planets do not follow the simple orbits described by Newton.

7

8 The principle of equivalence on which Einstein based his general theory of relativity states that gravity cannot be distinguished from acceleration. An astronaut is pulled to the floor of his stationary craft by gravity [bottom], in the same way as the floor is pushed towards him when the craft acclerates [top]. The effects are identical; if he let go of an object it would "fall" in either situation.

9 A ray of light passing a rotating wheel bearing a line of people would appear to be straight to an outside non-rotating observer [A]. As it passes, the people are carried away from it by the movement of the wheel. To them the ray appears to bend [B]. This analogy shows that light bends in an accelerating system and therefore, by equivalence, in a gravitational field.

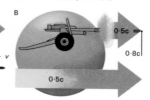

9 A

B

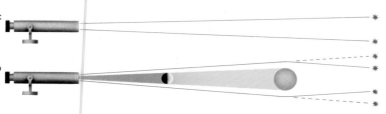

10 A

B

10 The bending of light by gravity was detected by photographing two stars normally [A] and in a solar eclipse [B]. As the light rays pass the Sun, they are bent by its field of gravity. As a result, the two stars appear to be farther apart [D] than usual [C].

C

D

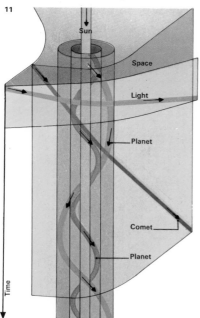

Sun

Space

Light

Planet

Comet

Planet

Time

11 The dimension of time is as necessary to describe the location of any body as are the three dimensions of space. Einstein realized that if light always travels at the same speed, then space and time must therefore be equivalent. This diagram shows the Sun, planets and a comet moving in time as well as space. The varying velocity and widely changing path of the comet demonstrates the effects of the various gravitational fields on its motion, as Einstein correctly predicted in his general theory of relativity.

Light energy

Light is energy and, in systems of constant mass, energy cannot be created but only changed from one form to another. Light can therefore be produced only from the conversion of some other form of energy. Electrical energy is changed into light in an electric lamp or discharge tube; heat is converted into light in a fire or a red-hot poker; chemical energy is changed into light in luminous animals such as glow-worms. The conversion may also go the other way – light produces electrical energy in a photoelectric cell.

Radiation and quantum theory
The conversion of energy involving light puzzled scientists at the end of the 1800s. A perfectly black object absorbs all light falling on it and all the invisible radiations such as ultra-violet rays and infra-red rays. However, when it is heated, the object gives out radiation, but only at certain definite colours or frequencies. Like the poker in the fire, it first gives out infra-red rays (which can be felt as heat rays), then it glows red, yellow and finally white as it gets hotter. If it could be heated hot enough, it would glow blue-white

and emit ultra-violet rays, as do the hottest stars. The wave theory of light [1] could not explain why a black object should differ in the radiation it produces when heated. According to the wave theory, all frequencies should be produced when the object is heated, not different ranges of frequencies at different temperatures.

In 1900, the German physicist Max Planck (1858–1947) put forward a convincing, although revolutionary, theory. He suggested that all energy, including light, consists of whole units of energy: an object can have one unit or a million, but not 0.8, 2.5 or 354.67 units, for example. Each energy unit is called a quantum of energy, from the Latin for "how much". The amount of energy in a quantum is minute and we are unable to make out the individual quanta in light rays as they strike the eye. A quantum of light energy is called a photon.

The quantum theory therefore explains why a poker glows in the way it does. As more heat is applied to the poker, the light produced has more energy, and this is shown by a change in colour – a "blue" photon has more

energy than a "red" photon. Planck explained that the energy content of each photon of light depends on its frequency; the higher the frequency (more towards the blue or ultra-violet), the greater its energy.

Particles and waves
The idea of light existing in indivisible units was a return to the particle theory of light. A light quantum is included in the fundamental particles that make up matter. If light consists of streams of particles, then it will cross empty spaces with no need for the medium of "ether" that scientists had sought in vain. But such effects as diffraction and interference could be explained only if light behaved as waves. Scientists solved this by assuming that light can behave both as particles and as waves, depending on the situation. This was not just an easy way out of a difficult problem, because the duality can be shown to exist – both experimentally and mathematically. Also, fast-moving particles were found to have wave-like properties. A beam of electrons acts as a wave in an electron microscope, for example.

1 **The wave nature of light** is demonstrated by the way light passes through openings. If light consisted of particles, it would produce a pinpoint of light on passing through a pinhole [A]. In fact diffraction gives a larger image [B]. Particles passing through two slits would give a broad band of light [C], but in fact, interference fringes are seen [D]. Diffraction and interference are properties of waves.

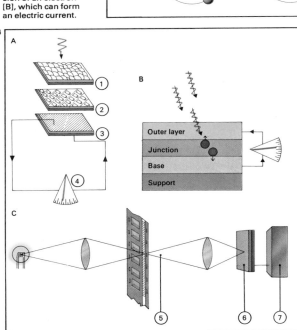

3 **The photoelectric effect** was explained by Einstein in 1905 as the absorption of a quantum of energy [A] by an atom and the resulting emission of an electron [B], which can form an electric current.

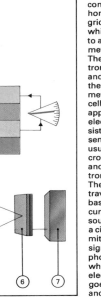

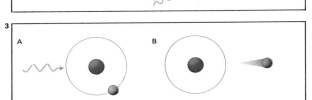

4 **A light meter** [A] contains a glass honeycomb [1] and a grid [2] through which light travels to a light-sensitive metal element [3]. There it causes electrons to be excited and these pass round the circuit to the meter [4]. The solar cell [B] is another application of photoelectricity. It consists of layers of semiconductor, usually silicon. Light crosses the outer layer and produces electrons at the junction. These are emitted and travel towards the base, producing a current. The optical sound track [C] on a cine film transmits a varying light signal [5] to a photoelectric cell [6] which produces an electric signal that goes to an amplifier and loudspeaker [7].

2 **Light exerts pressure** on any object it encounters. A vane [1] struck by light [2] moves under this pressure, and the movement is counterbalanced by a horizontal mirror [3] onto which light is directed

by a vertical mirror [4]. The torsion heads [5] first level the balance arm. The mirror [6] reflects light from a lamp [7] onto a scale [8]. A timer [9] detects motion of the light across the scale and adjusts a

power source [10] to change the intensity of the lamp [11] illuminating the mirror [4] and thus keeps the arm in balance. Deflection of the mirror is detected by the torsion head [12].

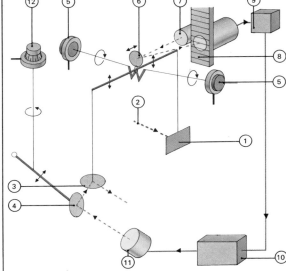

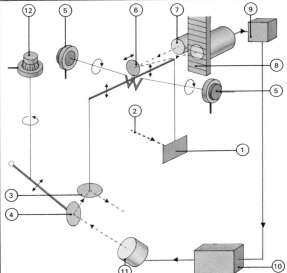

5 **A photographic exposure meter** measures the light coming from a scene. The light strikes a photoelectric cell which produces an electric current that varies in strength according to the intensity of the light. The current is low but sufficient to move a needle across a dial and give a value for the light. Many single-lens reflex cameras have built-in exposure meters that measure the light entering the lens of the camera.

The quantum theory – especially its application to light – finally resolved the problem that had divided scientific thought for centuries. Isaac Newton (1642–1727) had championed the particle theory and Christiaan Huygens (1629–95) had maintained that light travels as waves. With Max Planck's proposal the dilemma ceased to exist – light can be regarded as behaving as particles *or* waves, depending on the phenomenon being investigated.

Certain metals emit electrons when light falls on them – a phenomenon known as the photoelectric effect [3]. It had been observed that brighter light produces more electrons than dim light but not electrons of greater energy; whereas blue light always gives electrons of greater energy than red light, regardless of the intensity of the light. In 1905, Albert Einstein (1879–1955) explained that each electron is released by one photon of light; a bright light has more photons of the same energy than a dim light, but a blue light has photons of greater energy than red light.

Changing the frequency of light, or converting an invisible frequency into a visible one, has several uses. Fluorescent substances take up light of several frequencies and immediately radiate them at a different frequency, making the resulting colour very bright because extra light has been transformed into it [9]. The fluorescent paints and inks used in some advertisements work in this way. Some washing powders contain optical brighteners that convert invisible ultra-violet rays into blue light and thus making the washing look brighter.

Effects of phosphorescence

Phosphorescence is similar to fluorescence, but the production of light continues for some time after the initial radiation has ceased. Television screens contain phosphors that glow for a short time after being struck with the electron beams inside the cathode ray tube and give a picture on the screen. Many instruments make use of the light produced by phosphors to detect invisible rays such as X-rays and fast-moving particles such as cosmic rays. Some phosphorescent paints store light for a long time after being exposed to it and glow in the dark.

The radiometer, invented by William Crookes (1832–1919), measures radiant energy. When sunlight falls on it, the vanes move round, seemingly pushed by the light. In fact, heat is absorbed by the black side and the few gas molecules left in the bulb's vacuum rebound faster, exerting pressure on that side.

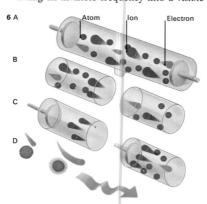

6 A   Atom   Ion   Electron

B

C

D

6 A discharge tube contains a gas at low pressure through which electricity is passed [A]. Electrons (negative particles) and ions (positive particles) move towards the electrodes [B]. Ions strike electrodes to produce more electrons [C]. Light is produced as electrons collide with gas atoms [D].

7 Advertising signs contain a gas such as neon or are coated with phosphors to give various colours.

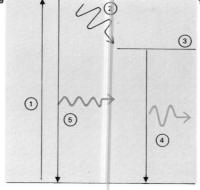

9 Fluorescence occurs when an atom receives light energy [1] and divides emission of the energy into two stages: a small energy change producing infra-red [2] to an intermediate energy state [3] and a large change giving light at a lower frequency than that received [4]. Normal light production occupies one change [5]. Phosphorescence is similar but stage two takes some time.

8 An infra-red view of the Colorado River and Lake Powell taken by a satellite shows vegetation in various shades of red and water as black. Diseased plants can be detected by colour.

10 Automatic letter checking utilizes ultra-violet light. Invisible phosphor codes printed on the stamp glow as the letter passes an ultra-violet scanner, classifying it as first or second class.

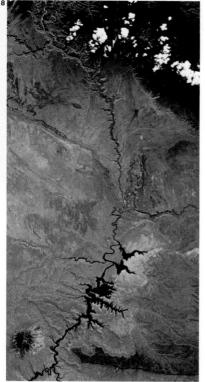

11

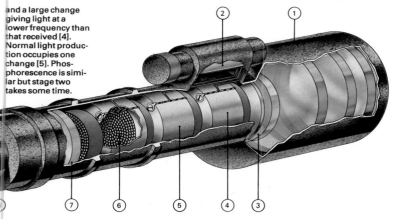

1 Night lens system
2 6·75 volt mercury battery
3 Photocathode
4 15kV channel plate
5 30 kV channel plate
6 45kV channel plate
7 Fluorescent screen
8 Ocular lens system
9 Eyepiece

11 An image intensifier gives a bright picture of a dimly lit scene. Light from the scene is focused onto the photocathode, which emits electrons. These pass to the electron multipliers that contain tubes lined with electron-emitting substances and produce more electrons from each electron entering the tubes. The resulting intensified electron beam is focused on a fluorescent screen which is viewed through an eyepiece.

Energy from lasers

From boring holes through diamonds to performing delicate eye operations, from spanning space between the Earth and the Moon to detecting the smallest movement, the laser has found an amazing range of uses during its short life. Its future looks no less extraordinary with the promise of three-dimensional television and cheap nuclear power. Clearly the laser is no ordinary source of light.

What is a laser?

A pulse laser is basically a device for storing energy and then releasing it all at once to give a very intense beam of light. The heart of the laser is a crystal or tube of gas or liquid into which energy is pumped [1]. This is usually done by surrounding it with a device to produce a powerful flash of light or an intense beam of radio waves or electrons. As pumping occurs, more and more of the atoms inside take up energy and are excited to high energy states. Suddenly an atom spontaneously returns to its first energy state and gives out a particle of light (a photon). This photon strikes another excited atom and causes it to produce another photon. Very rapidly, a cascade of photons develops. The crystal or tube is closed at both ends by mirrors and the photons bounce to and fro between them, building up the cascade. A proportion of this light is able to escape through one of the mirrors, which is half-silvered, and an intense flash of light emerges from the laser.

The first pulse laser, invented by Theodore H. Maiman in 1960, contained a ruby crystal and produced a short flash of red light. Continuous wave lasers now produce continuous beams of many colours and some give out infra-red rays or ultra-violet rays.

The activities of photons

The atoms that discharge photons are stimulated to emit them by the arrival of other photons, which make up light radiation. The light that is pumped into the laser consists of many frequencies but what emerges is a far more intense light at a single frequency.

The result - *light amplification* by the *stimulated emission* of *radiation* – gives the laser its name.

Each photon triggers the production of another one and so they all travel together and produce light waves that are exactly in step. This light is said to be in phase, or coherent. (In ordinary light, the waves are all out of phase.) Because the waves are all in step, they reinforce each other and laser light is very bright. The construction of the laser produces a narrow beam that hardly spreads at all – even at the distance of the Moon, a laser beam directed from the Earth is only 3km (2 miles) wide [3]. A narrow beam of intense, coherent light is extremely concentrated in energy and, if a laser beam is focused to a point by a lens, it will heat the air to a state of incandescence (bright and glowing with heat) or burn a hole in a steel plate. A straight, narrow beam of laser light can be used for precise alignment in the construction of tunnels and pipelines, for example. The beam is directed along the proposed route and can be seen by the construction engineers only when they are directly in line with it.

Other uses of lasers

Laser beams can also be used to measure distances and speeds. These have included firing a laser beam at the moon to reflect it from a

CONNECTIONS

See also
108 Light energy

1 Normal emission of light occurs when an electron in a high-energy orbit falls to low orbit [A]. Stimulated emission [B] is triggered by light emitted from another atom. In a laser [C], most atoms are brought to a high-energy state by pumping in energy. Some begin to produce light by normal emission and mirrors at each end reflect the light to and fro, producing stimulated emission until all the atoms are in a low-energy state. The light leaves the laser through one of the mirrors. Ordinary light [D] is a mixture of different frequencies moving in various directions, whereas laser light [E] has a single frequency and moves in the same direction with all waves in phase. The first laser [F] contained a synthetic ruby crystal surrounded by a flash tube (to pump in light energy) and a pair of reflecting mirrors.

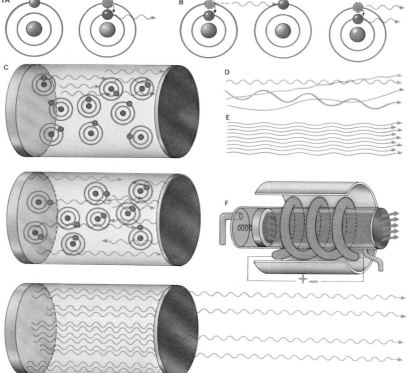

2 The helium-neon gas laser contains two gases and operates continuously. The helium atoms are first excited and transfer their energy to the neon atoms, which then produce laser light [A].

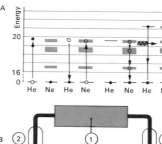

The laser itself [B] consists of a glass tube containing the gases and electrodes [2, 3] connected to a power supply [1] to excite the gas. The beam emerges through the half transparent mirror [4].

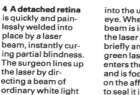

4 A detached retina is quickly and painlessly welded into place by a laser beam, instantly curing partial blindness. The surgeon lines up the laser by directing a beam of ordinary white light into the unaffected eye. When the white beam is in place, the laser is fired briefly and the green laser light enters the other eye and is focused on the affected retina to seal it in place.

3 Laser communication would be ideal for interplanetary missions because the narrow, powerful beam [B] can reach a small, distant target. A laser carries a television or other signal by modulating the beam at the transmitting end, focusing the beam on a detector at the receiving end and demodulating the signal produced [A]. Ordinary light is unsuitable for communication because it has many frequencies that interfere with each other; only the laser can be used.

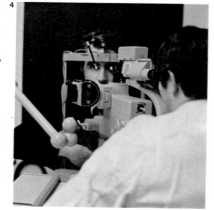

special mirror placed there by the Apollo astronauts and thus give a very accurate measure of the moon's distance.

In meteorology, laser beams are used to detect invisible air layers and movements as well as clouds, and they are useful in studies of air pollution.

The intense heat of lasers gives them all kinds of uses in medicine and industry. A laser beam directed into the eye at insufficient power to damage the lens is focused by the lens onto the retina, where it can painlessly weld a detached portion back in place and restore failing sight [4]. Laser beams can burn away skin growths without surgery, by firing the beams along fibre-optic tubes inserted into the body, and painlessly drill decayed teeth. In industry, lasers cut out patterns, drill holes in diamonds to make dies for wire manufacture and shape and weld parts for microelectronic circuits [Key].

Communication by laser beams instead of radio waves is desirable because light beams can carry many more channels of information than can radio. Data, sound and pictures can be transmitted by a laser beam, which routes it along an enclosed path of some kind to avoid loss of signal strength from having to pass through fog or mist in the air.

One of the most amazing consequences of producing coherent light in lasers is the development of holography, with which three-dimensional images can be made [5, 6, 7]. Although three-dimensional colour television and motion pictures may one day result from it, holography has several uses now. Double-exposure holograms record any movement of the object between the exposures and so readily picture the vibrations in a surface. Vibration analysis is essential to the design of components such as aircraft and engine parts which must perform faultlessly for long periods at high speeds and stresses.

Another field that may be revolutionized by the laser is nuclear energy. Research is being carried out to see if thermonuclear fusion (the reaction that takes place in a hydrogen bomb and in the stars) can be initiated by a laser beam instead of producing a high-temperature plasma by means of a powerful electric discharge.

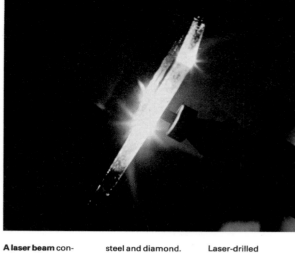

A laser beam contains sufficient energy to "burn" a hole in hard materials such as steel and diamond. Here a laser beam is drilling a hole in a sheet of toughened glass. Laser-drilled diamonds are used as dies for drawing metal into extremely fine wires.

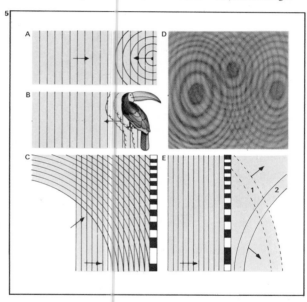

5 Holography reconstructs light waves. An illuminated point produces spherical wavefronts [A] and the surface of an object makes a complex wavefront [B]. When a curved wavefront combines with a plane one from the original light [C], a pattern [D] is formed on a photographic plate. When laser light is passed through this hologram [E], the original wavefront is reconstructed [1, 2].

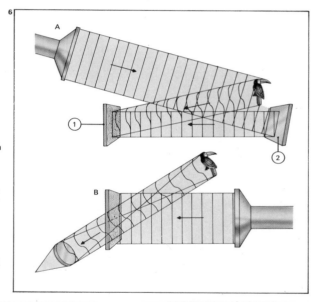

6 A hologram is made [A] by placing a photographic plate [1] near an object lit with laser light and reflecting some of the laser light onto the plate from a mirror [2]. The image is re-created [B] by illuminating the hologram with a laser beam from behind.

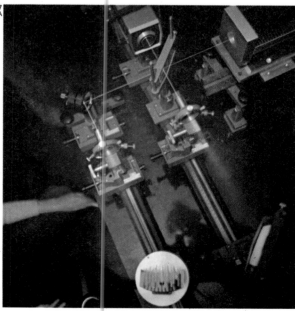

7 Images in 3-D can be obtained by holography. A hologram of an object is made by the technique described in illustration 5, using apparatus such as that shown in A. It is arranged so that laser light reflected from the object travels about the same distance as the reference laser beam reflected onto the photographic plate by a mirror. When the hologram is illuminated from behind with laser light, an image of the object is seen which is not only 3-dimensional but also demonstrates parallax: as the head is moved, a different view of the object is obtained, as in real life. This is evident in two views of chessmen [B, C]. The image has depth because the hologram completely reconstructs the light rays coming from the object and striking the plate. The colour of the object can also be produced by using several lasers of different colours. It is possible that holography may one day give us a totally realistic cinema or television picture.

What is electricity?

To the man in the street, electricity is the cause of a lightning flash [1–3] or the form of energy that powers his television set and washing machine. He knows that electric trains use electrical power and he is reminded of his dependence on it by the network of power lines criss-crossing the countryside, or by a power cut, when he has to read by candlelight. But there are other less well-known everyday processes that involve the use of electricity. A beating heart, a running athlete, a dreaming baby and a swimming fish all generate a form of electricity just as surely as a power station does.

Electrons and protons
To a scientist, electricity results from the movement of electrons and other charged particles in various materials. A scientific understanding of electricity therefore depends on a knowledge of atoms and the sub-atomic particles of which they are composed. The key to this understanding is the tiny electron – tiny even when compared with the minute atom in which it may be found.

Atoms of all materials have one or more electrons circling in orbits of various sizes – much as the planets move round the sun. Normally the number of electrons equals the number of protons in the nucleus. The protons, however, being much heavier than the electrons, are virtually stationary in the atom's centre. This extremely simplified model of the atom is sufficient to explain the basis of electricity.

The electrons and protons each have an electric charge (but of opposite polarity) and attract each other. Charges of the same polarity repel each other. To distinguish the proton's charge from that of the electron the former is called positive and the latter negative. An atom that has more, or fewer, electrons than normal is called an ion. If it is deficient in electrons, it is called a positive ion; if it has an excess of electrons, it is called a negative ion.

When an electron moves away from an atom the atom is left with a net positive charge. The electron, deprived of its positive counterpart in the atom's centre, eventually moves about another atom or possibly returns to the atom it has left.

Why do electrons move at all? There are a number of possible causes. A common one is simply that, if an incoming electron or light pulse hits an atomic electron, the latter can be knocked out of its orbit. Heat makes atoms dance faster, causing the electrons to move so energetically that they may shoot away from the parent atoms. Chemical activity will also cause electrons to move out of atoms.

A good example of the relationship between chemical and electrical activity is found in the muscles. Muscle fibres contract when they are electrically stimulated [4]. Normally, this is caused by the release of a chemical from an associated nerve, following the receipt of an electrical signal from the nervous system. When part of this system is damaged and muscles become weak or fibres are destroyed, it is possible to apply external electrical signals to stimulate muscle activity and strengthen their fibres.

Conductivity
The electrons of some materials move more freely than others. This characteristic is known as conductivity. Most metals, hot

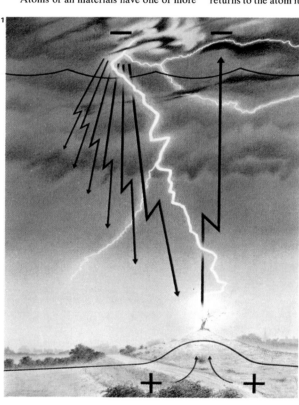

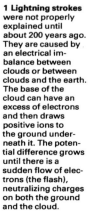

1 **Lightning strokes** were not properly explained until about 200 years ago. They are caused by an electrical imbalance between clouds or between clouds and the earth. The base of the cloud can have an excess of electrons and then draws positive ions to the ground underneath it. The potential difference grows until there is a sudden flow of electrons (the flash), neutralizing charges on both the ground and the cloud.

3 **Benjamin Franklin** (1706–90) was the first man to recognize the true nature of lightning. During a thunderstorm he induced a flash of lightning to flow along the string of a kite to the earth.

4 **Galvanism** was the term used to describe the twitching effect produced by an electric current on a pair of frog's legs. Luigi Galvani (1737–98) used this to show the connection between muscle activity and electricity.

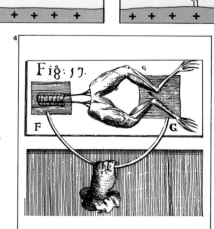

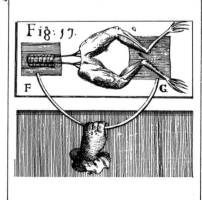

2 **Formation of lightning** starts with a big storm cloud [A], within which there is a significant temperature difference. Electrons move downwards and positive ions move upwards within the cloud, causing positive ions to gather on the earth below. When there are sufficient electrons, a sudden breakdown of the air occurs and a stream of electrons shoots earthwards [B] to be met by an upward stream of ions [C].

gases and some liquids are good conductors. Air, rubber, oil, polythene and glass are bad conductors so that they can be used to cover good conductors without themselves taking part in electron flow.

These bad conductors are called insulators. No insulator is "perfect". Under certain circumstances the electrons of any atom can be forced out of it. But the conditions required are generally so unusual and difficult to arrange with these materials that, for practical purposes, they can be considered inactive.

There is also a group of materials – the semiconductors – that behave partly as insulators and partly as conductors. Among these are germanium, silicon and copper oxide [6]. Their properties can be exploited for many purposes. For example, using one of the semiconductors it is possible to make an electric "valve" that, like the valve on a bicycle tyre, allows easy electron movement in one direction only. This device is called a rectifier; it is used in both tiny radio sets and large power stations to change an alternating current to a direct current.

Heat is simply a chaotic form of molecular activity or electronic motion and temperature is a measure of its vigour. When the temperature of most metals is reduced, it is easier for electrons to move freely; that is, the electrical resistance (to free electron movement) falls as the temperature drops and the conductivity of the metal increases.

Superconductivity
If in certain materials the temperature drops low enough, the resistance to electron flow ceases completely and electrons, once started on their journey, continue to move indefinitely provided the temperature is kept sufficiently low. The condition of zero resistance is called superconductivity. It occurs in metals such as tin, lead, aluminium and niobium [7] at a few degrees above absolute zero (−273°C or −460°F).

Electricity, therefore, is simply the movement of electrons or other charged particles. These particles are among the smallest components of matter, and yet the way in which they move and interact has a great influence on every aspect of life.

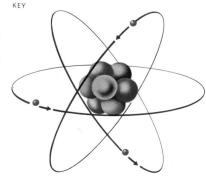

The electron is the basic unit of electricity; it is also a fundamental particle found in all kinds of atoms. In this simple model of an atom of the metallic element lithium, three electrons [red] can be seen circling the central nucleus. Larger particles called protons [blue] and neutrons [grey] make up the nucleus. Each electron carries a negative electric charge and each proton carries a positive one, so that the three electron charges are exactly balanced by the three proton charges, making the whole atom electrically neutral. In a conductor – ie most metals – an external electromotive force (voltage) causes electrons to "drift" from atom to atom and it is this flow of electrons that constitutes an electric current. Electron movement occurs because in a conductor the outermost electron is not tightly bound to its nucleus. In a non-conductor or insulator the electrons are too tightly bound to leave the nucleus easily and so such substances do not conduct electricity. In some situations atoms can completely lose or gain one or more electrons to become permanently charged. Such charged atoms, called ions, can also act as current carriers.

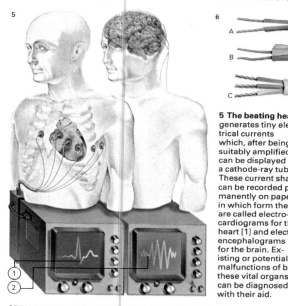

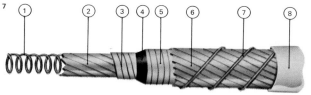

5 The beating heart generates tiny electrical currents which, after being suitably amplified, can be displayed on a cathode-ray tube. These current shapes can be recorded permanently on paper, in which form they are called electrocardiograms for the heart [1] and electroencephalograms [2] for the brain. Existing or potential malfunctions of both these vital organs can be diagnosed with their aid.

6 Domestic wiring uses copper wires encapsulated in rubber or plastic [A, C]. In fire-proof wiring [B], the wires are embedded in a noninflammable powder surrounded by a copper tube.

1 Helical conductor support
2 Strips of inner conductor
3 Inner conductor screen
4 Lapped tape dielectric
5 Outer conductor screen
6 Strips of outer conductor
7 Helical skid wires
8 Helium pipe

7 Perfect conductors can be made out of alloys of metals (eg tin, lead and niobium) at temperatures close to absolute zero (−273°C). Once electrons become detached from their parent atoms they move through the supercooled conductor, without slowing down or coming to rest, for indefinitely long periods. At such low temperatures the atoms in the metal vibrate only very slightly.

9 Extremely high voltages can be produced with the Van de Graaff generator [B]. If a body having an excess of positive ions is placed inside a container, the inside acquires electrons [A] and the outside an equal number of positive ions. If the charged body touches the inside, all the free electrons flow into it, thus making it neutral. The outside of the container still retains its positive ions. In the Van de Graaff generator, positive ions are sprayed from a suitable source [1] onto an endless conveyor belt which carries them inside a metal sphere. The belt connects to the inside wall through a conductor in the form of a comb [2], thus permitting an electron flow to the belt. This causes positive ions to form on the sphere's outside wall [3]. The effect may be enhanced by using two generators connected as shown [C].

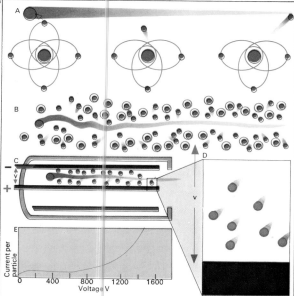

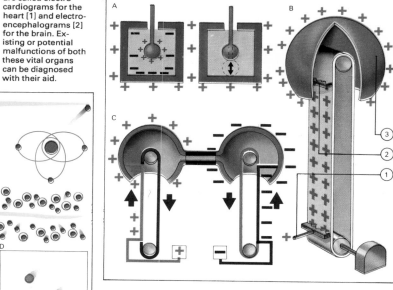

8 A high velocity particle passing through a gas knocks electrons off the otherwise neutral atoms [A]. As it sweeps through the gas, it leaves behind a stream of free electrons [B], shown here in blue. If these break free with sufficient energy, they can knock other electrons out in a Geiger counter [C, D]. Subatomic particles enter the chamber where they are accelerated to produce more free electrons. These are attracted to the positive plate and are led from there to drive a meter or earphones. The current flow depends on the voltage v between the plates [E]. The meter reading or frequency of clicks indicates the amount of radioactivity of the source.

113

What is an electric current?

Electricity flowing along a wire is known as an electric current. The wire is the conductor. When an electric lamp is connected across a battery and switched on, current flows along a wire from one terminal of the battery to the lamp, through its filament, making it glow white hot, and back along a second wire to the other battery terminal. If the switch is turned off, the circuit is broken, current flow stops and the lamp is extinguished.

Movement of electrons

The current carriers in most circuits are electrons from the metal making up the conductors. In all conductors, and a few other materials as well, there is always a random movement of electrons (minute charged particles), even when no current flows. The electrons may be relatively free to move or more tightly bound. Good conductors have freer electrons and hence more electron movement than do bad conductors, or insulators, in which most of the electrons are too tightly bound to their parent atoms to move easily. Sometimes, through natural or contrived processes, there can be a net movement of electrons in a specific direction. This concerted flow is the electric current and it is measured in amperes, generally given by the abbreviation A. Other current carriers include ions (charged atoms or molecular fragments) in gases and solutions, and "holes" (a deficiency in electrons in some types of semiconductors – the holes behave as positively-charged carriers of electric current).

A force has to be applied to upset the random character of electron motion in a conductor. In nature this can be derived from a number of sources such as sunlight, magnetic action or chemical activity. Some of these have been exploited to generate electric current. Two common devices designed for this purpose are the generator [9], which utilizes magnetic effects, and the cell (sometimes called a battery) [6], which depends on chemical action. Both force electrons to move in one direction round a circuit by virtue of the electromotive force (emf) they generate. The emf is measured in volts, using a voltmeter.

The voltage of an emf and current flow are related in a similar way to water pressure and water flow. In a household, all the pipes are full of water at a certain pressure. But there is no movement until a tap is opened, allowing water to run.

An electrical circuit may be connected to a source of emf without causing any specific electron flow (current) until a path is provided through which the electrons can move. This may be a light bulb [Key] or a vacuum cleaner; an electric switch is like a tap that turns on the current.

Relationship between voltage and current

As the voltage in a circuit increases, so does the current. An electrical circuit is, however, made up of a number of different parts. There is normally a switch, conductors and the appliance that is being supplied with electricity. All these taken together have a resistance to current flow, which is constant (provided the temperature remains the same) for that particular group of components. Therefore although the same voltage may be applied to a light bulb and an electric iron, the actual current flow is different in each, because each has a different resistance. So it

1 When two metals with different energies of free electrons are joined, the electrons redistribute themselves partially on both sides of the junction [A]. Joining the other ends of the metals stops this action because electrons cannot move two ways at once [B]. A temperature difference between the junctions can upset this [C], and electrons start moving in one particular direction. An arrangement of many junctions forms an "electrical" thermometer [D]. This also occurs at junctions between two kinds of semiconductors, known as n-type and p-type. Current flows from the hot to the cold end of n-type [E] and the other way in p-type [F].

2 Circuit-breakers can be used either in conjunction with, or in place of, a fuse to interrupt dangerously high currents. When too large a current flows through the coil [1], a magnetic field is produced which activates the catch [2], causing the spring-loaded contact [3] to rise. The current is interrupted, thereby protecting the circuit of which the breaker is a part. No further current can flow until the circuit-breaker is closed again, which is achieved by pressing the reset button [5]. To test the circuit-breaker, or to operate it manually, a push button [4] is pressed to move the contact.

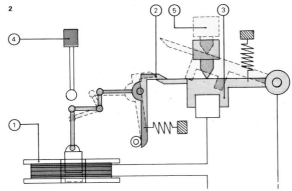

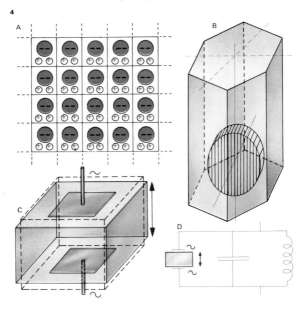

3 A gas discharge lamp gets its light from the energy changes in gas atoms. A positive and negative electrode (anode and cathode) at opposite ends of a gas-filled glass tube [A] attract electrons and positive ions. Reducing the pressure [B] speeds this up. As ions hit the cathode [C] they dislodge electrons which speed towards the anode, colliding with gas atoms on the way. The atoms absorb the energy for a moment, then release it in the form of light [D].

4 Certain crystals, including quartz and sapphire, exist in the form of "cells' in a delicate state of electrical equilibrium [A]. Applying a varying voltage across the crystal disturbs the delicate balance and causes the crystal to vibrate and emit sound or ultrasonic waves, generally at a specific angle to the direction of the applied voltage [B]. Conversely, when such a crystal is vibrated, it generates a voltage. This effect – which is known as piezoelectricity – is made use of in a gramophone pick-up [C] and in crystal microphones. In the pick-up, the movement of the stylus in the record groove rapidly vibrates a piezoelectric crystal and generates a small electric current. In a crystal microphone, sound waves vibrate a diaphragm coupled to a crystal and generates a current that can be amplified and fed into a tape recorder or sound system. Most crystals respond strongly to only one frequency, depending on their dimensions. Radio transmitters make use of this property to "hold" a particular fixed frequency [D]. Vibrating quartz crystals are able to keep almost perfect time. They are used in quartz clocks and watches, which are accurate to within a few seconds over a period of of several years. Piezoelectric crystals are also used to generate the electricity for the spark that ignites the gas in certain kinds of "electronic" cigarette lighters, which do not need any batteries.

is not only the magnitude of the voltage that determines how much current flows through a particular piece of equipment, it is also the resistance of that equipment and the conductors. This property of electrical resistance is measured in ohms (Ω). For any conductor, or system of conductors and equipment, the relationship between voltage, current and resistance is given by the formula: voltage = current × resistance. This is the mathematical expression of Ohm's law, named after Georg Ohm (1787–1854) who was the first person to specify the interdependence of these three factors in a precise way.

The resistance of electrical conductors depends on their dimensions and on the materials they are made of. As the cross-sectional area increases, the resistance falls; but as the length increases, the resistance rises. A long thin conductor therefore has more resistance than a short thick one with the same volume of material. Silver has a lower resistance than copper, whereas aluminium and iron have higher resistances.

In an electric circuit, the amount of electron movement is always the same past any two points. According to a convention adopted before the nature of electricity was properly understood, a direct current (that is, current from a battery or dynamo) is assumed to flow from a positive point to a negative one. As it happens, electrons move from negative to positive, so that electron movement is opposite to the assumed current flow.

Effects of current flow
Three phenomena that typically occur when a current flows (and by which it can be detected) are heating, chemical and magnetic effects. Its heating effect is used to provide warmth in electric fires, cookers and industrial furnaces. Such heating can also be unwanted; large cables carrying thousands of amperes have to be cooled to prevent the current-generated heat melting the insulation or even the wires themselves.

The chemical effect of current is used in electroplating and in energy storage, particularly in cells, the most familiar of which is the lead-acid accumulator or battery [7]. The magnetic effect is used in motors, electromagnets and many other devices.

KEY

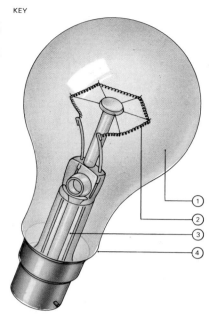

Heat generated by the passage of electric current is the source of light in a so-called "filament" or "incandescent" lamp. Because it is enclosed in a vacuum, or inert gas [1], the filament [2] cannot oxidize when the current passes through, causing it to become hot. It is made of a tungsten alloy, combining mechanical and thermal strength. Though it is extremely thin, it glows white hot when enough current passes through it. It is supported on two glass columns [3] through which the connecting wires pass. The whole assembly is enclosed in a thin glass envelope [4]. Only about two per cent of the electrical energy is converted into light.

5

Unbroken fuse

Burnt-out fuse

7

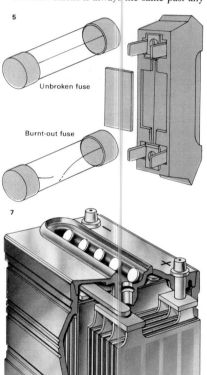

5 The heating effect of an electric current is used in a fuse, which consists of a thin wire that melts when excessive current passes through it, thereby cutting off the electricity supply.

6 The Léclanché cell [A] consists of a leakproof jacket [1] containing a porous pot [2] in which there is a paste of manganese dioxide and carbon granules [3] surrounding a carbon rod [4]. The top can be sealed with pitch [5]. A zinc rod [6] stands in a solution of ammonium chloride [7], and is connected to the carbon rod via a circuit and a light bulb [8]. The zinc dissolves in the solution, setting up an electromotive force. The ammonium ions migrate to the carbon anode and form ammonia (which dissolves in the water), and hydrogen ions. Torch dry batteries [B and C] use wet paste cells of the Léclanché type.

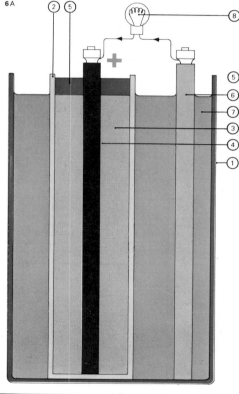

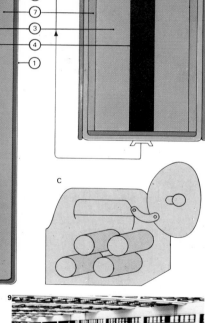

7 A 12-volt car battery has six two-volt cells connected in series. The cells have anodes of brown lead oxide and cathodes of porous grey lead immersed in sulphuric acid. An electric current flows if the electrodes are connected through a conductor. When the battery supplies current the sulphuric acid converts the anode to lead sulphate, thus reducing the strength of the acid. This process is reversed during recharging. Each cell of the battery is made of several anodes and cathodes separated by porous insulators. The cells are housed in a hard rubber case and the various cells are interconnected with lead bars.

8 Tiny electric motors, such as this one, which is used for driving a miniature tape recorder, emphasize the enormous range of size and applications of electrical equipment.

9 The heart of a power station is the generator, which sends electricity over hundreds of kilometres through a wide network of transmission lines. It is the point at which mechanical energy is converted to electrical energy.

115

Magnets and magnetism

Magnetism and electricity are not two separate phenomena. The error of thinking they were arose from the fact that their interrelation was not appreciated until 1820. In that year the Danish scientist Hans Christian Oersted (1777–1851) showed that an electric current flowing in a wire deflects a compass needle close by. Whenever an electric current flows, whether from cloud to ground in the form of lightning or through a muscle in the body, a magnetic field is created.

Thousands of years before electricity was recognized and used, magnetism was observed and applied – mainly for navigation. Eventually, when science became aware of the atomic nature of matter, it was finally realized that the properties of magnetism and electricity are both bound up in the nature of the physical structure and arrangement of atoms and their electrons.

Whenever magnetism can be detected there must be a current of electricity. Those materials that appear to be magnetic without any external source of electricity depend on electron movements within their atomic structure to provide the electric current; this is the class of magnetism dealt with here.

The property of attracting iron and iron-based materials occurs naturally in a mineral called lodestone [3], itself a chemical compound of iron. It is likely that some form of lodestone was used in the first magnetic compasses that the Chinese are thought to have made [1]. It is relatively easy to transfer deliberately magnetic properties between various materials, of which iron and steel are the best known common examples.

Permanent magnets

Iron-attracting materials form a class of so-called permanent magnets, although they may retain their magnetic properties for only a limited time. In the form of a bar, a permanent magnet experiences a force due to the earth's magnetism such that, if it were free to move, one end would point roughly in the direction of the earth's North Pole and the other to the South. The two ends are named the north-seeking (or north) pole and south-seeking (or south) pole.

Unlike magnetic poles attract each other. A magnet that attracts other material does so by first turning the material into a weak magnet. Like poles repel each other (although this is not so obvious as attraction) because whenever an iron or steel object comes within the influence of a magnet, and itself becomes a magnet, it acquires the opposite polarity. As a result it is automatically attracted. But when two identical magnets of equal strength are positioned with their *like* poles close to each other, each experiences a repulsive force equal to the attractive force that results when the two unlike poles are placed close to each other.

It is not only ferrous (iron-containing) materials that are affected by magnetism. But its effects are easiest to observe in pure metals such as iron, nickel and cobalt.

Domain strengths

In general the metals affected by magnetism consist naturally of tiny magnets within the structure of the material, all of them aligned in a random manner. These magnets occupy areas known as domains [6], which can be seen with an electron microscope. In the unmagnetized material, the result of these

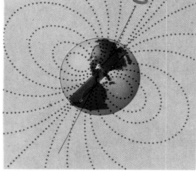

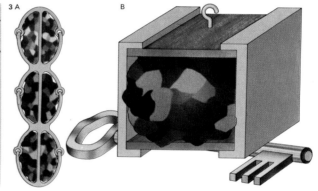

1 The Chinese were probably the first to realize the directional properties of magnetic materials and built compasses to help them navigate at sea as well as overland. By the 12th century magnetic compasses were in use in the West. This 13th-century one consists of a disc of lodestone (meaning "way stone") marked with the compass points, mounted on a block of wood and floating in water.

2 A simplified model of the earth's magnetic field may be made by picturing a long bar magnet lying in the centre of the earth. Magnetic materials on the globe's surface tend to align themselves so that their north-seeking poles point to what is called north (actually the south pole of the imaginary magnet) and their south-seeking poles point to the south (north pole of our magnet).

3 In an early "magnet" the magnetic properties of lodestone were intensified by placing lumps of the material within a soft iron structure [A]. This provided a path of low resistance to the magnetic flux of the lodestone, and had the effect of concentrating the flux. The attractive force depends on the square of the flux density, so that by directing the flux through a small section the lifting power of the lodestones was increased. Further improvements to lodestones' attractive qualities were gained using iron pole-pieces [B]. The flux lines follow the iron path to produce the two poles.

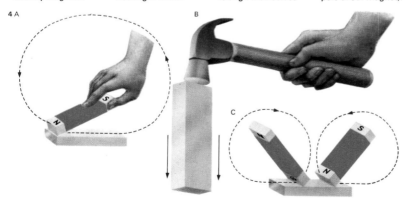

4 A simple way to magnetize materials such as iron and its alloys is to stroke it with a bar magnet [A], the nearness of which, coupled with its movement, tends to align the magnetic domains within the material. They then reinforce each other rather than keeping their normally random arrangements. The south-seeking ends of the domain try to follow the movement of the original magnet's north pole so that the right-hand side of the new magnet becomes a south pole. The domains lie with their south poles to the right, and so their north poles are to the left. Another way to magnetize a bar of suitable material is simply to hit it [B]. The domains receive a mechanical shock and the earth's field tends to align them with itself. Adapting the technique shown in [A] it is possible to make a bar magnet of suitable material using two magnets [C]. In this case, the right-hand side acquires a south pole and the left-hand side a north pole.

5 William Gilbert (1544–1603), an English physician and philosopher, demonstrated magnetic phenomena to Elizabeth I. His *De Magnete* was the first major work in Europe to describe the characteristics of magnets and magnetism in an organized way. Some of the theory is now known to be incorrect but it was still the most important contribution to the subject for many years. He suggested that the earth itself is a large magnet. The compass was already at that time used as a navigational aid – an important tool for the long commercial and military ventures of the 16th century – but the process by which the compass worked had never been explained. This painting is by A. Ackland Hunt.

millions of tiny magnets, acting in different directions, is to produce a neutral field – one with no magnetic properties whatsoever. It is as if hundreds of children were all tugging equally at a maypole from different positions; the result of their combined efforts is that the pole does not move.

The process of magnetizing consists of causing all the domains to assist each other by lining them up in the same direction. As they all come into line the total effect is additive and the whole of the material begins to display the properties of a magnet. If all the domains become perfectly aligned then the material has reached the limit of its magnetic capability. As a result, the magnetic strength of a material depends ultimately on its domain strength, and this is determined, in turn, by the way its individual atoms are structured within the domains.

Earth's magnetic field
The magnetic field of the earth has been accurately measured and charted but it still cannot be adequately explained [2]. In very simple terms, it is as if a single bar magnet lies

between the geographic North and South Poles to produce some of the observable effects. But this does not explain the very unusual variations of strength, and even change of direction, of the magnetic forces over the earth's surface. Nor does it explain why millions of years ago the magnetic poles were oppositely aligned to their present direction, nor why they are slowly but constantly moving.

Both terrestrial magnetism and that exhibited by small pieces of iron can be better understood by considering that lines of magnetic force (often called flux lines) leave the north pole and enter the south. But this is an entirely arbitrary concept, in the same way as lines of latitude and longitude on a map are drawn merely for the sake of convenience.

In a simple bar magnet, lines of flux [Key] are pictured as forming an approximate cylinder stretching in air from one pole to another and enclosing the magnet itself. The flux lines are of the same polarity so they repel each other. They all start from and end at the same poles, but they each follow unique paths that can never cross.

KEY

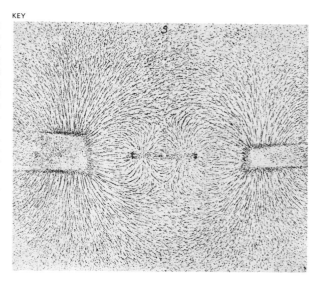

The pattern of iron filings in a magnetic field demonstrates how the lines of magnetic flux are distributed. The lines never cross and their mutual repulsive effects push each other away.

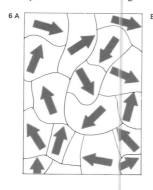
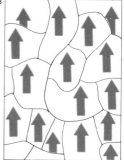
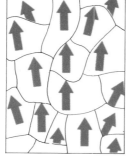
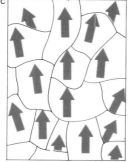

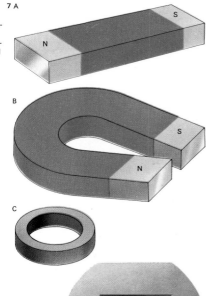

6 A

D

E

Magnetization M

① ②

−H

Magnetizing force H

−M

F

6 The random arrangement of domains in unmagnetized materials [A] becomes highly ordered in a strong external magnetic field [B]. On removing the field the domains do not revert completely but retain some degree of alignment [C]. Domains on each side of any breaks ensure that large bar magnets always split up into smaller replicas of themselves [D]. Increasing the magnetizing force beyond a certain limit [E] cannot increase perfect alignment, and the material "saturates". Reversing H causes demagnetization. Different materials have similar shaped curves [1 and 2]. Removing H leaves the material partly magnetized [F].

7 A

B

C

D

N S

7 Magnets can be made in almost any form, from the bar [A] to a horseshoe [B], a ring [C] or a shape like that of D, used in an electrical measuring instrument. The poles are marked as N and S.

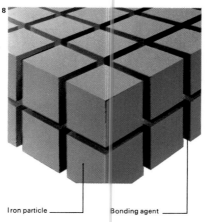

Iron particle ____ Bonding agent

8 Magnets of complicated shapes can be made using powdered iron, mixed with a suitable bonding agent and cast into the required form. In granular form the structure is similar to bricks and mortar, the bricks being the particles, the mortar the bonding agent. Each magnet is separated from the next by non-magnetic material. For this reason the whole structure makes a weaker magnet than it would be if made from solid material.

9

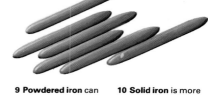

9 Powdered iron can also be bonded as needle-like particles magnetized in such a way that their poles correspond to their points. The flux lines tend to run along the axes so that the bonding agent has a limited weakening effect.

10 Solid iron is more easily magnetized than any of the forms of powdered iron (such as the grains of iron in illustration 8 or the needles of illustration 9 which are joined with a bonding agent) as the domains form an uninterrupted array.

10

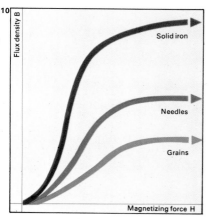

Flux density B

Solid iron

Needles

Grains

Magnetizing force H

Electromagnetism

Electromagnetism is the effect by which electrical currents produce magnetic fields. Occasionally the process is unwanted, such as when a current flowing through a piece of equipment or cable on a ship produces magnetism that deflects the ship's magnetic compass. Often the effect passes unnoticed because it is very weak. But sometimes electricity is deliberately used to produce magnetic fields of great strength, as in the electromagnets that lift scrap iron [5].

Current flow and magnetic flux
The intensity of a magnetic field is measured in flux lines or webers (Wb). These lines are produced whenever a current flows; and in air there is a simple proportional relationship between electric current flow and magnetic flux. A straight wire carrying a current can be looped to form a single turn. Provided the radius is reasonably small, the effect of forming a loop is to increase the concentration of magnetic flux without having to increase the current.

This concentrating effect can be further intensified by using more turns of wire to form a coil [4D]. At the point of maximum flux density – that is, maximum flux lines per unit area – the relationship between electric current A, turns of wire T and magnetic flux B is such that AT is proportional to B. Additional turns are simply a way of making the same current pass the same way more than once, and 12 amps flowing through three turns has precisely the same magnetic effect as three amps flowing through 12 turns.

Solenoid is the name given to a coil of wire wound to produce a magnetic field. Solenoids may be wound on iron (iron-cored) or on a non-magnetic support (air-cored). As far as flux is concerned, any non-magnetic core has the same properties as air, which means that the relationship connecting current, turns and flux holds good.

The presence of iron influences the magnetic field in two ways. It enhances the magnetic effect of the current, often by a factor of a thousand or more, but it also destroys the simple relationship applying to air-cored coils. Both these effects are a result of the structure of iron.

Microscopic regions called domains in the iron tend to align themselves with the magnetic field produced by the current. The iron provides an easy path for the magnetic flux passing through it. As a result, a given current produces more flux per unit cross-sectional area – that is, there is a high flux density. When all the domains have been aligned, further increase in current (or in the number of turns of wire in the coil) increases the flux density only negligibly.

Limiting characteristics
An iron-cored solenoid has a vastly stronger magnetic field compared with that of an air-cored one but is limited by the characteristics of iron. Theoretically there is no maximum to the magnetic field produced by an air-cored solenoid. But generally the enormous currents required to make them comparable to iron-cored ones are too expensive and technically difficult to produce.

A changing magnetic field can produce a current just as a current can produce a magnetic field. As a magnet moves towards a conductor the flux lines sweeping past cause an electromotive force (voltage) to be

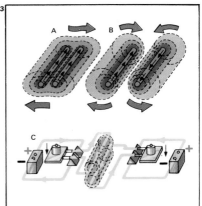

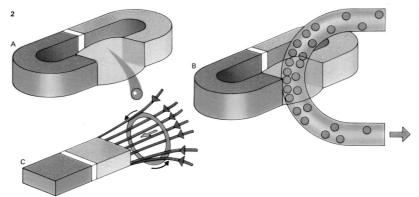

1 An electricity generator devised in 1883 by a Frenchman, Hippolyte Pixii, consisted of a horseshoe magnet set on end between two coils. The magnet was rotated through a gear system driven by a hand crank. As the magnet rotated, an alternating voltage was induced in the coils. A commutator, added later, enabled positive voltage to be picked off one side, and negative off the other, to produce direct current.

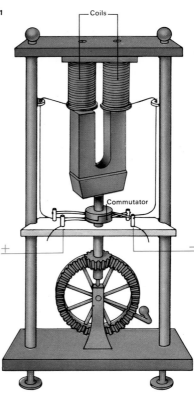

3 An electric current carried by a wire sets up a magnetic field around itself. As a result when parallel wires carry currents in the same direction [A] they attract each other, but when the currents flow in opposite directions [B] they repel. The ampere balance [C] measures the force of attraction or repulsion between two electric current-bearing conductors. Such forces can be very destructive when the currents are high and the conductors close.

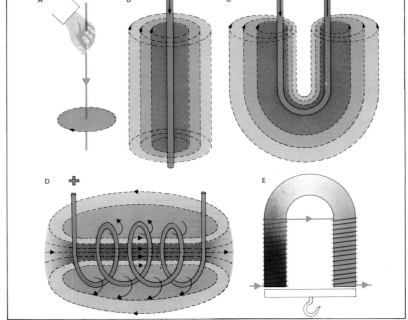

2 Pixii's generator is based on the principle that moving electrons [A], having their own magnetic field, are affected by a moving magnet, which attracts or repels them according to their relative polarities: if a path [B] is provided for them, electrons tend to move along it. Electrons will flow in a closed conductor [C] if there is relative movement between it and the magnetic field, as in a dynamo.

4 An electric current in a conductor produces a magnetic field in a plane at right-angles to its flow. The direction of the field can be found using the "right-hand rule". Holding the wire in the right hand [A] with the thumb pointing in the direction of the current flow, the curled fingers indicate the field's direction. The field pattern round the wire is symmetrical [B] even if the wire is bent [C]. The magnetic effect is increased by winding the wire into a coil or solenoid [D]. An electromagnet [E] can be made by winding a solenoid on an iron core.

induced. The polarity of the induced voltage depends on the polarity and the direction of flux movement. The effect is greater in a coil than in a single wire, and is in proportion to the number of turns of wire in the coil. Similarly if the coil is iron-cored the induced voltage is more than in an air-cored coil because the flux changes are larger.

In inducing a voltage in this way there must be relative movement between the flux and the conductor (or coil). If not, flux lines will not move relative to the conductor and no voltage will be indicated.

How power is produced

Electric generators produce current using precisely these principles [1]. In their basic form a magnet is rotated between coils. A voltage is induced depending on the factors outlined previously – that is the strength of the magnet and its speed of rotation (since this determines the rate of flux change). The voltage in a conductor is directly proportional to the rate at which flux sweeps past it.

In many generators the magnet is replaced by a solenoid that must be energized or "excited" with current to produce the magnetic field necessary for the generator to function. It is the combination of voltage and current that constitutes the electrical power output from a generator.

Another aspect of the interrelationship between current in a conductor and magnetic flux makes use of the flow of an electric current in a magnetic field to produce physical movement. This is the principle on which motors and some electrical measuring instruments operate [6] but electrical power must be supplied to cause movement against a mechanical force.

Magnetic fields far stronger than ever before are now created by means of superconductivity, the zero-resistance effect in some metals at temperatures approaching absolute zero. As a result, current can flow without losses or heating, and it is possible to use vast currents in air-cored coils, avoiding the limitations of saturation imposed by iron. These enormously strong magnetic fields open up prospects for electromagnetic levitation and new forms of motors and generators capable of high outputs at reduced costs.

The "lift" obtainable using magnetism has been applied to make a working model of a magnetic levitation train, which is seen here end-on. The train has no wheels, but instead "floats" over a long magnetic strip that takes the place of a conventional track, below which a series of electromagnets generate the necessary magnetic field. The electric current can be supplied to these electromagnets in such a way that they behave as a linear motor, so driving the train along the track. Such "maglev" trains are frictionless, pollution-free and virtually silent; a full-scale experimental prototype train has been built and successfully tested in West Germany.

5 **Electromagnets** are often used in scrap metal yards to lift ferrous metals. This method not only reduces manhandling but also provides a means of separating the iron from other scrap materials.

6 **In electrical indicating instruments** a coil [1] turns when energized in a magnet's field [2]. The pointer [3] shows the strength of current on the scale [4]. The hairspring [5] returns the pointer to rest.

7 **Relays** are used to allow a low-power source to switch a high voltage circuit. When the coil [1] is energized by a small current, magnetic flux appears between the poles of the core [2] attracting the armature [3]. The moving contact [4] then engages the fixed contact [5], closing the high-voltage circuit. When the coil is de-energized, the balance weight [6] overcomes the weakened magnetic field and opens the circuit.

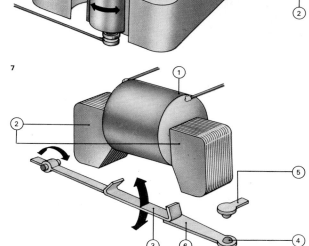

8 **In an electric bell,** the magnetic field of an electromagnet is effectively turned on and off rapidly to make a hammer strike the bell. Pressing the switch [A] allows current from a battery to energize the magnet, pulling over a spring-loaded armature and with it the hammer [B]. This action also breaks the circuit at a point contact and "turns off" the electromagnet. The armature springs back, re-makes the circuit [C] and the whole sequence is repeated.

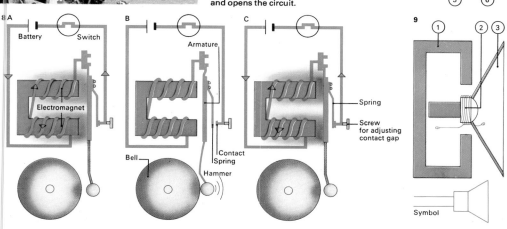

9 **A loudspeaker** commonly has a permanent magnet [1] to provide a magnetic field in which a coil [2] attached to a fibre cone [3] is held balanced, but is free to move backwards and forwards. A varying electrical current is fed to the coil from the amplifier output, resulting in a varying magnetic flux in the coil. This reacts with the permanent magnet's field, causing the coil and hence the cone to move back and forth, producing sound.

Using magnets

The two main types of magnets used in apparatus such as electric bells, motors, dynamos, speedometers and the like are permanent magnets and electromagnets. Permanent magnets, chiefly made from iron-based alloys, retain their magnetism all the time. An electromagnet consists of a coil of wire, sometimes wound round a soft iron core, and behaves as a magnet only when an electric current flows in the coil.

The magnetic field

Those processes that generate motion by creating a strong and then a weak magnetic field (as in an electric bell) expend more electrical energy in the electromagnet than is gained from the resulting mechanical motion. But weakening a magnetic field (or alternatively "turning it off") is in fact the basis of many important devices that have characteristics that would be difficult or expensive to obtain in other ways. In permanent magnets the field cannot be turned off without destroying it. But it can be diverted.

The best example of diversion of the magnetic field is the magnetic chuck [Key].

This is a device for holding ferrous metals tightly onto a work table. The chuck is used almost exclusively in grinding machines because a vice could distort the metal, or not hold it level relative to the grinder.

The chuck comprises a number of small bar magnets embedded on a movable metal plate so that north and south poles are pointing vertically. The metal plate is of a material with low magnetic properties and the poles of the magnets are positioned alternately north–south and slightly separated. A second metal plate is placed above the magnet assembly. This incorporates soft iron pieces that correspond to the position of the magnets fixed in the base plate.

When a workpiece is placed on the upper plate it provides a flux circuit for the embedded magnets through the soft iron pieces of metal and is attracted to them, holding the workpiece in position. However, the operation of a simple lever allows the lower plate to be moved horizontally in such a way as to bring the magnet poles out of alignment with the pieces of soft iron in the upper plate. This causes the magnetic flux to

be diverted from the workpiece and it goes instead through the metal of the upper plate to link the embedded magnets north to south. The workpiece is then free to move.

Magnets in railway service

A system in which permanent and electromagnets complement each other for reasons of safety is commonly used in railways. A strong permanent magnet is attached to the track at a set distance from the signals. As a train passes over the magnet it causes a pivoted permanent bar magnet in the cab to swing (like a see-saw) through a small angle and rest in this new position. The bar magnet's movement closes a switch to bring into action an alarm bell or hooter signal. A few seconds later the cab passes over an electromagnet connected to the signals. If they are set at "clear" the electromagnet is energized and the pivoted magnet in the cab is repelled so that it returns to its first position, turning off the alarm.

But if the signals are set at "stop" or "caution", the electromagnet is not energized and, after a short pre-set delay, the brakes

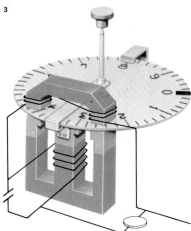

1 Magnetic mines took a great toll of merchant ships during World War II. Placed in busy shipping lanes, the strong magnets built into the devices were attracted to metal-hulled ships and on impact they exploded. Countermeasures such as electrical cables to reduce the ships' magnetic fields were devised, and are sometimes used in places where mines may still be located.

2 A system of permanent magnets attached to movable metal plates can be suspended above a fixed set of magnets. Provided there is a guide-rail to stop sideways motion, it can be used to move heavy loads around a factory or within an area where it is convenient to build a magnetic "track". The advantage over comparable methods such as rail systems is the absence of friction and moving parts.

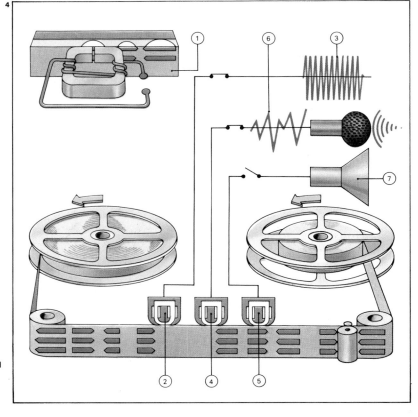

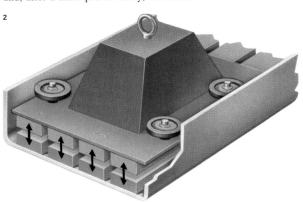

3 A domestic electric meter – a watt-hour meter for recording the amount of electricity consumed – makes use of electromagnets. When current is being used, it flows in coils that energize magnets and make a disc rotate. The disc is coupled to a counter reading in kilowatt-hours.

4 Magnetic metal oxide development has led to a revolution in the sound-recording industry. Metal oxides in powder form are bonded to flexible plastic tape [1], usually PVC, forming a moving surface on which magnetic patterns can be imposed corresponding to sound, visual or other signals. These tapes are used in machines that consist of an erase head [2] using high-frequency input [3] to demagnetize the tape as it is driven past, and a record [4] and replay head [5]. These either magnetize the tape according to the signal input [6] or reconvert the previously imposed magnetic patterns into the signals that formed them (ie plays back) [7]. Stereo or twin-track recording combines two record/replay heads in one. The erase head is taken out of circuit when playing back and, to prevent accidental erasure, most tape recorders have a built-in fail-safe arrangement.

are automatically applied if the driver fails to apply them. The brake-time circuit (like the audible alarm) is energized from the moment the pivoted magnet moves. If this magnet is returned to its original position (within the pre-set time) the brakes are not applied.

Meters and medicine

A phenomenon associated with magnetic fields is the eddy current. When there is relative motion between an electrical conductor (not necessarily one with magnetic properties) and a magnetic field, currents called eddy currents are induced. These, in turn, produce another magnetic field of opposite polarity. There is a tendency, because of the attraction between the opposite fields, for the conductor and original magnetic field to move together while the motion exists.

This principle is the basis of the car speedometer. A permanent magnet within the instrument housing rotates at a speed related to the crankshaft speed. It turns within a specially shaped aluminium disc which itself can rotate, but only through about 270° because of the restraint of a spring. As the magnet rotates, eddy currents are induced in the disc, which tries to follow the magnet. The strength of these currents is proportional to the speed of rotation and therefore the disc moves according to the speed of the car. A pointer attached to the disc moves over a scale calibrated in kilometres or miles per hour.

An electricity meter [3] works on similar principles. Current used by the consumer passes through an electromagnet, which induces eddy currents in an aluminium disc. In this case the disc can rotate freely through 360° and its movement is coupled to a gear train that drives the indicating dials.

Medical science also benefits from the use of powerful magnets. Experiments are taking place in the application of magnetically guided "pills" within the body. The "pill", which may be swallowed or inserted in a vein, is a minute radio, capable of transmitting information about such factors as temperature and salinity. It incorporates a "nose" of suitable metal and may be guided to particular organs by a magnet operating outside the body, thus aiding doctors in diagnosis.

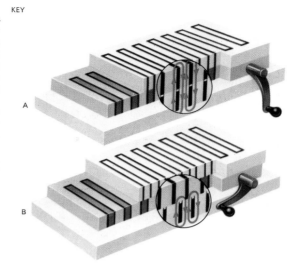

A **permanent magnet** cannot be switched off as electric currents can. Instead, its magnetic flux can be diverted. A magnetic chuck holds a workpiece in position on a grinding machine. Flux enters the workpiece and holds it in place [A]. When the flux is cut, the work is released [B].

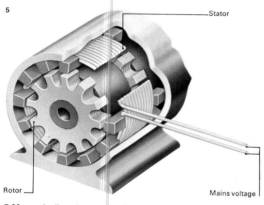

Stator
Rotor
Mains voltage

5 Magnetically polarized rotators are essential elements of modern timing devices and similar equipment. Because the mains frequency is fixed precisely at the power station, motors whose rotation is tied to this frequency have a highly accurate speed. Using magnets in the rotor (often permanent for ease of manufacture and reliability) ensures that once it is "locked" to the stator frequency, the rotor follows the speed exactly. The rotor is made from thin sheets of silicon steel cemented together to form a cylinder. The ridges are then magnetized to form poles.

6 Electricity meters are often difficult to read. The cyclometer type shown here suffers from the disadvantage that the digits representing the high numbers (on the left) rotate slowly, to give ambiguous readings. A version developed by Ferranti has a small magnet attached to every wheel behind the figure 7. A bar magnet is fixed over the wheels as shown. When the wheels move from 9 to 0 magnetic attraction ensures a quick changeover of figures. Only the highest number wheel is excluded from this arrangement because when it changes to 0 all the other figures will change to 0.

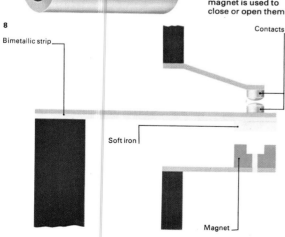

Bimetallic strip
Contacts
Soft iron
Magnet

7 The reed switch, used as a safety switch and in electronic counters, has its contacts enclosed in a glass envelope to protect them from corrosion. A magnet is used to close or open them.

8 A thermostat often incorporates a bimetallic strip, which moves to open or close a circuit according to the temperature. One way of ensuring a quick "on" or "off" is to use a fixed magnet and a soft iron pellet (armature) attached to the moving contact arm. As the bimetal starts to curve it is either attracted swiftly to the magnet or else is released suddenly, providing a fast "snap" action.

9 The electromagnetic clutch, often used on ships, has a soft iron cup [1] attached to the propeller shaft and has electromagnets on the engine shaft. Energizing the coils produces a strong field and transmits the drive motion.

10 Magnetic catches are used on doors of furniture and refrigerators to hold them shut. A common type on a refrigerator uses "magnetic rubber" (rubber with ferrous particles magnetized to form a convenient pole pattern) on the edge of the door. The type shown here, on a cupboard, uses metallic magnets in a plastic housing fixed to the frame. It "catches" on an iron plate on the door.

Transformers, motors and dynamos

The transformer, one of the most essential and efficient of electrical machines, has wide uses in the supply of electricity. It is used in power stations and at sub-stations – in the former to boost voltages for transmission over power lines and in the latter to reduce voltages to levels suitable for industrial or domestic use. Transformers are also used in many electrical appliances – such as radios, television sets and battery chargers – wherever alternating voltages different from the mains supply are required.

Motors and generators

A transformer, with its two main elements, magnetic and electrical, linked by a laminated soft iron core [Key, 1], has no moving parts and is up to 98 per cent efficient. This is not true of electric motors and dynamos (seldom more than 60 per cent efficient), both of which have rotating elements.

Motors and dynamos (or generators, as they are now more usually called) are basically the same in construction, although their functions are different. Motors are supplied with electrical power to provide mechanical power; generators are supplied with mechanical power to give electrical power. But it is important to remember that they are so similar that some machines can act as motors or as generators, depending only upon whether they are supplied with electricity or mechanical power.

The two essential elements of each of these machines are the field and the armature. The field is a magnetic field, which may be derived from permanent magnets or electromagnets. The former are cheaper but the latter are more convenient because with an electrically energized field it is easy to increase or decrease their strength. It is, however, a convenience gained only at the expense of having to provide the coils for the electromagnets that form the field (field "windings") and to do all the work that goes into insulating and installing them.

How the armature works

The armature is also a winding but is arranged differently from the field. It is essentially a conductor (or conductors) arranged to cut the field's magnetic lines of flux at right-angles. The armature conductors may be wound onto a cylinder that rotates in a field. Or they may be fixed to the inner walls of a cylinder, within which the field windings rotate. The static part of the machine is called the stator and the revolving part the rotor. Both the field and the armature may be on either the stator or the rotor.

By a basic principle of electromagnetism, a voltage is induced in a conductor that moves in a magnetic field and a conductor in a magnetic field experiences a force and tends to move when a current flows in it.

To make the best use of this basic effect, the magnetic and electrical elements in electrical machines (the field and the armature) have to interact in the most efficient manner possible. The armature is generally wound on a soft-iron core so as to be in the path of the maximum number of magnetic flux lines. The field coils are also wound on soft-iron cores, to produce the maximum flux for a given current. The soft iron for both the field and the armature is laminated – that is, made up of thin slices [1]. This is to prevent the currents from circulating and "eddying" in the iron

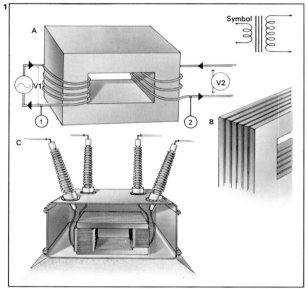

1 In the transformer [A], the input or primary current [1] causes lines of magnetic flux to form in the iron core, linking it to the output or secondary [2]. As the supply alternates the flux lines collapse and reform in the same pattern, but with different polarities. They cut the output coil, generating a voltage. The ratio of input to output voltage (V1 to V2) is the ratio of turns on the output and input coils. The iron core is laminated [B] to reduce eddy currents. The high-voltage transformer [C] has its terminals insulated to prevent "flashover".

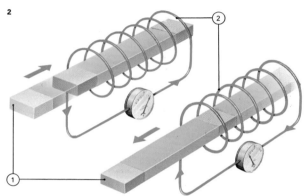

2 Electricity can be generated by a magnet, an electrical conductor and relative movement. Moving the magnet [1] causes the flux lines surrounding it to cut the conductor of the coil [2] and induce a voltage in it – coinciding with the movement; the faster the movement the higher the voltage induced. Opposite movements produce opposite voltages – any current in a circuit between the coil ends flows first one way then the other as the magnet is moved in and out. These are alternating currents and the generators are called alternators.

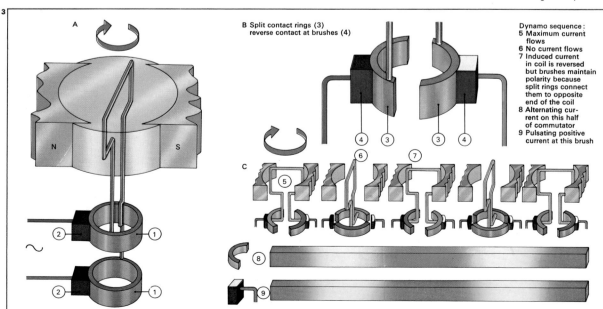

B Split contact rings (3) reverse contact at brushes (4)

Dynamo sequence:
5 Maximum current flows
6 No current flows
7 Induced current in coil is reversed but brushes maintain polarity because split rings connect them to opposite end of the coil
8 Alternating current on this half of commutator
9 Pulsating positive current at this brush

3 Electricity is generated mechanically when a coil is rotated in a magnetic field [A]. An AC voltage is induced in the coil and is connected to the external circuit by contact rings [1] and carbon brushes [2]. A current flows when a circuit is made between the brushes. To produce direct current – as in a cycle dynamo – the generator is modified [B]; the contact rings are split [3] and the halves insulated from each other. A fixed pair of brushes [4] contacts the ring segments in turn. The arrangement of split segments is called a commutator; the sequence [5–9] describes how it works in a cycle dynamo.

itself and generating wasted heat, as a result of the magnetic flux in the machine.

The armature must be supplied with current (through rotating contacts) if it is the rotor in a motor and there must be a way of taking current from it if it is in a generator. The same applies to the field if it is electrically energized. The rotating contact arrangement can be either a set of slip-rings or a commutator [3]. These rotate under fixed contacts, called brushes, made of carbon and held in place by springs. From time to time the brushes have to be replaced when the carbon wears away.

Varying the field

The armature and field of a motor are supplied with current and, by using a resistor in the field circuit, it is possible to vary the field strength. Weakening it causes the motor to revolve faster – provided the armature current is constant – but with less torque (turning force), and vice versa [4]. In a generator driven at constant speed, strengthening the field increases the voltage output; weakening the field decreases it.

The motors and generators already described are generally suitable either for AC (alternating current) or for DC (direct current) supplies. There is a group of machines, however, that is suitable only for alternating current supplies. These are induction motors in which the armature is the rotor and can be wound or made in the form of a "squirrel cage" [5]. It obtains its current from the constantly changing flux gained from the field, which is energized by alternating current. No slip-rings are necessary.

The simplest field consists of two coils. As one becomes a north magnetic pole, the other becomes south, unlike DC machines whose field polarities are constant. The rotor experiences a pulsating and reversing field, but not one that appears to rotate; without a starting force the armature will not rotate either. Given a push in either direction, it begins to follow the field as it alternates.

Some induction motors have a wound rotor and slip-rings may be used. The purpose of the rings is not to supply current to the armature but to alter its characteristics with the use of external resistances.

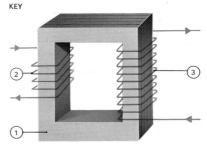

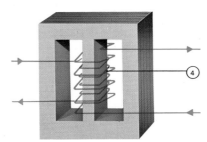

The transformer, a simple and efficient means of raising or lowering AC voltage, consists of three basic elements: an iron core [1], which provides a magnetic link between the primary or input [2] coil and the secondary, output coil [3]. The turns ratio between the input and the output determines the ratio between AC voltages "transformed"; fewer turns on the output side give a proportional decrease in voltage and vice versa, transforming the flow of current. The core may take a number of shapes; sometimes more than two coils are used and sometimes the two coils are wound on each other [4].

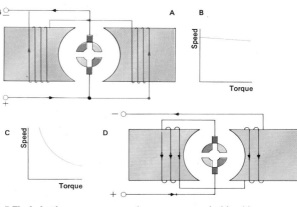

4 Electric motors are similar in principle to generators. Current supplied to the armature or coil, and to the electromagnetic field, causes the armature to rotate. Connecting the field coils and armature in parallel [A] gives an almost constant speed for any torque [B]; connecting them in series [D] produces high torque [C] at low speeds, as in motors [E] for starting electric trains.

5 The induction motor is the one in most widespread use; the coil forming the armature of a simple DC machine is replaced by a "squirrel cage". This consists of aluminium or copper bars connecting each end to a ring, the whole embedded in a laminated soft-iron rotor. The field, at least two coils inside the motor body, is shaped to allow the rotor to revolve inside with a small clearance. Flux lines caused by an alternating current passing through the field cut the cage bars, inducing a current in them – hence the "induction" motor.

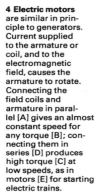

6 The "rotor" of a linear motor moves length-wise rather than revolving. It is a flat plate, either sandwiched between two long field windings or resting solely on one. Energizing the fields with alternating current causes the plate to move in a linear direction, making use of exactly the same principles as an induction motor [A], from which it can be pictured as being "made" by cutting [B] and opening out [C].

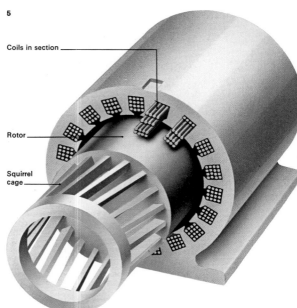

Coils in section

Rotor

Squirrel cage

7 A large linear motor, the field windings of which are shown here, could be used to drive a silent, non-wheeled train or other vehicle. Smaller motors are already used – for example for opening and closing sliding doors. The armature or metal plate, usually aluminium, is fixed to the upper part of the door and the windings attached to the door frame. When the field is energized, the plate passes horizontally along the field, thus moving the door.

Basic DC circuits

An electrical circuit is the system by which an electric current is directed, controlled, modified, switched on or switched off. Circuits can contain from two or three to many hundred different components, according to the way in which the current is to be controlled, but all share certain characteristics.

The formation of a circuit
The primary requirement of a circuit is that it must form a complete path; electrons must be able to flow round the whole system so that as many electrons pass back into the source of the current as leave it. Certain occurrences, such as lightning strikes or electric shocks, seem to deny this first requirement, but are nevertheless examples of electrical circuits. This apparent contradiction can be resolved by considering the earth and all the structures on it as a vast electron bank. If clouds develop an electron imbalance, the earth makes it up with a flash of lightning and the net result is that the numbers of electrons leaving the earth and arriving at it are equal.

Electric currents can also be "carried" by charged atoms, or ions. Ions of dissolved salts

and other chemicals conduct current through the electrolyte in an electroplating bath and gas ions conduct electricity in a fluorescent strip-light. But whatever the current carriers, all circuits share three characteristics: a current (I), a voltage (V) and a resistance (R).

An illustration of electron movement (and the use of high-voltage direct current) is the arrangement of the transmission line from the Cabora Bassa dam in Mozambique to the South African town of Apollo some 500km (310 miles) away. There are two lines to carry the current, one taking electrons to Apollo, the other returning the same number to Cabora Bassa. If one of the lines breaks, the earth itself "replaces" it and carries the electrons in the appropriate direction.

In a similar way, the chassis of vehicles are used as the so-called earth or return circuit, although this is a loose and generally inaccurate term. One terminal of the battery is connected to the bodywork and a single wire is brought from the other terminal through a switch to each piece of electrical equipment. These in turn are also connected firmly to the chassis. The circuit so estab-

lished allows the number of electrons leaving the battery to be matched exactly by the number of those arriving. Using the chassis to complete the circuit in this way makes a second wire to each component unnecessary.

Direct and alternating current
One major practical difference in circuit components (although there are no differences in principle) is determined by whether they are used for direct current or alternating current. Direct current is unidirectional; the electron flow is always in the same direction and although it may stop and start, grow or diminish in quantity, it never reverses.

Current flow (as opposed to electron flow, which is always against the conventional direction of current flow) is assumed by convention to be from a positive to a negative terminal. In direct current (DC) generators, batteries and some other sources, the terminals are determined by the nature of the machine or equipment and are irreversible. The most common example of direct current source is a chemical cell or battery, in which the nature of the chemicals themselves

CONNECTIONS

See also
114 What is an electric current?
122 Transformers, motors and dynamos

1 A cell or battery sends electrons round an electrical circuit as a result of chemical action. This unidirectional character of the electron flow is called direct current (DC), even if it varies greatly or stops altogether from time to time. In a primary cell current flow stops as soon as one of the chemicals or electrodes is consumed. A secondary cell or accumulator can be recharged, often forming these gas bubbles.

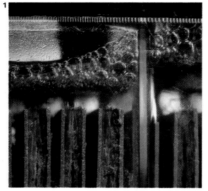

3 A method of examining direct current is to measure how it varies over a fixed period. Graph A shows a current, such as that taken by a lamp supplied by a generator, which does not alter over the time it is measured (the vertical axis represents current and the horizontal one time). Graph B shows a direct current typical of a welding circuit. It varies with time, although its value is always positive.

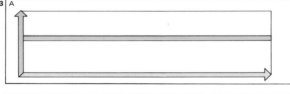

2 Electrical generators can be AC or DC machines. The output of a DC machine does not depend on an exhaustible chemical process, so it usually gives a steadier direct current than chemical sources. Large generators like these supply alternating current (AC), characterized by a rapid periodic reversal of electron flow. In an AC system the current falls to zero every time the direction of its flow is reversed.

4 The drawings used by engineers and other workers in the electrical industry have to be clear to the reader irrespective of his language. These symbols are just a few of the hundreds used.

5 It is often economical to use direct current at high voltage when transmitting large amounts of electrical power. Only in the last few years has reliable equipment been available to switch large DC voltages on and off. This large thyristor "switch" operates on similar principles to transistors used in radios. Because power stations are being sited farther away from population centres, high voltage DC is increasingly used instead of AC.

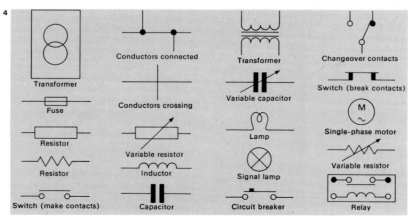

Transformer

Fuse

Resistor

Resistor

Switch (make contacts)

Conductors connected

Conductors crossing

Variable resistor

Inductor

Capacitor

Transformer

Variable capacitor

Lamp

Signal lamp

Circuit breaker

Changeover contacts

Switch (break contacts)

Single-phase motor

Variable resistor

Relay

fixes the polarity of the system. Although the output of the cell may vary, the current flow is always in the same direction. The same applies to a DC generator (dynamo), because the structure of the machine determines the polarity. Another device, known as a rectifier, also has fixed polarity. A rectifier is used to convert alternating current to direct current – irrespective of how the input varies, the current direction at the output terminals is always the same.

Alternating current (AC) is the more common mode, although in certain instances direct current is particularly appropriate and alternating current cannot be used. In electroplating, for example, direct current is used because it is vital that the current always flows in the same direction. If it did not, material would pass back and forth from the coating metal to the coated surface and no plating would take place. The recharging of a battery [1], which is a specialized form of "electroplating", can also be carried out only by direct current and mains-powered battery chargers must contain a rectifier.

In systems for transmitting power over long distances, direct current may be chosen because it requires less insulation and uses fewer and narrower conductors for the amount of power transmitted than does alternating current. Direct current, provided it is steady, uses its conductors fully throughout the whole of the transmission period whereas the conductors of an alternating current are not always fully utilized. As a result, a smaller conductor can be used for the same effective power transmission and if the current is to be transmitted over very long distances the savings may be considerable.

Interrupting the flow
Despite some advantages, the use of direct current is beset by one major problem: switching it off quickly. Interrupting a current that is flowing tends to produce a spark between the contacts providing the interruption. Sometimes this can take the form of a spark so large that it melts not only the contacts but the switching device itself. Alternating current, by its very nature, falls to zero many times a second, so that the spark also falls to zero, limiting the damage it can do.

Electroplating of metals was one of the earliest uses of direct current. The inherent electrical properties of certain chemicals in solution make it possible to coat metallic surfaces with a thin durable coating of another metal. In this automatic plating plant car components are given a protective coating of chromium.

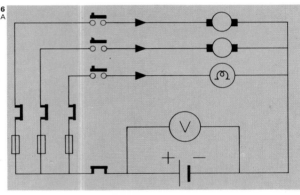

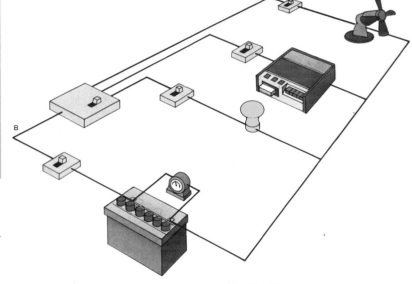

6 This diagram of a circuit [A] shows a typical but simplified electrical system [B] of the kind that is used for equipment powered by a battery. Switches control the current flow into three leads to a lamp and two appliances containing electric motors. The voltmeter records the electromotive force (that is, the voltage) of the battery, which remains virtually constant. The current flowing through the lamp will also remain virtually constant, but those through the appliances may vary, depending on the demands made by the motors in them. Fuses can protect the circuit from current surges.

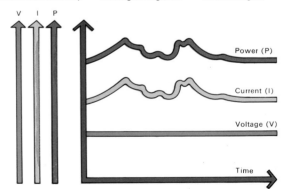

7 The circuit shown in illustration 6 has many interdependent electrical variables. Representing them in the form of a graph indicates how they are related. (There are different scales on the vertical axes but the horizontal axis, time, is the same for all the other variables.) Three of the most important variables – voltage, current and power – are shown. The voltage (V) is fixed by the battery. If the battery is in good condition the voltage will not vary significantly. The current (I) depends on the voltage and the resistance of the appliance it is feeding. This may vary (as it does in the motors) or may be fixed (as it is in the lamp). The power (P) is a product of voltage and current and is measured in watts.

8 Welding is a process of joining metals together with a bond so strong that it is often superior to the parent materials. A large electric current – perhaps 2,000 amps – at low voltage can be used to provide the necessary heat, so that the metal in a particular area is melted. The current passes in the form of an arc from an insulated electrode in the welder's hand and enters the workpiece across a short air gap. The object that is to be welded is connected to one terminal of the current source and the welding electrode to the other. Electric welding generally uses direct current, generated by special equipment because of the very large currents needed.

125

Basic AC circuits

The physical processes that take place in electrical circuits carrying alternating current (AC) differ from those in direct current (DC) circuits, reflecting the differences between the two types of electricity. Alternating current regularly reverses its direction, becoming zero before each reversal (100 times a second in European countries and 120 times a second in North America, corresponding to 50Hz and 60Hz supply frequencies). With reference to zero, the current is negative and positive alternately. Direct current always flows in the same direction.

Current wave and circuit components
The shape of the "current wave" (the curve representing its change of value with respect to time) can take an infinite variety of forms. For most purposes it is sinusoidal (like a sine wave [Key]).

The number of times the curve repeats the whole alternating cycle in a second is called the frequency and is measured in hertz (Hz) – one cycle per second equals 1Hz. A sinusoidal voltage (V) applied to a circuit produces a sinusoidal current whose value at any instant in the cycle is equal to V/Z, where Z is called the impedance (which depends on the resistance, capacitance and inductance of the circuit and the supply frequency); Z is measured in ohms (Ω). The equation is analogous to that used to express Ohm's law: the direct current flowing through a conductor is directly proportional to the electromotive force (voltage) that produces it and inversely proportional to the resistance.

The three main types of circuit components are inductors, capacitors and resistors. A resistor behaves in the same way in either an AC or a DC circuit; inductors and capacitors, however, do not. In these devices, currents are out of phase with the applied voltage in parallel circuits (in which there is more than one path for the current) and voltages are out of phase with the current in series circuits (in which the source and output devices are connected by only one path).

Phase lead and lag
A simple analogy to the phase differences in alternating current and voltage is the action of a yo-yo, where the hand from which the spinning mass derives its energy can move in the opposite direction to that of the mass. The current taken by a capacitor is out of phase with the applied voltage; it is zero when the voltage is maximum and vice versa. Sine waves may be represented by rotating vectors (a vector is a quantity that has both magnitude and direction) and on a vector diagram the capacitor current is 90° out of phase with the voltage and is said to be leading.

For a pure inductance the reverse applies – that is, the current lags the voltage by 90°. This can be explained in another way by saying that for a capacitor the voltage lags the current by 90° and for an inductor the voltage leads. With a resistance, the current and voltage are in phase [3].

In a circuit that has both capacitance and inductance, one current leads by 90° and is equal in magnitude to another lagging by 90°. The overall effect is subtractive – they cancel each other out. When this happens the circuit is said to experience current resonance. In effect, the capacitor's current feeds the inductor and vice versa, because each needs

1 An inductor [A] is a circuit component consisting of a coil of wire. When current flows through a coil a magnetic field is set up whose lines of magnetic flux thread through the coil. Their number and distribution depend on the design of the coil. As the field changes in strength with the changing current, flux lines increase or decrease, cutting the windings of the coil. This is the principle of a generator and an emf (electromotive force) is generated in such a way as to oppose current changes. The effectiveness of an inductor can be changed by screwing a threaded iron core into or out of the coil [B].

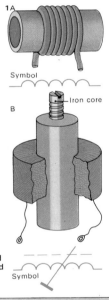

2 A capacitor or condenser [A] is a component used primarily in alternating current circuits (particularly in electronic applications). The two parallel plates become electron deficient or rich as the current varies. The capacitor holds the electrons in balance and releases them at the same rate as the supply current, but out of phase with it. Shown here are (top) a variable condenser consisting of parallel metal plates separated by air gaps, commonly used as the tuning control in a radio set, and an electrolytic condenser (bottom) consisting of a roll of aluminium foil. A large industrial condenser is shown at B.

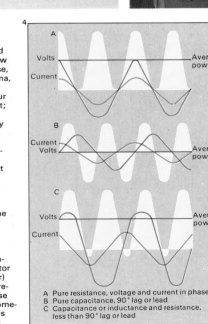

3 In an AC circuit that has only resistance in it [A], the voltage applied and the current flow are exactly in phase, that is, their maxima, minima and zero points always occur at the same instant; this is true no matter how quickly the voltage fluctuates. With only inductance in the circuit, the voltage and current are out of phase. In an inductive circuit [B], current is said to lag the voltage, or the voltage lead the current. In a capacitive circuit, the reverse applies – the current leads the voltage. The inductor and capacitor (unlike the resistor) store energy and release it out of phase with the input – something like flywheels on steam engines.

A Pure resistance, voltage and current in phase
B Pure capacitance, 90° lag or lead
C Capacitance or inductance and resistance, less than 90° lag or lead

4 Power in an AC circuit is the instantaneous product of voltage and current averaged over a fixed period. In a resistive circuit, the voltage and current are in phase [A] and the power dissipated is given by the formula VI (voltage × current). In a capacitive circuit [B] the current and voltage are 90° out of phase and the circuit returns as much power to the source as it absorbs. This also applies to a purely inductive circuit. Circuits that have a mixture of resistance, capacitance and inductance [C] take power according to the formula VI cos ø, where cos ø (the power factor) varies from 0 to 1 depending on the phase angle ø.

the current at different times in the cycle. It is a repeated borrowing and lending action.

The same applies to a circuit in which a 90° leading voltage is of the same magnitude as a 90° lagging one. The two "cancel out" and the circuit is then said to exhibit voltage resonance [7].

The actual value of the current taken by a capacitor or inductor depends on three factors: the voltage, the frequency and the value (size) of the capacitor or inductor. The higher the capacitance the higher the current, but for an inductor the current is smaller as the inductance rises. Current or voltage vectors may be added or subtracted using special mathematics to give a resultant vector, combining all the individual ones. The process is similar to that applied to mechanical vectors. Pulling a barge with ropes at an angle from each side of a canal, for instance, gives a resultant forward motion.

Root mean square values
In an AC circuit the magnitude of the alternating voltage or current is defined as the rms (root mean square, or "effective") value. It is used because the average value is zero, since during any short interval of time the number of half cycles in one direction equals those in the other. The rms value can be derived using simple mathematics and is that quoted in all descriptions of electrical equipment. On an electric iron, a plate reading "230V, 2A" refers to rms values. The voltage and current in this case vary constantly in the form of a sine wave, reaching ±325V and +2.828A – both 50 times every second.

All electrical appliances that are essentially resistive in structure – such as incandescent lamps, heaters and irons – operate perfectly well in either alternating or direct current circuits (provided they are of the same voltage). But equipment that depends on inductive or capacitive properties, such as some motors, transformers and fluorescent lamps, can operate only with alternating current. Alternating current is preferred for the domestic electricity supply because it can be transmitted efficiently and easily from the power station to the domestic consumer and, unlike direct current, is safer when it is switched on and off.

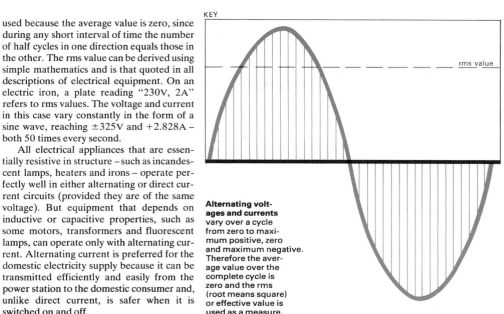

KEY

rms value

Alternating voltages and currents vary over a cycle from zero to maximum positive, zero and maximum negative. Therefore the average value over the complete cycle is zero and the rms (root means square) or effective value is used as a measure.

5 A vector diagram shows the relationship between the three branches of current in a capacitance, resistance and inductance. Three currents and their phase relationships with the applied voltage are depicted, by convention, by lines whose lengths represent the various values of current (they are vectors): A is said to lead, B to be in phase and C to lag behind the voltage.

6
A

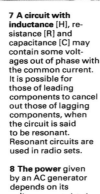

B

6 Electric clocks connected to the AC network depend on the system frequency for keeping accurate time. This is why the kitchen clock [A] always shows the same time as a master clock in the power station [B]. A reduction in frequency makes the clock read slow and vice versa. Because many essential timing devices depend on the mains frequency, it is maintained at a constant value (normally 50Hz in Europe). Any fall in frequency (when demands are heavy) can be gradually made up by adjusting the frequency. The frequency rarely varies by more than 4 per cent.

7

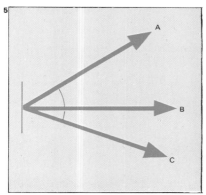

Voltage

Current

Current I

C

Voltage V

H

R

H

7 A circuit with inductance [H], resistance [R] and capacitance [C] may contain some voltages out of phase with the common current. It is possible for those of leading components to cancel out those of lagging components, when the circuit is said to be resonant. Resonant circuits are used in radio sets.

8

8 The power given by an AC generator depends on its voltage, current output and frequency. The power to weight ratio becomes more favourable as the frequency increases, which is why aircraft generators operate at 400Hz. Here an auxiliary generator is being used to power electrical circuits in an aircraft while its engines are idle.

Semiconductors

Metals such as copper and aluminium are good conductors of electricity. Glass, rubber and most plastics are non-conductors, or insulators. But there are some materials, such as germanium and silicon, which are neither good nor bad conductors, and they are called semiconductors. They are used for making transistors and other solid-state devices.

Current carriers
The atoms in a semiconductor easily lose one of their electrons, allowing another from a nearby atom to replace it. Although this electron exchange process goes on, the overall charge of the material is nil; in other words, it is electrically neutral. But by adding some different atoms in the form of a slight impurity, for example with one more electron per atom than the atoms of the material itself, an entirely new material is created. Just one of these new atoms (such as phosphorus, arsenic or antimony) for every one hundred thousand million germanium or silicon atoms can make a semiconductor called an *n*-type material, in which a few extra electrons are available for carrying current.

The opposite situation can be produced, making a material deficient in electrons, by adding atoms with one fewer electron per atom than those of the original material. In this case, aluminium, gallium or indium in the same remarkably small proportion is added to the germanium or silicon to produce *p*-type material [Key]. In both *n*-type and *p*-type materials, the electrons involved in the creation of the particular type of semiconductor are known as the valence electrons – the ones in the outer shell of the atom.

In *n*-type materials, the surplus electrons provide the means for current flow, whereas in *p*-type surplus "holes" are created for the electrons to settle into. And as a hole exerts forces of attraction on the surrounding electrons, it can be thought of as if it were a positively charged particle. The most numerous – electrons or holes – are called majority carriers, or current carriers, in contrast to the minority carriers which are the few residual electrons or holes [1].

The simplest form of semiconductor device, a *p-n* junction diode [3], is made by joining pieces of opposite types of semicon-

ductor material together, attaching a wire to each and enclosing the combination in a metal or plastic shield, with the leads protruding. By connecting a battery, so that the positive terminal is connected to the *n*-type material, a very small current flows consisting of minority carriers only. But if the battery is reversed [4], there is a large current flow, because it consists of majority carriers.

Forming a transistor
When a layer of one type of semiconductor material is sandwiched between two layers of the opposite type, a conventional two-junction, three-layer semiconductor device is formed, known as the junction transistor [5]. This arrangement can be used to form either a *p-n-p* or an *n-p-n* device. Apart from voltage polarities, both can be connected in a circuit to provide current amplifying devices. The voltages in each instance are low – for example, a voltage of 6 volts DC between the collector and base of an *n-p-n* transistor.

When the base/emitter voltage is increased from, say, 600 to 620 millivolts, the collector current might increase from 0.995

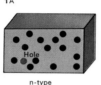

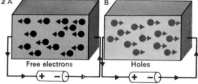

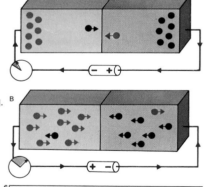

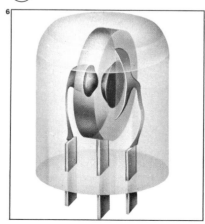

1A　　　**B**

n-type　　　p-type

1 Electrons and holes deliberately introduced into the intrinsic (pure basic) material are seen here to be much more numerous than the carriers (holes [A] or electrons [B]) made by thermal activity.

2A　　　**B**

Free electrons　　　Holes

4 Wiring the terminals of a battery across a piece of joined *n*-type and *p*-type semiconductor material, in which a barrier free from carriers has been produced, simply widens the barrier if the positive pole of the battery is connected to the *n*-type [A]. Little current flows. But if the battery is reversed, forward bias is created together with the breakdown of the barrier and a large current flows [B].

6 Most alloy junction semiconductor devices are of the *p-n-p* type, with germanium as the *n*-type base material in the form of a wafer. A pellet of "impurity" material, such as indium, is placed on each side and heated along with a piece of nickel attached to the base (called the base tab). The indium melts, and after recrystallization, *p*-type areas form. Lead-out wires are connected and the whole device is encapsulated.

2 In an *n*-type material [A], electrons are attracted by the oppositely charged (positive) terminal of the battery. But for the fact that electrons are also flowing from the negative terminal into the other end of the material, it would be left with a net positive charge. With *p*-type material [B] a similar action ensues, but initial attraction is between positive holes and the negative terminal. The invention of the transistor in 1948 revolutionized the world of electronics. Before this no one had ever considered how vitally important semiconductors would become to the electronics industry.

5 A transistor is a sandwich of particular crystal types, produced by taking two pieces of the same semiconductor material and sandwiching the opposite type between them. The crystal structures behave as if the *p*-type is carrying positive electrical charges and the *n*-type carrying negative charges [A]. If a voltage is applied, the charges drift in the directions shown [B]. Amplification of a signal with a negative bias can be achieved [C], the oscillations of the signal influencing the drift of the charges. Also shown is the actual size of the device [D], its symbol [E] and a transistor amplifier circuit [F].

3

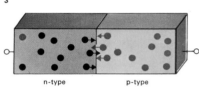

n-type　　　p-type

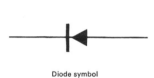

Diode symbol

3 Joining together two different semiconductor materials (*n*-type and *p*-type) causes carriers to start drifting across the junction area. As soon as a few holes and electrons have crossed the junction, they make a thin section of each material oppositely charged from the rest. A barrier which is free from carriers is therefore produced. The combination acts as a semiconductor diode.

5A

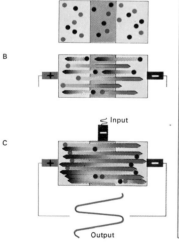

B

C

Input

Output

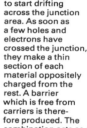

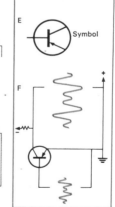

D

E　Symbol

F

7

7 Where more power is needed than would be possible from an ordinary alloy junction transistor, a totally different type of construction is used, with larger pieces of material. The collector is cut and bonded to a piece of material called the header. To enable the device to handle high powers continuously, it is attached to a piece of metal known as a heat sink. With the larger junction transistor more heat is produced and has to be dissipated by the heat sink.

to 1.990 milliamps (mA), whereas the corresponding base current would probably increase only from 0.005 to 0.010mA. Therefore the gain of 0.005mA in the base current has caused a gain of 0.995mA in the collector current, a current amplification of 200 times.

The earliest transistors were of the point contact type, but this method of manufacture was quickly replaced by the alloy junction [6]. It involves the use of heat to form two regions of *p*-type material in an *n*-type germanium wafer. The resulting device is a *p-n-p* semiconductor, although it can handle only low currents. When more power is needed, larger pieces of each material are used to form an alloy junction power transistor [7]. To ensure cool running, a large heat sink is essential, bolted to the metal plate which is bonded to the collector.

Thyristors and photocells

Thyristors [8] are used to control voltage in light dimmers for the home and in electric trucks for railway stations, factories and warehouses. Also known as a silicon con-trolled rectifier (SCR), a thyristor has four layers, three *p-n* junctions and three electrodes – the cathode, anode and gate. When reverse biased (cathode positive relative to the anode), an SCR blocks the flow of reverse current, just like an ordinary diode. When forward biased (cathode negative relative to the anode), an SCR also blocks forward current flow until a trigger signal is applied to the gate. The device then switches to a highly conductive state. It remains in this state until the anode current is reduced below a given maintenance level, or the anode voltage is reversed. When this happens the gate must be triggered again to make the thyristor conduct. A photocell-actuated circuit [9] provides automation for street lamps, making them switch on whenever daylight fades.

Integrated circuits [10] are a development from transistors and are finding increasing applications in all types of electronic apparatus. These circuits incorporate hundreds or even thousands of transistors, resistors, capacitors and interconnecting circuitry, on a single piece of silicon up to 20mm by 20mm (0.78in), and about 0.2mm thick.

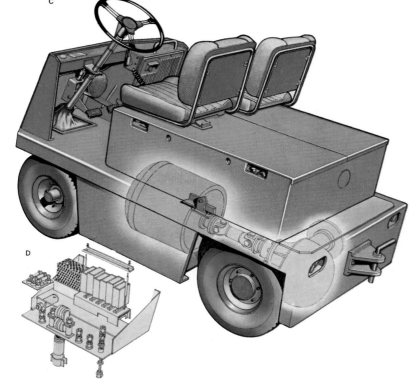

KEY

Electron

○ ● ○ Impurity atom ☼ Hole
○

If a "foreign" or impurity atom with three electrons in its outer shell enters an array of atoms that have four electrons in their outer shells, a "hole" is created in the lattice. This hole acts as an attraction to the surrounding electrons; it can be considered as a virtual positive charge.

8 The thyristor family includes reverse blocking triode thyristors (also known as silicon controlled rectifiers [SCR], or just simply as thyristors), diacs and triacs. A thyristor [A] has four layers, three *p-n* junctions and three electrodes – the cathode [1], the gate [2] and the anode [3]. Thyristors are used to achieve controlled conduction in apparatus such as temperature controllers, motor speed controllers, lamp dimmers, power supplies and inverters. Extremely small thyristors are made to fit within a domestic light dimmer [B], which is based on the action of a thyristor to control the degree of illumination, from zero to full power. Much bigger thyristors capable of handling very high powers are used in layers, or banks [D], in electric vehicles such as this small tractor [C] to give precise control of the speed whatever the load it may have to pull.

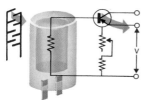

9 A photocell-actuated circuit incorporated into a street lamp [A] turns on the lamp when natural light falls to a pre-set level. At that point, the output voltage falls to trigger another circuit and operate the lamp. A photoconductive cell is superior to a time-switch because no time-switch can be programmed to anticipate fogs or sudden storms. The basic construction of a photoconductive cell is very simple [B] and modern units are most reliable.

10 Integrated circuits make feasible the miniaturization of all kinds of electronic devices. Such circuits are used in this electronic watch [A]. The tiny size of the circuits can be compared with surrounding crystals of common salt [B].

Basic electronic principles

Everything in electronics begins with the electrons that are part of every atom. Scientists have painstakingly built up the modern picture of the atom, but no one has ever seen one because they are so small that even the most powerful electron microscopes have difficulty revealing them. Even smaller are the minute, negatively charged electrons, which can be thought of as orbiting at a distance round the atom's central nucleus where most of its mass is concentrated.

The movement of electrons
Although atoms are normally neutral they can acquire an extra electron and so become negatively charged, or lose an electron and become positively charged. It is this ability of certain atoms easily to "exchange" electrons that enables a stream of them – an electric current – to flow in a conductor [1]. By using a battery, or a generator, a surplus of electrons can be provided at one terminal and a deficit at the other, to produce an electromotive force (emf). If a conductor is connected between these terminals the emf causes electrons to flow (or rather "drift" – the rate is

seldom more than 2cm [0.75in] a minute) from the "surplus" terminal (negative) to the "deficit" terminal (positive). This is opposite to the adopted convention that assumes that electric current flows from positive to negative. Unfortunately this convention was firmly established before anyone knew anything about electrons and has been allowed to remain ever since.

In electronic circuits, conductors (in the form of wires or thin copper strips on an insulating material such as paxolin) act as paths for the free flow of electrons from one part of a circuit to another. But elements are needed to control the flow, to allow precise currents of electrons to pass through various circuit components such as valves and transistors. These elements are known as resistors [2], and are available in a wide range of values from a fraction of an ohm (the unit of resistance) up to tens of millions of ohms.

Valves and their components
The diode valve is the simplest form of vacuum tube [3A] and can change an alternating current (such as that at the mains) into

a series of pulses – direct current – by a process known as rectification. A diode with a single anode produces half-wave rectification [3B]. The efficiency of the process is improved by full-wave rectification [3F]. To obtain a direct current that has virtually no pulses in it at all the pulsating current can be fed into additional circuit elements, such as capacitors and chokes, which "smooth" it.

Nowadays valve diodes are usually rejected in favour of their solid-state semiconductor equivalents. Virtually all modern domestic circuitry is made up of solid-state components, apart from a few television sets that still employ one or two valves as well as semiconductors.

Although the solid-state diode [3C] is much smaller than the vacuum tube equivalent it performs exactly the same rectifying function when used in a similar type of circuit [3D]. But it does not have a filament (heater) and so a semiconductor diode does not consume a large amount of power. It thus makes cooling the equipment unneccessary as well as enabling the size and cost of any associated transformer to be reduced.

CONNECTIONS

See also
114 What is an electric current?
126 Basic AC circuits
128 Semiconductors

In other volumes
102 Man and Machines

1 By applying an electromotive force (emf) to the ends of a wire made of a metal such as copper a flow of electrons can be maintained. Such a current is possible because the atoms of good conducting materials allow their electrons to escape easily under the influence of an emf. And just as easily these atoms pick up other electrons that have drifted away from neighbouring ones.

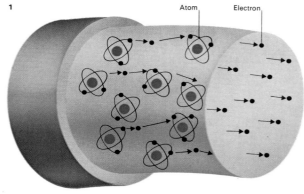

Atom Electron

2 Carbon resistors usually have their resistance value in ohms (ranging up to millions of ohms) marked on them in an internationally recognized colour code. This resistor has a value of 470 ohms with a tolerance (accuracy) of 10%. The key to this simple code is as follows: end ring [A] has no significance; ring B gives the first figure; ring C gives the second figure; ring D gives the numbers of noughts to be added to the digits in B and C. If there is a fourth ring E it indicates tolerance: silver 10%, gold 5%. An absence of a fourth ring would indicate a tolerance of 20%.

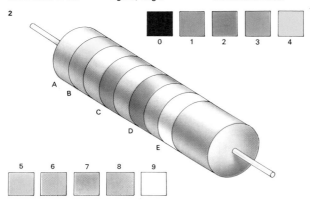

0 1 2 3 4

5 6 7 8 9

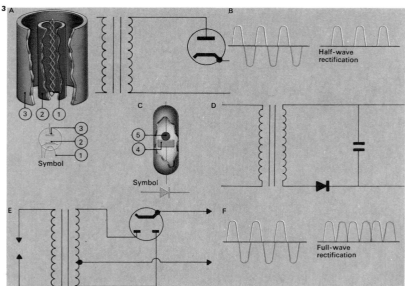

Half-wave rectification

Symbol

Symbol

Full-wave rectification

3 When a voltage is applied [A] to the filament (heater) [1] of a diode valve it causes the cathode [2] to emit electrons that are instantly attracted to the anode [3]. As the anode cannot emit electrons, current flow is possible only in one direction. Therefore the diode valve is ideal as a rectifier of alternating current, either as a half-wave rectifier [B] or a full-wave rectifier [F] using a different circuit with a pair of anodes to produce a "double diode" valve [E]. The solid-state component, known as a semiconductor diode, is more efficient for it does not need a heater. This device [C] has a layer of *p*-type semiconductor material [4] and a bead of *n*-type [5]. This combination has a low resistance in one direction and a high resistance in the other, allowing current to flow in only one direction and therefore producing rectification. A half-wave rectification circuit using such a diode is shown in D.

4 By adding an electrode called a control grid [3] between the cathode [2] and the anode [4] the current in the valve can be controlled by the voltage on the grid. As with the diode valve, a heater [1] is essential to start current flow in the resulting triode valve [A]. In practice, it is found that a small change in grid voltage results in a large change in anode current. By using this effect in the circuit [B] a varying anode current can be converted to a voltage in the resistor R – the result is a signal amplified relative to the input at the control grid [C].

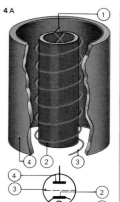

Symbol

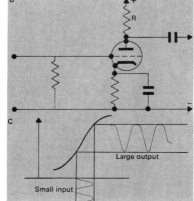

R

Large output

Small input

By adding an extra electrode to the diode valve in 1906 the American inventor Lee De Forest (1873–1961) controlled the flow of electrons between the cathode and the anode [4A]. And by adding other basic circuit elements, in this case resistors and capacitors, the triode valve could be used as a voltage amplifier [4B]. Later, other grids were added to the triode to improve performance, especially in early radio receivers.

The transistor

Following a successful research programme directed by William Bradford Shockley (1910–) at the Bell Telephone Laboratories, Murray Hills, New Jersey, USA, the world of electronics was suddenly presented with the first solid-state three-electrode device, which was destined to end the supremacy of the electronic valve. Shockley, John Bardeen and Walter Houser Brattain were awarded the Nobel prize in physics in 1956 for work on the development of the transistor in 1948.

As a result of this work the world now enjoys the benefits of small, inexpensive

portable radios that run on torch batteries [5]. Inspecting a circuit shows that transistors are even smaller than many of the other conventional components, despite the fact that these are now miniaturized. A transistor consists of a layer of one type of semiconductor material between two layers of a different type. These materials are called p-type and n-type and either of them can form the inner layer. Taking up even less room, an integrated circuit has one or more transistors and other circuit components formed within a single "chip" of semiconductor material.

An interesting feature of practical valve and transistor circuits is the use of "feedback" to improve the quality of the sound produced by an amplifier. This is a refinement to overcome the distortion of the signal that can take place between the input and the output circuits of a valve or transistor or of several such devices. Feedback may be voltage feedback or current feedback.

In small portable transistor radios, there is seldom any output transformer at all because a "single-ended" push-pull output stage is used, reducing cost and weight.

The mysteries of the atom have fascinated many artists. "Theme on Electronics" by the English sculptor Barbara Hepworth (1903–75) was created in 1957 for the Mullard Electronics Centre in London. It symbolizes the world of the electron. Every atom can be pictured with one or more electrons orbiting a central nucleus (with one or more protons and sometimes one or more neutrons as well). The electron always carries a negative charge and the central nucleus a positive one. Normally these charges balance each other out exactly so that neither the negative nor the positive charge predominates – in other words an atom is normally electrically neutral.

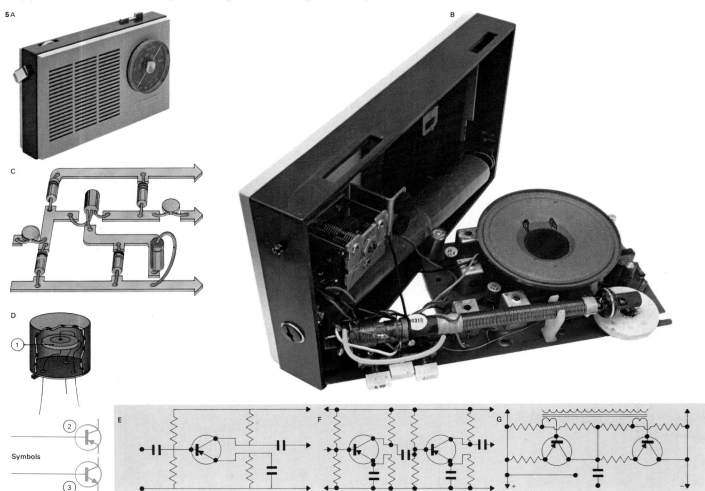

5 A modern portable radio [A] owes its existence to the invention of the transistor in 1948. Without transistors such devices as this one were almost out of the question as far as mass production was concerned because valves, and

most other parts, were too large. Looking inside the set [B] it is possible to see how many transistors and other components can be packed into such a small space and yet leave room for the batteries and loudspeaker. Within a

portable transistor radio there are several types of amplifying circuits. In a typical one-stage transistor amplifier [C], several components are grouped around the transistor, which emphasizes its small size even more.

The upper wires of the various capacitors and resistors are cleverly folded back and joined onto the printed circuit board. Basically, the transistor itself [D] comprises a centre layer called the base [1] of one type of semiconductor

material sandwiched between two other layers of the opposite type of semiconductor material (the collector and the emitter). The two types are n-type and p-type and the composition of the sandwich is denoted by the names of the tran-

sistor types: pnp [2] and npn [3]. The essential difference is in the polarities of the three terminals, and the two types of transistors are distinguished by the directions of the emitter arrows in the symbols. The circuit [E] of the

transistor amplifier [C] is used in a portable radio with a negative feedback circuit [F] for improving sound quality. Where high output is needed a "single-ended", push-pull circuit [G] is used, each amplifying half of the signal.

131

What is chemistry?

Dyestuffs, drugs, synthetic fibres, photographic products, detergents, fertilizers – these are just a few examples of the very many products that have been made through chemistry. But what is chemistry? Every science looks at the world in its own special way; the basic building block with which chemistry is concerned is the atom. Chemistry deals with the properties of different atoms, the ways in which they join together to form molecules, and the interactions of various kinds of molecules with one another.

The stuff of atoms

To the chemist, an atom is made up of three kinds of sub-atomic particles – the proton, neutron and electron. The only difference between a neutron and a proton is that a neutron has no electric charge, whereas a proton has a unit of positive charge. An equal unit of negative charge is carried by the electron, which is much smaller.

In any atom, the protons and neutrons are packed closely together in the central nucleus. Surrounding this, but much less closely packed, are the atom's electrons. The radius

of a neutral atom – that is, one in which there are as many electrons as protons – is about 10,000 times larger than the radius of its nucleus. An atom is composed largely of empty space. Because of this, it is much more likely, when two atoms collide, that their electrons will interact with one another than that the two nuclei will ever come into contact. Consequently, chemists are concerned primarily with the electrons in atoms.

Different kinds of atoms result from the combination of different numbers of protons, neutrons and electrons. The number of protons in an atom is its atomic number, and the total mass of all the sub-atomic particles (protons, neutrons and electrons) is its atomic weight. The simplest atom, that of hydrogen, consists of a single proton and a single electron. If a neutron is added to the nucleus of hydrogen, a different kind of atom, called deuterium, is formed. In many ways, the behaviour of hydrogen and deuterium is the same – as one might expect of two atoms each with only one electron. As a result, the hydrogen atom with no neutron (sometimes called protium) and the

deuterium atom are classed by the chemist as being different isotopes of the same element. There is also a third isotope of the element hydrogen. This is tritium, with two neutrons and one proton in its nucleus. But if a second proton is added to a tritium nucleus and, to balance the electric charge, another electron is placed round this nucleus, the situation is quite different. The atom shows no resemblance in its chemical behaviour to any of the hydrogen isotopes. It is an atom of an entirely different element – helium.

How elements are built up

The element to which any atom belongs is defined by the number of protons in its nucleus. The number of neutrons can usually vary slightly, to give a range of isotopes. The isotopes of an element have different masses but identical chemical properties. There are many stable isotopes in nature, but among the elements with about 90 or more protons, such as uranium, the isotopes tend to be unstable and the nuclei break down to form atoms of other elements [4]. Nuclear reactors and atomic bombs depend on this instability.

1 This iron rosette, dating from the 7th century BC, once decorated an Etruscan chariot. Techniques of working iron and, earlier, copper and bronze go back thousands of years. They allowed men to make tools and implements that advanced both agriculture and warfare. Extraction and purification of metals probably developed accidentally, but led to the founding of present-day chemistry.

2 Ancient Greeks, such as Democritus, believed that matter is made from tiny particles, which they called atoms. But "atomism" did not become a useful part of chemistry until John Dalton (1766–1844) proposed that atoms have different weights, and that the weight of any particular atom is constant. By 1808 Dalton had drawn up this list of symbols for the different types of atoms (elements).

3 John Dalton, the son of a weaver, is one of chemistry's most distinguished figures as a result of his work on the atomic theory. He was born in Cumberland, England, in 1766, into a Quaker family. At the age of 12 he became a schoolteacher. He moved to Manchester in 1793 and stayed there until his death in 1844, earning his living by teaching mathematics and natural philosophy (physics). Dalton was colour-blind and in addition to the atomic theory, which he developed during the first decade of the 1800s, he also investigated colour-blindness (sometimes called Daltonism) and meteorology.

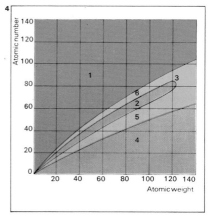

4 The nucleus of an atom is made up of protons and neutrons. Only certain combinations of these sub-atomic particles are stable [2]. Atomic nuclei containing a large excess of protons [1] or of neutrons [4] are both highly unstable. Beyond the stable region is a group of atoms [3] where nuclei break apart to form smaller, more stable nuclei. Below [5], and above [6], the stable regions are unstable nuclei, which decay into stable elements.

5 Light of specific wavelengths is emitted by an element such as hydrogen when it is "excited" by passing an arc discharge through it.

If light of these wavelengths is passed through the vapour of the same element, it is absorbed to form an absorption spectrum.

In the 1800s men wondered why elements should have such discrete spectra, and why the spectral lines could be divided into four classes – called sharp, principal, diffuse and fundamental. Only when modern atomic theory was developed was an explanation found.

As the elements build up, their nuclei are surrounded by more and more electrons which are arranged according to definite rules. The positive charge of the protons in the nucleus attracts the negatively charged electrons, whose momentum prevents them from "falling" into the nucleus.

Electronic orbitals
The electron in a hydrogen atom spreads out round the nucleus in a spherical shell. It is possible to assign to any point only a probability that the electron is there at an instant in time. The region round a hydrogen nucleus where there is the highest probability of finding the electron is the electron's orbital.

In 1925, the German physicist Wolfgang Pauli formulated rules for electronic orbitals. His major rule was that no two electrons in the same atom can be in exactly the same quantum state. The quantum state of an atom is defined by four different numbers. The first of these, known as the principal quantum number, describes the average distance between the electron and the nucleus. The second and third quantum numbers are related to the shape of the orbital, which is not always spherical. The final quantum number is the spin of the electron, which can only be $+\frac{1}{2}$ or $-\frac{1}{2}$. The possible values for the second and third quantum numbers depend on the value of the principal quantum number in such a way that, when it is 1, there are only two possible quantum states (and orbitals); when it is 2, there are 8; when it is 3, there are 18; and so on, according to the formula $2n^2$, where n is the principal quantum number.

The energy of any electron depends upon the first two quantum numbers and, because its behaviour depends largely on its energy, chemists have developed a kind of shorthand for describing an electron's energy level. Each electron in an atom can be described by a number followed by a letter. The number is the same as the principal quantum number, but the values 0, 1, 2 and 3 for the second quantum number are represented (for historical reasons) by the letters s, p, d and f [7]. By knowing the numbers of electrons in different orbitals, a chemist can predict the behaviour of any particular atom.

Flasks, beakers and the simple test-tube have long symbolized the chemist's search for new elements and compounds.

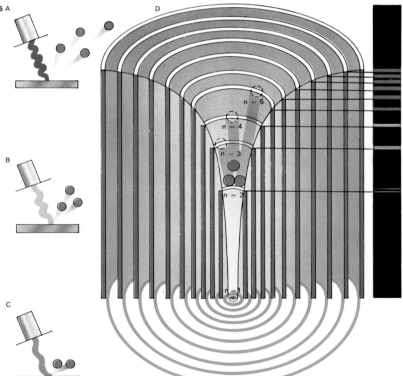

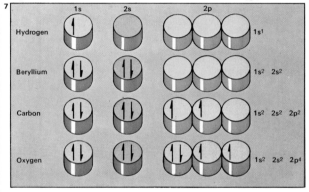

7 The positions of electrons in a neutral atom of any element can be shown in a "pigeon-hole" notation. The orbitals are signified by a number – the principal quantum number that indicates the average distance of the electron from the nucleus – and a letter that indicates the shape of the orbital. The letters used for identification are derived from the initial letters of the four types of spectral lines, s, p, d and f and then proceed alphabetically. Because electrons – shown by half arrows – can spin in opposite directions, each pigeon-hole can hold two electrons.

6 The photoelectric effect was another puzzling phenomenon for 19th-century physicists, Some metals give off electrons when light is shone on them, but the number of the electrons depends on the intensity of the light [A–C] and not its wavelength, and demonstrates the quantum nature of light [D]. Electrons occupy orbitals that have specific energies associated with them and they absorb or emit energy when they "jump" between orbitals.

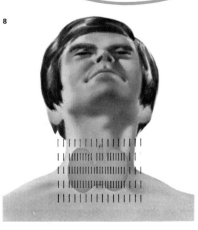

8 The breakdown of unstable isotopes of different elements can be detected and sometimes utilized. Radioactive isotopes (also called radioisotopes) have widespread applications in medicine. For example, a dose of radioactive iodine is taken up preferentially by the thyroid gland in the neck. Its presence can be detected and used to map the gland, to see if it is diseased, cancerous or (as in the one pictured here) enlarged.

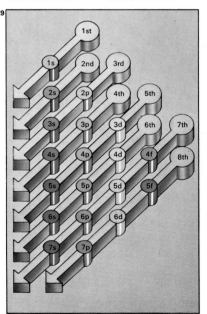

9 Atoms have more and more electrons as they build up to form the heavier elements and the order in which they occupy orbitals depends on the particular binding energy. In general, the closer an electron is to the nucleus, the greater is the energy. As electrons get farther away from the nucleus, the energy relationships become more complex. So that, for example, the 4s orbital is more strongly binding than the 3d orbital, and is filled with electrons before it. The chart shows the general order for the filling of orbitals as atoms get larger. However, even this is an approximation and there are a few exceptions to it.

Classification of elements

Since man first began to purify metals from rocks thousands of years ago, he has been learning how different substances behave and trying to detect a pattern in that behaviour. But the major breakthrough in discovering the pattern of chemistry did not come until just over 100 years ago when the Russian chemist Dmitry Mendeleyev [Key] proposed his periodic system of the elements.

Earlier, the French chemist Antoine Lavoisier (1743–94) had revived Robert Boyle's use of the word "element" for substances that could not be broken down into anything simpler. During the next 75 years, many new elements were discovered [4] and substances previously thought to have been elements were shown to be compounds – two or more elements combined.

Atomic weights
As more elements were discovered, and more of their properties catalogued, it became clear that some elements were similar to others. Sodium and potassium, for example, first isolated in the early 1800s by Humphry Davy (1778–1829), are both soft metals that react violently with water to produce alkaline solutions. It gradually became obvious that there must be a way of tabulating the elements so that those with similar properties were grouped together.

One property of elements that was being catalogued at that time was atomic weight. An atom is extremely small – a toy balloon might easily hold a quadrillion of them (that is 1 followed by 24 zeros). Nevertheless each atom does have a definite mass. Most of this comes from the neutrons and protons in its nucleus. For example, a deuterium atom – with one proton and one neutron in its nucleus – is almost twice as heavy as a hydrogen atom, with its one proton. An oxygen atom, with eight neutrons and eight protons, is about 16 times as heavy as a hydrogen atom.

Using various analytical skills, nineteenth-century chemists gradually catalogued the comparative atomic weights of the elements with increasing accuracy. The weight of a hydrogen atom was formerly taken as 1 and the weights of other atoms related to it; atomic weights are now based on a value of 12 for carbon-12 (six neutrons and six protons), making hydrogen 1.008.

By organizing the elements in tabular form in order of increasing atomic weight, Mendeleyev produced a periodic table. Unlike those proposed by other chemists, his had gaps in it. Where the difference in atomic weight between two neighbouring elements seemed exceptionally large, he assumed that there was an element of intermediate weight that had yet to be discovered. Because his scheme also arranged elements in families, he was able to predict the properties of some of these undiscovered elements [6]. Before the end of the nineteenth century, his assumptions were vindicated by the discovery of some of these "missing" elements and by the close agreement between Mendeleyev's predicted and their actual properties.

Atomic numbers
It is known that the physical basis underlying the periodic classification is not the atomic weights of the elements, but their atomic numbers – that is, the numbers of protons in their nuclei. And the regularities observed in different families of elements result from

CONNECTIONS

See also
132 What is chemistry?
136 Survey of groups of elements
138 Joining atoms

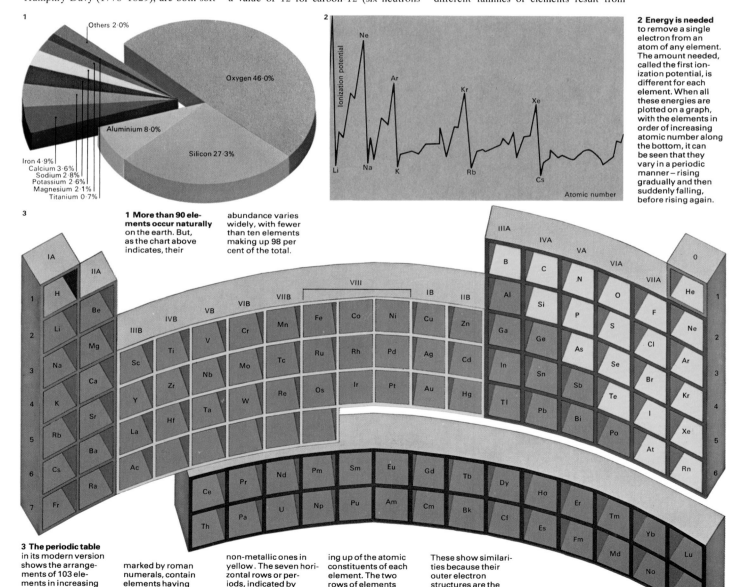

2 Energy is needed to remove a single electron from an atom of any element. The amount needed, called the first ionization potential, is different for each element. When all these energies are plotted on a graph, with the elements in order of increasing atomic number along the bottom, it can be seen that they vary in a periodic manner – rising gradually and then suddenly falling, before rising again.

1 More than 90 elements occur naturally on the earth. But, as the chart above indicates, their abundance varies widely, with fewer than ten elements making up 98 per cent of the total.

Others 2·0%
Oxygen 46·0%
Aluminium 8·0%
Silicon 27·3%
Iron 4·9%
Calcium 3·6%
Sodium 2·8%
Potassium 2·6%
Magnesium 2·1%
Titanium 0·7%

3 The periodic table in its modern version shows the arrangements of 103 elements in increasing atomic number. The vertical columns, or groups, marked by roman numerals, contain elements having similar properties. Metallic elements are in blue boxes, non-metallic ones in yellow. The seven horizontal rows or periods, indicated by arabic numerals at the sides of the table, relate to the building up of the atomic constituents of each element. The two rows of elements shown separately are the lanthanide and actinide series. These show similarities because their outer electron structures are the same; they are built by filling inner orbitals.

similarities in electronic arrangement [5].

Lithium (Li) has three electrons and potassium (K) 19. According to the rules for the occupancy of orbitals (the regions in space where electrons are most likely to be found), lithium has as many electrons as are allowed (two) in the orbital with the principal quantum number 1. It also has one electron left over: this occupies the next lowest energy orbital, called 2s. Potassium, on the other hand, has electrons filling all possible orbitals in the first three levels, and one more. This last electron is in the 3s orbital. Thus, in both elements, the outermost occupied orbital has a single electron in it.

If two atoms have the same number of electrons in the outermost orbital layer, or "shell" as it is usually called, it is reasonable to expect that their chemical properties should be similar. All the elements beneath lithium in a modern periodic table [3] have only one electron in their outermost shell.

Man-made elements

For many years, it was believed that element 92 (uranium) had the heaviest atoms occur-ring naturally on earth. It was believed that as atomic size increases, the atoms become less stable and any atoms of heavier elements that might once have been present on earth had broken down. Since 1940, chemists in the United States and the Soviet Union have been making "transuranium" elements artificially [7]. Glenn Seaborg (1912–), who has been involved with much of this work, has used the periodic table to predict the likely properties of transuranium elements. His predictions go as far as element 168, although, for physical reasons, few of these elements would be stable for long enough to check whether or not their properties did coincide with Seaborg's predictions.

But another of his theories based on the periodic table has been proved. Because of their chemical similarities, members of the same family of elements often occur in the same minerals. Seaborg predicted that, if there were any traces of transuranic elements left on earth, they would be found in minerals rich in other elements of the same family. In 1971, he discovered naturally occurring plutonium in a sample of uranium ore.

The Russian chemist Dmitry Mendeleyev was responsible for proposing the periodic table of the elements in 1869. This major theoretical breakthrough provided the necessary classification system for making the similarities in the properties of certain elements understandable. Mendeleyev was born in Siberia in 1834 and trained as a teacher in St Petersburg. Subsequently he became professor of chemistry at St Petersburg University, a post which he held until 1890. He died in 1907.

4 The elements, their symbols and atomic numbers

Actinium	Ac	89	Hafnium	Hf	72	Promethium	Pm	61
Aluminium	Al	13	Helium	He	2	Protactinium	Pa	91
Americium	Am	95	Holmium	Ho	67	Radium	Ra	88
Antimony	Sb	51	Hydrogen	H	1	Radon	Rn	86
Argon	Ar	18	Indium	In	49	Rhenium	Re	75
Arsenic	As	33	Iodine	I	53	Rhodium	Rh	45
Astatine	At	85	Iridium	Ir	77	Rubidium	Rb	37
Barium	Ba	56	Iron	Fe	26	Ruthenium	Ru	44
Berkelium	Bk	97	Krypton	Kr	36	Samarium	Sm	62
Beryllium	Be	4	Lanthanum	La	57	Scandium	Sc	21
Bismuth	Bi	83	Lawrencium	Lr	103	Selenium	Se	34
Boron	B	5	Lead	Pb	82	Silicon	Si	14
Bromine	Br	35	Lithium	Li	3	Silver	Ag	47
Cadmium	Cd	48	Lutetium	Lu	71	Sodium	Na	11
Caesium	Cs	55	Magnesium	Mg	12	Strontium	Sr	38
Calcium	Ca	20	Manganese	Mn	25	Sulphur	S	16
Californium	Cf	98	Mendelevium	Md	101	Tantalum	Ta	73
Carbon	C	6	Mercury	Hg	80	Technetium	Tc	43
Cerium	Ce	58	Molybdenum	Mo	42	Tellurium	Te	52
Chlorine	Cl	17	Neodymium	Nd	60	Terbium	Tb	65
Chromium	Cr	24	Neon	Ne	10	Thallium	Tl	81
Cobalt	Co	27	Neptunium	Np	93	Thorium	Th	90
Copper	Cu	29	Nickel	Ni	28	Thulium	Tm	69
Curium	Cm	96	Niobium	Nb	41	Tin	Sn	50
Dysprosium	Dy	66	Nitrogen	N	7	Titanium	Ti	22
Einsteinium	Es	99	Nobelium	No	102	Tungsten	W	74
Erbium	Er	68	Osmium	Os	76	Uranium	U	92
Europium	Eu	63	Oxygen	O	8	Vanadium	V	23
Fermium	Fm	100	Palladium	Pd	46	Xenon	Xe	54
Fluorine	F	9	Phosphorus	P	15	Ytterbium	Yb	70
Francium	Fr	87	Platinum	Pt	78	Yttrium	Y	39
Gadolinium	Gd	64	Plutonium	Pu	94	Zinc	Zn	30
Gallium	Ga	31	Polonium	Po	84	Zirconium	Zr	40
Germanium	Ge	32	Potassium	K	19			
Gold	Au	79	Praseodymium	Pr	59			

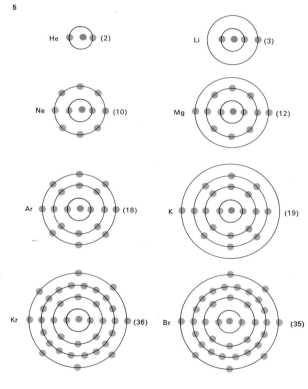

5 The periodicity of the elements is now known to be due to their electronic structures. The electrons in any atom can be envisaged as occupying "shells", each of which can hold only a certain number of electrons. For the first three shells, the numbers are 2, 8 and 18. If the outermost shell is either completely filled, or holds 8 electrons, the element is chemically unreactive: the gases helium (He), neon (Ne), argon (Ar) and krypton (Kr) all fulfil this condition. Lithium (Li) and potassium (K) are similar because they both have one outer electron, and both differ chemically from magnesium (Mg) and bromine (Br).

6 Germanium, a metal "grown" in this form for making transistors, was unknown when Mendeleyev proposed his periodic table. However, in 1871 he predicted the existence of an element with the properties of germanium and called it eka-silicium, from silicium (the old name for silicon) and the Sanskrit word *eka*, meaning "one". Germanium was discovered in 1886, 15 years after he had made his prediction.

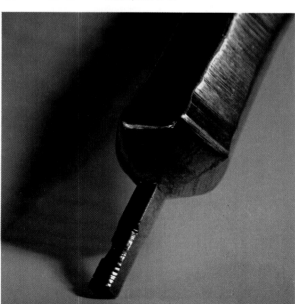

7 The first samples of element 94, plutonium, were produced in 1940 in an American laboratory. One of 11 "transuranium" elements that have now definitely been synthesized, it is used as a fuel in some types of nuclear reactors.

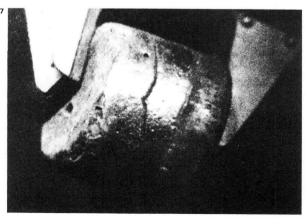

Survey of groups of elements

Atoms of all elements consist of a central nucleus surrounded by a "cloud" containing one or more electrons. The electrons can be thought of as occupying a series of well-defined shells. The behaviour of a particular element depends largely on the number of electrons in its outermost shells. Other factors, such as the total number of electron shells, also play a part in determining behaviour but it is the dominance of the outer electron configuration that underlies the periodic law and justifies the grouping of the elements into groups or families.

The s-elements and their reactions

Each electron shell is made up of various volumes in space called orbitals, known as s-, p-, d- and f-orbitals, and each is at a higher energy than the one below it. Those elements in which the outermost shell can have only one or two electrons can be grouped together as the "s-elements" (because it is only the s-orbital that is occupied in the outer shell). These are (in addition to hydrogen and helium) lithium, beryllium and the elements directly below them in the periodic table [1].

All but helium readily form positive ions by the loss of their outer electrons and they are therefore mostly found as components of ionic compounds, commonly called salts. Many common substances contain these elements – for example, soda (sodium, Na), potash (potassium, K), gypsum (calcium, Ca) and carnallite (magnesium, Mg).

The s-elements with only one electron are more reactive than those with two. Thus if dropped on to water, sodium reacts so violently that it catches fire; if magnesium is dropped into hot water the reaction (release of hydrogen) gives off light, but is less violent.

The energy levels of electrons in the heavier elements are slightly complicated and, for this reason, many elements have a filled s-level but only a partly filled d-level below it. The part of the table ten elements wide beginning with scandium (Sc) includes the "transition" elements in which a d-level is successively filled. The outermost electron shell of these elements has an s-configuration, but it is the underlying layers of d-electrons that determines the element's chemical behaviour. The lanthanide and actinide series form further sub-groups: in their cases an f-group on a lower level begins to fill while the levels beyond contain one d-electron and two s-electrons.

The d-elements: "rare earths" and metals

For the elements scandium, yttrium (Y) and lanthanum (La), as well as the entire 14-element lanthanide series, the chemical behaviour is dominated by the presence of the single d-electron. All these elements tend to form positive ions by the loss of this electron and the two s-electrons – giving ions with a charge of +3. The elements are all fairly reactive and all are rare, but some of these "rare earths" have found commercial uses, for example in the manufacture of specially tinted spectacles.

The other sub-series, the actinides, is of greater importance as it contains the nuclear reactor fuel elements uranium and plutonium. But the importance of these elements is based on their nuclear instability rather than their chemical properties.

The other transition elements [4] are also important because they are all metals and

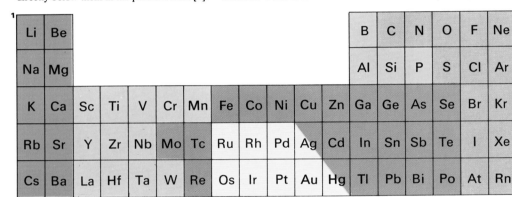

1

Li	Be												B	C	N	O	F	Ne
Na	Mg												Al	Si	P	S	Cl	Ar
K	Ca	Sc	Ti	V	Cr	Mn	Fe	Co	Ni	Cu	Zn		Ga	Ge	As	Se	Br	Kr
Rb	Sr	Y	Zr	Nb	Mo	Tc	Ru	Rh	Pd	Ag	Cd		In	Sn	Sb	Te	I	Xe
Cs	Ba	La	Hf	Ta	W	Re	Os	Ir	Pt	Au	Hg		Tl	Pb	Bi	Po	At	Rn

1 Elements are won from minerals in different ways, according to their position on the periodic table. In the red group are reactive metals usually extracted by electrolysis. In the orange group, the elements frequently occur in ionic compounds, often combined with oxygen. These elements are often prepared by electrolysis. The third group (coloured green) is commonly found as sulphides and the elements are obtained by roasting these and reducing the resultant oxides. Group four (coloured yellow) are relatively unreactive elements, found free or as compounds that give the element when they are heated to a certain temperature. The fifth group (light brown) are non-metals that occur free or as negative ions that can be converted to elements by electrolysis.

2 Magnesium is one of the reactive s-elements. It burns with an intense white flame, which made it useful in early "flash" photography. It is now used in flames that burn under water.

3 Tungsten was discovered in 1783 by the brothers Fausto and Don Juan d'Elhuyar. A fine wire made from tungsten becomes white hot when carrying an electric current, as in an electric lamp.

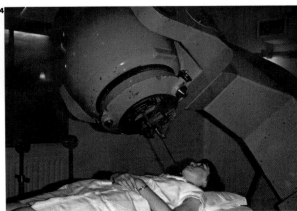

4 Cobalt, one of the transition metals, is the basis of various blue pigments. It is also an important constituent of some biological molecules. However, a most important use in recent years has been in the medical "cobalt bomb". A radioactive isotope of cobalt (cobalt-60), which gives off high-energy gamma-rays, is used to direct this cell-destroying radiation at tumour sites in humans. It is used extensively in hospitals to treat cancers and arrest their growth.

5 A compound exhibits a specific colour when it absorbs white light – a mixture of all colours – by selectively reflecting a few wavelengths. Many salts of transition metals, such as iron and nickel, are coloured. The exact colour depends on which other atoms are associated with the metal. Thus chromium compounds have for centuries been used to provide painters with yellow pigments. Vincent van Gogh's "Sunflowers", shown here, is a good example of their use.

many of them have large-scale industrial uses. All of these elements – except copper, silver and gold [6] – have two electrons in an outer s-orbital and between two and ten electrons in the underlying, but more energetic d-orbital. Silver, copper and gold could be expected to have two s-electrons and nine d-electrons. But, because the complete filling of a d-shell increases stability, they have only one s-electron and a full complement of ten d-electrons.

The transition elements are characterized by the ability to form several different ions, because of the complex behaviour of electrons in the d-orbitals. Thus iron is found in ionic compounds as both Fe^{2+} (ferrous, with two electrons lost) and Fe^{3+} (ferric, with three electrons lost from the atoms).

The p-elements and their grouping
To the right of the transition elements in the periodic table are the p-elements. In the groups headed by boron, carbon, nitrogen, oxygen, fluorine and helium, it is the three p-orbitals (capable of holding a maximum of six electrons) that are the most important in determining chemical behaviour. From left to right through this group the p-orbitals become increasingly filled, until the "completed octet" (two s- and six p-electrons) of the noble gases – the group headed by helium. These elements are unreactive, and only recently have chemists made compounds including them.

The group headed by fluorine – the halogens – all need only one electron to complete the octet and they readily do this to form ions carrying a single negative charge. The oxygen group can form double-negative charged ions, but tend more to link with other atoms through covalent (non-ionic) bonds. The tendency towards covalent bonding is even more marked in the groups headed by nitrogen and carbon. The group headed by boron, where there are two s-electrons and one p-electron, like the group headed by scandium, shows more of a tendency to form ions. Aluminium, for example, readily loses its three outer electrons to form Al^{3+}, but aluminium chloride ($AlCl_3$) is not an ionic compound. The bonds joining the chlorine and aluminium atoms are partly ionic and partly covalent.

Elements such as iron and lead have been known since antiquity.

Others were first purified by alchemists such as Hennig Brand

(died c. 1692), shown praying after discovering phosphorus.

6 Precious metals often occur as the free elements in nature. Not only gold and silver, which for centuries have been used in jewellery and ornamentation, such as this Fabergé egg, but others such as platinum and iridium also come into this category. Silver and mercury are borderline cases. Mercury, for example, occurs as its sulphide but heating releases the metal, which readily alloys with other metals.

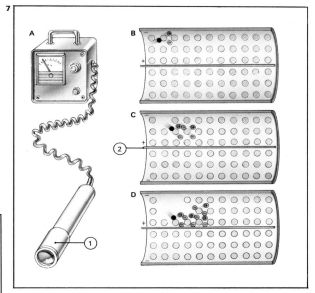

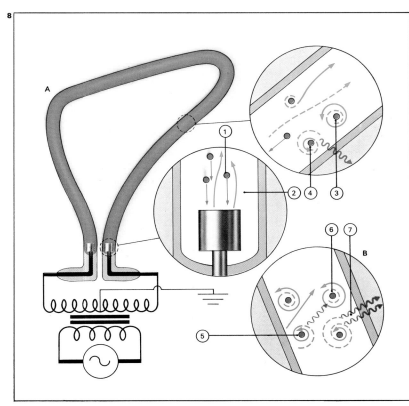

7 A Geiger-Müller counter [A] detects radioactivity in a chamber [1] containing neon atoms at low pressure [B]. The chamber has neon atoms [grey] that are ionized by beta particles [black] to form positive ions [red] and electrons [blue]. [C] An electric field round an anode [2] accelerates electrons, which collide with other neon atoms [D] and split them to release more electrons, to be recorded by the anode.

8 Neon is an unreactive element yet it can be used to produce coloured lights [A]. Positively charged ions [1] strike a negative electrode, causing emission of electrons [2] which "energize" electrons in neon atoms [3]. When the energized electrons return to their stable "ground" state [4], red light is emitted. Similarly, a mercury vapour lamp [B] emits ultraviolet light [5]. This may be absorbed by a fluorescent substance [6] which can then release the energy in stages [7].

9 Marie Curie (1867–1934) was the first person to win two Nobel prizes; one in physics, the other in chemistry. She discovered polonium and radium and was responsible for much of the early research on radioactive elements.

Joining atoms

Human beings and the world in which they live exist because atoms of elements join together to form compounds or molecules [Key]. Such compounds may have as few as two atoms or they may contain thousands linked together. From the basic building blocks – consisting of fewer than 100 elements – natural processes have produced hundreds of thousands of compounds and chemists have synthesized many more.

Every atom consists of a small central nucleus surrounded by a "cloud" containing one or more electrons. When two atoms approach closely their clouds of electrons interact. Each electron has a negative electric charge and for this reason the electron clouds repel one another. But when two atoms are close together the electrons of each, located in space in "orbitals", are attracted by both the atomic nuclei.

Forming a bond
The net result of these dual forces of repulsion and attraction can be a rearrangement of the electron orbitals to form new orbitals that encompass and hold together both nuclei.

When this happens a chemical bond [1] has been formed and a molecule created. The electrons involved in the chemical bond now occupy molecular, not atomic, orbitals.

The rules that apply to electrons occupying atomic orbitals also apply to those in molecular orbitals [4, 5]. Using the normal "pigeon-hole" notation for showing how electrons occupy atomic orbitals it is also possible to show how electrons are distributed in chemical bonds.

In the commonest form of chemical bond, in which a single "pigeon-hole" is involved, the molecular orbital contains two electrons. Such a bond, in which two electrons are shared between two nuclei, is called a single, covalent bond [3]. In some cases an atom may have more than a single electron available for bonding. If it meets up with a similar atom, two (or more) covalent bonds may form between the two nuclei.

The number of electrons in an atom that are available for forming chemical bonds depends on its outer electronic structure. (All the inner electrons are in completely filled atomic orbitals, which generally cannot

accept any more.) For this reason elements such as neon and helium that have completely filled outer shells are highly unreactive and form practically no compounds.

Complete electron shells are very stable structures, and there is a tendency for atoms to borrow or share as many electrons from other atoms as they need to complete a shell. If two oxygen atoms meet, for example, to fill their outer shells they each take a share of two electrons from the other. As a result, each nucleus has eight (instead of six) electrons.

Hydrogen needs – and can accept – one electron to complete its shell (it has only one) [4]. This means that if oxygen and hydrogen atoms come together, two hydrogen atoms each form a single bond to oxygen. In the resultant compound, H_2O (water), the oxygen shares two of its own electrons and has a share in two others, so it again completes its outer octet in the molecule.

Exchanging electrons
When two electrons are shared between atoms the bond is said to be covalent [5, 6]. But some atoms have a stronger affinity for

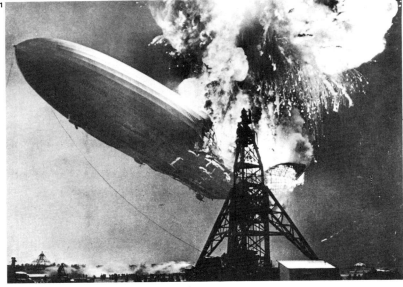

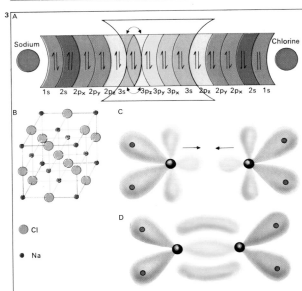

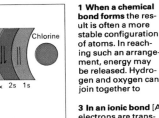

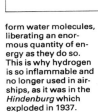

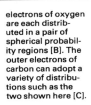

1 When a chemical bond forms the result is often a more stable configuration of atoms. In reaching such an arrangement, energy may be released. Hydrogen and oxygen can join together to form water molecules, liberating an enormous quantity of energy as they do so. This is why hydrogen is so inflammable and no longer used in airships, as it was in the *Hindenburg* which exploded in 1937.

2 The nucleus of any atom has its complement of electrons distributed in orbitals, or volumes of probability. For hydrogen's single electron the orbital is spherical [A] while the two available electrons of oxygen are each distributed in a pair of spherical probability regions [B]. The outer electrons of carbon can adopt a variety of distributions such as the two shown here [C].

3 In an ionic bond [A] electrons are transferred, eg from the 3s orbital of sodium (Na) to the 3p orbital of chlorine (Cl). Ionic compounds often have distinct geometries in which ionic charges balance [B]. But in a covalent bond two electrons in a molecular orbital are shared by both nuclei. Such orbitals have specific shapes as in the "double bond" (four electrons) linking the carbon atoms in ethylene [C, D]. The other carbon orbitals are bound to hydrogen.

4 The simplest molecule, H_2, is made up of two covalently bonded hydrogen atoms. The two nuclei share each other's single 1s electrons [A]. As two hydrogen nuclei come together [B], and electron sharing commences, energy is released. The positively charged nuclei repel one another and partly counteract the binding force of the electrons. This keeps the nuclei a roughly constant distance apart – called the bond length. Such two-atom molecules are called diatomic.

electrons than others and completely gain or lose electrons to combine by means of an ionic bond. The resulting atoms are no longer electrically neutral; they carry a positive or negative electric charge (depending on whether they have lost or gained electrons) and are known as ions. Chlorine, for example, with seven electrons in its outer shell needs only one more to achieve the stability of a completely filled shell (with eight electrons). When it gains this electron it becomes a chloride ion (Cl^-) and carries a negative charge. Sodium, by contrast, has only one electron in its outer shell and as a result readily loses this to form a positively charged sodium ion (Na^+).

Sodium chloride, or common salt, is an example of a compound with an ionic bond. When it is dissolved in water the chloride ions each carry an extra electron borrowed completely – not shared – from the sodium ions.

The co-ordinate bond
A third possibility exists for joining atoms in a simple way. This is an extension of the sharing arrangement of covalent bonding.

But in this case both electrons in the bond, called a co-ordinate bond, come from a single atom and use a totally empty orbital in the second atom involved in the bond. Nitrogen, for example, has five electrons in its outermost shell. Three of these may be involved in covalent bonds with three hydrogen atoms, to form the compound ammonia, NH_3. Nitrogen now has shares in eight electrons – a complete outer octet. In a molecule of ammonia, however, there remains a pair of electrons in the nitrogen atom that have not formed covalent bonds with hydrogen atoms. These can be donated to an empty orbital of various metal atoms, such as copper, to form a third type of bond called a co-ordinate bond. The resulting compounds are known as complexes or co-ordination compounds.

When more than two atoms are involved in joining together, more complicated arrangements are possible. Molecular orbitals may spread out over more than two nuclei to give compounds in which the average number of electrons per bond is less than two, but in which the molecules nevertheless hold together.

Atoms join together to form molecules. Only since the discovery of the wave-nature of matter in the 1930s has it been possible to describe this bonding of atoms clearly. Before this time the way in which atoms held together was a mystery, often shown fancifully, as in this "model" of benzene. Today benzene is pictured as a hexagonal ring of six carbon atoms all linked together by equivalent bonds.

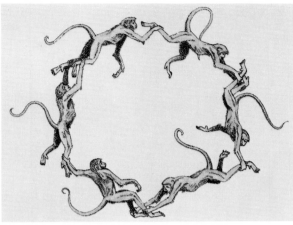

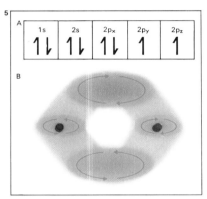

5 The "pigeon-hole" notation for the electron structure of an oxygen atom [A] shows that it has two partially filled orbitals ($2p_y$, $2p_z$). To complete its outer octet it therefore needs to take over, or share, two electrons from other atoms. In nature oxygen is usually found as a diatomic molecule (O_2) in which two oxygen atoms (red) share a total of four electrons – enough to form two covalent bonds [B].

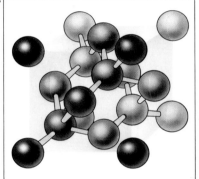

6 In a diamond any one carbon atom is covalently bonded to four others, orientated at the corners of a tetrahedron. This strain-free covalent configuration accounts for the diamond's hardness.

7 By mimicking the strain-free structure of diamonds chemists have been able to "engineer" even tougher compounds, such as the tungsten carbide on the tips of drills used in oil wells.

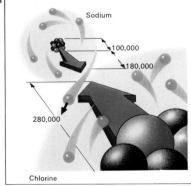

8 A sodium atom here appears as it would look seen from the nearest chlorine atom in a crystal of sodium chloride. One electron is shown moving from the sodium to the chlorine, resulting in an ionic bond between them. The dimensions of the sodium nucleus (made up of protons [red] and neutrons) and of the two complete atoms are given in femtometres – a unit equal to a million-millionth of a millimetre (10^{-15}m).

Sodium

100,000

180,000

280,000

Chlorine

9 When two atoms spontaneously form a bond the release of energy stabilizes the compound. The process may be reversed by supplying energy. Colourless silver salts used in automatically dimming sunglasses (A) and an astronaut's helmet (B) take the form of silver ions in a glass matrix. In sunlight the glass darkens because light energy breaks the ionic bonds and reconverts the silver into metallic atoms. Colourless ions re-form when the light is removed.

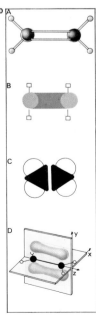

10 Chemical bonding can be illustrated in many ways. The four diagrams shown here depict the simple molecule ethylene, made up from two carbon atoms and four hydrogen atoms. As in the diamond structure, each carbon atom can form four bonds: two of these link the two carbons together and each carbon also has two hydrogens attached to it by single covalent bonds. The simple ball-and-stick model A shows the atoms and the number of bonds linking them. Model B gives similar information but is designed specially for biological molecules while models C and D reveal more about the actual molecular shape, and the distribution of orbitals, respectively.

Simple chemicals and their structures

Mankind's interest in chemistry derives from the useful information that the subject can provide about the properties of different substances. By understanding the structure of molecules and the way they interact with one another it is possible to invent new compounds that are useful, for example, as drugs, or building materials or fibres for clothing. It is also possible to acquire a better understanding of how the earth on which we live came to its present form and how it is still developing. Geochemistry (as this sub-branch of the subject is called) can lead to the discovery of new supplies of fuels and ores and new ways of processing them.

Structures of molecules

All substances are made up of molecules that, in turn, are composed of individual atoms. One of the basic aspects of chemical knowledge is an understanding of the structure of molecules. An atom is a relatively insubstantial entity: a small, hard nucleus surrounded by an electron occupying a volume of probability. Molecules composed of such atoms are similarly made up, in terms of volume, largely

of electron orbitals (the areas in space occupied by the electrons). Nevertheless, despite the fact that an orbital can contain a maximum of only two electrons, they often have definite directions in space, so that molecules have particular shapes. In complex molecules such shapes may be crucial to the behaviour of the substance. This is particularly true of biological molecules, which contain thousands of individual atoms linked together, but even simple molecules have shapes that can determine their properties.

The structures of ionic compounds [1] depend on the electric charge of the ion and on its ionic volume. An ion of sodium, for example, can be regarded as a small sphere having a particular diameter. In general, provided ions of different elements have the same electric charge and similar radii, they can substitute for one another in different materials. As the earth settled down after its formation, rocks gradually solidified from molten material. Many of these original rocks contain crystalline compounds but the compounds are nearly always impure. For example, the common mineral potassium

felspar, an ionic compound containing potassium, silicon and oxygen, is always contaminated with rubidium because rubidium forms a singly charged ion of approximately the same radius as potassium. Germanium forms a similarly charged, similarly sized ion to silicon so it is also a contaminant of felspars.

In covalent (non-ionic) compounds the bonds between the different atoms are separate. Covalent compounds can still form crystals, but this is because the molecules as a whole can arrange themselves in geometric arrays, whereas with ionic compounds it is the geometric arrangement of ions that leads to crystallinity. With a crystal of a covalently bonded compound each component molecule has its own characteristic shape.

Shapes of covalent compounds

The shape of a covalently bonded molecule [3] depends on the shapes of the orbitals occupied by the electrons – both those involved in chemical bonding and any others in the outermost shells of the individual atoms. A water molecule, for example, in which two hydrogen atoms are each singly

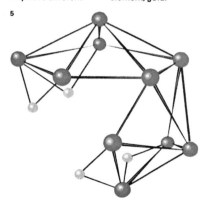

1 Sodium chloride, common salt, has a simple cubic crystal structure in which each sodium ion (Na⁺) is surrounded by six chloride ions (Cl⁻) and vice versa. Around each ion, the six ions of opposite electrical charge are situated at the corners of an imaginary octahedron. The main force holding the ions in place is the balanced electrical attraction of the neighbouring ions.

Sodium ion   Chloride ion

2 The shape of any crystalline, ionic solid depends on the size and the number of ions that make it up. However, different chemical substances may look alike, even though they have different compositions, molecular structures and chemical properties. Iron sulphide, shown here, is called "fools' gold" because of its resemblance to the much more valuable element, gold.

A Nitrogen dioxide NO_2
B Carbon dioxide CO_2
C Phosgene $COCl_2$
D Sulphur dioxide SO_2
E Water H_2O
F Ammonia NH_3

3 Covalently bonded molecules have individual molecular orbitals holding different atoms together, unlike the balanced geometric cluster of an ionic compound. Because of the forces of repulsion between the electron clouds surrounding adjacent nuclei, even small molecules have distinct shapes, which may differ quite markedly from one another as in the half-dozen common molecules here.

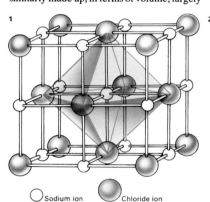

4 In hydrogen peroxide two hydrogen atoms and two oxygen atoms are joined in a single molecule. It readily releases an OH group whose oxygen can, for example, bleach hair (as here) or kill germs and act as a mild disinfectant.

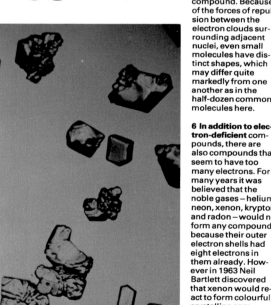

5 In most covalent bonds two electrons hold two nuclei together. By adopting tightly geometric patterns electron-deficient compounds, such as this decaborane, can hold together with a smaller number of electrons.

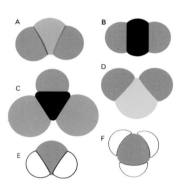

6 In addition to electron-deficient compounds, there are also compounds that seem to have too many electrons. For many years it was believed that the noble gases – helium, neon, xenon, krypton and radon – would not form any compounds because their outer electron shells had eight electrons in them already. However in 1963 Neil Bartlett discovered that xenon would react to form colourful crystalline compounds. An extension of chemical bonding theory has shown how the formation of such compounds does not violate any chemical law and since 1963 a large number of compounds of xenon (shown here) and krypton have been made in laboratories throughout the world.

bonded to a central oxygen atom, might be visualized as the three atoms joined in a straight line. Electrons, all being negatively charged, repel each other, so this in-line arrangement might seem best. It would mean that the volumes of electron probability around the two hydrogen atoms are as far apart from each other as they possibly could be, thus reducing repulsion to a minimum. But, in addition to the two electrons from the oxygen's outer shell, which are involved in bonding, there are four other electrons in this shell, situated as two "lone pairs" in filled orbitals. Their effect on neighbouring electron clouds has to be taken into account. When this is done it is found that a shape almost like that of one segment of a diamond [3E] is adopted. It is a tetrahedral structure in which the hydrogen atoms and the two pairs of oxygen electrons (not involved in bonding) all lie as far away from each other as possible.

Methane [8], in which one carbon atom is surrounded by four hydrogen atoms and all the outermost carbon electrons are involved in bonding, adopts a regular tetrahedral configuration. Intermediate in structure be-tween methane and water is ammonia (NH_3), in which there are three nitrogen-hydrogen bonds and one filled orbital containing a lone pair of nitrogen electrons [3F].

Co-ordination compounds

The lone-pair electrons in water and ammonia molecules sometimes form bonds to metal atoms that have empty orbitals. Anhydrous copper sulphate, for example, is white. When water is present the molecule turns blue as a number of water molecules "co-ordinate" to empty copper orbitals through the unbonded oxygen lone pairs. Because the empty orbitals in such metals have definite shapes most "co-ordination compounds" generally have highly geometric structures [7].

Some atoms can form several bonds with other atoms, so it is often possible for a number of different molecules to be made up from the same mixture of atoms [10]. One of the major advances in chemistry since the 1940s has been the development of sophisticated instruments that analyse exactly how the atoms of a compound are linked together.

KEY

Washing soda (Na_2CO_3)

Baking soda ($NaHCO_3$)

Vinegar (containing CH_3COOH)

Common salt (NaCl)

Epsom salt ($MgSO_4$)

Everything around us is made up of chemicals, most of them complex mixtures. But these familiar substances from the kitchen are comparatively simple chemical compounds.

7 Where a central atom in a molecule can bond to a number of other atoms, or groups of atoms, a variety of different molecular geometries is possible. In the particular case of co-ordination compounds, in which the central atom is sharing electrons from other atoms and not contributing any of its own electrons to the bonding, the commonly found geometries are those shown here. Whether a compound made up of, say, a central cobalt atom and four outer atoms adopts a square planar or a tetrahedral form depends on the influence of non-bonding electrons associated with the different atoms involved.

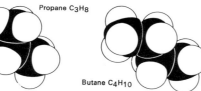

7

Square planar

Octahedral

Tetrahedral

Tetragonal pyramidal

Trigonal bipyramidal

8

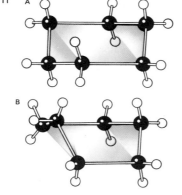

Methane CH_4

Ethane C_2H_6

Propane C_3H_8

Butane C_4H_{10}

8 Carbon atoms can combine covalently to form long chains. Consequently carbon compounds occur in series such as the alkanes (paraffins); the first four (commencing with methane) are shown here.

9

Ethylene C_2H_4

Acetylene C_2H_2

9 As well as forming chains joined by single bonds carbon atoms can be joined to one another by two or three molecular orbitals: ethylene and acetylene are, respectively, the doubly and triply bonded analogues of ethane, all of which contain two carbon atoms.

10

A

Carbon

Carbon

Oxygen

OOO
OOO Six hydrogens

B

10 Where the same numbers of the same kinds of atoms are joined to one another in different ways the resultant compounds behave differently. Because the gross structures of the molecules are different they are called structural isomers; examples are dimethyl ether [A] and ethyl alcohol [B].

11 Some compounds such as cyclohexane, can exist in two forms differing only in the way their chemical bonds are arranged in space. This illustration shows the "chair" [A] and "boat" [B] forms of cyclohexane. Such compounds are known as conformations: they may "flip" from one form to the other, and a sample of the compound may contain both.

11 A

B

12 During the 20th century hundreds of thousands of carbon compounds hitherto unknown in nature have been synthesized in chemical laboratories. Some of the simplest such molecules of commercial significance are insecticides. Developed mainly to protect food crops, they are also a help to the home gardener. The active ingredients in many insecticide sprays and powders – dichlorvos and BHC for example – are molecules containing only one or two dozen atoms.

12

141

Complex chemicals and their structures

Carbon atoms have the ability to link together in large numbers to give an infinite variety of different substances [2]. Most complex chemicals are carbon-based although there are some important complex materials, such as glass, that contain no carbon.

At one time it was believed that most carbon compounds could be made only by living processes. For instance urea was discovered in the urine of mammals. Consequently, this and similar compounds were called organic chemicals, a name that has stuck. Then organic chemicals were synthesized in the laboratory – for example urea (organic) was made from ammonium cyanate (an inorganic compound). Today organic chemicals account for 50 per cent of the total output of the chemical industry.

The basis of an organic chemicals industry
The discoveries that led to the development of an organic chemicals industry came towards the end of the nineteenth century [3]. It was found that a wealth of useful substances could be obtained from coal tar, a by-product of the manufacture of domestic gas

from coal. At that time many of the substances extracted from coal tar were too complex in structure to be made in the laboratory. Once purified, however, it was possible to use them as starting materials from which to produce a range of commercial substances. Dyes, aspirin, saccharin, and explosives such as TNT were all made before the end of the nineteenth century from coal-tar chemicals.

A basic constituent of many of the coal-tar products, and of many other complex organic chemicals, is the group formed of a ring of six carbon atoms joined together. As in many other molecules, the chemical bonds between the carbon atoms take the form of electron-carrying orbitals. These orbitals are not located strictly between adjacent carbon nuclei but spread over all six, so that each carbon atom is effectively joined to the next by one-and-a-half bonds.

The simplest of such compounds, in which each carbon atom is also linked to a hydrogen atom (giving a formula C_6H_6), is benzene – originally a coal-tar product but now made mainly from petroleum [1]. In some compounds several benzene rings are fused

together to give what are called polycyclic structures such as naphthalene (used in mothballs) benzpyrene (a cancer-inducing chemical) and the hallucinogenic drug LSD.

Many complex organic chemicals found in nature do not contain benzene-type rings. Instead, they are made up of long chains of carbon atoms, with other atoms attached. The other atoms nearly always include hydrogen and often oxygen and nitrogen as well. These compounds include useful natural products such as fats, waxes, sugars and proteins.

Synthetic polymers and their products
Organic molecules have been the basis of one of the major industrial developments of the twentieth century – the widespread manufacture and use of synthetic polymers as plastics, rubbers and fibres. Polymer is a general term for any large molecule that is made by repeatedly linking together the same small molecular unit, which is called the monomer.

Most organic polymers are formed either by addition reactions or by elimination reactions. In the first type the monomer molecule has a double bond and links with others by

1 The six-carbon benzene ring is one of the most important molecules in organic chemistry. Each carbon atom is attached to a single hydrogen atom, so it can form three other bonds with its two neighbouring carbon atoms. It was once thought that single and double bonds alternated around the benzene ring [A, B], although if this were so not all the carbon atoms would have the same chemical reactivity. It is now known that all the carbon-carbon bonds in the ring are equal, because the molecular orbitals (blue [C]) spread out over the whole ring. Below, [D] is a more convenient representation of the actual shape of the molecule.

2 In benzene other atomic groups can replace hydrogen atoms. In aniline [A] one hydrogen is replaced by an amino (—NH₂) group, while phenol [B] has a hydroxyl (—OH) group in place of one hydrogen. Styrene [C], from which polystyrene is made, has a small carbon chain attached to the benzene ring. More than one hydrogen can be replaced: catechol [D], for example, has two —OH groups, while aspirin [E] has two different groups attached to the ring. The shared orbital structure of benzene is sometimes retained even when carbon atoms are replaced, as in pyridine [F].

3 The English chemist William Perkin (1838–1907) tried to make quinine from aniline. Instead his experiments of 1856 accidentally produced the dye aniline purple. This was the first synthetic dye: until that time all dyes were natural compounds extracted from plants or animals.

4 Modern pigments brighten this train with contrasting colours, and paint makers are no longer dependent on natural substances for their products.

5 The chemical heart stimulant digitalin was first obtained by herbalists from leaves of the foxglove (*Digitalis purpurea*).

using electrons from the double bond. In the second type atoms at the ends of two monomer molecules are "eliminated" and a bond forms between the remaining parts of the monomer molecules.

Natural rubber is an addition polymer, as are many of the man-made rubbers developed in the past few decades. Many synthetic fibres on the other hand are elimination polymers or condensation polymers (in which water is eliminated), as are their natural counterparts such as wool and cotton. A synthetic polymer can have characteristics built into it by the careful choice of starting materials. As a result, different synthetic fibres can have widely differing properties.

Plastics [6] can be fibrous, rubbery, clear, hard, opaque or flexible. The possibilities are almost endless, as the widespread use of synthetic polymers in everyday life confirms.

Linking silicon atoms
Silicon falls directly below carbon in the periodic table of the elements and so logically should have similar chemical properties. But because of their larger size silicon atoms cannot link to form long chains by themselves. Nevertheless, the important polymers called silicones are based on long chains of alternating atoms of silicon and oxygen. The commercially available silicones are often partly organic, and their forms range from plastics for replacement parts such as artificial heart valves in surgery, through greases and lubricating fluids, to insulating materials for electrical cables in submarines.

In silicones, carbon groups are linked to the silicon atoms and stick out on each side of the polymer chain. Like carbon, silicon can form four bonds, whereas oxygen forms only two. This means that, in a repeating —Si—O— chain, there must be other atoms joined to the silicon atoms. These do not have to be organic; each silicon atom can be bonded to four oxygen atoms. If each of these oxygens has two silicons attached, the result is a three-dimensional polymer matrix with the overall formula SiO_2. Sand and glass are composed largely of this inorganic polymer [7], which is often known as silica, and it is also used in ceramics and paint manufacture.

A modern chemist deals in hundreds of drugs. The old chemist, called an apothecary, displayed in his shop rows of glass jars filled with tinctures and essences extracted from botanical plants. Many balms and medicines made from the contents of such jars are still manufactured today, not from natural extracts but from substances that are produced in chemical plants. Increasing knowledge of the complex molecules found in nature has enabled man to copy and in some cases improve nature's own efforts. Synthetic drugs, rubber and pigments are only a few of the products that man's ingenuity has created to replace natural materials.

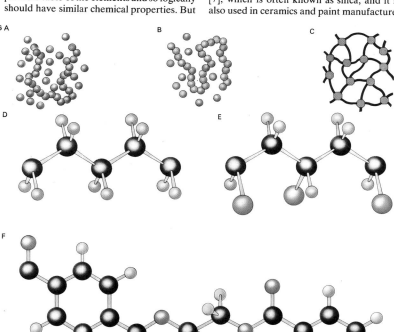

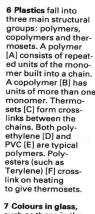

6 Plastics fall into three main structural groups: polymers, copolymers and thermosets. A polymer [A] consists of repeated units of the monomer built into a chain. A copolymer [B] has units of more than one monomer. Thermosets [C] form cross-links between the chains. Both polyethylene [D] and PVC [E] are typical polymers. Polyesters (such as Terylene) [F] cross-link on heating to give thermosets.

7 Colours in glass, such as those in the windows of Coventry Cathedral, come from added metallic compounds. Cross-linked polymers are not all based on organic molecules and glass is probably the oldest synthetic polymer.

Lead
Carbon
Oxygen
Hydrogen
Nitrogen

8 Modern plastics are replacing metals in many applications, such as the bodywork of motor vehicles.

9 Metal-organic chemistry bridges the gap between organic and inorganic chemistry. It can lead to important new products, for example, poison antidotes. A chelate, such as EDTA (containing carbon, hydrogen, oxygen and nitrogen atoms) can surround ions of metals and remove them from unwanted places; it may be used, for instance, to treat lead poisoning.

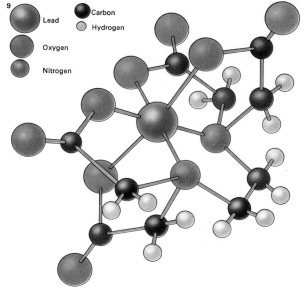

Chemicals in solution

Every day millions of people make solutions. Many start at breakfast, when they dissolve sugar in tea or coffee. So common is this action that it is taken for granted. But where does the sugar go when it dissolves? Why, if a spoonful of sand is stirred into hot liquid, does it not dissolve?

In chemical terms, a solution is a homogeneous mixture of different sorts of molecules. The criterion of homogeneity is that the two or more types of molecules involved are thoroughly mixed up with one another – as water and sand clearly are not.

Solutes and solvents
Solutions are usually thought of as solids dissolved in liquids – coffee in water, sugar in the coffee, salt in water in the sea [6], detergent in washing water, iodine in alcohol in tincture of iodine, and many more. But there are other kinds of solutions. Gases can dissolve in liquids, as in soda water. Many solutions are made by dissolving liquids in other liquids. Gases can also dissolve in some solids, and solutions of solids in solids are found in metal alloys, for example.

For a solution to form, there must be an interaction between the substance which dissolves (solute) and the material it dissolves in (solvent). Sugar, for example, usually occurs as crystalline arrays of sucrose molecules. To dissolve it, energy is needed to break apart the crystal lattice so that sucrose molecules can disperse evenly through the solvent. Where water is the solvent, the attraction for solute molecules comes from the "polar" character of the water molecule. The central oxygen atom in H_2O is electrically slightly negative, while the hydrogen atoms are slightly positive. Molecules of water tend to attract one another – which is why water is a liquid at room temperature whereas most such small molecules are gases [1]. Water molecules also tend to attract (ie dissolve) other "polar" compounds, such as sugar, with many OH groups in their molecules.

Molecules and compounds
For similar reasons, but via weaker attractive forces, non-polar molecules such as hydrocarbons will dissolve other non-polar compounds, such as fats. Modern detergents work by a compromise between both types of attraction: part of the molecule dissolves in grease, the other part dissolves in water, so that the detergent molecule acts as a bridge between them and disperses grease in water.

Some compounds, such as ethyl alcohol, dissolve completely in either water or hydrocarbons – substances that will not dissolve in one another. Other compounds may show a preference for polar or non-polar solvents, according to their chemical structures.

The attraction between individual molecules of different types can be seen easily where it results in a reduction of the overall volume of material. If, for example, equal parts of water and ethyl alcohol are mixed together, the total volume of solution is only about 97 per cent of the sum of the separate components.

When a substance will not dissolve, it is because the solvent does not overcome the intermolecular forces which hold the molecules together. With any substance, there are limits to solubility, which may vary with temperature – hot solvents dissolving more solute than cold solvents.

CONNECTIONS

See also
136 Survey of groups of elements
146 Key chemical reactions
148 Electrochemistry
140 Simple chemicals and their structures

In other volumes
26 The Natural World

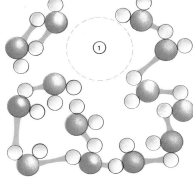

1 **The atoms and molecules** of all materials are in motion, the energy of this motion depending on the temperature. In a liquid, the motion prevents any permanent intermolecular structure from forming, but forces of attraction govern the overall volume. In water, there are many temporary linkages (shown blue) between the H_2O molecules; very small enclosed cavities [1] form and disappear continually.

2 **The slight positive charge** on the hydrogen atoms and slight negative charge on the oxygen atoms of water make it particularly suitable for dissolving inorganic salts. Trichloroethylene, widely used in dry cleaning, and chloroform both dissolve many organic compounds. Acetone, with behaviour intermediate between water and trichloroethylene, is able to dissolve both organic and inorganic compounds.

4 **The solubility of a substance** is different in different solvents. When immiscible solvents are shaken together, any compounds present are divided between the two liquid phases. If a solution of iodine and salt in water is shaken with benzene, iodine dissolves in the benzene layer, but salt does not. Thus it is possible to separate the two. Many different compounds can be purified in this way.

3 **Ordinary washing** removes water-soluble dirt and also non-soluble dirt that can be emulsified in water by the addition of detergents. (Milk is a typical example of an emulsion where one liquid mixed and joined with another does not separate.) In dry cleaning, an organic solvent is employed; most of those used commercially are small, halogen-containing molecules, such as trichloroethylene. These solvents dissolve grease; however, because their fumes are unpleasant they have to be used in special machines, as shown here. Similar compounds are also used in industrial degreasing operations, such as cleaning pieces of metal prior to machining, or in the manufacture of electronics components.

5 **Different compounds** can also be separated by partitioning them between a solid and a liquid. In thin-layer chromatography, the process shown in this illustration, a spot of mixed compounds is placed on a plate coated with absorbent powder, which is put into a development tank containing a solvent (ensuring the tank is saturated with its vapour). As solvent ascends the plate, the mixture separates out into its component chemicals.

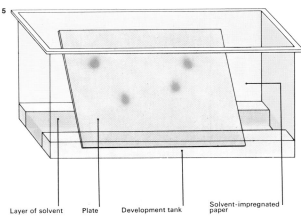

Layer of solvent | Plate | Development tank | Solvent-impregnated paper

A solution containing the maximum possible amount of solute (dissolved substance) is called saturated. A hot liquid generally has a greater capacity for holding solid in solution than a cold one (with gases in liquids, the reverse is true). Consequently, if as much solid as possible is dissolved in boiling water, some of it may crystallize out when the water cools. If it does not, the solution is said to be supersaturated. This is the basis of many experiments for growing giant crystals in school laboratories.

Dissolving a solid in a liquid affects the liquid in a variety of ways. For example, pure water freezes at 0°C (32°F) and boils at 100°C (212°F). The freezing-point of a solution of common salt, however, is lower than 0°C; thus, salt sprinkled on roads in winter prevents ice from forming.

Osmotic pressures

One of the more important properties of solutions is their ability to exert osmotic pressure. If a solution encased in certain types of membrane is brought into contact with pure solvent, the solvent molecules pass through the membrane into the solution, making it more dilute. The molecules of solute, on the other hand, are unable to pass through the membrane, which is called "semi-permeable" in consequence.

Osmosis is crucial to many living organisms. For example, the absorption of water by the root hairs of plants depends on it. If the concentration of dissolved matter in the plant cells is greater than in the water surrounding them, water is drawn in. On the other hand, if the reverse is true – in very salty soils, for example – water may be drawn out of the plants, so that they die.

Plants and animals are also responsible for producing complex substances that form so-called colloidal solutions. These are part way between a solution and a suspension. The jellies eaten by children are one type of colloidal solution, and non-drip paint is another. Jellies set on cooling and "melt" again on heating, but a non-drip paint becomes more liquid as it is stirred and more solid when it is left standing. These differences are accounted for by the different types of colloidal substances used.

Rivers and streams dissolve small quantities of minerals. Under certain conditions, these minerals can be precipitated from solution – sometimes in impressive forms, such as these stalactites and stalagmites formed over thousands of years in limestone caves in various parts of the world.

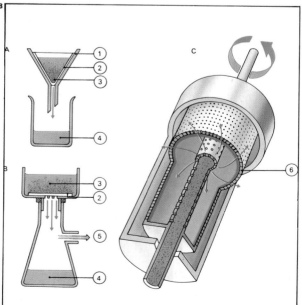

6 There is a limit to the amount of any particular compound that will dissolve in a solvent. When this limit is reached, the solution is said to be saturated. If solvent is evaporated, or if the temperature falls, the amount of compound that can be held in solution drops and solid precipitates. The Dead Sea is one of the most concentrated naturally occurring solutions of minerals, some of which crystallized out when the water level dropped in the past. Here Israeli industry evaporates large quantities of Dead Sea water in nearby "pans" to obtain important minerals for making various inorganic chemicals.

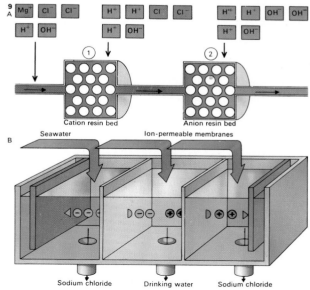

7 Although two-thirds of the earth's surface is covered with water, in many countries there is a lack of pure water for drinking. In this modern desalination plant, pure water is evaporated from solutions such as seawater and then condensed in large tanks. The concentrated brine left over from the process is usually returned to the sea, where it is diluted by tidal mixing.

8 Filtration removes solid particles from solution. A mixture placed [A] in a filter funnel [1] drains through the filter paper [2] by gravity, leaving sediment [3] trapped while the filtrate [4] flows through. In the Buchner funnel [B], the same principle is enhanced by suction [5]. In the industrial rotary filter [C] the spinning action of the drum drives the mixture onto the fine mesh filter [6].

9 Ion exchange removes ionic compounds from solution [A]. A cation resin [1] exchanges cations in solution for H^+, whereas an anion resin [2] exchanges anions for OH^-. An extension of the exchange process is electrodialysis [B].

Key chemical reactions

In an enclosed space, a mixture of air and an organic vapour composed mainly of hydrocarbons are pressed together. The hydrocarbons consist of molecules in which several carbon atoms are chemically bonded to each other and to a number of hydrogen atoms. A spark is generated. In an instant, many of the bonds between carbon and hydrogen and between carbon and carbon are broken and replaced by chemical bonds that combine these atoms with oxygen from the air. An explosion results, and a chemical reaction has taken place. This particular reaction takes place millions of times each day in most parts of the world, wherever people use petrol-driven internal combustion engines in cars and other motor vehicles.

Molecular energy

Chemists are interested in how individual compounds react: how quickly they will react, what the products are and how much "persuasion" is needed to make a reaction take place.

Why, for example, is a spark needed before the hydrocarbons that compose petrol will react with oxygen? Why, however many sparks are provided, will exhaust gases such as carbon dioxide and steam not burn? Why does the spark cause an explosive reaction in a petrol/air mixture, whereas a spark landing on this page would probably only char a small portion of the paper; and, if the paper did catch fire, why would it burn steadily rather than explode?

The answers to all these questions are related to the energies of different molecules. The world is full of molecules, rather than unlinked atoms, because the formation of chemical bonds releases energy and makes the resultant product more stable. The same is true of molecular reactions. If a reaction occurs spontaneously, there is usually a release of energy and the production of more stable molecules.

It is theoretically possible to get water and carbon dioxide to react to form petrol and oxygen. But, because petrol and oxygen are less stable molecules, large quantities of energy would have to be put into the reactants (starting materials) to succeed. A better example is, perhaps, metallic corrosion.

Many metals are more stable as compounds, such as the oxide or sulphide, than as pure metal. Using processes that supply energy to metallic compounds, scientists can refine their ores to make metals for steel girders or silver teaspoons. But if unprotected, the metals gradually corrode – that is, spontaneously form compounds such as iron oxide (rust) or silver sulphide (tarnish) that are more stable than the pure metals.

Kinds of reactions

Whereas certain reactions may occur spontaneously, they do not necessarily do so. A petrol/air mixture needs the spark to set it off because, in between the reactants and the products, there is a transition state which is of higher energy. It is necessary to give many reactants an energy "lift" to help them over the activation barrier. The larger the amount of energy arising from a reaction, the more molecules can be lifted over the barrier. And if the reaction happens fast enough, the result is an explosion.

The energy to initiate a reaction can come from a variety of sources. Heat is commonly

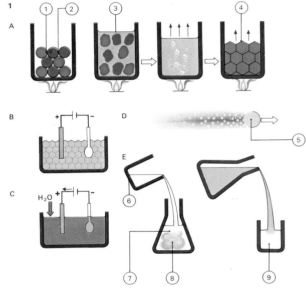

1 The application of heat [A] alone to aluminium sulphate [1] and potassium sulphate [2] produces no reaction, but they "combine" when dissolved in water [3]. If heating is continued to evaporation, alum [4] forms. Dry copper sulphate crystals [B] do not conduct current; but dissolve them in water [C] and electrolysis can proceed. Metals may react with a liquid [D]: a grain of sodium [5] dropped into a water bath melts, generating hydrogen. Solutions and other liquids react readily [E]. Phenolphthalein [6] added to a solution of alkali [7] produces a pink solution [8]. When this is added to an acid solution [9], the pink disappears.

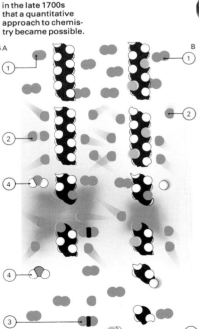

2 In a chemical reaction, matter is neither created nor destroyed. This conservation of mass can be shown by a classic experiment in which a candle is burned inside a weighed bell jar [A]. At the end of the experiment, the weight of the jar and its contents [B] are the same as at the beginning, although a part of the candle – made up largely of carbon and hydrogen – has "disappeared" as volatile reaction products (water and carbon dioxide). It was only after scientists accepted the principle of conservation of mass in the late 1700s that a quantitative approach to chemistry became possible.

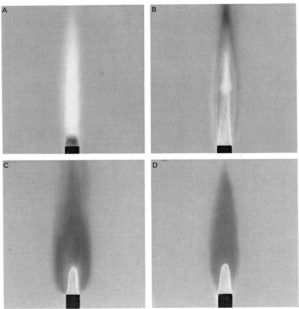

3 Unmixed ethylene burns with a luminous diffusion flame [A], reacting with oxygen drawn in from round the flame. If mixed with a little air, the ethylene gives a flame with three distinct layers – an inner cone of unburnt gas, a blue-green layer of reacting premixed gas and an outer cone where partially oxidized products of the premixed layer are burned by a diffusion flame [B]. Addition of nitric oxide to the mixture cuts down the amount of oxygen available for immediate combustion and the resultant flame [C] shows a complex series of reactions, as in B. But if more air is added to the gas mixture, the diffusion layer disappears [D].

4 A hydrocarbon fuel [A] will burn more readily than similar molecules in which some of the hydrogen atoms have been replaced by chlorine [B]. When mixed with oxygen [1] and subjected to a spark, both types of molecules burn, but the chlorinated one burns more slowly. The spark breaks oxygen molecules into reactive oxygen atoms [2]. These combine with carbon and hydrogen in the hydrocarbon to give carbon dioxide [3] and water [4]. Sufficient heat is produced to keep the reaction going rapidly. With the chlorinated material, more complex reactions take place more slowly. These produce, in addition, hydrogen chloride [5] but generate less heat.

used by chemists to help reactions along. But other forms of energy, such as light, are sometimes enough to initiate reactions.

There are a great many types of reactions, but they can all be broken down into simple categories. There are reactions in which a single substance rearranges its chemical bonds to produce a different single substance (rearrangement); alternatively, it may break into two or more different parts (decomposition, fragmentation). Conversely two, or occasionally more, compounds can combine to form a single compound (addition). More often, there are a number of products.

Reaction requirements

When there is only one starting material, the activated state is achieved by molecules absorbing sufficient energy to initiate a change. For example, when the visual pigment rhodopsin in the retina of the eye absorbs light, one of its electrons is energized and the molecule is catapulted into a more energetic state. As a result, the molecule changes shape to form a different geometric isomer (substance whose molecules are made of the same numbers of the same atoms, but whose atoms are differently arranged)

When different starting materials are involved, not only must they have enough energy to form the activated intermediate state, but they must also collide physically before the rearrangement of bonds can be completed. This is why some reactions take place under high pressure, as in a car engine, because packing molecules more tightly together makes collisions more likely.

Another way to improve the chances of a successful collision is to provide a surface that has an appropriate geometry for bringing molecules together. Substances that provide such surfaces are among various types of catalysts. They speed up reactions by lowering the activation barrier. Other, negative, catalysts slow or even prevent reactions.

To follow reactions as they occur and identify the various products, modern chemists use a large range of different techniques [6, 7]. Some of them, such as nuclear spin resonance and Mossbauer spectroscopy, are based on physical phenomena discovered during the past 30 or 40 years.

A reaction takes place as drops of alkali fall into a dilute solution of a copper salt. This is a precipitation reaction – one of many different kinds that can take place in chemistry. One of the chemist's major tasks lies in discovering how various materials react with each other.

5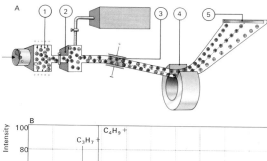

C_6H_6

$C_6H_5NO_2$

Nitrogen
Carbon
Hydrogen
Oxygen

5 The nitration of benzene produces a reaction that goes through several stages. Initially [A], the entering group approaches and associates weakly with the benzene ring [B]. Then, rearrangement produces an unstable high energy intermediate [C] which breaks down to a complex [D] in which the leaving group is weakly associated with the ring. It ends with the departure of the leaving group [E].

6 In a mass spectrometer [A], outer electrons of a compound are removed in an ionization chamber [1]. Positively charged ions pass into an adjacent chamber under vacuum [2] and are focused by electric [3] and magnetic [4] fields. The way they are deflected by these fields is characteristic for each ion, which can be identified by its position on a photographic plate [5].

A molecule such as n-dodecane breaks down into a number of fragments that produce various "peaks" on a graph [B]. From the position of these, the parent molecules can be precisely identified.

7 Mixtures of compounds can be separated by gas chromatography. Gaseous molecules travel down columns of liquid-impregnated solids at different rates. Special detectors reproduce on a graph the peaks produced by the molecules as they leave the column.

Each can frequently be identified by the time it takes to go through the column – the "retention" time. The chromatogram shows the separation of hydrocarbons: isobutane [A], n-butane [B], n-butene [C], isobutene [D] trans-but-2-ene [E], cis-but-2-ene [F], cis-1-3-butadiene [G].

6

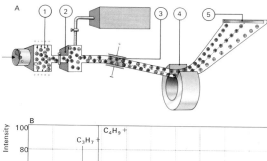

A

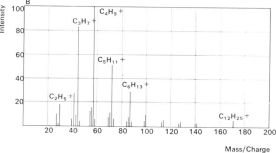

B

Intensity

$C_4H_9{}^+$
$C_3H_7{}^+$
$C_5H_{11}{}^+$
$C_6H_{13}{}^+$
$C_2H_5{}^+$
$C_{12}H_{25}{}^+$

Mass/Charge

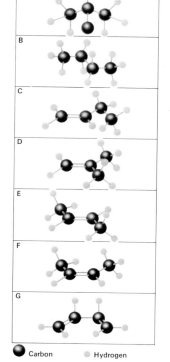

7 A
B
C
D
E
F
G

Carbon   Hydrogen

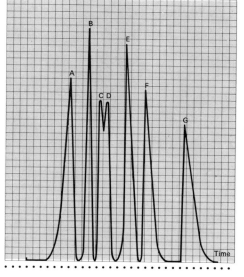

A
B
C D
E
F
G

Time

147

Electrochemistry

Electrons are negatively charged particles that form a part of every atom and it is with the interactions of electrons from different atoms that chemistry is mainly concerned. An electric current is no more than a flow of electrons. Consequently, it is not surprising that electricity and chemistry are connected.

Early research
Studies of electricity and chemistry went hand in hand long before anyone knew of the existence of electrons. During the eighteenth century, there was much interest in static electricity, leading to the development of the Leyden jar (for storing "electric fluid" generated by friction) and the lightning conductor. However, it was not until the Italian physiologist Luigi Galvani (1737–98) found, towards the end of that century, that frogs' legs would contract if different metals were applied to nerve and muscle that current electricity was discovered. In 1795, another Italian, the physicist Alessandro Volta (1745–1827) [1], showed that this "animal electricity" could be produced without living tissue. He separated two pieces of metal by a

cloth moistened with salt solution and thus made the first electrical battery. Within five years, it was discovered in England that current from such a battery could decompose water into hydrogen and oxygen gases. Thus, the foundations of electrochemistry were laid. By a chemical reaction involving two metals, a flow of electrons can be produced; such flow can bring about other reactions.

Batteries soon became important equipment in every laboratory and led to many new discoveries, such as the isolation of the elements sodium and potassium in the first decade of the nineteenth century by Humphry Davy (1778–1829).

Chemical reactions
When a metal such as zinc forms compounds, it does so in many instances by losing two electrons to form a doubly positive zinc ion (Zn^{2+}). Metals differ in the ease with which they lose electrons, so that if a piece of zinc metal is placed in a solution of copper sulphate (which contains Cu^{2+} ions), the zinc gives up electrons to the copper. The net result is that zinc is converted to zinc sulphate

and copper ions become metallic copper.

When an element gains electrons to form a negatively charged ion, it is said to be reduced; if it loses electrons to form a positively charged ion, it is said to be oxidized. A reaction where reduction and oxidation cancel each other out, as in the zinc/copper sulphate case, is called a redox reaction. Redox reactions can be tapped to supply electric currents by preventing the reduction and oxidation from occurring at the same place. A battery can be made by suspending zinc in zinc sulphate and copper in copper sulphate and linking the two solutions by a porous partition and the two pieces of metal by a length of wire.

Each of the reactions in such an arrangement is called a "half cell". When two "half cells" are added together, a cell is completed and the voltage it produces depends on the particular half cells that make it up.

That different batteries produce different, but specific, voltages depending on their chemical composition is not surprising, in view of the differences in reactivity between different elements. The reverse also seems

CONNECTIONS

See also
114 What is an electric current?
144 Chemicals in solution
138 Joining atoms

1 Alessandro Volta, professor of natural philosophy at the University of Pavia, Italy, constructed in 1800 an "artificial electrical organ", an apparatus he described as like the electrical organ of the electric eel. Made by piling alternate discs of copper and zinc, each pair separated by a piece of brine-soaked cloth, his electrical organ was one of the first scientific batteries ever made.

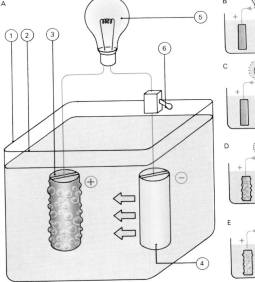

3 Heart pacemakers like the one shown here, and miniature hearing aids, can be powered by batteries. These are examples of primary cells. Secondary cells or storage batteries can be recharged. Early batteries all had metallic plates separated by solutions of salt-like chemicals. The dry, or Leclanché, cell replaced the liquid with a paste. Such batteries, which use zinc and carbon (with a manganese dioxide depolarizer) as electrodes, are the type used in transistor radios, torches, and many other everyday appliances. In recent years, battery technology has led to the production of very small but highly reliable batteries which also use zinc and a metal oxide, in this case mercury.

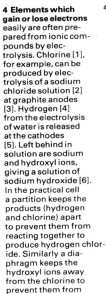

2 The voltaic cell, [A] named after Volta, consists of a jar [1] containing sulphuric acid [2] in which are suspended a copper anode [3] and a zinc cathode [4]. These are connected by a circuit containing a light bulb [5] and a switch [6]. When the switch is open [B] no current flows. When it is closed [C] an electromotive force (emf) is set up in the circuit and the light bulb glows weakly but the light gradually decreases [D] while the cathode is steadily eroded. Hydrogen bubbles are seen on the anode. The hydrogen bubbles then polarize the anode [E], setting up a "back emf", breaking the circuit.

4 Elements which gain or lose electrons easily are often prepared from ionic compounds by electrolysis. Chlorine [1], for example, can be produced by electrolysis of a sodium chloride solution [2] at graphite anodes [3]. Hydrogen [4] from the electrolysis of water is released at the cathodes [5]. Left behind in solution are sodium and hydroxyl ions, giving a solution of sodium hydroxide [6]. In the practical cell a partition keeps the products (hydrogen and chlorine) apart to prevent them from reacting together to produce hydrogen chloride. Similarly a diaphragm keeps the hydroxyl ions away from the chlorine to prevent them from reacting to form sodium hypochlorite.

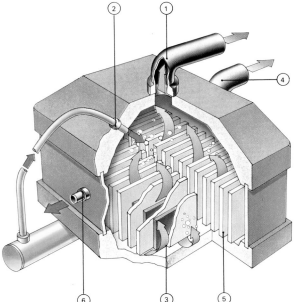

logical: that a particular quantity of electrons should produce a particular amount of change in a substance. The quantitative relationships between electricity and chemical reactions were stated during the nineteenth century by Michael Faraday [Key]. The extraction and electroplating of metals, and the production of reactive electronegative elements such as chlorine [4] and fluorine, are often done electrolytically.

Electrolysis

The products from electrolysis reactions [2] sometimes depend on the conditions used, as well as the amount of electricity. If fused (molten) sodium chloride is electrolyzed, sodium metal forms at one electrode and chlorine gas at the other. However, when a solution of sodium chloride is electrolyzed, using a graphite anode (positive electrode) and an iron cathode (negative electrode), chlorine and hydrogen gases are produced, leaving behind sodium hydroxide.

The ions of different elements may be positive (cations) or negative (anions). In a solution that is being electrolyzed, cations are attracted to the cathode and anions to the anode. If aluminium is made into an anode in an acid solution, a very thin layer of aluminium oxide forms on it. This anodization protects the aluminium from corrosion and is used on a wide range of articles.

Many oxidation reactions are used in everyday life: for example, the burning of petrol in a car is such a reaction. Instead of releasing the energy from such a reaction as heat, it can be converted into a flow of electrons in a "fuel cell". These cells are theoretically much more efficient energy converters than heat engines. However, difficulties in designing suitable fuel cells for everyday purposes have meant that their use has been limited largely to applications where cost is not an important factor.

The chief commercial application of electrochemistry is electroplating by means of electrolysis. For example, decorative metals such as gold and silver are electroplated onto articles of jewellery, whereas chromium is electroplated onto steel (preferably over base layers of copper and nickel) to provide resistance to corrosion.

As a youth, Michael Faraday (1791–1867) attended Humphry Davy's lectures at the Royal Institution in London. He copied these out and sent them to Davy with a request for employment. From Davy's assistant, he rose to become professor of chemistry at the Royal Institution, a post he held for more than 30 years. Most of Faraday's work was in physics – particularly in the field of electromagnetic induction. He also founded the science of electrochemistry through his discovery of the quantitative relationships between the amount of electricity passed through a solution and the amount of substances deposited as a result.

5 Electrolysis can easily be shown. A current is passed [A] between platinum electrodes [1] through a dilute hydrochloric acid electrolyte [2]. Positively charged cations [3] move towards the negative electrode (cathode) and anions [4] move to the positive electrode (anode). The hydrogen ions combine with water to form hydronium ions, H_3O^+; when two hydronium ions [B] reach the cathode, they each receive an electron, thus forming atoms of hydrogen which combine to form a molecule of gas. In copper plating [C], copper from the copper sulphate electrolyte is deposited onto the object to be plated [5], while copper from the anode [6] is drawn into solution. A modern barrel electroplating machine [D] can plate many small objects [7] simultaneously.

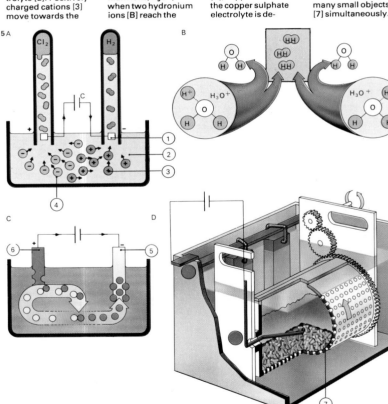

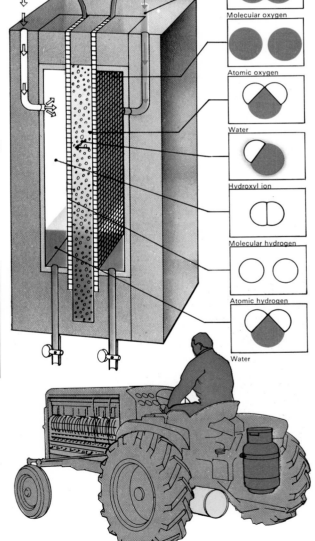

Molecular oxygen

Atomic oxygen

Water

Hydroxyl ion

Molecular hydrogen

Atomic hydrogen

Water

6 The fuel cell, like a battery, uses the energy generated during a chemical reaction to produce electrical power. One simple cell uses oxygen and hydrogen as "fuels" to produce electricity, the water formed being a by-product. The electrolyte in this case is located in a very thin, water-saturated membrane. This allows ions to pass through it, but does not allow the passage of atoms or molecules. The electrodes are of wire mesh, coated with platinum. Molecular hydrogen and oxygen are fed to them from gas chambers. The cathodic platinum converts oxygen to hydroxyl ions, which move across the membrane and react with hydrogen at the anode to form water. Electrons released by the hydrogen traverse the external circuit – as an electric current – to help form hydroxyl at the anode. A battery of such cells can power a tractor, with fuel tanks holding liquid oxygen and hydrogen either as a liquid or as a solid hydride.

Chemical analysis

One of the main branches of chemistry – chemical analysis – is concerned with determining the composition of a substance or a mixture of substances. Identifying the ingredients is termed qualitative analysis, whereas determining their precise proportions is called quantitative analysis. Organic chemicals (the large class of compounds containing the element carbon) and inorganic chemicals (all other compounds) require different analytical techniques in the laboratory.

Methods of inorganic analysis
Qualitative inorganic analysis is generally carried out on a semi-micro scale [1] using small quantities – less than a gramme. Chemists make preliminary tests on a dry sample of a substance and these give general information about its composition. The effect of heat may cause a colour change, sublimation or the evolution of a gas. A reagent is added to a solution of the sample and the resulting mixture examined for the evolution of a gas, a precipitate or a coloration.

Metal ions are identified by a systematic separation into "groups". A variety of techniques exists, but metals are usually split into groups by adding a series of reagents and collecting any precipitate produced. Group precipitates are, in turn, separated and identified by characteristic reactions.

Quantitative inorganic analysis may be carried out using either volumetric or gravimetric methods. Volumetric analysis involves reacting a solution of known concentration, referred to as a "standard" solution, with a solution of the substance to be determined. After preparing the standard solution the chemist carries out a titration [Key] in which one solution is slowly added to the other. From the concentration and volume of the standard solution, the concentration of the "unknown" can be calculated.

Gravimetric analysis involves preparing a solution with a known mass of the sample. This solution is then reacted with a chosen reagent so that the desired component is completely separated, generally as a precipitate. The product, which must be pure, is isolated and weighed and the amount of the component calculated.

Merely identifying the elements present is insufficient to make a definite description of an organic compound. The ability of carbon compounds to exhibit isomerism (in which two different substances can contain the same chemical elements, in the same proportions, but combined in a different way) means that the arrangements of the elements present must also be analytically determined.

Methods of organic analysis
Identification of the elements in an organic component involves systematic elimination of all possible elements one by one. Carbon and hydrogen are nearly always present and tests for them are rarely carried out. But tests are made to identify other elements. An example is the Lassaigne sodium fusion, which reveals the presence of nitrogen, the halogens (chlorine, bromine and iodine) and sulphur. A knowledge of the elements allows a chemist to allocate the compound to a main group. This is followed by the application, within the group, of classification tests for functional groups, which determines the types of organic compound present.

1 Analysis on a semi-micro scale saves time and money. Small test tubes are used for reactions and tapered tubes are used for the centrifuge. Solutions do not mix well in these small tubes, so a stirring rod is necessary. Solutions are handled in a teat-pipette and each reagent bottle should be fitted with its own "dropper". Solids require the use of a semi-micro spatula. To avoid the dangers of "bumping" during heating, solutions are heated indirectly in a metal block, although evaporation to dryness requires the use of a small crucible. Identification of gases is usually carried out with a bubble-cap fitted to a test tube.

2 The melting-point is a characteristic of an organic compound and can be measured, using this automatic apparatus, to help to identify it. A sample is first heated quickly to get an approximate value and then slowly melted to obtain a more accurate reading.

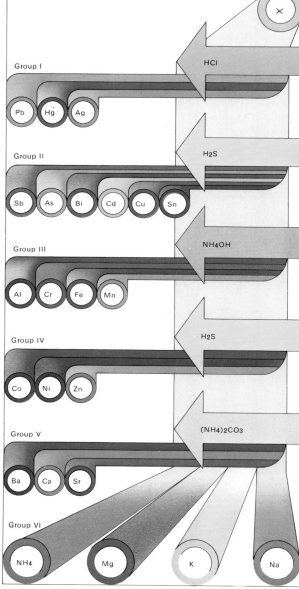

3 **Metal ions** can be detected in qualitative chemical analysis by a systematic separation into groups by means of a series of characteristic precipitation reactions. The chemist tries to dissolve a sample of the substance to be analysed in dilute hydrochloric acid. Metals with insoluble chlorides – lead, mercury and silver – constitute Group I. Next hydrogen sulphide is bubbled through the acid solution; a sulphide precipitate indicates one of Group II metals – antimony, arsenic, bismuth, cadmium, copper or tin. The addition of ammonia solution precipitates the hydroxides of the Group III metals – aluminium, chromium, iron or manganese. The Group IV metals have sulphides that are precipitated from alkaline solution by bubbling in hydrogen sulphide gas; they are cobalt, nickel and zinc. The addition of ammonium carbonate to the remaining solution at this stage of the analysis precipitates the carbonates of barium, calcium or strontium, the Group V metals. Group VI of the analysis table contains the metals magnesium, sodium and potassium as well as the "metallic" ion ammonium, left after eliminating all other possible metal ions. This analysis scheme can be carried out on the semi-micro scale and it can be enlarged to include some of the less common metals. It reveals only the presence of a metal and not its quantity.

Group I — HCl — Pb, Hg, Ag

Group II — H₂S — Sb, As, Bi, Cd, Cu, Sn

Group III — NH₄OH — Al, Cr, Fe, Mn

Group IV — H₂S — Co, Ni, Zn

Group V — (NH₄)₂CO₃ — Ba, Ca, Sr

Group VI — NH₄, Mg, K, Na

Quantitative organic analysis also involves estimation of the elements present, followed by purification and a determination of molecular weight to give the empirical and molecular formulae. The amounts of carbon and hydrogen are found by completely oxidizing a known mass of the organic compound and weighing the carbon dioxide and water formed. Then any other elements are estimated by a variety of methods.

These results allow the chemist to calculate the percentage composition of the substance (the proportions of each element present) and to determine its empirical formula. The molecular formula is found by comparing the empirical formula with the molecular weight. Dissolving a sample of the substance in a solvent affects the physical properties of the solvent. The lowering of vapour pressure, the elevation of boiling point and the depression of freezing point are all proportional to the mole fraction (concentration in terms of molecular weight) of the dissolved substance present in the solution. The concentration also affects the osmotic pressure, viscosity and light-scattering properties of solutions. Careful measurement of one of these effects is then followed by a calculation of the molecular weight before the chemist has enough information to complete the analysis.

Modern instrumental analysis

Various forms of chromatography are based on the fact that different substances diffuse or are absorbed at different rates. Spectroscopy [7] makes use of the fact that each species of atom has a unique characteristic spectrum. A spectrum is produced when atoms, ions or molecules are excited by absorbing energy and may be observed by using a prism or, preferably, a diffraction grating.

In mass spectroscopy [5] a substance is bombarded with low-energy electrons and fragmentation produces a number of positive ions. Ions of the same mass/charge ratio are focused by magnetic or electrostatic fields and detected photographically or electronically. The highest mass/charge ratio can give the molecular weight. Investigation of the fragmentation pattern determines the molecular structure.

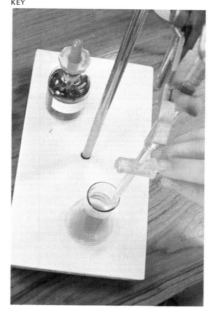

A titration is performed to estimate the unknown concentration of a solution by reacting it with a "standard" solution of known concentration. This is usually added from a burette to a fixed volume of the "unknown". The end-point of the reaction is shown either by a visible change in the reactants or by the addition of a chemical indicator. Acid/base indicators have different colours according to the hydrogen-ion concentration (pH) of the solution and change as the pH of the solution changes. The pH at which colour changes occur varies so an indicator can be selected that shows a colour change at a pH close to the end-point.

4 Molecular weight may be determined by the depression of freezing-point. When a substance is dissolved in a solvent the freezing-point is depressed. If dilute solutions are used then the depression is directly proportional to the number of molecules of the solute in unit mass of the solvent. The molecular depression constant of a solvent is the depression of freezing-point produced for one mole of solute in 100g (3.5oz) of solvent. Experimentally this quantity would need too high a concentration so dilute solutions are used and the constant calculated by proportion. The most convenient freezing agent is ice and water.

5 The masses of atoms can be compared with great accuracy using a mass spectrometer. A vaporized sample is ionized by electron bombardment and the beam of positive ions produced is accelerated to a constant speed. Application of a strong magnetic field deflects the beam – the lightest ions being deflected most and the heaviest least. The field is adjusted so that ions of a particular mass fall on to a detector, either a moving photographic plate or an electrometer. From an electrometer the signal is amplified and recorded on a graph. Further adjustment of the magnetic field allows ions of different mass to be recorded.

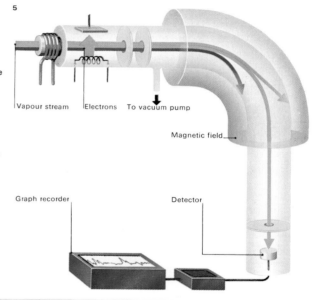

Vapour stream   Electrons   To vacuum pump

Magnetic field

Graph recorder   Detector

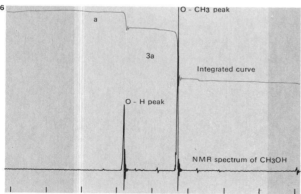

O - CH3 peak

a

3a

Integrated curve

O - H peak

NMR spectrum of CH3OH

Infra-red spectrum of CCl4

Transmittance (%)

C-Cl peak

Frequency (cm⁻¹)

6 Nuclear magnetic resonance (NMR) spectroscopy of a substance with a molecular formula of CH_4O shows two peaks. The areas under the peaks are integrated automatically and indicated by the upper curve. This shows a ratio of one to three. The inference is that the four hydrogen atoms in the molecule are arranged so that three are in the same environment, and the other is different. The larger peak is produced by those in the O-CH₃ group and the smaller by that in the O-H group – indicating the structure CH₃OH: methanol. Normally a "standard" is added and all lines are measured relative to it. Tetramethylsilane or a sodium salt of 4,4-dimethyl,4,silapentanesulphonic acid is a suitable standard.

7 Infra-red radiation is absorbed by the chemical bonds in an organic compound and can cause vibrational effects in them. The frequency of the absorbed radiation is characteristic of the bond concerned, so that measuring these frequencies provides a means of determining the bonds present and analysing the compound. This is the basis of infra-red spectroscopy. The absorption frequencies are measured electronically and plotted as a series of peaks on a graph. In this example the main absorption peak at a frequency of 750cm⁻¹ is due to stretching of a carbon-chlorine bond, indicating that the substance producing it is probably tetrachloromethane (carbon tetrachloride, CCl_4). The minor peak at 1,550cm⁻¹ is probably a harmonic of the main one. The bending and rocking of chemical bonds after infra-red absorption also produce characteristic peaks on the spectrograph and aid the analysis.

Towards the chemistry of life

Only half a dozen of the 93 or more chemical elements that occur naturally on earth make up the bulk of living matter, and life's diversity is due largely to the combining properties of just one element: carbon. Carbon atoms can form chemical bonds with each other to produce an extensive range of basic structures. These can be modified by the addition of the atoms of other common elements of life – hydrogen, oxygen, nitrogen, phosphorus and sulphur – to produce the enormous diversity of chemical substances found throughout the living world.

Isomers and polymers
Many naturally occurring carbon compounds have another distinctive property. A single atom of carbon can form chemical bonds to four different atomic groupings. The bonds can be arranged in space in two distinct ways to produce two different molecules that are as similar to and as different from each other as a pair of gloves [1]. They are called optical isomers and where two of these are possible, usually only one form occurs naturally.

Many key substances in living organisms are polymers – giant molecules containing thousands or even millions of individual atoms linked together. Carbohydrates, proteins and nucleic acids are all polymers. But they are all made by joining together small molecular building blocks rather than individual atoms.

Carbohydrates, for example, are all made from small molecules called sugars, or saccharides. Common table sugar is not one of the simplest: it is made by linking two smaller saccharides. Chemists call it a disaccharide. It is an ingredient of many proprietary foods, such as pickles and ketchup [3].

Like all carbohydrates, saccharides are composed of carbon, hydrogen and oxygen. These elements are generally linked together in such a way that a loop is formed between the ends of the molecule by an oxygen atom bridging two carbon atoms.

Sugars and fats
Sugars [2] not only supply living organisms with energy, but they also make up a broad range of polymeric substances, such as starch and cellulose. Starch, for example, is the chief carbohydrate in potatoes, rice and bread [3]. Human beings cannot digest cellulose, but the polymer can be broken down chemically to form molecules of glucose, which is an example of a simple sugar (monosaccharide). Sugar from beet or cane is sucrose.

Virtually all food has sugars or polysaccharides in it. Almost as common in food are the lipids, composed solely of carbon, hydrogen and oxygen. Lipid is a general term that includes oils, fats and waxes, all of which have similar chemical structures. In a simple lipid, the same sort of carbon-oxygen-carbon bond that holds a disaccharide together joins three fatty acids to a small molecule called glycerol (also known as glycerine) [4B]. The resultant triglyceride may be a liquid (oil) or solid (fat) at room temperature, depending on the structures of the fatty acids, which can all be the same (as in olive oil) or all different. All meats contain some fat [3].

The so-called unsaturated fatty acids contain carbon-carbon double bonds. In recent years, margarines containing them have been available in many countries because a link has been suggested between high consump-

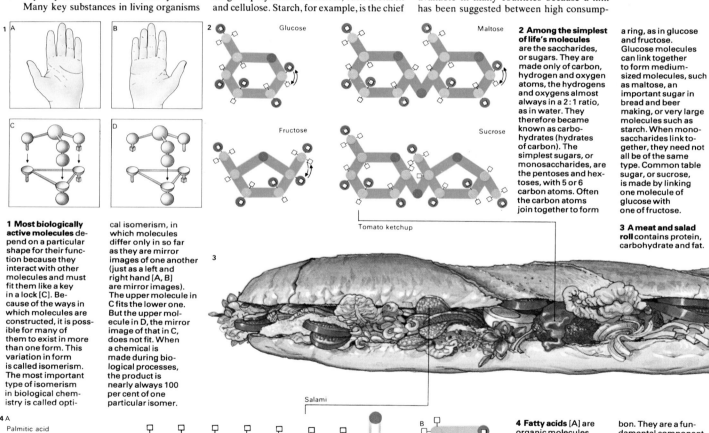

1 Most biologically active molecules depend on a particular shape for their function because they interact with other molecules and must fit them like a key in a lock [C]. Because of the ways in which molecules are constructed, it is possible for many of them to exist in more than one form. This variation in form is called isomerism. The most important type of isomerism in biological chemistry is called optical isomerism, in which molecules differ only in so far as they are mirror images of one another (just as a left and right hand [A, B] are mirror images). The upper molecule in C fits the lower one. But the upper molecule in D, the mirror image of that in C, does not fit. When a chemical is made during biological processes, the product is nearly always 100 per cent of one particular isomer.

2 Among the simplest of life's molecules are the saccharides, or sugars. They are made only of carbon, hydrogen and oxygen atoms, the hydrogens and oxygens almost always in a 2:1 ratio, as in water. They therefore became known as carbohydrates (hydrates of carbon). The simplest sugars, or monosaccharides, are the pentoses and hextoses, with 5 or 6 carbon atoms. Often the carbon atoms join together to form a ring, as in glucose and fructose. Glucose molecules can link together to form medium-sized molecules, such as maltose, an important sugar in bread and beer making, or very large molecules such as starch. When monosaccharides link together, they need not all be of the same type. Common table sugar, or sucrose, is made by linking one molecule of glucose with one of fructose.

Glucose
Fructose
Maltose
Sucrose

Tomato ketchup

Salami

3 A meat and salad roll contains protein, carbohydrate and fat.

4 Fatty acids [A] are organic molecules that all have a carboxyl group (–COOH) at one end of a chain of carbon atoms that may vary in length and in the number of hydrogen atoms attached to each carbon. They are a fundamental component of lipids, compounds found in materials as diverse as bacon fat and olive oil. In the most common lipids, three fatty acid molecules are linked chemically to one molecule of glycerol [B]. In some important biological molecules, one of the fatty acids may be replaced by a phosphorus-containing molecule, such as choline phosphate [C], to form a type of compound known as a phospholipid.

4 A
Palmitic acid
Stearic acid
Oleic acid
Erucic acid

tion of saturated fats (such as butter) and the incidence of heart disease. Lipids also provide energy for animal cells but, in more complex forms, they play other roles, such as "insulators" for nerve fibres.

Triglycerides are the simplest lipids. An important and more complicated example is cholesterol, widely found in dairy foods such as cream and cheese [3]. It is a major constituent of gall-stones and also implicated in heart disease. Cholesterol, with a complex chemical structure related to that of human sex hormones, is a type of molecule called a steroid. Steroids may be synthesized in living systems from a molecule called squalene, a kind of terpene. Some terpenes are made from only carbon and hydrogen, but others also contain oxygen and nitrogen. They include not only substances such as turpentine, vitamin A and cholesterol, but many flavours and fragrances as well.

Amino acids

From small groupings of carbon, oxygen, hydrogen and occasionally sulphur, all joined in a particular pattern, come the amino acids [5], the building blocks for proteins. Proteins are major constituents of some of the most important foodstuffs, such as meat, fish and eggs. They are usually made from only 20 different amino acids and yet they have a wide range of valuable properties. In addition a number of other, less common amino acids are known. These can combine to form compounds such as the antibiotic valinomycin, which is made by linking together a small number of amino acids, and the extremely poisonous seven and eight amino-acid rings occurring naturally in toadstools of the genus *Amanita,* such as the death cap and fly agaric.

More complex than the amino acids are the nucleotides, the basic building blocks of the nucleic acids, which carry genetic information. Each nucleotide is made from one of the five types of base [7] which is joined to a sugar and this, in turn, is joined to a phosphate grouping. Some foods are very rich in nucleic acids and too much of them can cause an illness, such as gout, in which the mechanism that usually deals with the bases falters or fails completely from overwork.

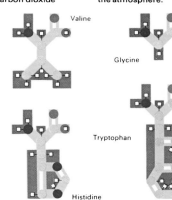

Atmospheric carbon dioxide

Carbon, the basic element of organic chemistry, undergoes a natural cycle in the environment. It exists in the form of carbon dioxide in the atmosphere.

From there it is absorbed by plants [1] to build carbohydrates in green leaves. When plants burn [2], and animals breathe out [3], carbon dioxide

passes back into the air. Also in decaying plant and animal remains [4], carbohydrates are broken down to release carbon dioxide into the atmosphere.

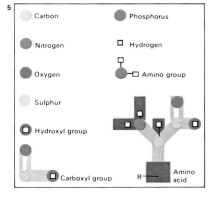

5 About 20 amino acids play essential roles in the structure and function of living organisms. All contain carbon, hydrogen, nitrogen and oxygen atoms; a few also contain sulphur. All except proline have a free unsubstituted amino group and a free carboxyl group. In addition, each amino acid has a characteristic "R" group attached to this carbon atom. This key applies to both of these pages.

Carbon
Nitrogen
Oxygen
Sulphur
Hydroxyl group
Carboxyl group
Phosphorus
Hydrogen
Amino group
R — Amino acid

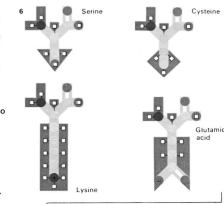

Serine
Cysteine
Valine
Glycine
Lysine
Glutamic acid
Tryptophan
Histidine
Proline

6 Amino acids usually occur in living organisms linked together in hundreds to form complex molecules. But monosodium glutamate, a simple derivative of glutamic acid, has been used as a food additive to enhance the flavour of meat. The "R" group attached to the central carbon atom in an amino acid can vary widely – from the simple hydrogen atom in glycine, to the complex groups found in the

other amino acids illustrated here. In proline, the end of the "R" group is linked also to the nitrogen atom of the amino group, so looping round on itself to form a ring structure.

Meat

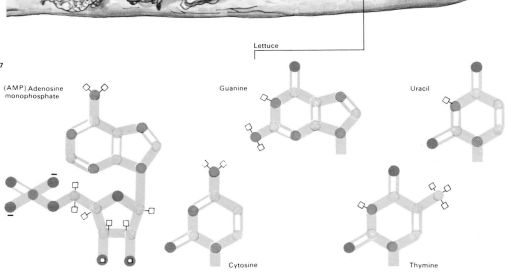

Lettuce

(AMP) Adenosine monophosphate

Guanine

Uracil

Cytosine

Thymine

7 All living matter, be it in the form of liver or lettuce, contains substances called nucleic acids. These are the giant molecules that ultimately are in control of all living processes. Protein made by bacteria, a possible source of synthetic food for man, is rich in nucleic acids. But an excess of these substances or foods containing them can produce unpleasant side-effects in human beings. They are formed by linking together large numbers of a few simpler molecules

called nucleotides. A single nucleotide is, in turn, made by linking together three even simpler chemical groupings: a phosphate, which is joined to a 5-carbon monosaccharide (either ribose or deoxyribose), to which a nitrogen-containing base is attached. Only five of these bases are common: adenosine (shown here linked to phosphate), cytosine, guanine, thymine and uracil. The nucleic acids DNA and RNA contain chains of four such bases linked through a sugar and phosphate.

The chemistry of life: biochemistry

All living things – plants and animals – build up and break down different chemicals. These chemical processes ensure that an organism has an adequate supply of both the basic materials and the energy it needs for survival. A person eating a salad roll derives energy from it, which he may use up by running several kilometres; but there is no obvious, direct link between the salad roll and the exercise. Biochemistry, through interlocking reactions, provides that link.

Fundamentals of biochemistry

The complexities of biochemistry can be reduced to two fundamental processes. The first is the way in which living cells develop an energy currency. This, like ordinary money, can be used to exchange one vital commodity for another. The second is the use of substances called enzymes [2] as go-betweens to reduce the amount of energy needed to make many chemical reactions essential to life take place fast enough.

The "currency" used by living cells is a chemical called adenosine triphosphate (ATP) [1]. Closely related to one of the units from which nucleic acids are built, ATP can break down to form adenosine diphosphate (ADP) and phosphate. In doing so, it can supply energy for a biochemical reaction – either one in which simple molecules are built up into more complex ones, or one that controls an activity such as muscle contraction [Key]. On the other hand, where a biochemical reaction gives off energy, as in the breakdown of sugars, that energy can be used to re-form ATP. Consequently, an organism can balance energy inputs against energy outputs by recycling ATP.

Although some biochemical reactions ultimately give off energy and others use it, all need an energy push to get them started. The strength of this push can be decreased if the reacting molecules are close together and are lined up in the correct relationship to each other. Substances called catalysts introduce reacting molecules to one another more efficiently and so reduce the amount of energy needed to get the reaction moving. The overall effect is to make chemical reactions proceed much more quickly.

Life relies heavily on a special class of catalysts, the enzymes [4]. These are made mainly from protein, but they may also include metal atoms or small non-protein organic molecules called coenzymes. Many of the vitamins included in a balanced diet are used by the body as coenzymes.

The shape and activity of enzymes

Enzymes [3] are very large molecules whose activity is governed by their shape. By changing their shapes it is possible to inactivate them, and thus stop certain reactions from occurring at a noticeable rate. For example, the important protein-digesting enzyme chymotrypsin occurs in an inactive form called chymotrypsinogen. Only when a few of the amino acids that make up this protein are removed does it adopt the catalytic shape of chymotrypsin. This change is triggered by the presence of food in the digestive tract. If the chymotrypsin were active all the time it would rapidly digest the intestine wall while waiting for food to arrive.

In many biochemical processes a molecule is passed from enzyme to enzyme before it becomes an end product. At each

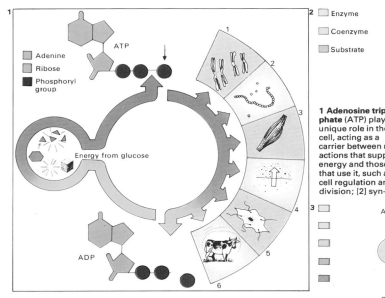

1 Adenosine triphosphate (ATP) plays a unique role in the cell, acting as a carrier between reactions that supply energy and those that use it, such as [1] cell regulation and division; [2] synthesis of important biochemicals (proteins); [3] muscle contraction; [4] transport of materials into cells; [5] conduction of nerve signals; [6] regulation of the body temperature.

2 Enzymes make molecules split up or join together much faster than they would otherwise. To work effectively, a good physical fit is needed between the enzyme and the other molecule(s) – the substrate(s). Only a part of the enzyme, the active site, often containing a coenzyme, comes into contact with other molecules. The rest of the enzyme is needed to give the active site the correct shape.

3 The shapes of enzymes can be changed by small non-substrate molecules. A molecule of similar shape to a true substrate may compete for the active site [A], thereby slowing down the desired reaction. A substance at a different site [B] changes the shape of the active site, affecting the fit of the substrate. Some enzymes possess more than one active site. The occupation of one or an activator may change the shape of the second site so that it can also accept the substrate [C]. An inhibitor molecule may prevent either of the sites from being used [D].

4 Enzymes can lower the "energy hump" that must be overcome in a biochemical reaction. They act as catalysts and are found in many common reactions, such as the rotting of fruit.

5 Pyruvic acid, a key biochemical, is formed during the breakdown of glucose and some amino acids. Further breakdown varies, depending on the biochemical system. In yeast, for example, the end product can be alcohol or gas to raise dough.

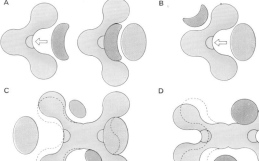

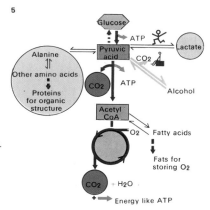

stage, an intermediate compound is formed. Sometimes the final product, or one of the intermediates, can combine with an enzyme farther back along the chain and switch it off. This feedback is like automation in a factory that ceases production when enough of a particular material has been made. Other small molecules may combine with an enzyme molecule to increase its activity.

Energy from combustion
The many-step processes involving enzymes are a necessary adaptation to circumstances. Most organic chemicals are combustible. Common sugar, for example, can be burnt completely to produce carbon dioxide, water and heat. But heat energy from total combustion is of no use to living cells. To use such energy a large temperature difference is needed, as in a car engine. Living organisms have roughly the same temperature throughout – the temperature of a healthy human being rarely deviates far from 37°C (98.6°F). Consequently, chemical energy is extracted from the "combustion" process by breaking it down into a large number of small

steps, each of which produces energy measurable (and removable) in one or two units of the cell's energy currency.

A molecule of glucose, which contains six carbon atoms, can be broken down into two three-carbon pyruvic acid molecules [5]. The process takes ten different steps, uses up two molecules of ATP and produces four new molecules of ATP. The net result of the process is therefore the production of two units of energy currency.

Pyruvic acid becomes involved in one of the key cycles of biochemistry, the citric acid (tricarboxylic acid) or Krebs cycle. It is converted into carbon dioxide and other chemicals [8]. At the same time, energy is transferred to another type of "currency" molecule, but this is soon exchanged for more ATP. Another example of energy-generation is photosynthesis, the primary process taking place in the leaves of green plants. Sunlight is absorbed by complex molecules, particularly chlorophyll, to produce "excited" molecules. These power chemical cycles that, after a number of steps, produce the ATP necessary to pay the biochemical cost of living.

The energy that a strong man uses to lift his weights comes ultimately from the sun. Plants use light to make energy-rich chemicals. Animals eat these and make even more complex substances. Human beings eat animals and plants as food, which is broken down by the digestive system into its component chemicals. These are reassembled to make various body tissues and chemicals for essential life processes. In this way, these biochemicals are employed to supply both the weightlifter's muscles and the energy necessary to flex them. The detailed study of these substances and their circuitous interactions is the science of biochemistry.

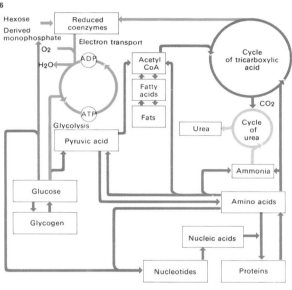

6 Every cell is the site of a complex series of chemical reactions that includes both synthesis (anabolism) and breakdown (catabolism). These processes are known collectively as metabolism. Pathways that are basically catabolic are shown in blue, those basically anabolic in green. Paths directly concerned with energy production and use are in brown. Many thousands of separate chemical reactions are involved, each controlled by a different enzyme. Overall balance and control is ultimately maintained by the genetic material of the cell, which governs the production of enzymes.

8 All cycles interlink so that the energy from breaking down one type of chemical can be used to make another. If a man overeats, he will put on excess weight – in other words he becomes fat. Yet as the dieter is frequently warned, it is carbohydrates such as sugar, more than fats, that increase weight. Pyruvic acid is the key to this apparent paradox. When it is broken down, in addition to carbon dioxide, a substance called acetyl coenzyme A can be formed. The acetyl part (the triangles) is the basic building block for the fatty acids which, together with glycerol (E-shape), make up oils and fats such as the one pictured at the bottom.

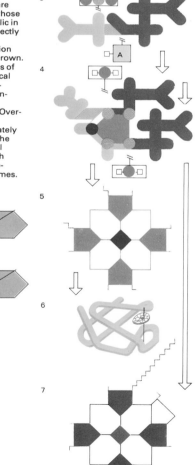

KEY
- Oxygen
- Carbon
- Nitrogen
- Hydrogen
- A Coenzyme
- Iron
- Magnesium

7 Nature's wide diversity is achieved with a remarkable degree of chemical economy. Large molecules are built from a small range of simple ones such as glycine [1] and succinyl coenzyme A [2]. Also certain key structures can have widely different functions. Both haemoglobin [6], which transports oxygen in the bloodstream, and chlorophyll [7], with which plants trap the energy of sunlight, need metal atoms in order to function. In haemoglobin iron is used; in chlorophyll, magnesium. In both cases the metals are attached to an organic molecule called a porphyrin [4], made from a compound such as d-aminolaevulinic acid [3]. Thus the red pigment haem [5] and the green pigment chlorophyll are synthesized from the same smaller molecules, although one of them occurs in animals and the other in plants. Similar compounds play key roles in other biochemical processes. Vitamin B_{12}, a porphyrin-like compound with a cobalt atom at its centre, is needed to prevent pernicious anaemia.

Polymers: giant molecules

The metals that make up a car's body and engine are chemically quite different from the oil products that power and lubricate them. Nature is more economical; the same few elements used to build living organisms are also those that trap and transport the energy, all originating from sunlight.

Proteins: polymers of amino acids

The important structural molecules of plants and animals are polymers, very large molecules known as macromolecules, made by joining together a succession of simple chemical building blocks. Proteins, for example, are polymers of amino acids which are small molecules, each containing an amino group and a carboxylic acid group. These two groups can react with one another to form a chemical bond. As a result, different amino acids can be linked through these groups in very large numbers. A small protein, such as insulin [1], may be made up of only 50 or so amino acids, but on the other hand many proteins contain hundreds of individual amino-acid units.

Animals employ proteins [2] both to build tissues and in the biochemical processes which take place in them. Collagen, for example, is a common structural protein. One of its jobs is to provide materials for tendons, which are essential to movement. A tendon "rope" of intertwined collagen molecules can have the strength of light steel wire. Another structural protein, keratin [8], occurs in hoof, hair, horn and feathers, and actin and myosin are important constituents of muscle. Proteins also supply the major (in some cases the sole) component of enzymes, the cell's catalysts which speed up biochemical reactions, and the antibodies which fight infective micro-organisms.

Essential to these differing roles are the various physical structures of proteins. When amino acids link together, they do not just form long chains. According to the shapes and the chemical properties of the side-chains of the individual amino acids used in a protein's make-up, it can be long and thin, or compact and globular. The structures contain electrically charged groups and, in addition, sulphur atoms can form bridges between amino acids. In insulin, sulphur atoms bridge adjacent chains of amino acids. In cytochrome C, a sulphur atom attaches a non-protein organic molecule (in this case haem) to a protein molecule.

Polysaccharides: polymers of sugars

The essential structural components of plants are polysaccharides, polymers of the small sugar molecules which provide most of the energy for cells. Not surprisingly, some polysaccharides are also used as a convenient means of storing energy.

It has been estimated that up to 50 per cent of the carbon atoms incorporated in plant tissues are in molecules of cellulose, a structural glucose polymer. In some forms cellulose has commercial value; cotton, for example, is 98 per cent cellulose.

The main energy reserve polymer of plants is starch, which is composed of two polymers – amylose and amylopectin. Like cellulose, amylose is a straight-chain polymer of glucose. The only difference between the two molecules is in the shape of the chemical bond that links the units together. This single difference is enough to make starch a readily

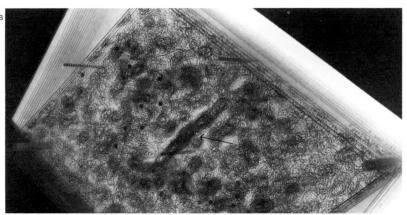

1 **The small protein insulin** is formed in the body from a simpler substance called proinsulin, a single chain with 84 amino acid molecules linked together. As well as the chemical bonds linking each amino acid chain, there are bridges between sulphur atoms [S] from the amino acid cysteine. When insulin is formed by removal of the 33 amino acid units [yellow], two chains are formed, held together by bridges.

2 **Proteins** are not simply amino acid molecules strung together like beads. Superimposed on the primary structure is a secondary structure, either an alpha-helix [A] or a pleated sheet [B]. Both arise from the formation of weak hydrogen bonds between similar parts of amino acid units. Although weak, hydrogen bonding is important in many biological macromolecules.

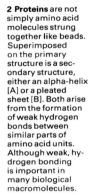
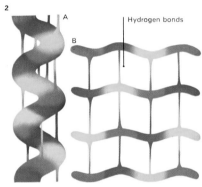

Hydrogen bonds

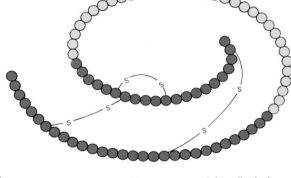

Enzyme

Substrate molecule

3 **As well as hydrogen bonds**, other interactions, such as those between the side-chains of similar amino acids, make the shape of some proteins even more complex. This may cause some disruption of the secondary structure, to produce the "tangled spaghetti" shape exemplified by this model of an enzyme, which has a substrate molecule fitted into its active site.

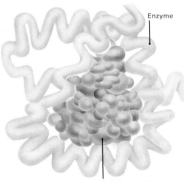

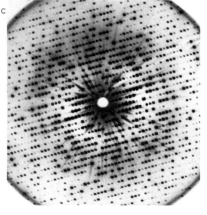

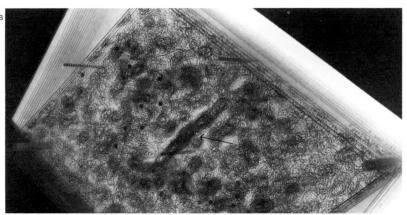

4 **Some proteins** can be purified sufficiently to form crystals. Myoglobin, a protein involved in carrying oxygen in muscles, occurs in many species, including sperm whales from which the crystals [A] are obtained (seen here enlarged 40 times). If a beam of X-rays is directed at a single crystal, it is possible to obtain a diffraction photograph [C]. Analysis of such photographs, often with the aid of computers, provides information from which electron density maps can be drawn. Made from sheets of Perspex stacked on top of one another, these maps show where particular atoms in the protein molecule are located [B]. It is then possible to construct an accurate 3-dimensional model of a protein. Much of the early work in determining protein structures by this method was done in Cambridge, England, by John Kendrew and Max Perutz.

digestible dietary ingredient and cellulose completely indigestible to human beings. In amylopectin, there are chemical links at more than one point on some of the glucose units, so that a branched-chain polymer is produced. The same sort of structure occurs in glycogen (animal starch), the glucose polymer used by animals for energy storage.

Many other sugars, apart from glucose, can form polysaccharides. Chitin, the hard shell material of insects, crabs and lobsters, is a polymeric aminosugar. Alginates, important food additives which keep the head on beer and give dehydrated soups their thickness, are polysaccharides from seaweed, while the pectins, which are widely used in jam-making, occur notably in apples. Natural adhesives, such as gum arabic, are polysaccharides, as is heparin, an important substance that prevents blood clots; it is often used in the treatment of thrombosis.

Nucleic acids
Although not present in such large quantities in most cells as proteins and polysaccharides, the most important macromolecules are the

nucleic acids [7]. These make up the genetic material which controls each cell, making it not only a man-cell or a mouse-cell, but a man-liver-cell or a mouse-tail-cell. Nucleic acid polymers are able to reproduce themselves accurately, therefore allowing any species to produce more of its own kind. They also control the chemical building of proteins. As the latter effect includes production of various enzymes, nucleic acids control all other chemical building up and breaking down in living tissues.

Complex polymers [7] can be thousands of units long. The basic repeating unit is made up from a nitrogen-containing base and a phosphate group. Both of these are attached to one of two types of sugar: ribose (in RNA) or deoxyribose (in DNA). Because of the chemical properties of the bases, particularly those in DNA, it is possible for two strands of nucleic acids to fit together readily to form the "double helix".

Synthetic rubber and plastics are also polymers – often man-made copies of natural molecules. Glass and similar substances are inorganic polymers.

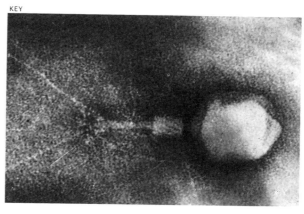

On the margin of life lie the viruses, each made up of a few macromolecules, all of which can be defined in purely chemical terms. Yet, when placed in a living cell, a virus is able to take over that cell's biochemical machinery and make it reproduce virus components. This electron micrograph of a bacteriophage – a virus that attacks bacteria – shows the shape of its protein molecules: inside the diamond-shaped head is the bacteriophage's nucleic acid which directs the build-up of further examples of both itself and the protein components after it has infected a cell.

5 Carbohydrates, proteins and polynucleotides are the basic polymers of life. They are not, however, the only ones. Natural rubber, for example, is a polymer made up mainly of repeating units of the unsaturated hydrocarbon isoprene. From 1,000 to 5,000 such units join together in a single molecule of rubber. The cell walls of some bacteria are made from a combination of sugar and amino acid molecules, to form a mixed polymer. This photograph, taken at a magnification of 280,000 times with the aid of an electron microscope, shows part of the outer cell wall of the bacterium *Clostridium thermohydrosulfuricum.* At this magnification it is possible to see the individual subunits, arranged in regular rows, that make up the surface of the bacterium's cellular wall.

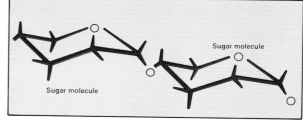

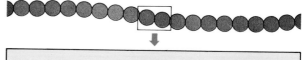

6 Monosaccharides, the simplest of the sugar molecules, can join together to form very large molecules such as starch. According to where the links form, the macromolecule may be a single chain (as shown) or branched.

Sugar molecule

Sugar molecule

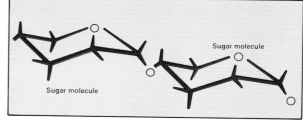

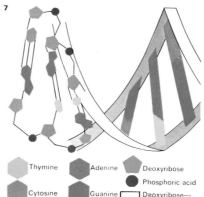

Thymine   Adenine   Deoxyribose

Cytosine   Guanine   Phosphoric acid

Deoxyribose—phosphate chain

7 Deoxyribonucleic acid (DNA) is the master molecule of life. It occurs as a double helix in which two complementary strands of polymer are held together by hydrogen bonding. This bonding occurs between the nitrogen-containing bases which form part of the nucleic acid unit.

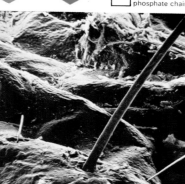

8 Hair is composed mainly of the protein keratin, which also occurs in feathers and skin. Although these hairs are magnified many times, it is still not enough to make individual keratin molecules visible. Whether a person's hair is straight or curly depends on the tertiary structure of the keratin molecules.

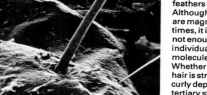

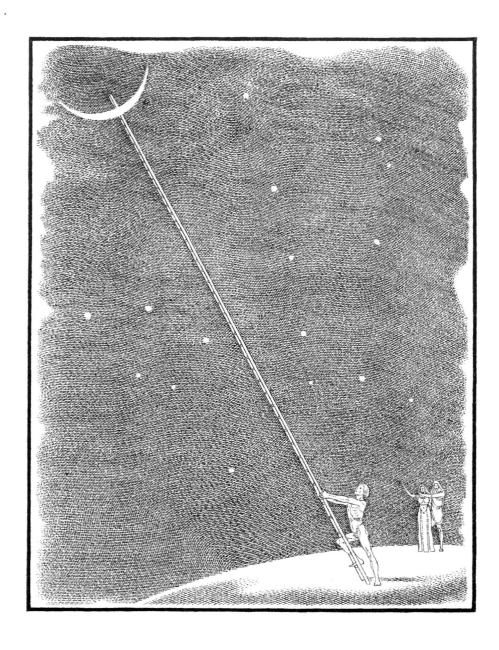

The Universe

To me, the vast emptiness of space is terrifying, is appalling.
Pascal, *Pensées* (1670)

An undevout astronomer is mad. **Edward Young,**
Night Thoughts (1742–6)

Observatory, n. A place where astronomers conjecture away
the guesses of their predecessors. **Ambrose Bierce,**
The Devil's Dictionary (1881–1911)

Comets are the nearest thing to nothing that anything can be
and still be something. National Geographical Society, press
release, 31 March 1955

Whoever starts out towards the unknown must consent to
venture alone. **André Gide,** *Journals,* 12 May 1927

Space flights are merely an escape, a fleeing away from
oneself, because it is easier to go to Mars or to the moon than
it is to penetrate one's own being. **Carl Gustav Jung,** quoted
in Miguel Serrano's "The Farewell", *C. G. Jung and Hermann
Hesse* (1966)

The Universe

An introduction by *Sir Bernard Lovell,* FRS,

Professor of Radio Astronomy, University of Manchester; Director of the Experimental Station, Jodrell Bank

Throughout the ages astronomical discovery has stretched the imagination of the human mind. The transference from the belief in a flat Earth to a globe that could be circumnavigated must have been exceedingly difficult. Indeed the direct visual and photographic evidence of the sphericity of the Earth awaited the high-flying aircraft and Earth satellites of our own age. In the sixteenth and seventeenth centuries the recognition that the Earth was not fixed at the centre of the universe created immense turmoil in the human mind. Aristotle in the 4th century BC and Ptolemy in the second century AD assumed that the Earth was fixed at the centre of the system of heavenly bodies. In the Ptolemaic system each planet moved in a small circle (or epicycle) whose centre was carried around the Earth in a larger orbit (the deferent), and for 14 centuries astronomers accepted this theory.

When the proposition that the Earth moves round the Sun was made by Copernicus nearly 500 years ago Luther declared "The fool will turn the whole science of astronomy upside down. But, as Holy Writ declares, it was the Sun and not the Earth which Joshua commanded to stand still". In 1508 Copernicus wrote an astronomical commentary in which he said "What appears to us as motions of the Sun arise not from its motion but from the motion of the Earth", and his heliocentric theory, published in the famous *De Revolutionibus Orbium Coelestium* in 1543, marks a vital stage in the development of human thought.

The Copernican theory dethroned the Earth from its hierarchical static position at the centre of the Solar System and with the invention of the telescope and its gradual improvement over the next few centuries man's interest turned to the stars. It is curious that for nearly another four centuries after the acceptance of the Copernican theory the Sun and the Solar System were believed to be at the centre of the stellar universe. The decade following 1918 was the critical epoch during which astronomical measurements finally eroded man's fundamental egocentric concept of his place in the universe. These years mark a period of revolutionary progress in our understanding of the structure of the Milky Way and of the larger scale organization of the cosmos. This emergence of a new understanding followed the discovery of a means for measuring the distances of stars far removed from the Solar System. For the past century it had been possible to measure the distance of the nearer stars by the straightforward trigonometric method of measuring their displacement against the background of faint stars with the Earth at opposite points of its orbit around the Sun. Even with the advent of photographic techniques this method had been extended only to distances of about 100 light-years by the end of the century and this enabled the distance of a few thousand stars to be determined. The discovery by Henrietta Leavitt (1868–1921) of Harvard in 1912, of the relationship between the period of variability and the apparent magnitude of the Cepheid variables, and the calibration by Harlow Shapley (1885–1972) in terms of absolute magnitude instantly led to the possibility of a great extension in the distance measurements. Immediately there was a revolution in our understanding of the larger scale structure of the universe.

Shapley studied the Cepheid variables in the globular clusters. By 1918 he had measured the distances of 25 of the 100 known objects of this type and found that they were at great distances from the Sun – 15,000 to 100,000 light-years. The clusters are unevenly distributed over the sky – a third are concentrated in the neighbourhood of the Sagittarius star cloud – and Shapley concluded that the Sun was far removed from the centre of the Milky Way system of stars. His work marked the final destruction of the age-old egocentric concept of man's place in the universe. The 100,000 million stars of the Milky Way are not arranged symmetrically with ourselves at the centre. They lie in a flattened disc extending for 100,000 light-years, with the Sun lying 32,600 light-years from the central region. The contemporary achievements of the radio astronomers who have been able to study the neutral hydrogen gas in the Milky Way soon led to a definite picture of the spiral structure of the Galaxy but these measurements have raised many new difficulties. It is accepted that the Galaxy is rotating with the arms trailing like a viscous fluid. At our distance from the centre we rotate once in 220 million years, whereas at a tenth of this distance the rotation rate is once in 28 million years. We know that of the total mass, equivalent to about 2×10^{11} Suns, only about 2 per cent is in the form of dust and gas. Most of this (nearly 99 per cent) is in the form of hydrogen gas but the irregular distribution of this gas and the recent discovery of small amounts of other complex molecules (including water) raises problems.

One of the remarkable features in the structure of the Galaxy is that in the central regions the mass is mainly composed of old red stars – within 2,000 light-years the gas contributes only about one per cent of the total mass. On the other hand in the spiral arms where the Sun is situated the ratio is markedly different – the stars are predominantly young blue stars and the gas represents about 20 per cent of the mass. We do not know why these differences occur.

We believe that the contemporary observations of the gas-clouds in the spiral arms provide good evidence that new stars are forming in these clouds. The results obtained by radio astronomers using radio telescopes working at very short wavelengths to study these gas-clouds has therefore been of special significance. Neutral hydrogen atoms emit a spectral line of a wavelength of 21cm. Similarly, various molecules have characteristic spectral line features but until recently no one envisaged the possibility of finding evidence for them in the interstellar medium. In 1963 the hydroxyl radical (OH) was detected and in the space of three years following 1969 the spectral lines characteristic of another 25 were found. Since that time the number of molecules now known to exist in space has steadily increased. A new subject of astrochemistry has arisen and a scientific basis has been established for the speculation that organic evolution may have occurred elsewhere in space.

Shapley's measurements on the globular clusters in 1918, which so quickly led to a radically new understanding about the Milky Way and also to a host of formidable problems unsolved today, was one of two major discoveries during the years immediately following World War I. The 100in (254cm) telescope of Mt Wilson had just been commissioned and with this instrument Edwin Hubble (1889–1953) obtained the first definitive evidence that the Milky Way galaxy did not comprise the whole universe. There had been speculation for a century that some of the nebulous objects to be seen in the heavens might be separate stellar systems outside of the Milky Way. Once more the ability to measure distances using the Cepheid variables provided the answer. In 1926 Hubble published the results on 400 systems in which he had measured the light variation of Cepheid variables. He found that they were at great distances from the Milky Way.

Hubble's proof of the extragalactic nature of these star systems is a major event in the history of astronomy. So, too, were the observations that he published on the relation between the distance of these systems and the shift towards the red end of the spectrum of their spectral lines. Interpreting the red shift as a Doppler effect, he found that the speed of recession increased linearly with distance and thereby established the observational foundation for the belief in the large-scale expansion of the universe.

When Hubble published these results he estimated that with the sensitivity of the 100in (254cm) Mt Wilson telescope it was possible to penetrate 140 million light-years into space, which encompassed two million extragalactic systems, and that the speed of recession at these limits of penetration was about 3,000km (2,000 miles) per second. Results published in 1975, obtained with the new Anglo-Australian telescope at Siding Spring in New South Wales, refer to systems at least five magnitudes fainter than the faintest objects available to Hubble. The number of observable extragalactic objects is estimated to be 100 million but at these limits the number was still increasing at about two or three times per magnitude.

Hubble found that the extragalactic nebulae were primarily of two types. The spherical and elliptical galaxies exhibiting little or no structure comprised about one-fifth of the nebulae he measured. Apart from a small percentage of irregular objects he classed the remainder as spiral galaxies and he believed that there was an evolutionary progression from the ellipticals to the spirals. Doubts about this evolutionary sequence arose with the discovery that the stars in the elliptical galaxies were mostly old, whereas the young stars were in the arms of the spiral galaxies. Discoveries made with

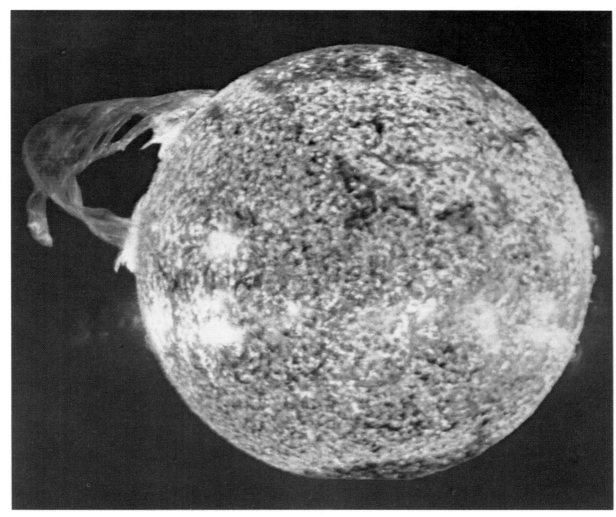

The violent Sun –
a solar prominence
reaches 400,000km
(250,000 miles)
out into space.

the radio telescopes since 1950 that have finally shattered belief in any such straightforward ordered sequence.

This new and confusing era began in 1951 when a strong radio source in Cygnus was identified with a peculiar image on a photograph taken with the 200in (508cm) telescope on Mount Palomar. The red shift measurements on this faint object established the distance as 700 million light-years away and the double nature of the image suggested two galaxies in collision. With the rapid discovery of more objects of this type, soon to be known as radio galaxies, and the realization that extremely large amounts of energy were involved, the idea of collisions was abandoned. Many of the radio galaxies consisted of two strong centres of emission straddling the optical image, implying that violent disruptive events had occurred in the nucleus of the galaxy. The strength of the radio signals from these radio galaxies encouraged the optical identification of more and more distant objects. In 1959 an important stage was reached with the radio and optical identification of a galaxy in Boötes with a red shift that implied a recessional velocity of 40 per cent of the velocity of light and a distance of 4,500 million light-years.

The attempt to find even more distant objects led to a surprising discovery. Objects, which from their radio properties were believed to be even more distant, were identified in 1960 with photographic images having a star-like appearance. They were characterized by their unusual intensity in the blue region of the spectrum and for two years were thought to be a new type of star in the Milky Way. Then in the spring of 1963 Schmidt at Palomar succeeded in the attempt to identify the spectral features of one of these objects. In fact they were not local to the Milky Way, but on the contrary possessed the largest red shifts of any objects known. They soon became known as quasars and since that time about 500 have been identified. Most of them have red shifts indicating recessional velocities of more than half the velocity of light and a few are known in which the implied recessional velocities are of the order of 80 per cent of the velocity of light. Their distances depend on the cosmological model adopted for the universe, but on the debatable assumption that the red

shifts are wholly associated with the cosmological expansion of the universe, these high red shift quasars are probably about 7,000 million light-years distant.

We do not understand how the quasars generate their vast output of energy – especially since in many the energy appears to be generated in volumes of space that are exceedingly small by astronomical standards. There has been speculation about the problem of gravitational collapse and the existence of superdense matter in the nuclei of the quasars. It is remarkable that radio galaxies and quasars, which were beyond imagination when Hubble established the existence of extragalactic objects, should now play such a vital role in our attempts to understand the universe.

The observed rate of expansion of the universe seems to imply that 10,000 million years ago the primeval material must have existed in a highly condensed state. In that case the possibility of studying quasars with a "look back" time of more than three-quarters of the time since the beginning of the expansion offered hope that the early history of the universe would be revealed. The attempts to interpret these observations within a cosmological framework led to great dispute – especially between the adherents of the steady state and evolutionary cosmologies. However, important and probably decisive evidence came unexpectedly from an entirely different source. In 1965 scientists at the Bell Telephone Laboratories in New Jersey, USA, tested equipment designed for communication tests using the American balloon satellite. They found that the signals from the sky exceeded by 100 times the noise level they had anticipated and furthermore that this signal was uniform from all parts of the sky. Their claim that the signals were the relic radiation from the initial hot and dense state of the universe 10,000 million years ago has been confirmed from many other tests by radio telescopes and high altitude equipment.

We now appear to have this direct evidence of the state of the universe only seconds after the beginning of the expansion when the temperature of the primeval material was thousands of millions of degrees. The possibility that the universe evolved from a dense initial condition was embraced by Einstein's general theory of relativity in 1915, but the contemporary observational evidence in favour of this initial state poses a severe problem. The singularity in the solution of the equations implying that at zero time the universe was of infinitesimal dimensions and of infinite density has often been regarded as a mathematical difficulty arising from the assumption that the universe is uniform. But the measurements of the relic radiation now indicate that the universe does possess a high degree of uniformity. Within the framework of the contemporary laws of physics it seems possible to envisage a physical state where the entire primeval material existed with a universe of dimensions 10^{-33}cms – a condition predicted to exist 10^{-43} seconds after the beginning of the expansion. In the approach to a physical description of the beginning of time we reach a barrier in contemporary theory at this point. The problem as to whether this really is a fundamental barrier to a scientific description of the initial state of the universe, and the associated conceptual difficulties in the consideration of a single entity at the beginning of time, are questions of outstanding importance.

Will the universe continue to expand for ever or will it eventually collapse upon itself to another state of high density? There are clear observational tests to be applied in the attempt to answer this question. For example does the strictly linear relation between red shift and distance break down at great distances, or is the density of the universe greater or less than 2×10^{-29} grammes per cubic centimetre? If it is greater, then the gravitational forces will eventually overcome the forces of expansion and the universe will collapse. These types of measurements are fraught with difficulties which have not yet been surmounted and with current technology we do not know the answer.

The immense advances in our knowledge of the universe have been related to our ability to study it over a wide range of the spectrum. The first great advance came immediately after World War II when the new techniques of radio astronomy emerged. Then with the launching of Sputnik 1 in 1957 another era opened when it became possible to dispatch scientific instruments into space and so evade the problems posed by absorption in the Earth's atmosphere. The entire spectrum embracing gamma-rays, X-rays and the long-wave radio waves has been thrown open for investigation. Even from the earliest results it is evident that only one prediction is safe. It is that our future description of the universe and all its component parts will change continuously as it has done for centuries past.

The style of the heavens – the Dumbbell nebula in Vulpecula, taken through the 200in (508cm) telescope at Palomar.

The restless sky

Astronomy is the oldest of all the sciences. It was a natural and long-held assumption that the Earth must be flat and lie at rest in the centre of the universe, with the entire sky revolving round it once every 24 hours; but at an early stage it was also clear that many celestial bodies had their own relative motions across the heavens.

Movements in our sky

The Moon was seen to shift quickly in position against the starry background and the Sun, of course, had its own motion. Then there were also the occasional spectacular phenomena; sometimes the Sun was blotted out during a solar eclipse and sometimes the Moon became strangely dim when full. It was not then known that a solar eclipse occurs when the Moon passes between the Sun and the Earth, casting a shadow on the Earth, and a lunar eclipse occurs when the Sun, Earth and Moon are in line and the Moon enters the Earth's shadow; but it has been suggested that some of the old stone circles were primitive eclipse computers [Key].

The Greeks recognized that the five bright planets – Mercury, Venus, Mars, Jupiter and Saturn – moved against the stars and were thus fundamentally different from them. The constellation patterns remained unchanged over long periods and it was originally thought that the stars were fixed onto a crystal sphere revolving round the Earth.

The planets were regarded as being closer to the Earth and it was thought that they, like the Sun and Moon, moved round the Earth between the Earth's surface and the sphere of the so-called fixed stars. The old system was perfected by Ptolemy (c. AD 90–168). In the Ptolemaic system [1], all celestial orbits were assumed to be perfectly circular; but as the observed movements of the planets did not conform to the idea of circular motion at uniform velocity, it was necessary to introduce complications – the epicycle that described the movement of a planet is a small circle the centre of which (the deferent) itself moved round the Earth in a circle.

A few of the earlier Greek philosophers, notably Aristarchus (310–230 BC), had believed that the Earth moved round the Sun, but the Sun-centred or heliocentric theory was generally rejected until the work of Copernicus (1473–1543), a Polish canon, made its impact in the sixteenth century. Copernicus took the drastic step of removing the Earth from its proud central position and put the Sun there instead [2]. However he retained perfectly circular orbits and was even compelled to retain epicycles. The modern phase of astronomy dates from the publication of his great book, *De Revolutionibus Orbium Caelestium*, in 1543.

Revolutionary outlooks

Inevitably, the Copernican system was strongly opposed. Tycho Brahe (1546–1601), the Danish astronomer who was the most thorough observer of pretelescopic times, believed the planets moved round the Sun while the Sun and Moon moved round the Earth [3]. When Tycho died his observations of the positions of the stars and the movements of the planets came into the possession of his last assistant, Johannes Kepler (1571–1630). After years of work, Kepler realized that the planets move round the Sun not in circles but in ellipses; and be-

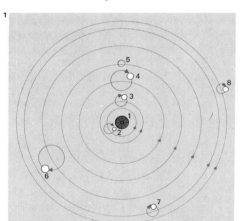

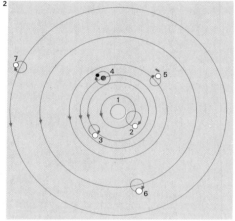

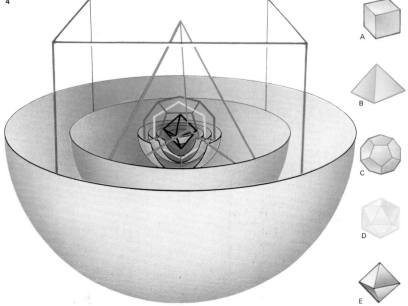

1 In the Ptolemaic system, the Earth [1] lies at rest in the centre of the universe. Round it move the Moon [2], Mercury [3], Venus [4], the Sun [5], Mars [6], Jupiter [7] and Saturn [8], each body moving in a small epicycle.

2 The Copernican theory places the Sun [1] in the centre of the Solar System orbited by Mercury [2], Venus [3], the Earth [4], Mars [5], Jupiter [6], and Saturn [7]. Copernicus's book was published in 1543. His theory met strong opposition from the Church, and religious persecution persisted for a century. Copernicus retained both circular orbits and epicycles.

3 Tycho Brahe believed that the Earth [1] was the centre of the Solar System, orbited by the Moon [2] and the Sun [3]. The planets – Mercury [4], Venus [5], Mars [6], Jupiter [7] and Saturn [8] – moved round the Sun, with the stars beyond.

4 Kepler's theory of "five regular solids" shows that his ideas provided a link between the past and the present. He believed that the five regular solids – the cube [A], tetrahedron [B], dodecahedron [C], icosahedron [D] and octahedron [E] – could be fitted inside the orbits of the various planets. He reasoned that there were only five such solids and exactly five spaces between the six planets known at that time: Mercury, Venus, Earth, Mars, Saturn and Jupiter. It was his brilliant work, based upon observations made by Brahe, that showed that the Sun, not the Earth, was the centre of the Solar System. Kepler was both a mathematical genius and an astrological mystic.

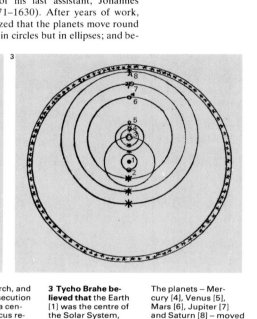

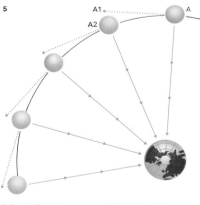

5 According to Newton, were it not for the Earth, the Moon would move in a given period from A to A1, but because of Earth's pull, the actual movement is from A to A2. The Moon keeps "falling" towards the Earth, although it comes no closer to us. This illustrates the law that a body will continue in a state of rest, or uniform motion in a straight line, unless acted upon by an outside force. This law was laid down in Newton's *Philosophiae naturalis principia mathematica* (1687).

tween 1609 and 1618 he published his three fundamental laws of planetary motion. The first law states that the orbit of a planet is an ellipse, with the Sun at one of the foci. The second law states that a planet moves at its fastest when it is closest to the Sun and the third law provides a definite relationship between a planet's sidereal period (that is to say, the time taken for the planet to complete one journey round the Sun) and its distance from the Sun. Using Kepler's laws it became possible to draw up a scale map of the Solar System. When one distance could be determined absolutely, the distances of all the rest could be obtained by calculation.

The revolution in outlook was completed by Isaac Newton (1642–1727), whose book – usually called the *Principia* [5] published in 1687 – laid the foundations of all subsequent work. By then the distance of the Sun from the Earth was known with reasonable accuracy and it had become clear that the Solar System was a very small part of the universe as a whole. The stars were known to be suns in their own right and to be so far away that their apparent individual or proper motions

were to an observer very slight indeed.

Edmond Halley (1656–1742), friend and contemporary of Newton, used ancient observations to demonstrate that a few of the bright stars had shown relative shifts over the centuries, so that even the constellation patterns could change gradually with time.

The scale of the universe

The scale of the universe was established only much later, when distances to stars began to be measured. In 1838 Friedrich Bessel (1784–1846) first measured such a distance to a nearby star (in Cygnus) and found it to be about 96 million million kilometres (60 million million miles) away. Since light takes about 11 years to cover this distance, the star is said to be 11 light-years away. Most stars are much more remote than this, but modern techniques enable astronomers to measure their proper motions from year to year. The old name of "fixed stars" is misleading; all the stars are moving at high velocities relative to each other. In our own century it has been shown that our Galaxy is itself one of many; there are millions of other galaxies.

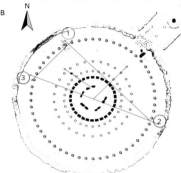

Stonehenge [A], the famous megalithic "stone circle" in Wiltshire, England, is made of standing stones with alignments that may have astronomical significance. Many of the alignments among the concentric rings [B] point to spots on the horizon where the Moon and Sun rise and set. For example stone 1 as viewed from stone 2 marks the point where the Moon sets in its most northerly position in midwinter. Viewed from stone 3, stone 1 marks the midsummer sunrise. Stonehenge may have been an early primitive computer predicting eclipses for religious ends or, more practically, fixing the solstices, which were important in the agricultural calendar.

6 The two planets whose orbits lie within that of the Earth – Mercury and Venus – show lunar-type phases and remain in the same area of the sky as the Sun [A]. An inner planet is at inferior conjunction [1]; its dark side is turned towards Earth so it appears "new". When on the far side of the Sun [2], it is full. The synodical period, or mean interval between successive inferior conjunctions, is 115.9 days for Mer-

cury and 583.9 days for Venus. The orbits of the Earth [4] and Venus [3] are shown [B] and the white line indicates the apparent motion of Venus in the sky. Mercury behaves in a similar way.

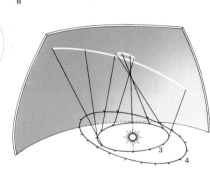

7 Orbits of superior planets [A], beyond Earth's orbit, reach opposition [1] and conjunction [2]. Apparent motion [B] of a superior planet [4] in relation to Earth [3] appears temporarily retrograde.

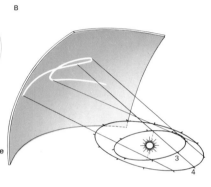

9 The real movement of a star in space includes actual motion [A] where the star moves from 1 to 2 in a given period. Radial motion [B] is when the star moves from 1 to 3 if receding (positive radial motion), or 3 to 1 if approaching (negative radial motion). Proper motion [C] is the term used for the transverse movement (1 to 3) against the background of more distant stars. [A] combines [B] and [C].

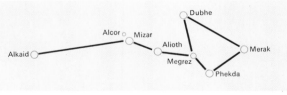

10 To the naked eye constellations appear unchanged over thousands of years, but over a sufficiently long period the proper motions show up. The seven main stars of Ursa Major, including the double star Mizar, are shown as they were 100,000 years ago [A], as they are today [B] and as they will appear 100,000 years hence [C]. It is evident that Dubhe and Alkaid are moving in an almost opposite direction to that of the other five.

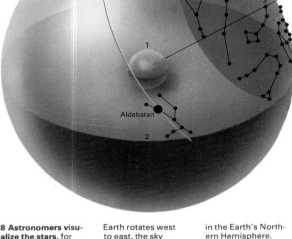

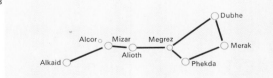

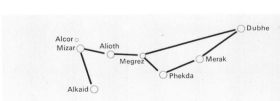

8 Astronomers visualize the stars, for the sake of convenience, as lying on the inside of a sphere centred on the Earth. To an observer on Earth [1], the horizon becomes a circle [2] when projected on to this celestial sphere. As the

Earth rotates west to east, the sky seems to move east to west, taking the stars, such as Aldebaran in Taurus, with it. The north pole of the sky [3], which is indicated approximately by Polaris, is stationary to an observer

in the Earth's Northern Hemisphere. Stars in the mauve area [4], drawn for an observer at latitude N50°, remain permanently above the horizon and are called circumpolar stars. The circumpolar area depends on the observer's latitude.

Measuring the restless sky

The size of the universe is almost unimaginable. It is easy to comprehend the distance from London to New York, or from New York to Australia; and the Moon does not seem impossibly remote, because its distance is only ten times greater than that of a journey right round the Earth. But any attempt to visualize what is meant by "a million kilometres" is doomed to failure – and a million kilometres is a very short distance on the cosmic scale.

Early estimates of distance
The ancients had no idea of scale (it was once thought that the diameter of the Sun was only 70cm [27in]), but they were able to measure the size of the Earth itself with remarkable accuracy [1]. As soon as the old concept of an Earth-centred universe was abandoned, distance estimates became much more realistic. Giovanni Cassini (1625–1712) gave the distance between the Earth and the Sun as 138 million kilometres (86 million miles), which was approaching the true figure. Astronomers then decided that the Earth–Sun distance was to be *the* astronomical unit.

and measuring it became a major task [Key].

The basis of any method of computing this distance was Kepler's third law, which established a definite relationship between the revolution period of a planet and its distance from the Sun. The revolution period of the Earth was known to be 365.25 days and the periods of the other planets could be found from observation – 687 days for Mars, and so on – with the result that a complete scale model of the Solar System could be drawn up. Thus if it were possible to obtain the distance from Earth to any planet (eg Mars or Venus), Kepler's third law would give the Earth–Sun distance [4].

The parallax principle
The obvious way to calculate the distance from the Earth to one of the planets was to use parallax [2], a method also used by surveyors. If a not-too-distant object is observed against a background of more remote objects, its position will seem to alter according to the position of the observer. If the distance between the two observation points is known, and the respective angles

formed by the object with the line are measured, an astronomer can calculate by trigonometry the height of the triangle thus formed, ie the distance of the heavenly body.

Edmond Halley (1656–1742), the second Astronomer Royal, proposed to make use of transits of Venus – the rare occasions when Venus passes in front of the Sun as seen from Earth, and appears as a black spot against the solar disc – to determine the planet's absolute distance. Attempts made during the transits of 1761 and 1769 were only partly successful. (Captain Cook's voyage to the South Seas in the latter year was for the express purpose of observing the transit.) The next transits were those of 1874 and 1882; all measurements agreed in making the astronomical unit about 150 million kilometres (93 million miles).

When Venus passed across the face of the Sun, however, it seemed to draw a strip of blackness after it, distorting its shape; when this so-called "Black Drop" vanished, the transit had already begun, which meant that the measurements were subject to considerable error. In 1877 efforts were made to determine the parallaxes of three asteroids –

CONNECTIONS

Read first
164 The restless sky
60 The scale of the universe

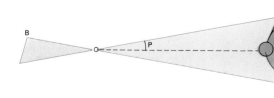

1 Eratosthenes calculated the Earth's circumference by noting that when the Sun was overhead at Syene [A], it was 7.2° from the zenith at Alexandria [B]. AB was known; as 7.2° is 1/50 of a circle, the circumference was 50 × AB.

2 Apparent parallaxes of stars are illustrated in these diagrams. The parallaxes are measured for the apparent movements of relatively nearby stars against a background of more remote stars, which are too distant for any detectable motion. With stars lying in the direction of the axis [Y] of the Earth's orbit, the parallactic motion over a year will be circular [A]. If a star is co-planar with the ecliptic [B],

the parallactic motion will take the form of a to-and-fro straight line. P is the angle of parallax and from this the distance of the star can be measured. The main difficulty is, of course, that the angle P is always very small. With modern photography and other techniques we can now "see" out to at least 200 light-years, but at greater distances the parallax shifts are swamped by observational errors.

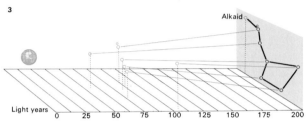

3 Stars in Ursa Major seem to the casual observer to be the same distance from the Earth; there is no "three-dimensional" effect. In fact

the stars in any particular constellation are not necessarily associated. The diagram shows the relative distances of the seven main stars in

Ursa Major; Alkaid (210 light-years) is easily the most remote. Mizar, lying next to it in the sky, is only 88 light-years away.

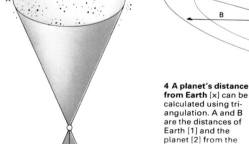

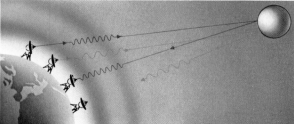

4 A planet's distance from Earth [x] can be calculated using triangulation. A and B are the distances of Earth [1] and the planet [2] from the

Sun. M1 and M2 are the planet's apparent positions as seen from W1 and W2 on Earth. The ratio of the angles at M1 and M2 is the same as the

ratio of the distances W1–W2 and x. Using Kepler's third law $(T1/T2)^2 = (A/B)^3$, where T is the period of orbit, A and B can also be found.

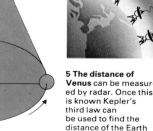

5 The distance of Venus can be measured by radar. Once this is known Kepler's third law can be used to find the distance of the Earth

from the Sun. Radar pulses of differing wavelengths are reflected from Venus to the receivers on Earth. The time-lag between transmitting

and receiving echoes gives the distance travelled, provided allowances have been made for the delaying of the echoes by the ionosphere.

Iris, Victoria and Sappho – which appeared as star-like points. In 1931, when the asteroid Eros passed within 24 million kilometres (15 million miles) of the Earth, there was a worldwide programme to determine its parallax precisely.

New methods involve the use of radar. Radar transmits a pulse of energy to a remote object and receives an echo from it: radio waves travel with the speed of light and, because this speed is constant, the time-lapse between the transmission and the arrival of the echo enables the distance of the object to be calculated. The planet Venus is contactable by radar [5]. This new method gives the length of the astronomical unit as 149,600,000km (92,957,000 miles).

The problem of mapping the stars
Star distance presented different problems and here again parallax was used. If a nearby star is observed over a six-month interval it will show a parallax shift against the background stars, because during the interim the Earth will have moved from one side of its orbit to the other, giving a "baseline" of 300

million kilometres (186 million miles). This method was first applied in 1838, when Friedrich Bessel (1784–1846) showed that a faint star in Cygnus lay at a distance of 11 light-years. (One light-year is the distance travelled by light in one year; that is, 9,460,000 million kilometres [5,880,000 million miles].)

The parallax methods work well for the nearer stars, but beyond a few hundreds of light-years the shifts become too fine to be measured and less direct methods must be used. Spectroscopic work will give the real luminosity of a star and this, compared with its apparent brightness (or visual magnitude), can give the distance. It is now known that the diameter of our Galaxy is about 100,000 light-years.

But our Galaxy is not the only one. The hazy patches known as nebulae are of two kinds: some can be resolved into stars, others cannot. In 1845 Lord Rosse (1800–67), using his 72in (183cm) telescope, found that many of the starry nebulae are spirals and it has now been found that spirals are external systems, millions of light-years away.

KEY

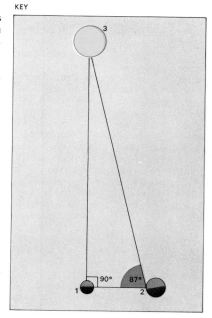

Aristarchus, the Greek astronomer of the 3rd century BC, is said to have been the first man to propose a heliocentric theory of the universe. He was also able to measure the relative distances of the Sun and Moon. When the Moon is at first quarter [1], the angle it makes with the Sun [3] is near 90°. By measuring the angle at the Earth [2] Aristarchus could determine from the triangle the relative distances from Earth. He found the angle to be 87°, instead of the true value of 89°52'; but a small error can lead to a large discrepancy in the ratio of the distances of the Sun and Moon respectively from the Earth. Aristarchus' ratio was 19 to 1; the true ratio is 390 to 1.

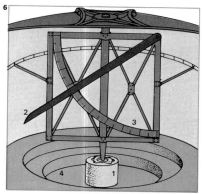

6 Tycho Brahe's quadrant was one of the instruments used for measuring star positions. This quadrant, used between 1576 and 1596, was mounted on a central pillar [1]; a pointer [2] with sights rotated against a graduated metal circle [3]. The well [4] accommodated the observer at various levels determined by the position of the pointer. Modern work depends on fundamentals like those established by Tycho.

7 Ancient astrolabes [A] had simple pointers and scales to measure the altitudes of stars and other objects in the sky. In the modern astrolabe [B] the light strikes a prism [1] and a mercury surface [2], forming a double image along the collimation line [3]. The images separate when the object moves and this separation is measured by the azimuth scale [4] giving the altitude of a moving celestial body.

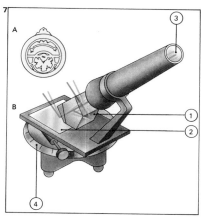

8 An orrery made in 1790 is illustrated here. The name originates from the Earl of Cork and Orrery, for whom an elaborate instrument was made. Orreries indicate the movements of the planets round the Sun. In the orrery shown here, the Sun is represented by a brass ball in the centre. Round it move the three innermost planets, Mercury, Venus and the Earth; an ingenious system of gears makes the planets move round the Sun in the correct relative periods, even though on a scale of this kind it is naturally impossible to give the correct relative distances. The lands and seas of Earth are shown and also the Moon in its orbit round the Earth, which is inclined at the correct angle. When, by turning a handle, the mechanism is moved, the planets revolve round the Sun and the Moon revolves round the Earth. More modern orreries are driven by clockwork, and in some the more distant planets than Earth and those discovered in the years after this model was made, are also shown.

Telescopes

The telescope is the main research instrument of astronomy. Without it our knowledge would be very limited indeed because other instruments – such as those based upon the principle of the spectroscope – depend upon telescopes to collect light that is to be analysed. George Ellery Hale (1868–1938), who was largely responsible for the building of the 200in (508cm) reflector at Palomar, in the United States, which was for many years the most powerful telescope in the world, once said that his call was always for "More light!" This is still true today, for modern astronomers are continually striving to investigate extremely faint objects that lie at immense distances from the Earth, and are gazing ever more deeply into the universe.

How refractors work

Telescopes are of two main kinds: refractors and reflectors [Key]. Each type has its own advantages and, unfortunately, its own drawbacks. Refractors, developed during the first decade of the seventeenth century, were first in the field and were used by pioneers such as Galileo (1564–1642). In a refractor the light

from the object to be studied passes through a specially shaped lens known as an object-glass or objective; the rays of light are brought to focus and the resulting image is magnified by a second lens known as an eyepiece or ocular. The larger the object-glass the greater the light-grasp of the telescope; thus a 6in (15.2cm) refractor (that is to say, a refractor with an object-glass six inches across) is twice as powerful as a small 3in (7.6cm) refractor.

The only function of the object-glass is to collect light; all the actual magnification is done by the eyepiece. Every astronomical telescope is equipped with several eyepieces, which can be used as desired. The limit depends upon the amount of light available. Thus if, say, an eyepiece giving a magnification of 500 were used with a 3in (7.6cm) refractor, the resulting image would be so faint that it would be useless; to make use of a magnification of 500 a larger object-glass would be needed.

All refractors have one defect in common: they produce false colour. This is due to the nature of light itself, which is a

blend of all the colours of the spectrum [1]. As the ray of light passes through the object-glass it is bent or refracted in order to be brought to focus; but the longer wavelengths are bent less sharply than the shorter ones. Thus the red part of the beam is bent less than the blue and so is brought to focus in a different place. The result is that a bright object such as a star is associated with false colour that may look beautiful but which, to the astronomer, is most unwelcome. This can be partly remedied by using compound object-glasses with one lens of crown glass and the other of flint glass; these have different refractive properties, and the false colour is reduced. It could be almost entirely removed (as it is in cameras) by adding more lenses, but that would significantly reduce the amount of light reaching the eye of the observer, and this is a fundamental consideration in astronomy [2].

Reflecting telescopes

The reflecting telescope, of which the first working example was made by Isaac Newton (1642–1727) in about 1671, works on an

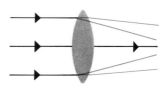

1 When white light, which contains all wavelengths of the visible spectrum, is passed through a glass prism, the beam is split up [A] (the colours bend unequally), into the spectrum ranging from the longest wavelength (red) to the shortest (violet). When one colour is passed through a hole in a screen and then through a second prism, there is no further splitting up [B]. By inverting the second prism the colours can be recombined [C].

2 The cause of the irritating false colour that is always present when a refractor is used is shown here. The light from the object passes through the object-glass and is split up to some extent so that the red rays are brought together from the lens at a different point from the blue rays [A]. The same is true if a different kind of lens is used [B]. The solution is to use a compound object-glass [C] made up of two lenses combined; the errors then tend to cancel each other out and the false colour is appreciably reduced, although for colour correction a refractor is always inferior to a reflector.

3 Reflectors are of various types. In the Newtonian pattern the light is collected by a parabolic mirror and sent on to a flat mirror at an angle of 45 degrees; the light is sent into the side of the tube, where the image is formed and magnified. To avoid the admittedly small light loss due to the flat, William Herschel tilted the main mirror and dispensed with the flat. However, this design is unsatisfactory. In the Cassegrain the secondary is convex and the light is reflected back down through a hole in the main mirror. In some designs the returning light is diverted by a second flat mirror into the side of the tube, which avoids making a hole in the main mirror.

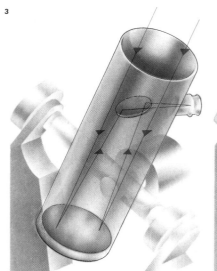

Newtonian reflector

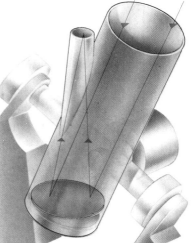

Herschel's reflector

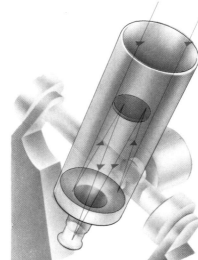

Cassegrain reflector

entirely different principle. On the Newtonian pattern [3] the light passes down an open tube until it hits a mirror at the far end. This mirror is curved; the shape is that of a paraboloid and the light is reflected back up the tube onto a second, flat mirror placed at an angle of 45 degrees. The light is then directed into the side of the tube, where it is brought to focus, and the image is magnified by an eyepiece as before. The presence of the flat mirror in the tube cuts out some of the light but the loss is not serious and with the Newtonian pattern there is no way of avoiding it.

Because a mirror reflects all colours equally there is no chromatic aberration – although a certain amount of false colour may be produced in the eyepiece. Modern mirrors are ceramic and are coated with a thin layer of some highly reflective substance such as aluminium or silver.

The Newtonian is not the only form of reflector. In the Cassegrain or Gregorian type [3] the second mirror is also curved and the light is reflected back through a hole in the main mirror. In the Herschelian type of reflector the main mirror is tilted and the secondary mirror is dispensed with altogether, but this involves distortions and as a result, Herschelian telescopes are now considered obsolete.

Advantages and disadvantages

Aperture for aperture a refractor is more effective than a reflector, but it is also more expensive because large lenses are harder to make than large mirrors. For this and other reasons all the world's largest telescopes are reflectors [5]. For the amateur astronomer the minimum really useful aperture is probably 3in (7.6cm) for a refractor and 6in (15.2cm) for a reflector.

The question of a mount is all-important. If the mounting is unsteady the telescope will be useless. It is highly desirable to use an equatorial stand in which the telescope is attached to an axis that is parallel to the axis of the Earth. With a driving mechanism added, the telescope can be driven in a way that compensates for the Earth's rotation and keeps the object under study perpetually in the field of view.

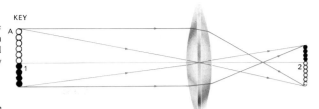

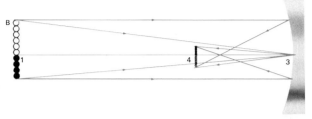

In a refractor [A] light from the object [1] passes through the lens to form an image [2]. The distance between the lens and the focal point is known as the focal length. Unless an extra lens system is used, the image is inverted.

In a reflector [B] light from the object [1] is collected by a curved mirror [3] and is brought to focus, forming an image [4].

4 A more modern type of reflector is the Coudé, which has a secondary mirror and an extra, rotatable mirror on the polar axis of the telescope. Since the light rays are reflected in a constant direction the image formed is stationary and the observer need not move as the telescope rotates. This has the great advantage that heavy and delicate equipment can be set up and need not be moved. Most modern reflectors allow for a Coudé focus and can be used according to several optical systems, which permits great versatility; in some telescopes the change-over to a Coudé system can be made very quickly. It can also be used with refractors.

5 The 200in (508cm) Hale reflector was for many years the world's largest and was unrivalled in its light-grasp. The diagram shows the primary mirror [1]; observer's cage [2]; Cassegrain focus [3]; the Coudé focus [4]; the southern end of the polar axis [5]; the Cassegrain and Coudé secondary mirrors [6]; the right ascension drive [7]; the declination axis [8]; the dome shutter, with an opening of 9m (30ft) [9]; the 42m (137ft) dome [10]; the primary focus, 16.5m (54ft) [11]; the northern pillar [12]; the southern pillar [13]; and the control panel [14], from which the telescope can be made to point to any part of the sky.

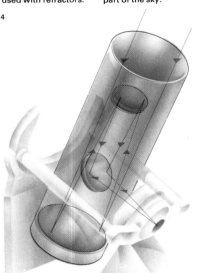

Coudé telescope

6 The cage for the observer can be set up within the tube itself in a telescope that is as large as the Hale reflector. This means that photographs can be taken at the prime focus. This is an obvious advantage because it means that no secondary mirror is needed – and every reflection from a mirror inevitably involves loss of light. The amount of light blocked out by the observer's cage is tolerated because of the benefits the cage provides.

7 One of the observers at Palomar loads a plate at the 200in (508cm) reflector, which was here being used at the Coudé focus. To change from one optical system to another takes a certain amount of time, but the operation is routine.

Great observatories

It is often thought that an astronomical observatory consists simply of a dome-shaped building containing a telescope. This is certainly true of some amateur observatories, but professional observatories are extremely elaborate and contain equipment of many kinds. Great care has to be taken in siting an observatory because a dark sky with no interference from artificial light and a clear, transparent atmosphere are absolutely paramount for good observation.

Today, little astronomical work is carried out visually – that is to say, by an observer sitting at the eyepiece of a telescope. Virtually all research is carried out by means of photography and the world's largest telescopes are used as giant cameras.

Observatory sites and equipment

To photograph a very faint object such as a remote galaxy it is necessary to make a time-exposure, which may last many hours. Consequently, stray light is probably the atronomer's worst enemy. Today, with the spread of cities and the consequent marked increase in light pollution, it is becoming more and more difficult to find really good sites that combine darkness with a high percentage of clear, cloudless nights. In addition the Earth's atmosphere is subject to turbulence and also absorbs light, so that it is desirable to set major telescopes upon mountains, above the densest atmospheric layers. This means that an observatory must be virtually self-contained, with accommodation for the observers as well as workshops, photographic laboratories and lecture halls. Even so, the usual procedure is for the observer to spend a limited time at the observatory itself and then return to his office with the photographic results of his work.

Observatories built during the late nineteenth century were equipped with large refractors. Pride of place must go to the 40in (102cm) refractor at the Yerkes Observatory in the United States. It is not likely that any larger refractor will be built, because a lens has to be supported round its edge and above a certain limiting size (about 40in [102cm]) the lens starts to distort under its own weight. Large lenses are also subject to severe chromatic and spherical aberrations that distort the image; these aberrations can be avoided with mirrors. For these reasons most modern telescopes are reflectors.

The Hale Observatories

The most famous planner of giant telescopes was George Hale (1868–1938) who designed observatories and persuaded friendly millionaires to finance them. At Mount Wilson, in California, Hale was responsible for the erection of first a 60in (152cm) reflector and then one of 100in (254cm). The latter, completed in 1918, remained the largest telescope in the world for more than 30 years and was instrumental in making fundamental advances in astronomy. It was then surpassed in 1948 by the 200in (508cm) reflector at Palomar, California, also masterminded by Hale, who died before the telescope was completed. Fittingly, Mount Wilson and Palomar are now administered jointly under the name of the Hale Observatories. The 200in (508cm) telescope can be used on three optical systems (prime focus, Cassegrain and Coudé), and is so large that the observer's cage can be placed inside the tube

CONNECTIONS

See also
168 Telescopes
172 Invisible astronomy

In other volumes
302 History and
Culture 1

1 The world's largest telescope is the 236in (600cm) reflecting instrument at Zelenchukskaya, in the Northern Caucasus. Its technical advantage over the Hale 200in (508cm) at Palomar is considerable, although observing conditions in this region are not quite as good as those in California. The 236in instrument is entirely Soviet made. It has an alta-zimuth mounting with a more complicated drive mechanism than the usual equatorial mount. The dome is made to the conventional pattern. The telescope will be used mainly to study remote star systems, exploiting its immense light-gathering power. Initial tests were carried out in 1974.

2 The Anglo-Australian telescope at Siding Spring Mountain near Coonabarabran, New South Wales, is a 153in (389cm) reflector. Four different optical systems can be used: prime focus, f/8 or f/15 Cassegrain and f/36 Coudé. The total mass of the telescope is 326 tonnes. Its design is similar to the 150in (381cm) Kitt Peak telescope.

3 Lick Observatory in California was founded with a donation from James Lick in 1874–5, and came under the direction of the University of California in 1888. The principal instrument is the 120in (305cm) reflecting telescope, shown here, that went into operation in 1959. Many of the design features are similar to the 200in at Palomar.

– where the 45-degree flat mirror of an amateur Newtonian telescope is sited.

The 48in (122cm) Schmidt telescope [4] at Palomar is used purely for photography and incorporates a spherical mirror together with a complicated correcting plate. The advantage of a Schmidt telescope is that it can photograph wide areas of the sky with a single exposure, whereas the field of a telescope such as the 200in (508cm) is by its very nature extremely limited.

The world's largest telescopes

The 200in Palomar telescope is no longer the largest optical telescope in the world, since the Russians have built a 236in (600cm) reflector in Zelenchukskaya [1]. There are also various telescopes in the 100-160in (254-406cm) range. Major observatories have been set up in the Southern Hemisphere, where the skies are clear and the important objects of the far south, such as the Magellanic Clouds, are accessible. There are sites in Australia [2], South America and South Africa, where the main telescopes of the Republic have been collected together at

a single site – Sutherland, in Cape Province – for particularly good observing conditions.

The largest telescope ever set up in Britain is a 98in (249cm) reflector, at Herstmonceux, Sussex, known as the INT or Isaac Newton Telescope. During the 1950s the instruments at the famous observatory at Greenwich [Key] were moved to Sussex but even there difficulties arose with cloud and scattered light and the 98in (249cm) is to be moved to a new site in the Canary Islands, where it will be joined by a new 160in (406cm) reflector.

Some observatories have special roles; for instance at Kitt Peak in Arizona there is elaborate equipment for studying the Sun, while the Lowell Observatory in Arizona [5] specializes in planetary work. New observatories are now being planned, many conceived primarily to overcome the limiting effects of the Earth's atmosphere. Thus several artificial satellites have been equipped with telescopes, culminating in the achievements of Skylab – the first manned orbiting observatory – and its Soviet counterpart, Soyuz.

Flamsteed House, in Greenwich Park, London, designed by Christopher Wren (1632–1723), is the site of the old Royal Observatory, set up in 1675. The instruments have been moved to Sussex and the old observatory is now a museum.

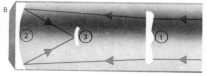

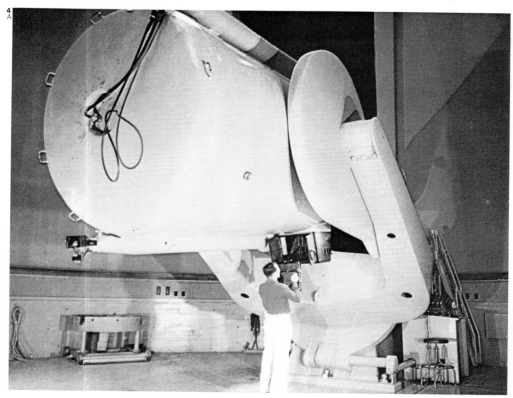

4 Modern telescopes have immense light-gathering power. But the conventional telescope is able to cover only a small area of sky with one photographic exposure. For studying individual objects such as galaxies this does not matter, but it means that to compile a photographic map of the whole sky would take too long. The principle of the Schmidt telescope [B], developed by Estonian optician Bernhard Schmidt (1879–1935) in 1932, enables large areas to be photographed with each exposure. There is a spherical mirror and a glass correcting plate over the end of the tube to compensate for optical distortion. The light passes through the plate [1] to the mirror [2] and is reflected onto a curved photographic plate [3] in the tube. The Schmidt telescope at Palomar [A] has a 48in correcting plate.

5 The Lowell Observatory at Flagstaff, Arizona, was set up by Percival Lowell (1855–1916) in 1895 mainly to study Mars; the observatory has been known for its planetary work, although much equipment, including a large reflector, has been added since Lowell's time. The photograph shows the dome of the 24in (61cm) refractor that Lowell used for his Martian studies from 1895 to 1916. The fine-quality optics are as good as when they were new.

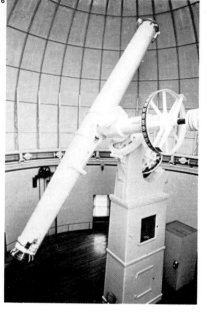

6 The 26in (66cm) telescope at Washington, DC, was one of the earliest of the great refractors. It was installed in 1862 and with it Alvan Clark (1832–97), who ground the object-glass, found the white dwarf companion of Sirius. Asaph Hall (1829–1907) used it to discover Phobos and Deimos, the two satellites of Mars in 1877. The photograph shows it as it is today with the telescope balanced by a counter weight, using a German-type mounting.

Invisible astronomy

Up to the 1920s, astronomers had to depend entirely upon the visible light coming from objects in space. This was a severe limitation, because visible light makes up only a small part of the whole range of wavelengths or "electromagnetic spectrum".

Light may be regarded as a wave motion and the colour of the light depends on its wavelength. The usual unit of wavelength is the ångström, which is equal to one hundred-millionth of a centimetre. Visible light extends from about 4,000Å for violet light up to 7,200Å for red. If the wavelength lies outside these limits the "light" does not affect our eyes. Below the violet end of the visible spectrum come ultra-violet, X-rays and the very short, penetrating gamma-rays; beyond the red end there are infra-red, microwaves and finally radio waves, whose wavelengths may amount to many kilometres.

Radio waves from space

The discovery of radio waves from space was made by Karl Jansky (1905–50), in the United States in 1931. The discovery was fortuitous: Jansky, a radio engineer, was investigating the nature of static when he found that he was picking up emissions from the sky. He tracked them down to the Milky Way. He published a few papers, but never followed the subject through. Before World War II, an American, G. Reber, set up a dish-shaped radio telescope and made the first radio map of the Milky Way. During the war a British team led by J. S. Hey found that radar equipment was being jammed not by transmissions from Germany, as was first thought, but by radio waves from the Sun.

Subsequently, radio telescopes were set up and a new branch of science was well under way. It was found that the Sun is a radio source, but by cosmic standards not a powerful one; it is obtrusive only because it is so close to the Earth. Jupiter is also known to be a source of radio waves. But most radio sources lie far beyond the Solar System. Those in our Galaxy include many supernova remnants, of which the Crab Nebula is the most celebrated example. At greater distances still are the radio galaxies, which are extremely powerful at long wavelengths, although why is still not definitely known.

Radio astronomy has added tremendously to our knowledge of the universe. Without it little would be known about pulsars, which are neutron stars, or quasars, which are extragalactic and may well be the most powerful objects known to man. Moreover, radio waves have been studied from greater distances than visible light waves, so that our information about the most distant regions of the universe is derived entirely from radio work.

Infra-red radiations

Beyond the longwave end of the visible range is the infra-red region of the electromagnetic spectrum. Most infra-red radiations are absorbed by the upper atmosphere and these are studied by means of equipment carried by satellites. However, there are a few "windows" through which infra-red radiations penetrate the atmosphere and these can be studied from the ground. This branch of research has provided a great deal of information about stellar evolution. For instance, there are extremely young stars, such as the variable V1057 in Cygnus, which

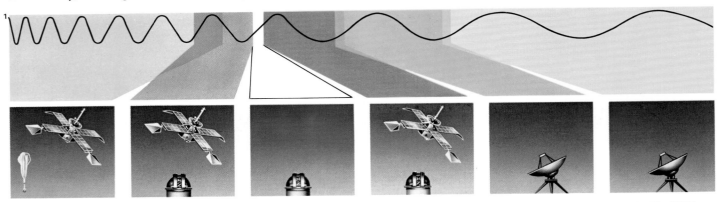

1 The electromagnetic spectrum shows the restricted "windows" in which radiations from space can reach the Earth's surface. Many of the largest wavelengths are blocked out, as are all the shortest. The illustration is not drawn to scale.

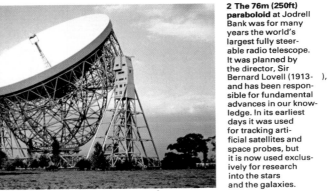

2 The 76m (250ft) paraboloid at Jodrell Bank was for many years the world's largest fully steerable radio telescope. It was planned by the director, Sir Bernard Lovell (1913-), and has been responsible for fundamental advances in our knowledge. In its earliest days it was used for tracking artificial satellites and space probes, but it is now used exclusively for research into the stars and the galaxies.

3 The Arecibo radio telescope, in Puerto Rico, has been built in a natural bowl. It is 300m (1,000ft) in diameter. Although the radio telescope is not steerable, some degree of direction can be obtained by moving the receiving aerial.

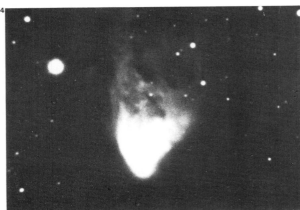

4 Hubble's variable nebula in Monoceros is 6,500 light-years away and is associated with a variable star, R Monocerotis. The infra-red radiation, of which it is a source, also varies.

seem to be surrounded by dust clouds, so that the dust heated by the star contained in the cloud sends out excess infra-red. There are even some objects that are detectable only by infra-red techniques, for example Becklin's Object inside the Orion Nebula. It may well be an immensely powerful star, perhaps a million times as luminous as the Sun, but it can never be seen because it is concealed by the nebulosity and only its infra-red radiation can pass through. The secondary component of the eclipsing binary Epsilon Aurigae is also detectable only in the infra-red, either because it is a very young star, not yet hot enough to shine in the visible range, or because it is a black hole and the radiations we receive come from dust just above.

Ultra-violet, gamma- and X-rays

Ultra-violet radiation, X-rays and gamma-rays lie beyond the shortwave end of the visible band of the electromagnetic spectrum. For high-energy ultra-violet, X-ray and gamma-ray studies (that is, radiations below 2,900Å), equipment carried by rockets or satellites has to be used since these radiations are absorbed by the upper atmosphere. There are many X-ray sources and there is a concentration at the main plane of the Milky Way, indicating that these sources are inside our own Galaxy – although X-radiations come from more distant sources as well. The celebrated galaxy M87 in Virgo, already known as a radio source, is also recognized as an emitter of X-rays.

The Crab Nebula, in our Galaxy at a distance of 6,000 light-years from Earth, contains a pulsar, an X-ray source. X-ray "binaries" are also known, each made up of an X-ray star orbiting a normal giant, and short-lived X-ray sources that may become obtrusive for a few weeks or months before fading away. Much research has been carried out by the British X-ray satellite Ariel 5, launched in October 1974.

Gamma-rays must also be studied from rocket-borne equipment and gamma-ray astronomy, although still in its infancy, has great potential. "Invisible astronomy" has caused a scientific revolution and has now become an established and vital part of astronomical research.

An improvised aerial was set up by Karl Jansky in 1931. He intended to study static on behalf of the Bell Telephone Company, and in the process discovered radio waves from the Milky Way, which led to the modern science of radio astronomy.

5 About 160 locations of X-ray sources along the main plane of the Milky Way have now been identified and their distribution indicates that these sources belong to our own Galaxy. More sources are being found each year. Not all are permanent; the British Ariel satellite found a highly energetic source in Taurus (near, but not associated with, the Crab Nebula) which lasted for some months in 1975.

6 This radio plot of emissions near the centre of the Galaxy also shows X-ray emissions superimposed as numbers. The higher the number the greater the emission. The highest (9) is in the galactic centre.

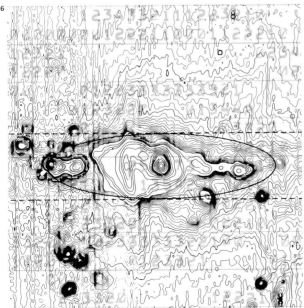

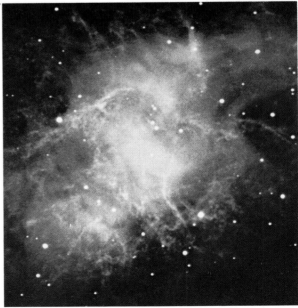

7 The Crab Nebula, in Taurus, is the wreck of the brilliant supernova seen by Chinese and Japanese astronomers in the year 1054. Today it is a cloud of expanding gas in which lies a pulsar – the only pulsar so far optically identified. The Crab is 6,000 light-years away, so that the actual supernova outburst took place in prehistoric times. As well as being a source of radio waves, the Crab also sends out radiation at virtually all wavelengths, so that it has been of the utmost value to astronomers; as far as we know there is nothing else quite like it. With a small telescope it may be seen as a dim, misty patch near the 3rd magnitude star Zeta Tauri.

Evolution of the Solar System

The question of how the Earth came into existence is one that has intrigued mankind for centuries. It was not until comparatively recently that plausible theories were advanced and even today it is impossible to be sure that the main problems have been solved, but at least some concrete facts exist.

Fallacies disproved
The concept of a central Earth, with the Sun moving round it, was abandoned during what is often called the "Copernican revolution", which began in 1543 with the publication of Copernicus's book *De Revolutionibus Orbium Caelestium* and was finally completed by the work of Newton in the latter part of the following century. Therefore it could be assumed that the Earth must have been formed in the same way as the other planets rather than being a special case.

Several centuries ago Archbishop Ussher of Armagh maintained, on religious grounds, that the world had come into existence at a definite moment in the year 4004 BC. Geological evidence soon disproved this but it was not until much more recently that any

reliable estimates could be made. The modern estimate for the age of the Earth is between 4,500 and 4,700 million years, or 4.5–4.7 aeons (one aeon being equal to a thousand million years), and this figure is as reliable as modern knowledge permits.

Further confirmation of the Earth's age has been obtained from analyses of the rocks brought back from the Moon by the Apollo missions and the Soviet automatic probes; it is now known that the Moon and the Earth are about the same age and no doubt the same is also true about the other planets. The Sun must be at least as old as the planets and probably rather older. There can be no doubt that the Sun has been responsible for the origin and formation of the whole of the Solar System.

The first scientific theories
The first serious attempt to explain the origin of the Solar System scientifically was made by the French mathematician Pierre Laplace (1749–1827) in 1796 (although earlier ideas, less purely scientific, had been proposed by Thomas Wright [1711–86] in England and

Immanuel Kant [1724–1804] in Germany). According to Laplace, whose "nebula hypothesis" [Key] elaborated an idea proposed by René Descartes (1596–1650) in 1644, the planets were formed from a rotating gas-cloud that shrank under the influence of gravitation. As it contracted, the cloud shed various rings, each of which condensed into a planet. The theory would mean that the outermost planets were the oldest and the innermost planets the youngest, with the Sun itself representing the remaining part of the original gas-cloud.

The nebular hypothesis was accepted for many years, but it was then found to have basic mathematical weaknesses and was abandoned. Next came a number of tidal theories, including the ideas proposed in America by Thomas Chamberlin (1843–1928) and Forest Moulton (1872–1952), who revived George de Buffon's original idea (1745), developed by James Jeans (1877–1946) in England [19]. It was assumed that the planets were formed by the action of a passing star, which came close to the Sun and pulled off a vast tongue of

CONNECTIONS

See also
176 Members of the Solar System
130 Earth's time scale
20 Life and its origins
72 Attraction and repulsion
76 Circular and vibrating motion
96 Extremes of pressure

1 According to modern theory, the Solar System began as a mass of gas without any definite form. There was no true Sun and no production of nuclear energy. Most of the gas was hydrogen.

2 As time passed by, this cloud, which may be referred to as a solar nebula, began to assume a regular shape and there was a certain increase in temperature, although the Sun was not yet recognizable.

3 The gas-cloud continued to contract under the influence of gravitation and the densest part of it was at the centre. This was the site of the Sun, which now began to radiate, and became a "star".

4 As the Sun became more luminous, the gas-cloud grew less uniform. Condensations appeared in it and were able to draw in surrounding material, so making up what may be termed proto-planets.

5 As the proto-planets increased in size and in mass, their gravitational pulls became stronger and they were able to pull in more and more material from the surrounding regions of the nebula.

6 As the solar nebula shrank, more and more material from it was absorbed into the proto-planets, while the radiation from the Sun continued to increase. The Solar System was not yet recognizable.

7 The main proto-planets continued to grow and to draw in more material by means of their own gravitational effects, so that the numbers of proto-planets became steadily less and less.

8 As the proto-planets grew, their forms became spherical and the Solar System began to assume its familar form. The Sun was now radiating energy from thermonuclear reactions.

9 During this long period of proto-planet formation, the Sun had completed its main contraction and had settled down to the start of its stable period on the main sequence which would last for 10 aeons.

10 By about 5,000 million years ago, the Solar System had assumed the form known today, with a stable central Sun surrounded by its planets.

11 In perhaps 5,000 million years from now the Sun will have exhausted its supply of available hydrogen and its structure will change. The core will shrink and the surface will expand considerably, with a lower surface temperature.

12 The next stage of solar evolution will be expansion to the red giant stage, with luminosity increased by 100 times. The size of the globe will increase, with the overall increase in energy output, and the inner planets will certainly be destroyed.

13 With a further rise in core temperature, the Sun will begin to burn its helium, causing a rapid rise in temperature and increase in size. The Earth can hardly hope to survive this phase of evolution as the Sun expands to 50 times its size.

14 By now the Sun will be at its most unstable, with an intensely hot core and a rarefied atmosphere. The helium will begin to burn giving the so-called helium-flash. After a temporary contraction the Sun will be 400 times its present size.

15 Different kinds of reactions inside the Sun will lead to an even greater increase of core temperature. The system of planets will no longer exist in the form we know today, but the supply of nuclear energy will be almost exhausted.

16 When all the nuclear energy is used up, the Sun will collapse, very rapidly on the cosmic scale, into a small, dense and very feeble white dwarf. It will continue to shine because it will still be contracting gravitationally.

17 The final stage of the Sun will be that of a black dwarf, devoid of any light or heat, still circled by its surviving but now dead planets. Black dwarfs may be common in the universe, but because they emit no radiation cannot be detected.

material. As the star receded, the tongue of matter was left whirling round the Sun, and broke up into drops, each drop becoming a planet. This was in agreement with the sizes of the planets, since the giants (Jupiter and Saturn) lie in the middle part of the system, where the thickest part of the cigar-shaped tongue would have been.

However, this theory too has serious mathematical objections and few modern astronomers support any form of tidal theory for the evolution of the Solar System.

It was also proposed, by Fred Hoyle (1915–) that the Sun used to be a binary star [18] and that the companion exploded as a supernova, producing scattered debris which formed the planets before the companion itself departed by a kind of recoil action. This has, however, met with little support among astronomers.

The future of the Solar System

Modern theories assume the existence of what may be termed a solar nebula, which contained the material that gradually built up into the planets by an accretion process. The exact details are still a matter for debate, but in essence the theory seems to be valid; if so, the Sun and the planets have a common origin and are made of the same material.

At the moment the Sun is a stable main sequence star, but it will not remain so indefinitely. In the far future – perhaps in 5,000 million years or so – it will have to change its structure, since the supply of available hydrogen "fuel" will be exhausted. What will happen is that the Sun will expand into a red giant star, and there will be a period when it will send out about 100 times as much energy as it does at present.

The effects of this expansion on the inner planets will be disastrous; even if they are not destroyed they will be stripped of their atmospheres and will become intolerably hot. Subsequently the Sun will collapse into a very small, feeble white dwarf star, still surrounded by the surviving members of its planetary system. The exact time scale is still a matter for debate, but one thing is certain: life on Earth cannot continue indefinitely and the Solar System in its present form must have a limited existence.

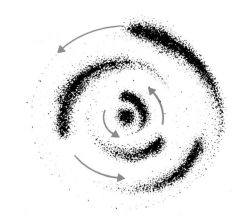

The nebular hypothesis, proposed by Laplace, assumed that before the birth of the planets the Solar System consisted of a gas-cloud which shrank, because of gravitational forces. This resulted in an increase in the speed of rotation and a ring separated from the nebula; the ring slowly condensed into a planet. Further rings were then thrown off, each producing a planet. The theory is mathematically weak.

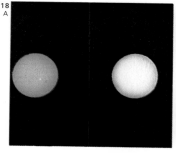  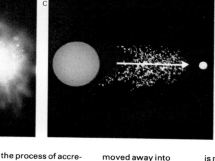 

18 The binary theory of the origin of the Solar System was proposed by Fred Hoyle. He argued that the Sun once had a binary companion [A] which exploded as a supernova [B] and was blown off, leaving a cloud of fragments [C] in orbit round the Sun; these fragments collected together by the process of accretion to form the planets [D], while the remnant of the supernova companion moved away into space and cannot now be identified. This theory is very difficult to substantiate and is not now generally favoured. If valid, it would mean that planetary systems would be very uncommon although on the modern solar cloud theories they are likely to be extremely common in our Galaxy.

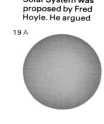

19 The tidal theory, as proposed by James Jeans, is shown here. The Sun [A] is approached by another star [B] which pulls a tongue of matter off the solar surface. After the wandering star recedes, the tongue of matter breaks up into drops which form planets revolving round the Sun. Jupiter, the largest planet, is in the position of the thickest part of the cigar-shaped tongue.

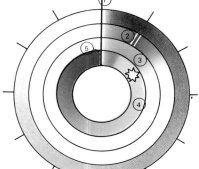

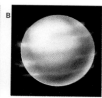   

20 The time scale of the Solar System can be represented on a 12-hour clock. From the inner circle outwards the life-span of the Sun, inner planets, Earth and outer planets are traced respectivley. At the 12-o'clock position [1] the Solar System was created. After 4,000 million years conditions on Earth are favourable for life [2]. As a red giant the Sun engulfs the inner planets [3] to collapse as a white dwarf [4] and finally end its life as a black dwarf [5].

21 The life-span of the Earth started from the material of the solar nebula [A] which at first had no regular form. When it reached its present size [B] the original hydrogen atmosphere had already been lost and had been replaced by a new one, caused by gases sent out from the interior. Life could begin and today the Earth is moving in a stable orbit round a stable star, so that it is habitable [C]. This state of affairs will not persist indefinitely. When the Sun enters the red giant stage, the Earth will be over-heated: the oceans will boil and the atmosphere will be driven off [D]; finally the Earth will be destroyed [E].

Members of the Solar System

The Solar System is made up of one star – the Sun – nine principal planets and various bodies of lesser importance, such as the satellites that attend some of the planets. It is entirely dependent on the Sun, which is by far the most massive body and the only one to be self-luminous. The remaining members of the Solar System shine by reflected sunlight and appear so brilliant in our skies that it is not always easy to remember that in the universe as a whole they are not nearly as important as they look.

Groups of planets
The planets are divided into two well-marked groups. First come four relatively small planets: Mercury, Venus, the Earth and Mars, with diameters ranging from 12,756km (7,926 miles) for the Earth down to only 4,880km (3,032 miles) for Mercury. These planets have various factors in common. All, for example, have solid surfaces and are presumably made up of similar materials, although the Earth and Mercury are more dense than Mars and Venus.

Their orbits do not in general depart much from the circular, although the paths of Mercury and Mars are considerably more eccentric than those of the Earth and Venus. Mercury and Venus are known as the "inferior planets" because their orbits lie inside that of the Earth; they show lunar-type phases from new to full and remain in the same region of the sky as the Sun. Mercury and Venus are unattended by any satellites. The Earth has one satellite (our familiar Moon) while Mars has two, Phobos and Deimos, both of which are very small and different in nature from the Moon.

Beyond Mars comes a wide gap, in which move thousands of small worlds known as the asteroids, planetoids or minor planets. Even Ceres, the largest, is only about 1,000–1,200km (600–750 miles) in diameter. This is much larger than was once thought, but is still small by planetary standards. It is not therefore surprising that the asteroids remained undiscovered until relatively recent times: Ceres was discovered in 1801, only one of this multitude of asteroids, Vesta, is ever visible from Earth without the aid of a telescope.

Far beyond the main asteroid zone come the four giant planets Jupiter, Saturn, Uranus and Neptune. These worlds are quite different from the terrestrial planets: they are fluid (that is, gas or liquid) rather than solid bodies with very dense atmospheres. Their masses are so great that they have been able to retain much of their original hydrogen; the escape velocity of Jupiter, for instance, is 60km (37 miles) per second as against only 11.2km (7 miles) per second for Earth. Their mean distances from the Sun range from 778 million km (483 million miles) for Jupiter out to 4,497 million km (2,794 million miles) for Neptune. Conventional diagrams of the Solar System tend to be misleading as far as scale is concerned; it is tempting, for example, to assume that Saturn and Uranus are lying next to each other when in fact the distance of Uranus from the Earth's orbit is about twice that of Saturn.

The giant planets compared
The giant planets have various points in common, but differ markedly in detail. Their densities are comparatively low and the

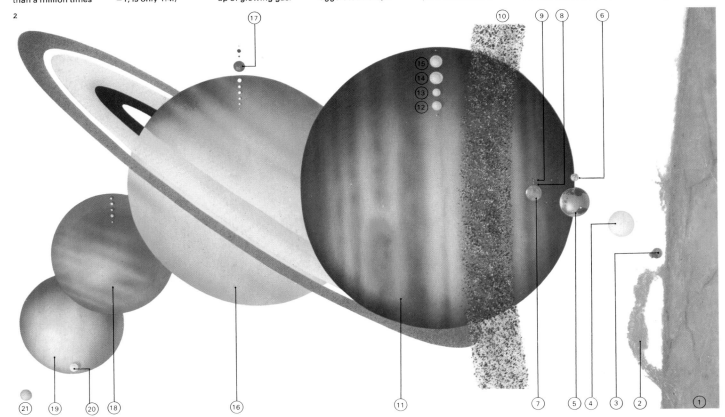

1 Shown here in cross-section, the Sun has an equatorial diameter 109 times that of Earth, or 1,392,000 km (865,000 miles). Despite the fact that its volume is more than a million times that of the Earth, its mass is only 333,000 times that of the Earth, because the density is lower. (The mean specific gravity, on a scale where water =1, is only 1.4.)

2 The planets of the Solar System are shown to the same scale. On the right is a segment of the Sun [1]; from its surface rises a huge prominence [2], made up of glowing gas. Then come the inner planets: Mercury [3]; Venus [4]; the Earth [5] with its Moon [6]; and Mars [7]. Mars has two dwarf satellites Phobos [8] and Deimos [9], exaggerated here; if shown to the correct scale, they would be too small to be seen without a microscope. Then come the asteroids [10], of which even the largest is only about 1,000–1,200km (600–750 miles) in diameter. Beyond lie the giant planets: Jupiter [11] with its four large satellites Io [12], Europa [13], Ganymede [14] and Callisto [15] two smaller moons are also seen in the picture; Saturn [16] with its retinue of satellites, of which the largest is Titan [17]; Uranus [18] with five satellites; Neptune [19] with its large satellite Triton [20]; and finally Pluto [21].

density of Saturn is actually less than that of water. Although Jupiter is seen solely by reflected sunlight, the planet does generate some heat of its own. However, even though the core temperature must be high, it is not nearly high enough for nuclear reactions to begin, so that Jupiter cannot be compared to a star like the Sun.

The outer planets
Five of the planets – Mercury, Venus, Mars, Jupiter and Saturn – have been known from ancient times, since all are prominent naked-eye objects. Uranus, which is just visible with the naked eye, was discovered fortuitously in 1781 by William Herschel (1738–1822), and Neptune was added to the list of known planets in 1846, as a result of mathematical investigations carried out concerning the movements of Uranus. All the giants are attended by satellites; Jupiter has 13 attendants, Saturn ten, Uranus five and Neptune two. Several of these attendants are of planetary size with diameters at least equal to that of Mercury.

The outermost known planet is Pluto, discovered in 1930 by astronomers at the Lowell Observatory, Flagstaff, Arizona. It is not a giant, being smaller than the Earth, and is usually ranked as a terrestrial-type planet, even though little is known about it. Whereas most of the planets have orbital inclinations similar to that of the Earth (the difference is 7 degrees for Mercury, much less for the remainder), the orbit of Pluto is tilted at the relatively steep angle of 17 degrees and the orbit is so eccentric that at perihelion, or closest approach to the Sun, Pluto will come closer in than Neptune. Pluto seems, in fact, to be in a class of its own and it may be a former satellite of Neptune which has achieved independence. Whether some more distant planets exist beyond the orbit of Pluto is a challenge to more advanced technology.

Comets are also members of the Solar System. They contain both dust particles and volatile material together with tenuous gas; most of them have eccentric orbits. Finally there are a great number of meteoroids, which may be regarded as the debris of the Solar System; some meteors are certainly associated with comets.

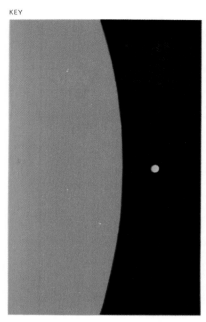

The Sun is an ordinary main sequence star with a magnitude of +5. It is the body on which the Solar System depends and its volume is more than a million times greater than that of the Earth. It is, in fact, far more massive than all the planets combined. However, the Sun is small when compared with a giant star. The diagram shows the Sun alongside a segment of the red giant star Betelgeuse which marks Orion's right shoulder. Betelgeuse is of spectral class M2 – a very cool star – but has an absolute magnitude of –5.5. Its diameter is 300-400 times that of the Sun and its globe is large enough to contain the Earth's orbit.

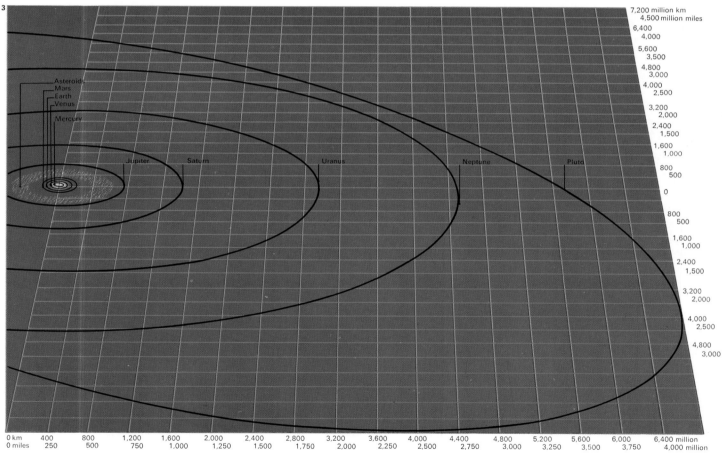

3 The map of the Solar System shows the approximate orbital inclination of each of the nine planets against a grid giving distances in kilometres and miles. The following measurements for diameter and rotation period refer to each planet's equator. The sidereal period is the time taken by each planet to orbit the Sun once.

Mercury
Distance from Sun,
mean 58 million km
(36 million miles)
Diameter,
4,880km (3,032 miles)
Rotation period,
58·7 Earth-days
Mass, 0·05 Earth.
Surface gravity,
0·37 Earth.
Escape velocity,
4·2km (2·6 miles)
per second.
Sidereal period,
88 Earth-days.
Venus
Distance from Sun,
mean 108,200,000km

(67,200,000 miles)
Diameter,
12,100km (7,500 miles)
Rotation period,
243 Earth-days.
Mass, 0·82 Earth.
Surface gravity,
0·90 Earth.
Escape velocity,
10·36km (6·4 miles)
per second.
Sidereal period,
224·7 Earth-days.
Earth
Distance from Sun,
mean 149,596,000km
(92,750,000 miles)
Diameter (equatorial),
12,755km (7,908 miles),

Polar 12,714km
(7,883 miles).
Rotation period,
23hr 56min.
Escape velocity,
11·2km (7 miles)
per second.
Sidereal period,
365·2 days.
Axial inclination : 23·5°.
Mars
Distance from Sun,
mean 227,940,000km
(141,323,000 miles)
Diameter,
6,790km (4,220 miles)
Rotation period,
24hr 37 min. 23 sec.
Mass, 0·11 Earth.

Surface gravity, 0·4 Earth.
Escape velocity,
5km (3·1 miles)
per second.
Sidereal period,
686·96 Earth-days.
Jupiter
Distance from Sun,
mean 778,300,000km
(483,600,000 miles)
Diameter (equatorial)
143,000km (89,000 miles)
Rotation period
(equatorial), 9hr 51min.
Mass, 318 Earth.
Surface gravity,
2·64 Earth.
Escape velocity,
60·22km (37·4 miles)

per second.
Sidereal period,
11·86 Earth-years.
Saturn
Distance from Sun,
mean 1,427 million km
(887 million miles)
Diameter (equatorial),
120,000km
(75,000 miles)
Rotation period
(equatorial), 10hr 14min.
Mass, 95 Earth.
Surface gravity,
1·16 Earth.
Escape velocity, 36km
(22 miles) per second.
Sidereal period,
29·46 Earth-years.

Uranus
Distance from Sun,
mean 2,869,600,000km
(1,780 million miles).
Sidereal period, 84 years.
Neptune
Distance from Sun,
mean 4,497 million km
(2,794 million miles)
Sidereal period, 164·8 years.
Pluto
Distance from Sun,
mean 5,900 million km
(3,658 million miles).
Sidereal period, 248·5 years.

The Moon

The Moon is much the closest natural body to Earth in the sky. Its distance from the Earth is, on average, only 384,000km (239,000 miles), which is about equal to ten times the distance round the Earth's equator. It is a small world [1] with a diameter of 3,476km (2,160 miles); its mass is only 1/81 that of the Earth and the escape velocity is 2.4km (1.5 miles) per second, which is too low for the retention of an appreciable atmosphere.

Movements of the Moon

It is not entirely correct to say that the Moon revolves round the Earth. More properly, the Earth and Moon revolve round the "barycentre", or centre of gravity of the system. But because of the discrepancy between the masses of the two bodies, the barycentre lies well inside the terrestrial globe, so that the simple statement that "the Moon goes round the Earth" is good enough for most purposes. The revolution period is 27.3 days and this is also the time taken for the Moon to rotate once on its axis. As a result, the same hemisphere is always turned towards the Earth.

The Moon's path is not quite circular, so the apparent diameter of the disc varies within narrow limits. The familiar phases are due to the fact that the Moon does not always turn its daylight side towards the Earth [2]. The boundary between the day and night sides is known as the terminator; it is rough and jagged because the lunar surface is irregular. A peak will catch the rays of the rising Sun while the land below is still in darkness. Before the flight of the circumlunar probe Luna 3, in 1959, nothing was definitely known about the Moon's far side. Effects known as librations [3] (irregularities in the Moon's movement) produced by the Moon extend the visible area to a total of 59 per cent of the whole surface (although never more than 50 per cent at any one time).

Theories of origin

Although the Moon is officially ranked as the Earth's satellite, it seems disturbingly large to be a truly secondary body. There are other satellites in the Solar System that are larger than the Earth's moon (three members of Jupiter's family, one of Saturn's and one of Neptune's), yet all these move round giant planets. Thus Triton, the senior attendant of Neptune, has only 1/750 of the mass of its primary, although it is possibly larger than the planet Mercury and certainly larger than the Moon.

This being so, it may well be that the Earth–Moon system should more properly be regarded as a double planet, which leads to the problem of the Moon's origin. The tidal theory proposed by George Darwin (1845–1912) in the nineteenth century was popular for many years. According to this hypothesis the Earth and Moon were once a single body, rotating quickly and therefore becoming unstable. Eventually the globe became so distorted that part of it broke away and moved off to form the Moon.

There are, however, mathematical objections to this theory which are so serious that few astronomers now support it in any form. It is much more likely that the Moon and the Earth were formed in the same way, from the solar nebula, either close together in space or else quite independently – in which case the Moon would have been "captured" by the

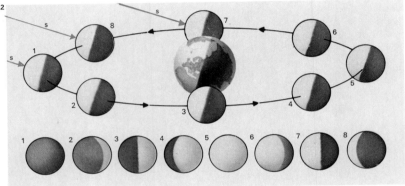

1 **The Moon is a small world** compared with the Earth. Its mass is much less and its specific gravity is lower. But the discrepancy between Earth and Moon is much less marked than with the satellites of other planets. With Neptune, for instance, the mass of its largest satellite, Triton, is only 1/750 of that of the planet. This is one reason why the Earth-Moon system may be regarded as a double planet.

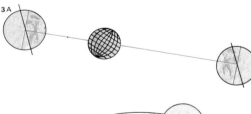

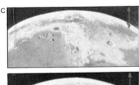

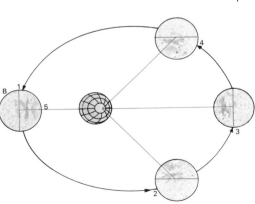

2 **The phases of the Moon** occur because the Moon has no light of its own. The daylight side reflects the Sun; the night side reflects "Earthshine". In the illustration, sunlight comes in from the upper left. In 1, the Moon's dark side is turned Earthwards (new) and the Moon cannot be seen – unless it passes directly in front of the Sun, producing a solar eclipse. Between 1 and 3 the Moon is crescent; at 3, half (first quarter); at 4, gibbous; and at 5, full. The Moon then wanes, through gibbous [6] to half or last quarter [7], crescent in the morning sky [8] and back to new.

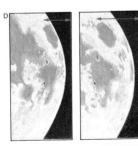

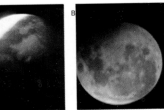

3 **Librations** are irregularities in the Moon's movement. Libration in latitude [A] occurs because the Moon's axis is tilted to its orbital plane, thus allowing views over the north [C] and south poles; libration in longitude [B] occurs when the Moon's speed of revolution changes slightly – it moves fastest at perigee [1] and slowest at apogee [3]. Its effect can be seen by tracing the position of point 5 through its locations in orbit from 1 to 4. This allows us to see a little farther round each limb (edge) [D].

4 **During a total eclipse** [A] the Moon does not vanish completely because a certain amount of light is refracted on to its surface by way of the Earth's atmosphere. The boundary between light and dark is never sharp. Partial eclipses [B] also occur. When the Moon passes through the area of "penumbra" that lies on either side of the main shadow cone, the visible effect becomes less striking.

5 **Eclipses** do not occur at every full Moon because the Moon's orbit is inclined at 5° to that of the Earth. At most full Moons the Moon passes either above or below the shadow [A]. A lunar eclipse [B] is produced when the Moon passes into the shadow cast by the Earth.

Earth later on. Based on our present state of knowledge, the former alternative seems to be the more plausible.

Surface features: seas and craters

The first telescopic maps of the Moon were drawn in 1609. Priority may well belong to Thomas Harriot (1560–1621), who drew a chart that shows many features in recognizable form. A longer and more systematic study was carried out from 1610 by Galileo, who described the mountains, the craters and the grey plains in some detail. The grey areas were named "seas" and the nomenclature has not been altered, although for centuries it has been known that there is no water in them. The names are usually given in Latin; thus the Sea of Clouds is Mare Nubium and the Ocean of Storms Oceanus Procellarum.

The walled structures commonly known as craters dominate the entire lunar scene. In size, they range from vast enclosures more than 240km (150 miles) in diameter down to tiny pits too small to be seen from Earth. A typical crater has a rampart that rises to only a modest height above the outer terrain, the

floor is sunken, and there may be a central mountain or mountain group. In some the wall may be more than 3,000m (10,000ft) above the deepest part of the floor. There have been endless arguments about the origin of the craters. The main controversy centres on one point: were the craters produced by external forces (meteoritic impact) or by internal ones (vulcanism)? No doubt, like the Earth, the Moon has both types.

Some of the waterless seas, like the regular craters, are more or less circular with mountainous borders. The huge Mare Imbrium or Sea of Showers, for example, is bounded by the Apennines, Carpathians and Alps, although the mountain boundary is not continuous and there are wide gaps. The Apennines [7] are the most spectacular of the ranges; their loftiest peaks reach heights of more than 4,570m (15,000ft).

Other lunar features include hills; domes, with gentle slopes and often a summit craterlet or craterlets; occasional faults; and many crack-like features called clefts or rilles. Some of these markings can be identified even with the naked eye.

The dark "seas" and bright uplands of the full Moon are seen here. The crater Tycho dominates the southern hemisphere.

6 The Sinus Medii or Central Bay, photographed from Apollo 10, is one of the relatively smooth mare areas.

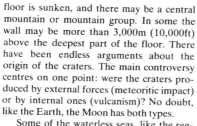

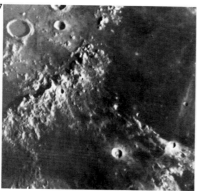

7 The lunar Apennines are here photographed with a 12in (30cm) reflector. The highest mountains rise to 4,570m (15,000ft) above the plains. They are by far the most spectacular of the lunar peaks.

8 The great crater Langrenus has massive terraced walls and a complicated central mountain group.

9 This small crater on the far side of the Moon was photographed by Apollo 10.

10 Clavius, photographed from Earth, is a walled plain 230 km (143 miles) wide with superimposed craters.

11 Part of the Sirsalis Rille, a telescopically visible collapse feature that is 32km (20 miles) in diameter.

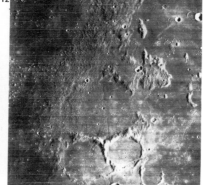

12 An Orbiter photograph of the connecting walls of three plains – Fra Mauro, Bonpland and Parry – is a typical example of damaged outer walls.

13 The Straight Wall, the Moon's best-known fault, is illuminated here by sunlight.

Moon missions

Flight to the Moon became a practical possibility after the opening of the Space Age in October 1957 when the USSR launched Sputnik 1, a satellite that circled the Earth. Two years later, she sent three vehicles on lunar missions. The first of these (Luna or Lunik 1) bypassed the Moon and sent back useful information, notably that there was no appreciable magnetic field. Luna 2 crash-landed on the Mare Imbrium in September 1959 and in October Luna 3 [1] went on a round trip sending back the first photographs of the Moon's far side. This proved to be just as mountainous, crater-scarred and sterile as the side seen from Earth, but there were no comparable seas.

Early unmanned explorations

The American Ranger programme [2] introduced a new phase in man's knowledge of the Moon. These Rangers were designed to send back close-range photographs before crashing onto the Moon. Ranger 7, in 1964, was the first successful probe of this kind. Two more followed; the last of them came down in the prominent crater Alphonsus, near the centre of the Moon's disc as seen from the Earth.

In January 1966 the USSR achieved a major triumph by soft-landing an automatic probe, Luna 9, on the surface of the Moon. Its cameras showed a landscape that looked remarkably like a lava plain, with hummocks and crater pits everywhere. Luna 9 was of special importance because it finally disproved a curious theory that the lunar seas were filled with soft, treacherous dust to a depth of several hundred metres. The landing showed that the Moon's surface layer, or regolith, is firm enough to support the weight of a spacecraft [13].

Lunar mapping was more or less completed during the two years following August 1966. Five American Orbiters [4] moved around the Moon in closed paths, sending back amazingly detailed pictures. The Americans also soft-landed several vehicles – the Surveyors. On 17 January 1968, Surveyor 7 came down near the crater Tycho, in the southern uplands, and sent back high-quality pictures of the outer slopes of the wall of the crater.

The USSR continued its programme of unmanned exploration into the 1970s. It achieved a major triumph with Luna 16 [6] in 1970, which landed in the Mare Foecunditatis and then returned, bringing samples of lunar rocks. Later in the same year Luna 17 landed in the Mare Imbrium. From it emerged Lunokhod 1 [7], an eight-wheeled craft powered by solar batteries and guided from Earth. After extensive exploration it ceased to function on 4 October 1971. Lunokhod 2 followed in 1972.

Men on the Moon

From the mid-1960s onwards the Americans had been concentrating on the Apollo programme of sending men to the Moon. This culminated in July 1969, when Neil Armstrong (1930–) and Edwin Aldrin (1930–) left Eagle [5], the lunar module of Apollo 11 [8], and made the historic "one small step" onto the lunar surface. After collecting samples of lunar material [14] and leaving recording instruments behind [12], the two astronauts returned to their module and rejoined the third member of the expedi-

1 Luna 3 was the probe that made the first circumlunar flight (October 1959). It sent back pictures of the far side.

2 Ranger 8 (1965) crash-landed near the crater Delambre, but sent back excellent pictures before impact.

3 Luna 13 made the first successful soft landing on 21 December 1966, and sent back pictures from the surface.

4 Orbiter 5 (1968) was the last of the series of Orbiters, which have provided full photographic surface coverage.

5 Eagle (1969), the lunar module of Apollo 11, took Armstrong and Aldrin to Tranquillity Base on the Moon.

6 Luna 16 (1970) landed in the Mare Foecunditatis and collected samples of lunar material to bring back.

7 Lunokhod 1, which landed with Luna 17 in November 1970, took photographs and crawled about the surface collecting data.

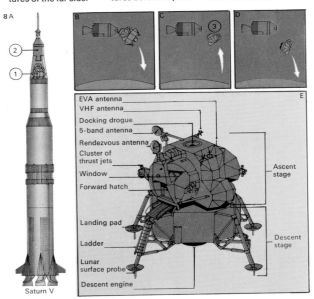

Saturn V

EVA antenna
VHF antenna
Docking drogue
5-band antenna
Rendezvous antenna
Cluster of thrust jets
Window
Forward hatch
Landing pad
Ladder
Lunar surface probe
Descent engine
Ascent stage
Descent stage

8 The Apollo Moon programme used a Saturn rocket [A] to carry the lunar module [1] and the command and service modules [2]. The lunar module, holding two astronauts, descended to the Moon [B]. Its upper part [3] later left, blasting off from the descent stage back into orbit [C] to rendezvous with the third member of the team, still in orbit round the Moon. The module was then jettisoned [D] and crashed onto the Moon. The lunar module is shown in detail [E].

9 The Lunar Roving Vehicle (LVR) was carried by the last three Apollos (15, 16 and 17) and was used by the astronauts to drive across the Moon's surface for considerable distances. Its speed ranges from 8 to 16km/h (5–10mph).

10 The ALSEP of Apollo 17 has a central station and thermal generator [1, 2] to provide the main power. The atmospheric composition experiment [3] analyses any residual lunar atmosphere. The ejecta and meteorite experiment [4] detects any impact from meteorite bodies and the gravimeter [5] measures any gravitational anomalies. The geophone experiment [6] involves artificial explosions that will help in studies of the Moon's surface layers. There are also seismic measuring devices [7] and core sampling equipment [8]. The solar wind experiment [9] was carried out on earlier Apollo missions.

tion, Michael Collins (1930–), who was orbiting the Moon in the command module.

Apollo 12 followed later in 1969; the explorers, astronauts Charles Conrad (1930–) and Alan Bean (1932–), landed near a previous automatic probe, Surveyor 3, and were able to bring parts of it home. With the next mission, Apollo 13, came the first real failure. An explosion in the service module of the spacecraft during the outward journey put the main power supplies out of action. The lunar landing was cancelled and it was only by a combination of courage, skill and luck on the part of both the astronauts and the operators at mission control that tragedy was averted. Since then there have been four more Apollo landings. The series ended in 1972 with Apollo 17, which landed in the Taurus–Littrow region, manned by Eugene Cernan (1934–) and Harrison Schmitt (1935–), a professional geologist.

Each successful mission has deployed what is called an Apollo Lunar Surface Experimental Package (ALSEP) [10]. Various investigations have been carried out and our knowledge of the Moon has improved immensely, even though astronomers still argue about the origin of the main craters. At the beginning of 1976 all the ALSEPSs, apart from that of Apollo 11, were still transmitting information back to the Earth.

Lunar landscape

Conditions on the Moon are unfamiliar. An astronaut has only one-sixth of his normal weight although his mass is unaltered. There is virtually no local surface colour and the lunar sky is black even when the Sun is above the horizon. The lunar day is long, because the Moon spins slowly. So far all the landings have been made in the early morning on a selected region of the Moon.

The Moon is not a welcoming world. The temperatures range between about 90°C (195°F) at noon on the equator down to well below −130°C (−200°F) at night. There is no air or water and we are now sure that there has never been any life. Yet the Moon is of tremendous importance to man. Before the year 2000, a permanent lunar base may well have been established there for scientific and astronomical research.

A Apollo
L Luna
O Orbiter (flight path)
R Ranger
S Surveyor

11 This lunar landscape photograph, taken on the Apollo 11 mission, illustrates several surface details encountered by the astronauts, although the area shown is smooth by lunar standards.

12 During the Apollo 11 mission Armstrong and Aldrin set up a lunar seismometer to measure ground tremors. The seismometer is similar to instruments used on Earth but can be more sensitive because the Moon is seismically "quiet". This particular one failed after a short period but similar instruments left on the Moon by later Apollos have shown that minor tremors do occur frequently.

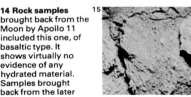

13 Edwin Aldrin's footprint on the lunar surface had a depth of penetration of less than 2.5cm (1in).

14 Rock samples brought back from the Moon by Apollo 11 included this one, of basaltic type. It shows virtually no evidence of any hydrated material. Samples brought back from the later Apollo and Luna missions show that there are numerous types of lunar rock; most of the minerals of which they are composed occur on Earth.

15 A section of the rock in illustration 14 is enlarged here; the fine structure is clearly shown. This was one of the first rocks to be analysed.

16 Microstructure photographs [A–E] of samples brought back by Armstrong and Aldrin from the Mare Tranquillitatis are shown here. Substances identified include plagioclase, ilmenite, pyroxene and microscopic rubies. There are many "glass marbles", well shown in E. These particles are small, the largest of them being only 0.5mm (0.02in) in diameter.

16 A

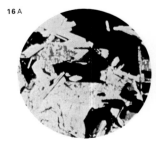

B

C

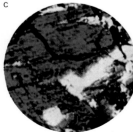

D

E

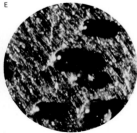

The Moon's structure

Analysis of the samples brought back from the Moon by the American Apollo missions and the Soviet unmanned probes have established that the Earth and the Moon are about the same age (between 4,500 and 5,000 million years, or 4.5 and 5 aeons). But because the masses of the Earth and the Moon are so different they have undergone different evolutionary sequences [1].

The surface of the Moon
The nature of the Moon's surface is intimately bound up with the problem of the origin of its craters and other features and this has led to endless arguments that have not been resolved even by the Apollo results. Some strange theories of crater origin have been advanced (ranging from coral atolls to atomic bombs), but the whole problem centres on whether the craters were formed by internal action or by external bombardment. The first of these rival theories is usually called the "volcanic" theory [3], the second is known as the "meteoric" or "impact" theory [2].

Both processes must, in fact, have oper-ated to some extent for both kinds of craters occur on Earth, and no doubt do so on the Moon as well. What must be decided is which of the two processes played the more important role. Opinion is sharply divided, although some of the features, such as the small chain craters, are undeniably "volcanic" in the broad sense.

Many efforts have been made to link the main lunar craters with terrestrial impact craters such as that at Arizona, USA, although the scales are different; if transferred to the Moon, the Arizona crater would appear insignificant. On the other hand, supporters of the volcanic theory point out that the distribution of the Moon's craters is not random [4]; for instance, the great crater plains tend to appear in distinct lines. When one crater formation breaks into another the smaller crater almost always intrudes into the larger. This is easier to explain with a theory of internal origin than with one of impact. It is also true that the lunar rock samples are essentially volcanic, although often affected by impact, and that the incidence of meteoritic material appears to be compara-tively scarce on the Moon's surface.

It is also questionable whether the circular maria, or lunar "seas", are essentially similar to the large craters in origin. At least there is some reliable information about their age and it seems probable that the chief maria (Imbrium, Serenitatis, Crisium and the rest) were formed about 4,000 million years ago: the Mare Orientale is probably the youngest of them with an age reliably estimated at about 3.8 aeons.

The filling of the lunar "seas"
Most experts agree that when the mare basins were formed they did not contain lava; the question of whether they were created by internal (endogenic) or external (exogenic) factors makes no difference to this conclusion. Between 3.8 and 3.2 aeons ago, the mare basins began to be filled by lava that poured out from beneath the crust and finally produced the aspect that is seen today. Because the eruptions were not confined during about a million years, the mare surfaces, although appararently simple, form a patchwork of overlapping lava flows [7].

CONNECTIONS

See also
178 The Moon
180 Moon missions
184 Moon maps
186 Moon panorama

In other volumes
28 The Physical Earth
30 The Physical Earth
100 The Physical Earth

1 The mare basins were formed, either by internal accretion or by impact, at an early stage in the history of Moon and Earth surfaces (4,000 million years ago). The general aspect of the Moon then must not have been dissimilar to that of today, although the basins were not filled. Little is known about surface details on Earth at this time. The surface of both the Earth and the Moon then remained the same for a considerable period. Some 2,000 million years ago the basins on the Moon were filled in; 1,000 million years after that, lunar activity was at an end. Geological techniques give some idea of the appearance of the Earth at that time, since when it has developed markedly into its present-day form.

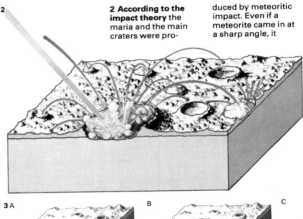

Present day

1,000 million years

2,000 million years

3,000 million years

4,000 million years

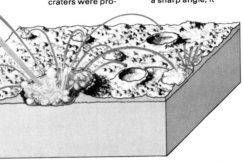

2 According to the impact theory the maria and the main craters were produced by meteoritic impact. Even if a meteorite came in at a sharp angle, it would still produce a circular formation. This theory is a popular one today.

3 The rival volcanic theory states that when the lunar surface was hot and plastic [A], domes were produced, by magmatic convection for example. On cooling the underlying material sank [B], leaving a void. The surface layer collapsed, forming a crater [C]. Central peaks are caused by penetration of magma.

3A   B   C

4

S

Clavius Group   Bailly
Janssen
Stofler
Petavius
Vendelinus
THEOPHILUS CHAIN
EASTERN CHAIN
GREAT
Mare Crisium
Hipparchus
PTOLEMAEUS-WALTER CHAIN
Sinus Medii
Archimedes
Mare Humorum
Grimaldi
WESTERN CHAIN
Sinus Iridum
Plato
Pythagoras
Mare Humboldtianum

E   W

N

4 The distribution of the lunar craters and walled formations is clearly crucial to any considerations of their origin. The small features tend to line up in chains and many of the so-called rilles are in part crater chains; there is no serious doubt that these are of internal origin. On the Earth-turned hemisphere the major formations also tend to line up. There are important chains, such as those including Vendelinus and Petavius (and also the Mare Crisium) in the east and the Grimaldi chain in the west. It has been argued that the important features formed along lines of crustal weakness produced by the gravitational influence of the Earth.

A certain amount of cratering occurred during this period. The ray-craters such as Tycho and Copernicus are probably the youngest of the major features and the age of Copernicus may be even less than one aeon. Subsequently the main activity ceased and only small (mainly impact) craters have been formed since that time.

Evolution of the Moon
Curiously, more information exists about the geological evolution of the Moon than that of the Earth. Unlike the Earth, which has a long history of continual erosion, the Moon has not suffered erosion for a long time. Two thousand million years ago the Moon may have looked much the same as it does today, whereas the Earth would perhaps have been completely unrecognizable.

The Apollo seismometers have been able to record "moonquakes" [5] and there is now no doubt that a certain amount of volcanic activity lingers on. Some of the moonquakes occur close to the crust, but others are deep-seated – up to half-way to the Moon's core [6]. It has also been established that the

Moon may have a hot core, so that the old idea of a globe that is cold throughout may be incorrect [Key]. Studies of moonquake records indicate that if a molten core exists it must be smaller than that of the Earth, both relatively and absolutely. Above this is the so-called asthenosphere, or zone of partial melting; above that the thick mantle, topped by a crust; and finally the rubbly regolith, which has a depth of 100m (328ft). There is virtually no general lunar magnetic field now, although some areas are locally magnetized; it seems that in the remote past the Moon had an appreciable general field that generally weakened and has now disappeared.

Earth-based lunar observers have recorded various minor "events" that may indicate the emission of gas from below the crust. These are known commonly as TLP, or Transient Lunar Phenomena [8]. They are thought to be commonest near perigee, when the Moon is at its closest to the Earth. The Earth's gravitational pull then produces the greatest strain on the lunar rocks and there may be a link between this and the positions of the epicentres of moonquakes.

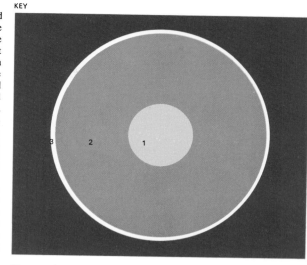

The core of the Moon [1] is believed to be much smaller than that of the Earth, both relatively and absolutely, and it is probably extremely rich in iron. Surrounding it is the lunar mantle [2]. This is overlaid by the crust [3], which is covered by a rocky "topsoil" (regolith).

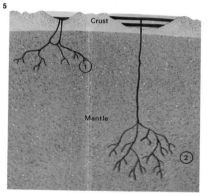

5 The lunar maria were formed when floods of lava erupted onto the Moon's surface, filling in basins previously excavated by planetesimal impact. They did not erupt in one burst of volcanic activity, but over a period of about 1,000 million years. The oldest basalts are thought to have been produced at a depth of 150km (95 miles) [1]. The more recent rocks were generated later at a depth of about 240km (150 miles) [2].

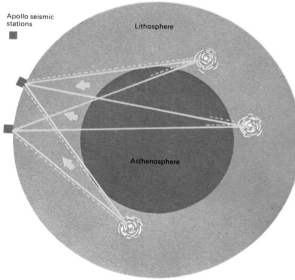

6 During the Apollo programme seismometers were set up on the Moon's surface and studies were made of moonquake waves. The force of even a major moonquake is slight by terrestrial standards, but valuable information about the structure of the Moon has been obtained from studies of the quakes, of which there are two types. Some occur just below the surface, but others have been recorded at a depth halfway between the Moon's surface and core. Shear waves [broken lines] are weakened when passing through a non-rigid medium. Compressional waves [solid lines] can pass through all material. The lunar lithosphere is solid; the asthenosphere is a near-melting zone.

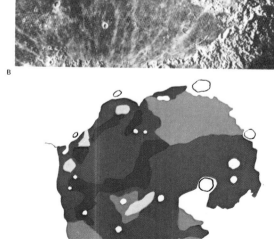

7 In the Mare Imbrium [A] the darkest red coloration on the map [B] represents the youngest of the lava flows, which is the area least affected by cratering. The lightest red represents the oldest flow. The most extensive flow dates back approximately 3.3 thousand million years.

8 Transient Lunar Phenomena (TLP) do exist on the Moon, as has been established in recent years. The TLP-prone sites are not distributed at random: they tend to congregate round the borders of the circular maria and in regions rich in rilles. The most active area on the Moon is that of the brilliant crater Aristarchus.

Moon maps

Even with the naked eye the Moon can be seen in considerable detail and binoculars or any telescope provide the observer with a seemingly inexhaustible panorama. Obviously the view depends on the angle of sunlight over the area being studied. A crater is at its most prominent when near the terminator (the boundary between the day and night hemispheres of the Moon), so that its floor is wholly or partially shadow-filled. Even a major crater may become difficult to identify under a high angle of illumination, unless its floor is either particularly dark or particularly brilliant. In general, the lunar surface has a low reflecting power, or albedo, of about seven per cent – that is to say it reflects only about seven per cent of the sunlight falling on it – but the brightest craters possess walls and central peaks that have albedos of more than 15 per cent.

The northern hemisphere
The northern hemisphere of the Earth-turned side of the Moon is dominated by two great seas, the Mare Imbrium (Sea of Showers) and the Mare Serenitatis (Sea of Serenity), both of which are approximately circular, although foreshortening makes them appear somewhat elliptical. The Mare Imbrium has mountainous borders for much of its outline, including the majestic Apennines with peaks rising to some 4,570m (15,000ft). Between the Apennines and the rather lower Caucasus Mountains there is a gap linking the Mare Imbrium to the Mare Serenitatis. The 95km (60 mile) dark-floored crater Plato lies in the region of the Alps and also in this range is the remarkable Alpine Valley, 130km (80 miles) long.

On the floor of the Mare Imbrium there are several major craters, including the 80km (50 mile) Archimedes and its two smaller but deeper companions called Aristillus and Autolycus. The Mare Serenitatis includes no crater of this size; the largest, Bessel, is only 39km (24 miles) across.

The Mare Tranquillitatis (Sea of Tranquillity), adjoining the Mare Serenitatis to the south, is less regular in form and is presumably older. It was in this sea that two of the astronauts of Apollo 11 made the historic pioneer lunar landing in July 1969.

The Mare Crisium (Sea of Crises), not far from the limb, is smaller but well marked and is easily visible with the naked eye. Of the other seas in this hemisphere the largest is the Oceanus Procellarum (Ocean of Storms), separated from the Mare Imbrium by the relatively modest Carpathian Mountains. Aristarchus, on the Oceanus Procellarum, is generally regarded as the brightest crater on the Moon and can often be seen shining prominently even when on the dark side of the terminator, lit only by light reflected from the Earth. The crater Copernicus, south of the Carpathian Mountains, is the centre of a bright ray system.

Another interesting feature in the northern hemisphere is a great bay, the Sinus Iridum (Bay of Rainbows), which leads off the Mare Imbrium. When the Sun is rising over it, illuminating the mountainous border, the effect is comparable to a jewelled handle.

The southern hemisphere
Slightly south of the equator are the great plains of which Ptolemaeus is the senior member. Ptolemaeus is almost 160km (100

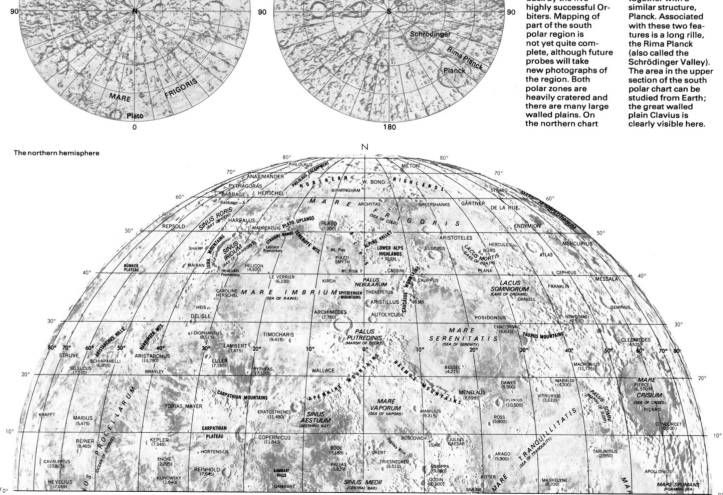

1 The polar zones of the Moon are difficult to study from Earth because of extreme foreshortening. Some of the areas in these maps cannot be seen at all and our knowledge of them is derived mainly from photographs sent back by the five highly successful Orbiters. Mapping of part of the south polar region is not yet quite complete, although future probes will take new photographs of the region. Both polar zones are heavily cratered and there are many large walled plains. On the northern chart Mare Frigoris, visible from the Earth, is shown; Plato appears near the bottom of the map. One of the interesting features in the south polar map is the large walled plain Schrödinger, on the far hemisphere, together with a similar structure, Planck. Associated with these two features is a long rille, the Rima Planck (also called the Schrödinger Valley). The area in the upper section of the south polar chart can be studied from Earth; the great walled plain Clavius is clearly visible here.

The northern hemisphere

miles) in diameter, with a relatively level, darkish floor. Adjoining it is the rather smaller Alphonsus, with a central mountain group and a system of rilles on its floor. In 1958, from Alphonsus the Russian astronomer N. A. Kozyrev recorded a reddish glow – one of the best authenticated examples of a Transient Lunar Phenomenon (TLP). This, he believed, indicated a certain amount of surface or sub-surface activity, which he interpreted as being of volcanic origin. The third member of the Ptolemaeus chain, Arzachel, is smaller than Alphonsus, but deeper, with a higher central peak.

The southern part of the Moon consists largely of rugged upland, although there are some sea areas – the Mare Nubium (Sea of Clouds) and the smaller Mare Humorum (Sea of Moisture). On the former, not far from Arzachel, is the Rupes Recta (Straight Wall), which is a major fault in the surface, with a length of 130km (80 miles) and with a height between the crest and the bottom of 240m (800ft).

Among other major walled plains are the dark-floored Schickard and the 230km (144

mile) Clavius, which has a chain of craters inside it. North of the Southern Highlands is the crater Tycho, called the "metropolitan crater" of the Moon because of its system of bright rays, which are brilliant and extensive. Near full Moon, as seen from Earth, they dominate the entire southern hemisphere and render virtually invisible even large walled plains in the vicinity. Tycho is 86km (54 miles) in diameter, with massive walls, and even under low illumination when the rays are not properly visible, it remains one of the most prominent craters on the Moon.

The far side of the Moon
The libration regions of the Moon were not well mapped until the age of space probes. Today we also have full information about the Moon's far side, although it has been seen "direct" only by the Apollo astronauts who have been round the Moon. Here there are no major maria, but there are walled structures of all kinds; of special interest is Tsiolkovskii, which has a very dark floor and was identifiable on the photographs sent back by Luna 3 in October 1959.

This Apollo 16 photograph shows parts of both the Earth-turned hemisphere and the far side of the Moon.

2 A map of the far side of the Moon shows no major maria similar to those on the familiar side of the Moon. The Mare Orientale is visible from Earth, although very foreshortened. On the far side there are craters of all kinds and the whole surface is "crowded uplands". The most striking feature is the dark-floored Tsiolkovskii, first seen on the Luna 3 photographs in 1959. It abuts onto another structure, Fermi, which has a light floor. On the Luna 3 photographs a bright streak showed up, thought to be a major mountain chain; the Russians named it the "Soviet Range", but subsequently it was found to be nothing more than a bright ray. The Mare Moscoviense (Moscow Sea) was also identified on the Luna 3 pictures. The question of naming the features on the far side caused some controversy. The nomenclature given here is that now officially adopted by a special committee set up by the International Astronomical Union. The Mare Marginis, Mare Smythii and Mare Australe may also be seen from the Earth.

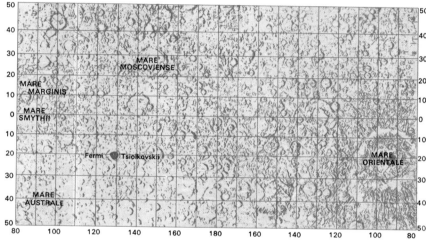

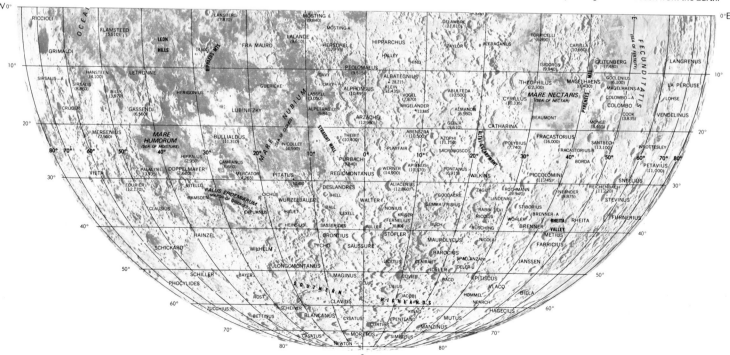

The southern hemisphere

Moon panorama

The Moon is a world of considerable variety. As well as the marked differences between the Earth-turned and the far hemispheres, there are also obvious changes of scenery on the familiar face of the Moon. For instance, the southwestern quadrant is dominated by rough uplands, with large craters and walled plains, whereas the northeastern quadrant contains vast stretches of mare surface.

In particular there is the region of Aristarchus [3], the brightest crater on the Moon, which is of special interest because so many local obscurations have been reported there by Earth-based observers. William Herschel (1738–1822), possibly the greatest practical observer of all time, on several occasions mistook Aristarchus for a volcano in eruption when he saw it shining conspicuously from that part of the Moon illuminated only by Earthshine.

Aristarchus is not the only feature in or near which activity has been suspected. Another is the walled plain Alphonsus [1], in the great Ptolemaeus chain near the centre of the Earth-turned hemisphere. Although Alphonsus and Aristarchus are so different, they do have one thing in common: they lie in regions rich in rilles or clefts. The same is true of most other regions with mild activity.

The Moon before Orbiter

Before the space probe era man's knowledge of the Moon was bound to be limited, despite the fact that the lack of a lunar atmosphere makes all the details more distinct. Measurements of the positions on the disc of various lunar features were possible and although work undertaken by S. A. Saunder and J. A. Hardcastle in 1907–9 was of value (the Saunder–Hardcastle measures are still referred to today) certain areas could still not be resolved. In particular, little was known about the limb regions, which are hopelessly foreshortened as seen from the Earth: when a crater is seen under such conditions it may be impossible to distinguish it from a ridge. Neither could there be any positive information about the Moon's far side. Research was carried out into the lunar rays that were seen to come from the far side; by "plotting back", it was possible to fix the positions of a few ray-centres with reasonable accuracy, but the distribution of the features on the far side remained unknown. It was, however, significant that none of the major maria on the Earth-turned side extends on to the far side – apart from the Mare Orientale, whose nature was not then realized.

Photographic missions

The original photographs sent back by the Russian probe Luna 3 were of immense value, but by modern standards they were blurred and indistinct and thus widely misinterpreted [Key]. In particular, a long feature seen stretching across the disc was taken to be a major mountain range and was dignified by the title of the Soviet Mountains; but later photographs showed it to be nothing more than a bright ray. It was therefore not until the advent of the Orbiters that lunar research made its greatest stride forward, although the three successful Rangers, which crash-landed but managed to send back many thousands of useful pictures during the last minutes of their flights, also played a significant part.

Despite the comprehensive coverage of

1 **Alphonsus,** the walled plain in which Ranger 9 landed and in which mild activity has been reported, is a member of the Ptolemaeus chain; Ptolemaeus itself is to the north [above] and Arzachel to the south.

2 **The Alpine valley,** 130km (80 miles) long, can be seen in this Orbiter photograph. Mont Blanc, the highest peak in the Alps, lies nearby. Note the delicate rille running along the valley floor.

3 **Aristarchus,** with a diameter of 38km (24 miles), is bright compared with the surrounding region. The walls are heavily terraced and the floor contains a central mountain of considerable height.

4 **Theophilus,** in a photograph taken from Orbiter 3 at an altitude of 55km (34 miles), shows the ramparts and the central mountain mass. Above and to the right can be seen the walls and peak of Cyrillus.

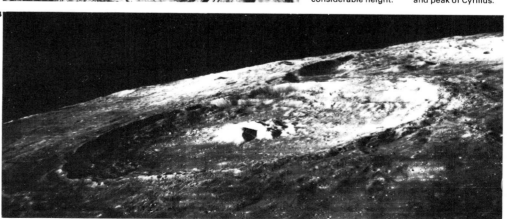

the Orbiter project, involving thousands of photographs – many of which still await complete analysis – there were still a number of outstanding questions. The Apollo programme (which was originally scheduled to continue to number 21) to a large extent supplemented the work of the Orbiters, particularly with more detailed coverage of future landing sites (for example, Apollo 10 the last pre-landing probe, photographed the Mare Tranquillitatis – the site selected for the Apollo 11 touchdown).

Ray-craters

The ray-craters, and the processes by which they may have been produced, also await examination. They must be the youngest of the major features of the Moon; the estimated age of Copernicus and Tycho, for example, is no more than one aeon (one thousand million years) but because samples have not been collected and analysed from those regions it is dangerous to be too definite. The last Surveyor landed on the outer wall-slopes of Tycho [5] and confirmed that the surface is extremely rough. This had been

expected, because it was already known from infra-red studies (carried out mainly by the American astronomers J. Saari and R. W. Shorthill) that Tycho cools down less rapidly during a lunar eclipse, or during a lunar night, than its surrounding areas; this implied a difference in surface covering or texture. Other ray-craters behave in the same way and have been called, rather misleadingly, "hot spots". This does not imply any internal source of heat; all it means is that during periods of darkness the "hot spots" record a somewhat higher temperature than other parts of the Moon.

It has been suggested that ray-craters were formed differently from other types of craters, but it seems unlikely. For instance there is no real difference in form between Tycho, which is the centre of the greatest ray-system on the Moon, and Theophilus [4], which is only slightly larger and which is not associated with a comparable system of rays. The surrounds are different, however; Tycho lies in a crowded area, while Theophilus is a member of a large chain of walled formations.

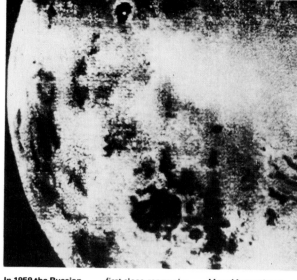

In 1959 the Russian Luna 3 sent back the first close-range photographs of the Moon. Mare Moscoviense can be seen.

5 Tycho, the great lunar ray-crater, was photographed from Orbiter 5. The terraced walls, the central elevations and the roughness of the floor are clearly shown. There is no evidence of lava flows of mare material; Tycho lies in the uplands and is probably one of the youngest of all the major craters.

6 The Hyginus Rille is one of the best known on the Moon. It is not, properly speaking, a rille or cleft; it is basically a crater-chain as is shown in this view of it taken from Orbiter 3. Hyginus itself, in the mid-position in the rille, is 6km (4 miles) across.

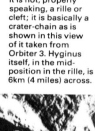

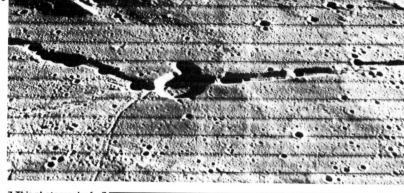

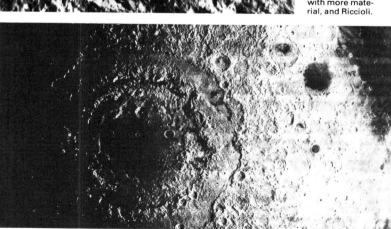

7 This photograph of the Mare Australe was taken from Orbiter 4 at a height of 3,500km (2,175 miles) above the Moon. The craters are flooded with mare material, but Australe is not a regular structure.

8 The Mare Orientale, photographed from Orbiter 4, is a complex structure with multiple ring-walls. To the right are seen Grimaldi, which is flooded with more material, and Riccioli.

The planet Mercury

Mercury can sometimes be seen with the naked eye but is much harder to sight than the other four planets that were known in ancient times – Venus, Mars, Jupiter and Saturn. It is the closest planet to the Sun and takes only 88 Earth-days to travel around it at a mean distance from it of about 58 million kilometres (36 million miles). In size and mass, Mercury is more like the Moon than the Earth [1]. Its diameter is 4,880km (3,032 miles). The escape velocity of only 4.3km (2.7 miles) per second indicates that its atmosphere is negligible.

Observation difficulties
The chief difficulty in observing Mercury is that it is never seen against a completely dark background because it remains in the same part of the sky as the Sun. Although quite bright (at its best exceeding magnitude 0, which is brighter than any star apart from Sirius, Canopus, Arcturus and Alpha Centauri), it is never conspicuous to the naked eye. An observer without a telescope sees it only on favourable occasions, either low down in the west after sunset or low down in

the east before sunrise. To make matters worse, the phase (illuminated surface) decreases as Mercury gets nearer to the Earth [3]. At its closest Mercury is at inferior conjunction and cannot be seen at all (except in a rare transit across the Sun) because its dark hemisphere is turned Earthwards.

Mapping the planet
The first serious attempts to map Mercury were made in Milan during the latter part of the nineteenth century by Giovanni Schiaparelli (1835–1910). Rather than study the planet at night when he could see it with the naked eye, Schiaparelli carried out his observations in broad daylight, with Mercury high above the horizon. He was able to see some dark shadings and brighter areas, but his chart was very rough. Later, between 1924 and 1933, a long study was carried out by E. M. Antoniadi. He used the 33in (84cm) refractor at the Observatory of Meudon, making his observations in broad daylight, and his chart [Key] remained the best – although we now know that it too was highly inaccurate – until the historic flight of

Mariner 10 which took place in 1973–4.

Antoniadi, like Schiaparelli, believed that the rotation period of Mercury was captured or synchronous. If that were so, then both the revolution period and the axial spin would be 88 Earth-days and, as a result, part of the planet would be permanently illuminated by the Sun while another part would be in everlasting night. Because the Mercurian orbit is decidedly eccentric, there would be an intermediate "twilight zone" between these two extremes over which the Sun would rise and set, always keeping close to the horizon. The effects would be analogous to the oscillations of the Moon. It is now known that Antoniadi was wrong. Radar measurements carried out since 1962 show that the true rotation period is 58.7 Earth-days so that all regions of the planet receive sunlight at some time or other [5]. Because of a curious relationship (which may or may not be coincidental), Mercury presents the same hemisphere to the Earth every time it is best placed for study; this is what misled the earlier observers.

Antoniadi believed that he had observed local obscurations or "clouds" on Mercury,

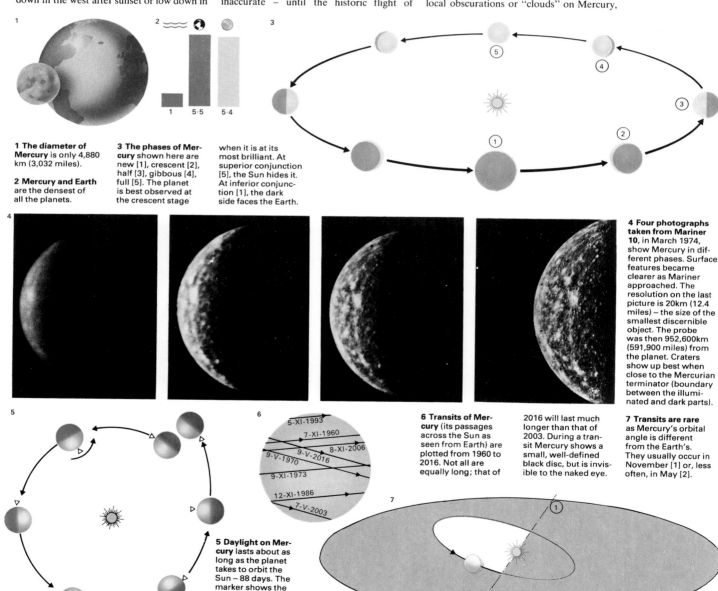

1 The diameter of Mercury is only 4,880 km (3,032 miles).

2 Mercury and Earth are the densest of all the planets.

3 The phases of Mercury shown here are new [1], crescent [2], half [3], gibbous [4], full [5]. The planet is best observed at the crescent stage when it is at its most brilliant. At superior conjunction [5], the Sun hides it. At inferior conjunction [1], the dark side faces the Earth.

4 Four photographs taken from Mariner 10, in March 1974, show Mercury in different phases. Surface features became clearer as Mariner approached. The resolution on the last picture is 20km (12.4 miles) – the size of the smallest discernible object. The probe was then 952,600km (591,900 miles) from the planet. Craters show up best when close to the Mercurian terminator (boundary between the illuminated and dark parts).

5 Daylight on Mercury lasts about as long as the planet takes to orbit the Sun – 88 days. The marker shows the rotation of a fixed point on the planet during this period.

6 Transits of Mercury (its passages across the Sun as seen from Earth) are plotted from 1960 to 2016. Not all are equally long; that of 2016 will last much longer than that of 2003. During a transit Mercury shows a small, well-defined black disc, but is invisible to the naked eye.

7 Transits are rare as Mercury's orbital angle is different from the Earth's. They usually occur in November [1] or, less often, in May [2].

which sometimes veiled the surface features. This was puzzling because it was obvious that the atmosphere must be extremely rarefied by terrestrial standards. It is now known that the atmosphere is much too thin to support clouds of any kind and there seems no escape from the conclusion that Antoniadi's observations were wrong. Because of Mercury's closeness to the Sun, its surface is extremely hot during daylight and a thermometer would register a temperature of more than 370°C (700°F) at maximum; but because of the virtual lack of atmosphere the nights are bitterly cold. No known form of life could survive on Mercury.

Mariner 10 probe
The first close-range information about Mercury's surface was received in 1974 from the flight of Mariner 10, the first "two-planet" probe. In February 1974 it bypassed Venus, sending back pictures, and then swung inwards towards a rendezvous with Mercury during the following month. The photographs revealed a landscape that was strikingly like that of the Moon [4]. Craters,

mountains, ridges and valleys showed up everywhere, although there were fewer broad dark plains like the lunar Mare Imbrium. On Mercury, the chief plain has been named the Caloris Basin [9].

In September 1974, after orbiting the Sun, Mariner 10 made a second rendezvous with Mercury and took more high-quality pictures. The third encounter took place in March 1975. Despite the incompleteness of the pictures, they revealed the same types of craters and mountains. Much of Mercury has now been mapped and for the first time there is reliable information on what this strange world is really like.

One discovery of great interest was the detection of a magnetic field. This field is weak compared with that of the Earth, but it is quite definite and produces a true magnetosphere.

Further probes to Mercury are being planned, but the chances of manned expeditions there are slight, at least in the foreseeable future. Mercury is overwhelmingly hostile, although from the scientific point of view it is of exceptional interest.

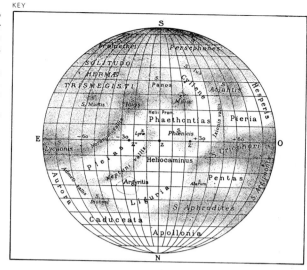

KEY

This map of Mercury was drawn between 1924 and 1933 by Antoniadi, who named the main features. There is poor correlation between the dark areas shown and the results from Mariner 10, and some recasting of names will be needed.

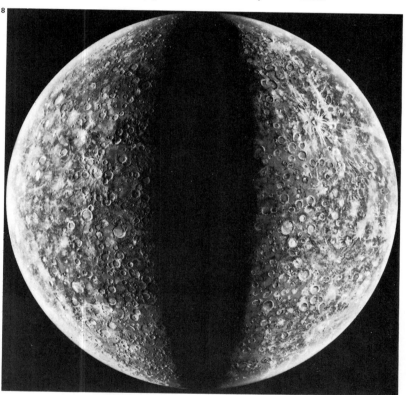

8 The whole surface of Mercury shown in this mosaic of Mariner 10 photographs is extensively cratered and distinctly lunar in appearance. There are also bright ray systems [upper right] like those of the Moon. The Mariner photographs were highly informative.

9 This mosaic of the Caloris Basin from Mariner 10 shows what is probably the most distinctive plain on Mercury. The mountain rim around the basin can be clearly seen and the interior is different from the surrounding region. The origin of the basin is still unknown.

10 Hills and ridges cut across many of the craters in the region shown here, which is unique to the planet. Their origin is obscure. They may be the result of shock waves caused by the formation of the 1,300km (800 mile) Caloris Basin, which lies diametrically opposite this region.

11 Another region of Mercury shows an area crowded with craters and craterlets. One interesting feature is the crater valley to the upper left, which resembles many of the crater chains on the surface of the Moon.

12 The similarity of Mercury to the Moon is clear in this Mariner 10 photograph. Some craters have centred peaks and all the well-marked craters are basically circular, although here they appear elliptical because of foreshortening.

189

The planet Venus

Venus, the second planet in order of distance from the Sun, is almost as large as the Earth and has more than 80 per cent of the terrestrial mass [1, 2]. Instead of being devoid of atmosphere, it has a deep, dense and cloudy surround which prevents us from ever seeing the true surface of the planet.

The mean distance of Venus from the Sun is 108,200,000km (67,200,000 miles) and this is practically constant because its orbit is more closely circular than that of any other planet. The revolution period is 224.7 days. Before the age of space probes and powerful radar, the rotation period of Venus was unknown; there are no markings on the disc persistent enough to be used for rotation measurements, as with Mars or Mercury. All that can be seen from Earth is the upper layer of cloud, and the shadings and bright areas on the disc are always vague and transitory.

Observing Venus from Earth
To the naked eye Venus is a splendid object and is far brighter than any other celestial body apart from the Sun and the Moon, which is why it was named after the goddess of beauty. Telescopically, however, Venus is a disappointment and it is not therefore surprising that until recent years it was often called the "planet of mystery"[4].

In the 1930s some positive information emerged. It was established that the atmosphere of Venus is made up largely of carbon dioxide, which tends to act as a "blanket" and shut in the Sun's heat. Two pictures of the planet were current. In one theory the surface was covered mainly with water, in which primitive life forms might have developed – just as happened on the Earth thousands of millions of years ago. In the other theory, Venus was a scorching-hot, arid dust-desert.

Information from early probes
The probe era began in 1962, when the US probe Mariner 2 bypassed Venus and sent back information that confirmed that the surface is extremely hot. It was also found that the axial rotation period is slow – about 243 Earth-days, which is longer than the revolution period of 224.7 days; therefore on Venus the "day" is longer than the "year", giving rise to a most peculiar calendar.

It has now been established that Venus rotates in a retrograde direction: east to west, instead of west to east as with the Earth and most of the other planets. To an observer on the surface of Venus, the Sun would appear to rise in the west and set in the east – although, in fact, the cloud-laden atmosphere would hide the sky completely.

Following Mariner 2, the USSR managed to soft-land various automatic probes on the surface of Venus, bringing them down by parachute through the dense atmosphere. A temperature of nearly 530°C (1,000°F) was recorded, with a ground pressure about 100 times greater than Earth's air at sea-level.

From the USA, Mariner 10 bypassed Venus in February 1974 and sent back the first pictures of the top of the cloud layer. Mariner 10 made only one pass of Venus; its main target was the innermost planet, Mercury. However, the pictures were of high quality, and showed the banded appearance of the clouds [9]. They also confirmed that the cloud tops have a rotation period of only four days, so that the atmospheric structure is unlike that of the Earth [5].

1 Venus and the Earth are almost equal in size, mass and surface gravity.

2 The density of Venus is less than Earth's, but it may have a heavy core.

3 Venus is brightest when at the crescent stage. When the planet is new [A] it is invisible (except at transit). At superior conjunction and full phase [B] it is on the far side of the Sun.

4 The apparent diameter of Venus changes according to phase. It is least at superior conjunction, when Venus is full, because the planet is then on the far side of the Sun and is at its most remote from the Earth. The apparent diameter increases as the phase shrinks, as is shown in these photographs which were taken with a (12in) 30cm reflecting telescope.

5 Mariner 10 provided the first really conclusive optical evidence of the four-day rotation of the upper clouds of Venus (although this phenomenon had been suspected from observations made from Earth) with these three photographs taken in February 1974. The cloud patterns are clearly shown (the arrow indicates the same area of Venus). The first photograph was taken on 2 February at 0 hours, the second at 7 hours and the third at 14 hours (note the movement of the arrowed marking). The axis of Venus is almost perpendicular to the plane of the orbit, but the rotation is retrograde in direction.

6 Transits of Venus (its passage across the Sun as seen from Earth) occur in pairs with an interval of eight years, each pair separated by more than a century. The diagram shows the transit paths of Venus for 1761–9, 1874–82 and 2004–12.

7 When Venus enters transit, it seems to draw a strip of blackness ("Black Drop") after it, an effect produced by the planet's atmosphere. The strip disappears only when the planet is well onto the Sun's disc, making measurements of Venus's entry difficult.

8 The Ashen Light is a faintly luminous patch that appears on the night side of Venus when it is in crescent. This is not the same as Earthshine on our Moon, and its cause is uncertain. Some authorities regard it as a contrast effect; others believe it to be due to electrical phenomena in the upper atmosphere of the planet. It is not easy to observe, but most serious students of Venus have seen it. It was first recorded in the 1790s. (This drawing is exaggerated.)

Meanwhile, US radar studies had shown that the surface contains large, shallow craters [11]. The origins of the craters are not known, but there must be intense erosion in so dense an atmosphere so that by "geological" standards they can hardly be very old. Volcanic action may be the cause of the cratering and the idea that vulcanism is in progress on Venus cannot yet be ruled out. It was also found that the clouds contain a great deal of sulphuric acid (perhaps even fluorosulphuric acid). Because this produces a highly corrosive "rain", Venus is clearly a most hostile world.

The transmission of pictures

The next major step came in October 1975, when two Soviet probes, Veneras 9 and 10, made controlled landings on the surface and transmitted pictures back. The pictures were relayed by the orbiting sections of the probes, which remained circling Venus at a height of about 1,500km (900 miles). It was an amazing triumph for the Soviet scientists – even though neither Venera 9 nor 10 could transmit for more than about one hour before

being put permanently out of action by the extreme temperature and pressure.

Surprisingly, the surface of Venus proved to be strewn with smooth rocks, many of which were about 1m (39in) in diameter [12]. There was plenty of light – in the Russian description, about as much as at noon in Moscow on a cloudy summer day – and the probes did not even have to use their floodlights. Neither was the atmosphere super-refractive, as had been expected, and all the details of the landscape were clear cut. A temperature of 485°C (900°F) was recorded and pressure was found to be 90 times that on Earth. It was also determined that the cloud layer ends at a height of some 30km (19 miles). Below this is what may be called a region of super-hot corrosive smog.

Venus is far from being the friendly world that was once thought. With its carbon dioxide atmosphere, its sulphuric acid clouds and its intense heat, it is wholly hostile to man. The knowledge has dashed some hopes: less than twenty years ago, Venus was widely regarded by many scientists as a more promising space target than Mars.

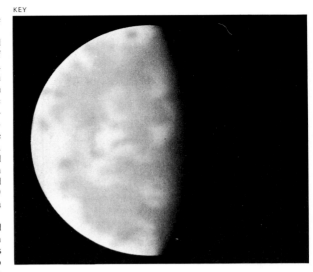

This drawing of Venus made by A. Dollfus at the Pic du Midi Observatory reveals more than any photograph taken from Earth, but the surface detail is so obscure it is hard to draw accurately; also the cloud patterns change quickly.

9 Venus is here photographed from Mariner 10. The light and dark zones of the cloud system, which account for Venus's brilliance, are distinctly shown.

10 In the equatorial zone of the planet the atmospheric conditions on Venus differ from those at the poles. This photograph was taken from Mariner 10.

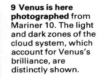

11 Craters on Venus can be detected only by radar. The planet's dense layers of brilliant cloud make telescopic observation of surface features impossible. Only a small part of the planet has so far been studied in detail, as shown here (the black strip has not been analysed).

Various features detected by means of radar are revealed in this picture [A] which covers an area in the equatorial zone of the planet, indicated in the circle [B]. The craters are of particular interest. They appear to be much shallower than those on Mars or Mercury, although their details are blurred in comparison with those obtained from other planets.

12 In October 1975 the USSR achieved a notable triumph when its probe Venera 9 soft-landed on Venus and transmitted a picture back. This showed a rock-strewn landscape. Part of the probe is seen at the bottom of the picture.

The planet Earth

The Earth is the largest member of the group of inner planets and is also the most massive. The difference in size and mass between the Earth and Venus is slight (the ratio for mass is 1 to 0.82), but Mars is much smaller, and Mercury is more comparable with the Moon than with the Earth.

When the Earth is compared with its planetary neighbours, marked similarities as well as marked differences are found. Of course, what singles the Earth out from any other planet is the fact that it has an oxygen-rich atmosphere and a temperature that makes it suitable for life of the kind we can understand. Were the Earth slightly closer to the Sun, or slightly farther away, life here could not have developed.

The ecosphere

What is termed the "ecosphere" [4], or the region in which solar radiation will produce tolerable conditions for terrestrial-type life, extends from just inside the orbit of Venus out to that of Mars. Until about 1960, it was commonly thought that such life might exist throughout the whole region. Although this possibility was far more remote for Mars, with a significantly lower mass than that of Earth, and hence a tenuous atmosphere, Venus was looked upon as the Earth's twin. Approximately equal in density as well as size and mass, Venus also absorbs about the same amount of solar energy as Earth because of the high reflecting power of its cloud. It was not until 1967 when the surface temperature of Venus was shown to register up to 485°C (900°F) that it was commonly accepted that advanced terrestrial life could develop only within a limited zone.

Another essential need is the presence of an atmosphere that will not only enable living creatures to breathe, but will also protect the planet from lethal short-wave radiations coming from space. On the surface of the Earth there is no danger, because the radiations are blocked out by layers in the upper atmosphere; but the Moon is unprotected, and so too is Mercury. Had the Earth been more massive, it might have been able to retain at least some of its original hydrogen (as Jupiter and Saturn have done) and the resulting atmosphere might have been unsuitable for life. A lower mass might have led to the escape not only of the hydrogen but also of all other gases, so that a fortunate combination of circumstances has produced terrestrial life.

Next, there is the question of temperature, which does not depend only upon the distance of the planet from the Sun or upon the composition of its atmosphere; there is also the axial rotation period to be taken into account. The Earth spins round once in approximately 24 hours, and the rotation period of Mars is only 37 minutes longer, but the situation with Mercury and Venus is very different – the periods are 58.7 days and 243 days respectively, leading to very peculiar "calendars". Were the Earth a slow spinner, the climatic conditions would, as a result, be both unfamiliar and hostile.

The Earth's magnetic field

The Earth's heavy, iron-rich core is associated with the magnetic field, and here too comparisons may be made with other planets. Venus again provides problems. With its comparable size and mass it should

CONNECTIONS

See also
174 Evolution of the Solar System
176 Members of the Solar System
130 Earth's time scale
66 The atmosphere

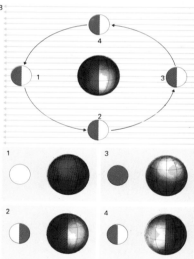

1 From space Earth will show phases, just as the Moon does to us. The five photographs shown [A] were taken from a satellite over a period of 12 hours. B shows the phases of the Earth as they are seen from the Moon. For this purpose it can be assumed that the Earth is stationary, with the Moon moving round it in a period of 27.3 days. When the Moon is full from Earth [1] a lunar observer will have "new earth"; at our new Moon [3] a lunar observer will see full Earth. Below the main diagram, the different phases of the Moon [left] and Earth are shown [1–4].

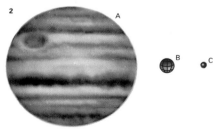

2 Relative sizes of Jupiter [A], the Earth [B] and Mercury [C]. Jupiter is the largest planet, Mercury the smallest; Earth is intermediate in size, but much more nearly comparable with Mercury. Earth is in fact the largest of the so-called terrestrial planets (Mercury, Venus, Earth, Mars, Pluto) but far inferior to the smallest of the giants (Uranus).

3 Apollo 10 sent back these pictures in May 1969; it was the second vehicle to take men round the Moon. The pictures show the Earth coming into view as the spacecraft comes from behind the far side of the Moon, from which the Earth can never be seen. In the first photograph the sharpness of the lunar horizon is particularly notable; there is no lunar atmosphere to cause the slightest blurring or distortion. Note the changing Earth phase.

have the same kind of core and hence an appreciable magnetic field; but space probes have so far failed to detect any magnetism, and it is now certain that even if a magnetic field exists it is very weak indeed. The same may be true of Mars, but Mercury has a perceptible field and even a magnetosphere. It is probably significant that Mercury and the Earth are the densest of all the planets, with specific gravities of about 5.5 (that is to say, their masses are 5.5 times greater than that of equal volumes of water).

The watery planet

The Earth is again unique in having a surface that is largely covered with water; thus although it is the largest of the four inner planets its land surface is much less than that of Venus and equal to that of Mars. There can be no oceans or even lakes on Mars, because of the low atmospheric pressure, and of course none on the Moon or Mercury, which are to all intents and purposes without atmosphere. On Venus the surface temperature is certainly too high for liquid water to exist, so that the old, intriguing picture of a "Car-

boniferous" type Venus, with luxuriant vegetation flourishing in a swampy and moist environment, has had to be given up.

Because the Earth is so exceptional, there have been occasional suggestions that it was formed in a manner different from that of the other planets; but this is certainly not so. The age of the Earth, as measured by radioactive methods, is approximately 4,600 million years (4.6 aeons) and studies of the lunar rocks show that the age of the Moon is the same; there is no reason to doubt that the Earth and all other members of the Solar System originated by the same process, and at about the same time, from the primeval solar nebula. It is often said that Mars is more advanced in its evolution than is the Earth and this may be true; but the absolute ages of the two worlds are probably the same, although Mars has "aged" more quickly.

The Earth's position in the middle of the ecosphere, and the particular size and mass that lead to its own kind of atmosphere, single it out. There is no other planet in the Solar System upon which men could survive except under artificial conditions.

The Earth's axis is inclined at 23.5° to the perpendicular to the orbital plane. This causes the seasons; the varying distance of the Earth from the Sun has only a minor effect.

4 This diagram shows the ecosphere, or the region round the Sun in which a planet will be at a suitable temperature for life to exist – assuming that the planet is of Earth type. The inner, yellow zone [1] is too hot. Beyond is the ecosphere [2] [orange], and beyond this [3] is the zone where temperatures will be too low. The Earth [4] lies in the middle of the ecosphere; Venus [5] is at the inner limit and Mars [6] at the outer.

The planet Mars

Mars, the first planet beyond the Earth in the Solar System, is of especial interest to man. In the earlier part of this century many astronomers believed implicitly in the existence of an advanced civilization on Mars. This belief has now been shattered. There are no Martians and it seems that the most advanced life the planet could support would be very primitive organic matter. It is more likely that the planet is sterile. Despite this, Mars is still more Earth-like than any other known world and it must surely be the first target for an expedition by a manned space probe after the Moon.

Through a telescope, Mars shows a red disc with white caps at the poles and prominent dark markings that are essentially permanent [Key]. Its mean distance from the Sun is 228 million km (141.5 million miles). The Martian year is equal to 687 Earth-days and a day is equal to 24hr 37min. Moreover the axial tilt is only slightly greater than the Earth's, so that the seasons are of the same basic type, although much longer. As with the Earth, the south pole is tilted Sunwards at perihelion (the point in orbit when a planet is nearest to the Sun); the effect of this on the climate of Mars is greater than its effect on Earth's climate because the orbit of Mars is more eccentric. The climates in the planet's southern hemisphere are more extreme than those in the northern. The summers are shorter and hotter and the winters longer and colder. The noon temperature on the Martian equator at midsummer may rise to more than 16°C (60°F). The nights are bitterly cold, because the thin atmosphere is inefficient in retaining warmth; however, Mars is certainly not a frozen world.

The Martian atmosphere

Because Mars is not only less dense than the Earth but much smaller – the diameter is a mere 6,790km (4,220 miles) – the escape velocity (the speed an object must reach to overcome gravity) is low: 5.0km (3.1 miles) per second, which explains the tenuous atmosphere. The main constituent is now known to be carbon dioxide (95 per cent), and the barometric pressure at ground level is less than 10 millibars. No advanced terrestrial creatures could survive without protection.

There is no free water on the Martian surface today. It appears, however, that the white polar caps are made up chiefly of water ice with some carbon dioxide (dry ice) also present. The size of the caps varies according to the Martian season; at their greatest extent they are conspicuous and can be seen with a small telescope [7].

Features shown by early maps

The first drawing to show any markings on the surface of Mars was made by the Dutch astronomer Christiaan Huygens (1629–95) in 1659 [1A]. It shows the V-shaped region now called the Syrtis Major Planitia, although in somewhat exaggerated form. Later observers using more powerful telescopes produced drawings that showed more detail. The first reasonably reliable maps date from the second half of the nineteenth century. What may be called the "modern" period of telescopic research was initiated by Giovanni Schiaparelli (1835–1910) in 1877 when Mars was at perihelion and opposition and was excellently placed for observation.

Schiaparelli, observing from Milan, drew

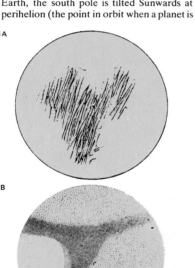

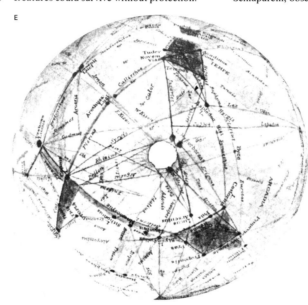

1 **Different stages** in the telescopic exploration of Mars are shown in these five drawings. The first [A], made in 1659 by Christiaan Huygens, shows only the Syrtis Major. In drawing B, made by the pioneer observer Johann Schröter (1746–1826) in 1800, the Syrtis Major is again shown reasonably accurately. The famous canal network appears in Schiaparelli's drawing [C] made in 1877. E. M. Antoniadi, who used the 33in (83cm) Meudon refractor from 1900–30, had no faith in the canals, but his drawing [D] was remarkably accurate. The drawing by Percival Lowell, made in about 1905 [E], shows the illusory canal network.

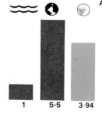

2 **The surface area of Mars** is 28% that of the Earth. Its diameter of about 6,790km (4,220 miles) is a little more than half that of the Earth and approximately twice that of the Moon. It has only one-tenth of the Earth's mass.

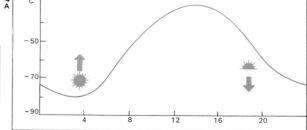

3 **The density of Mars** is appreciably less than the Earth's with a specific gravity of only 3.94, resulting in a fairly low escape velocity. The Martian surface gravity is 0.38 that of the Earth's. There is no detectable magnetic field and so presumably no heavy core.

4 **Temperature and wind speed** on the Martian surface were carefully monitored by Viking 1 in the area of Chryse Planitia. Despite a diurnal range of more than 70°C [A], the maximum temperature recorded during a Martian day was still below freezing-point. Chryse is, however, well to the north of the Martian equator. Speed (metres per second) and direction of winds were also recorded during the Martian day [B]: wind direction is indicated by the arrows. The pattern was repeated day after day, the winds remaining gentle, never attaining 24km/h (15mph) throughout the period. Viking's meteorology station was set up, however, when conditions in its subtropical location were very steady.

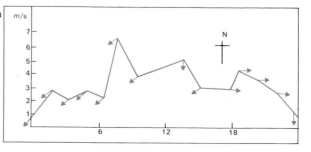

a map that was superior to any of its predecessors [1C]. On it he showed straight, artificial-looking features that he called *canali* or channels but which have since become known as the Martian canals. Inevitably, it was suggested that these features were artificial waterways constructed by the planet's inhabitants to form a vast irrigation system. According to this intriguing theory, water would be drawn from the ice-covered polar caps and pumped through to the arid regions closer to the equator. Where canals crossed each other, they did so with the formation of small patches called "oases", which were regarded as centres of population. Schiaparelli himself kept a reasonably open mind. But the American astronomer Percival Lowell (1855–1916), the founder of the Lowell Observatory at Flagstaff, Arizona, USA, was convinced that Mars supported a highly developed civilization.

The Martian surface

After it became obvious that dark patches on the surface of Mars could not be seas, it was generally believed that they must be tracts of vegetation growing in depressions. This was still thought to be true until Mariner 4, the first successful space probe to Mars, was launched in 1965. It has now been found that the dark regions are not depressions; some of them, including the Syrtis Major, are lofty plateaus sloping off to all sides. Evidently their colouring arises from some difference in surface texture and not from any vegetation.

Most of the Martian surface is reddish-ochre; these areas are generally called deserts, which may be an appropriate name although there is no real analogy with terrestrial deserts such as the Sahara. Dust-storms are not uncommon and there are winds in the atmosphere of the planet [6].

While the Martian "canals" do not exist, Viking spacecraft have given evidence that water may once have flowed on Mars in abundance. In the Chryse region cleanly cut channels meander and intertwine like dried-out river beds, and "islands" of ancient rock have tails extending downstream. According to a later theory, the mysterious channels may have been carved by molten lava flowing over the rock.

Martian volcanoes, three in the Tharsis Mts, plus Olympus Mons, were photographed by Viking 1 from 560,000 km (348,000 miles). Also seen is the impact basin Argyre.

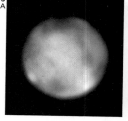

5 Martian clouds consist of high-altitude "white" clouds and much less common but more extensive dust clouds. These four views show Mars taken through different coloured filters. The area of cloud that forms every summer over the Syrtis Major is shown in A. B, C and D show the Syrtis Major to the left, covered by the same cloud. The bright area to the right is Elysium.

6 The Martian dust-storm of 1971 was one of the greatest ever observed. [A] 20 Sept 1971: before the storm; the dark markings show up clearly. [B] 12 Oct 1971: the dust covers the planet. Mariner 9 approached Mars during this period. [C] 8 Feb 1972: the dust is clearing and the most prominent surface features become visible.

7 Observations of Mars made in 1972 show a dark peripheral band well defined in A at the edge of shrinking polar caps [B to F]. Gerard de Vaucouleurs in 1939 attributed the shrinking to release of water vapour, which might support vegetation. Vegetation is now considered unlikely although the polar caps do contain water ice.

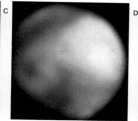

8 A

B

C

D

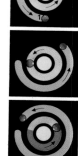

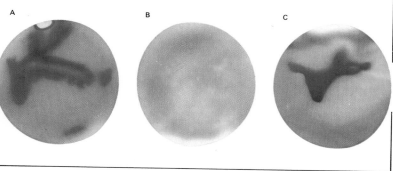

E

15 Dec 1975
22 Jan 1978
26 Feb 1980
15 April 1967
31 May 1969
10 Aug 1971
25 Oct 1973

8 The "opposition of Mars" describes the lining up in sequence of the Sun, Earth and Mars. Mars takes 687 Earth-days to complete one revolution round the Sun and this means that it comes to opposition once in approximately 780 days. As shown in E oppositions occurred in 1967, 1969, 1971, 1973 and 1975; the next will be in Jan 1978 and Feb 1980. The series of drawings [A to D] begin with Earth and Mars in opposition; by the time Mars has reached position 1 [A], the Earth will have moved to position 2. By the time the Earth has completed one full circuit [B], Mars has made only a little more than half a revolution. When the Earth has made 1.5 circuits [C], Mars has made almost one and after 780 days the two planets are again in opposition [D].

7 A

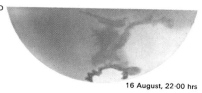

22 May, 02·10 hrs

B

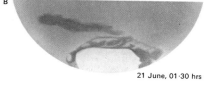

21 June, 01·30 hrs

C

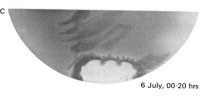

6 July, 00·20 hrs

D

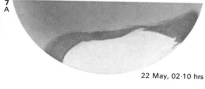

16 August, 22·00 hrs

E

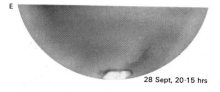

28 Sept, 20·15 hrs

F

10 Oct, 19·45 hrs

Mars missions

Mars is a fairly small world and can be properly observed from Earth for only a few months in every alternate year (the average interval between successive oppositions is 780 days). Before the Space Age, man's knowledge of Mars was bound to be incomplete; then in 1962 the Russians made a preliminary attempt to send a space probe past it [1]. They failed, because contact with the vehicle was lost at a relatively early stage and was never regained. But in July 1965 the American probe Mariner 4 [2] bypassed Mars at close range, and sent back the first detailed information about the planet.

What the Mariner programme revealed

The Martian atmosphere proved to be as thin as had been expected in that instead of being made up of nitrogen, with a ground pressure of 85 millibars, it was found to be composed mainly of carbon dioxide, with a pressure of less than 10 millibars – which at once reduced the possibility of any advanced life forms. But the most spectacular discovery was of craters, superficially similar to those of the Moon, but with some important differences. The craters

were large and some of them had lunar-type central peaks. This came as a major surprise, because scientists had expected that Mars would have a smooth landscape.

The next American probes, Mariners 6 [3] and 7, followed in the summer of 1969, only a few days after Neil Armstrong and Edwin Aldrin, in Apollo 11, made their landing on the Moon. They again showed craters, together with mountainous regions. When measurements were taken there were some more surprises. For instance, the circular white patch known as Hellas, just south of the V-shaped Syrtis Major, proved to be a depression instead of a raised plateau. Hellas is in fact the most depressed area on Mars and it seems to be almost devoid of craters or other important features.

These first three Mariners were fly-by probes, making one pass of Mars and then moving on to solar orbit. Mariners 8 and 9, launched in 1971, were different. They were intended to orbit Mars in order to send back data – including photographs – over a period of months instead of only a few days. Mariner 8 failed immediately after blast-off and fell

into the sea; but Mariner 9 [4] was a success, giving us a wealth of new data including 7,329 photographs.

The craft went into orbit around Mars on 14 November 1971, when almost the whole of the planet was surrounded by dust and little could be seen. Such huge dust-storms are uncommon although they have been recorded previously (in 1909 and 1956, for example). When the dust cleared, the spacecraft's cameras discovered volcanoes, similar in form to the terrestrial volcanoes of Hawaii, but much larger. For instance, Olympus Mons reaches up to about 25km (15.5 miles) above the general level of the surface and Arsia Mons to about 20km (12 miles). There were deep rift valleys, and features that looked like dried-out river beds.

Viking's search for life

But did some kind of life await discovery somewhere on this cold, harsh landscape? The United States sent the Vikings to find out. The first swung into orbit on 19 June 1976, the second on 7 August 1976. After a period of photo-reconnaissance to establish

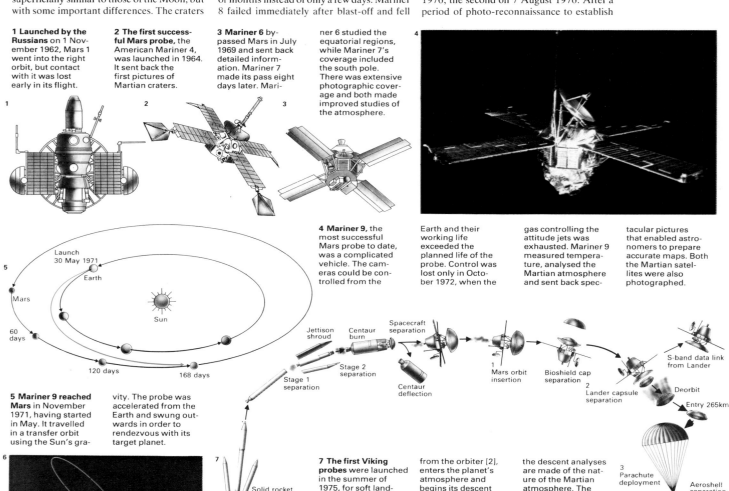

1 Launched by the Russians on 1 November 1962, Mars 1 went into the right orbit, but contact with it was lost early in its flight.

2 The first successful Mars probe, the American Mariner 4, was launched in 1964. It sent back the first pictures of Martian craters.

3 Mariner 6 bypassed Mars in July 1969 and sent back detailed information. Mariner 7 made its pass eight days later. Mariner 6 studied the equatorial regions, while Mariner 7's coverage included the south pole. There was extensive photographic coverage and both made improved studies of the atmosphere.

4 Mariner 9, the most successful Mars probe to date, was a complicated vehicle. The cameras could be controlled from the Earth and their working life exceeded the planned life of the probe. Control was lost only in October 1972, when the gas controlling the attitude jets was exhausted. Mariner 9 measured temperature, analysed the Martian atmosphere and sent back spectacular pictures that enabled astronomers to prepare accurate maps. Both the Martian satellites were also photographed.

5 Mariner 9 reached Mars in November 1971, having started in May. It travelled in a transfer orbit using the Sun's gravity. The probe was accelerated from the Earth and swung outwards in order to rendezvous with its target planet.

Launch 30 May 1971
Earth
Mars
60 days
120 days
168 days
Sun

6 As Mariner 9 approached Mars, it passed and photographed the two tiny Martian satellites, Phobos and Deimos. Their orbits are shown, together with the elliptical orbit of the probe after it had joined a closed path round the planet.

7 The first Viking probes were launched in the summer of 1975, for soft landing on Mars in mid-1976. Programmed steps in the exploratory sequence began with the first probe landing in the region of Chryse, the second in Utopia. Each probe is made up of two parts, an orbiter and a lander. Once the combined vehicle is in a path round Mars [1], the lander separates from the orbiter [2], enters the planet's atmosphere and begins its descent to the surface. At a comparatively low level, the main parachute opens [3]. Next, the lander separates from the parachute [4] and completes the descent to the surface under its own power. This turns off at about 15m (50ft) above the surface, and the lander falls gently on to its chosen site [5]. During the descent analyses are made of the nature of the Martian atmosphere. The orbiter meanwhile remains in a closed path round Mars, acting as a relay link for the lander which, once it has grounded, begins to carry out numerous experiments. The principal objective of the probes was to study the surface geology and chemistry of Mars and test its soil for signs of life.

Solid rocket separation
Launch
Jettison shroud
Centaur burn
Stage 2 separation
Stage 1 separation
Spacecraft separation
Centaur deflection
Mars orbit insertion
Bioshield cap separation
Lander capsule separation
S-band data link from Lander
Deorbit
Entry 265km
3 Parachute deployment
Aeroshell separation
4 Terminal propulsion 1,750m
5 Entry to landing 5–10 minutes

suitable landing sites, both released landers. That of Viking 1 touched down on the broad Chryse basin in Mars' northern middle latitudes on 20 July. The other landed on 3 September on Utopia Planitia roughly 7,400km (4,600 miles) from Viking 1 and 1,400km (870 miles) nearer the north pole.

Spectacular photographs received from the landers showed rock-strewn terrain in both places with an overlying reddish soil. The sky was pink from light scattered by red dust particles in the air. Major elements in the soil, as detected by Viking 1's X-ray fluorescence spectrometer, were silicon, iron, calcium, aluminium and titanium.

Both spacecraft dug into the soil under computer control to deliver samples for analysis to the biology laboratories, with results that both surprised and tantalized. The experiments had been designed on the basis that any life forms in the soil must eat and excrete certain basic chemicals.

The initial "gas exchange" experiment rapidly detected 15 times as much oxygen as had been expected. All terrestrial life forms with which science is familiar take time to

grow and reproduce, and these findings seemed more likely to be the result of a chemical reaction in the iron-rich soil.

Results of the experiments

The result of the "labelled release" experiment also at first looked interesting. If microbes were present, they were expected to take up carbon-14 and give off radioactive wastes such as carbon dioxide, carbon monoxide and carbon methane. A generous amount of carbon dioxide was detected but again the cause seemed to be chemistry.

The "pyrolytic" experiment indicated that something was taking carbon dioxide out of the air in the test chamber and incorporating it into other compounds within the soil, but whether that "something" was biological or chemical was open to doubt.

What worried scientists most was the absence of organic molecules. All three biology experiments showed many signs that could be interpreted as the result of living organisms, only if organic compounds were also present. But this did not detract from the success of the programme.

The Sampler scoop of Viking 1 is seen poised over the red-orange soil of Chryse — Planitia. Some of the rocks are dark and coarse-grained, while others have a lighter, — mottled appearance and may have come from lava flows or stream deposits.

8 The first colour photograph of the Martian surface taken by Viking 1 shows orange-red materials covering most of the surface with darker bedrock exposed in patches (lower right). The reddish materials may be limonite (hydrated ferric oxide). Such weathering products form on Earth with water and an oxidizing atmosphere. The sky has a reddish hue, probably due to the scattering and reflection of light from reddish dust particles suspended in the atmosphere. The scene was scanned three times by Viking 1's camera through a different colour filter each time. Colour balance was achieved with the help of a test chart on the spacecraft.

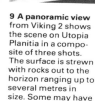

9 A panoramic view from Viking 2 shows the scene on Utopia Planitia in a composite of three shots. The surface is strewn with rocks out to the horizon ranging up to several metres in size. Some may have come from the nearby impact crater Mie which is about 1km (0.62 mile) across. The picture has been electronically rectified to remove the effect of the spacecraft's eight degree tilt toward the west.

10 This spectacular sunset over Chryse Planitia was photographed by Viking 1. The camera began scanning the scene from the left about 4 minutes after the Sun had dipped below the horizon, and continued for 10 minutes. The Sun had set nearly 3 degrees below the horizon by the time the picture was completed. The Martian surface appears almost black and the horizon line is very sharp.

Mars maps

Most of the Martian features were renamed by Giovanni Schiaparelli (1835–1910) after 1877. His system replaced the older nomenclature; for instance, the "Kaiser Sea" or "Hour Glass Sea" was renamed the Syrtis Major. Further revisions by the International Astronomical Union following the Mariner 9 results have assigned Latin qualifications to those of Schiaparelli's. These are given to topographical as opposed to albedo features (those features associated with the reflecting power of the planet). The new Latin names have been used on the maps below.

The western hemisphere

The western hemisphere (below left) includes most of the Acidalia Planitia. To most observers it is the more interesting of the two hemispheres because it also contains some of the greatest of the volcanoes, notably Olympus Mons, which can be seen as a tiny patch from Earth and is surrounded by an extensive, roughly circular area comparatively free from major craters. This area includes Amazonis Planitia to the west, Arcadia Planitia to the northwest and Tharsis Montes to the southeast. The relatively few small craters here were probably produced by meteoritic impact and are not the result of the extensive Martian vulcanism that took place in past ages. Of special significance in volcanic areas are the lava flows, which are particularly well marked around Olympus Mons. Ascraeus Mons, Pavonis Mons and Arsia Mons are also lofty volcanic cones which together make up the Tharsis Montes. During the great dust-storm of late 1971, when Mariner 9 reached Mars, these three peaks were some of the few features that could be identified with any certainty because their summits protruded through the top of the dust cloud.

The Vastitas Borealis extends around the north polar region and its southern border is seen across the top of the first map. During the northern winter the white deposits of the polar cap may extend as far south as Tempe Fossae and cover this dark region.

Chryse Planitia was selected as the landing site for the first Viking probe in the summer of 1976. (Viking 2 landed in Utopia Planitia in the eastern hemisphere.) About 5° south of the equator and on longitude zero lies the Sinus Meridiani, which used to mark the zero for Martian longitudes, and at a point 20° west and 20° south of that is the dark patch of Margaritifer Sinus. These names did not survive the IAU revisions (and are not on the map) because the markings do not correspond with any obvious topographical features.

The hemisphere is dominated by the tremendous system of rift valleys extending eastwards from the Tharsis Montes through Tithonius Chasma, Melas Chasma and Coprates Chasma through Simud Vallis. Immediately south of Tithonius Chasma lies Solis Planum, one of the most variable areas on Mars. Observers since 1877 have noted pronounced changes of shape and intensity. Such variations were easy to explain with the old vegetation hypothesis, but now that the dark areas are believed to be inorganic the changes are more puzzling.

The prominent dark patches around the furrows of Sirenum Fossae and the serpentine Nirgal Vallis are two of the classical mare that have disappeared in the new nomencla-

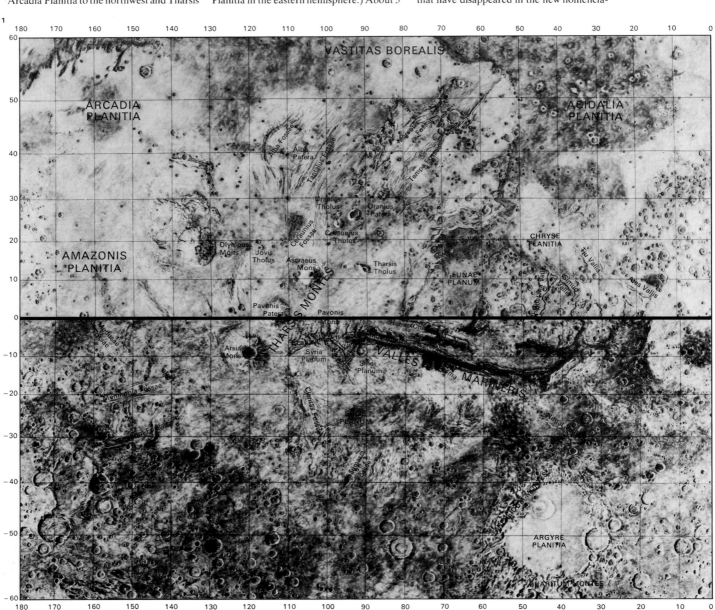

ture. Both are heavily cratered. Russia's Mars 3 probe landed south of Sirenum Fossae in December 1971, but it transmitted for only 20 seconds after arrival and nothing was learned from it.

The eastern hemisphere

The main feature in the eastern hemisphere (below right) is the Syrtis Planitia, which was recorded by Christiaan Huygens (1629–95) in 1659 and is the most prominent dark marking on Mars. Mariner 9 showed it to be a relatively smooth plateau sloping eastward towards the basin of Isidis Planitia and not, as had been believed, a sunken seabed. Surprisingly, there is little, apart from its colour, to distinguish it from the lighter Isidis Planitia and therefore the conclusion can be drawn alongside it that its prominence is simply due to the low albedo (light reflecting power) of its rocks.

Elysium Planitia to the east of the map is a volcanic province of intermediate geological age. It contains two large volcanic craters (calderas) as well as a clearly marked dome. The dark region north of latitude 55° is the other half of Vastitas Borealis. It may be partly responsible for the dark peripheral region of the polar cap formerly attributed to the visible effects of melting ice. The southern portion of this hemisphere is dominated by Hellas Planitia, which may appear extremely brilliant from Earth and can sometimes be mistaken for an extra polar cap. East of Hellas lie the two prominent dark features known as Mare Tyrrhenum and Mare Cimmerium respectively.

The polar caps

In the north polar region [Key] the white cap never vanishes completely. Part of the Acidalia Planitia is shown; this is the most prominent of the dark features in the northern hemisphere of the planet.

In the south polar zone [Key] the area within 10° of the actual south pole is seen to be smooth and laminated with the summer remnant of the polar cap offset at longitude 45°. The dark surrounding areas are heavily cratered with a prominent ridge, Argyre Dorsum. During the southern winter the polar deposit covers almost the whole area.

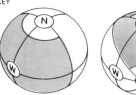

KEY

The yellow areas on these globes show [left] the western hemisphere of Mars, which is charted on the opposite page; [centre] the eastern hemisphere, charted on this page; and [right] the north and south poles of the planet, which are charted immediately below.

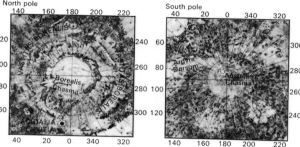

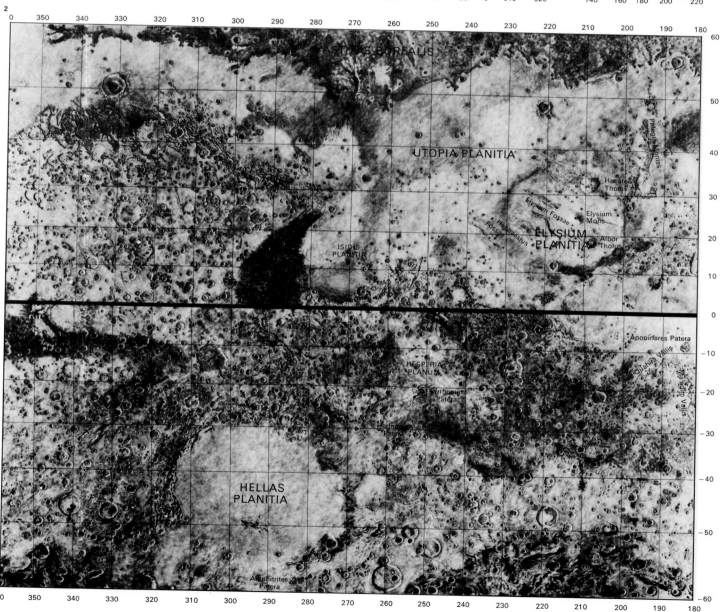

Mars panorama

Although Mars is almost certainly about the same age as the Earth (about 4,700 million years), it is so much smaller and less massive that it has evolved more quickly. This fact suggested that the surface features were likely to be more worn and eroded, because although the atmosphere of the planet is thin it is far from negligible.

Theories old and new

It is interesting to look back at what astronomers thought about Mars before 1965, when the first successful probe, Mariner 4, sent back data from close range. It was believed that the dark areas were depressions, probably old sea-beds, while bright regions such as Hellas Planitia and Argyre Planitia were plateaus; also the surface was thought to be gentle in relief with no lofty mountains or deep valleys anywhere. The reality was very different. The first pictures sent back from Mariner 4 showed signs of craters [Key] and, as the probe approached the planet, the photographs became clearer and the general nature of the landscape was no longer in doubt. Instead of being a world

with a level landscape, Mars proved to be extremely rough. Yet even from the Mariner 4 evidence it was clear that there were marked differences over various areas and that the surface of Mars was likely to be much more variegated than that of the Moon. Mariner 4 also showed that the atmosphere was much more tenuous than had been previously supposed and the theory of vegetation-filled sea-beds began to look less plausible.

Mariners 6 and 7, in 1969, produced a rather similar picture of Mars and it seemed that there were cratered areas and others that were described as "chaotic", with no particular pattern. Because of technical improvements, pictures were far clearer than those from Mariner 4; much had been learned during the intervening four years.

Mariner 9's discoveries

The most striking discoveries of Mars came in 1971 from Mariner 9, which proved to be a tremendous success – the more so because it also had to compensate for the failure of its predecessor, Mariner 8. After it had approached the planet and photographed the

two Martian satellites, Mariner 9 had to wait for dust-storms to subside. It appeared that the dust extended almost to the top of Olympus Mons and Arsia Mons, the highest-known points on Mars, with altitudes of more than 20km (12 miles). When the atmosphere became clear, these were revealed as towering volcanoes – and all ideas about the nature of Mars changed yet again. Few astronomers had expected to find volcanoes of a terrestrial type, yet the similarity between Olympus Mons and Arsia Mons on the one hand, and the Earth's Hawaiian volcanoes on the other, was unmistakable. The main discrepancy was in scale. Surveys of the height of surface features have been carried out both by radar and by measuring density of the carbon dioxide layer over different areas. As a result it was discovered that the Martian volcanoes are about three times as high as their Hawaiian counterparts and because Mars is a much smaller globe this makes them proportionately even higher.

The volcanoes had not been recorded by the earlier Mariners for two reasons. First the times of photography had been very limited.

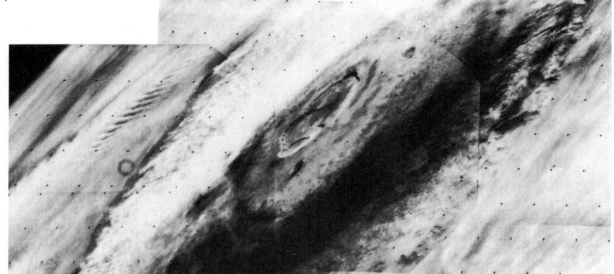

1 The great Martian volcano Olympus Mons was photographed by Viking 1 from 8,000km (5,000 miles) away. The 25km (15.5 miles) high mountain is seen in mid-morning, with clouds that extend up the flanks to an altitude of about 19km (12 miles). The multi-ringed calders, some 70km (43 miles) across, extend into the stratosphere. A well-developed wave cloud train stretches several hundred miles behind the mountain.

2 This oblique view across Argyre Planitia extends towards the horizon some 19,000km (12,000 miles) away. Argyre – surrounded by heavily cratered terrain – is the relatively smooth plain at left centre.

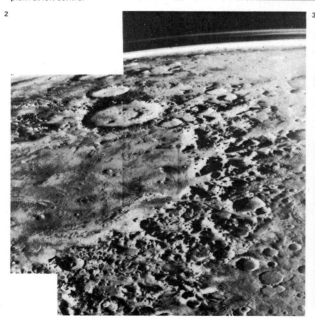

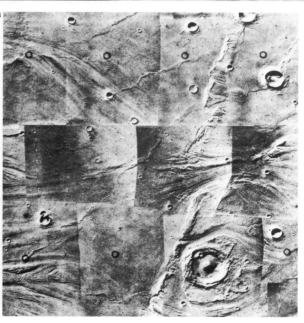

3 Eroded channels were photographed near the Viking 1 landing site in Chryse Planitia. They may be the remains of ancient stream beds that fade out near the landing ellipse, suggesting that the area is a sediment basin. The picture is a mosaic of 15 photographs taken by Viking Orbiter 1 from a distance of 1,680km (1,040 miles) and has an area of about 250 by 200km (155 by 120 miles). The lava flows are broken by faults that form ridges and are peppered by meteorite impact craters. A small stream flowed northward (toward upper right) from Lunae Planum, crossed the area and flowed toward the east. In places the water may have ponded behind the ridges before cutting through to form "water gaps".

Mariners 4, 6 and 7 each made only one pass of Mars and then moved on into never-ending orbits round the Sun, whereas Mariner 9 and the Vikings were put into orbit round the planet and were capable of sending back data until their power was exhausted. Second, there had been a great improvement in the computerized "cleaning-up" of Mariner pictures; it became possible to make vast improvements by electronic methods and the process took only a short time.

Photographic coverage
Our knowledge of the Martian surface has been revolutionized by Mariner 9 and the Vikings and a full analysis of their findings will take many years to accomplish. The most striking aspect is the diversity of the features in different areas. Regions that are thickly cratered are succeeded by relatively level areas; apart from great volcanoes there are drainage canyons and deep basins, of which Hellas and Argyre are the best examples. The equatorial canyons Valles Marineris cut deeply into the surface of Mars and stretch for nearly a third of its circumference. In

places the canyon walls appear to have been modified by huge landslides; in others by headward erosion to form integrated tributary systems. Elsewhere faulting seems to have predominated.

When the orbit of Viking Orbiter 1 was changed so that the whole planet revolved beneath it, the craft's water vapour and temperature mappers made another major discovery. It was found that the northern polar ice cap is mainly frozen water and not dry ice (frozen CO_2) as most scientists had believed. In some places the ice may be 100 to 1,000 metres (330 to 3,300 feet) thick. Photographs show that it virtually fills some craters.

Over large areas of the planet there is evidence of volcanic activity and water erosion. In Mars' geological past, when the atmosphere was very much thicker than it is today, torrential rains must have flooded the basin areas and cut channels through rock and desert. Some of the drainage channels may even have been caused by geothermal action. Another theory is that they were formed by flowing molten lava.

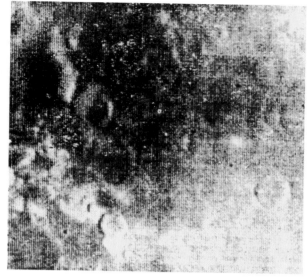

In 1965 Mariner 4 sent back the first close-range photographs of Mars. This picture took 8hr 35min to transmit.

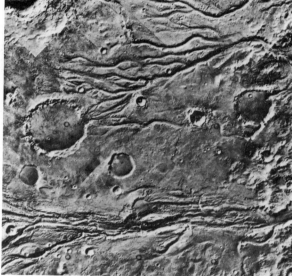

4 Large amounts of water still exist on Mars locked up in the north polar cap. In these overlapping pictures taken by Viking Orbiter 2 from 4,000 km (2,480 miles), the northern cap has receded to its smallest size as mid-summer approaches in the northern hemisphere. The solid white area near the top (north) is ice — mainly water ice with most of the frozen CO_2 evaporated off. The dark bands are regions devoid of ice.

5 The terrain in this Viking Orbiter 1 picture slopes from west to east with a drop of about 3km (1.86 miles). The channels are a continuation of those to the west of the Viking 1 landing site in Chryse Planitia. They are suggestive of a massive flood of waters from Lunae Planum, across this intervening cratered region, and into the general region from which Viking 1 took its soil samples.

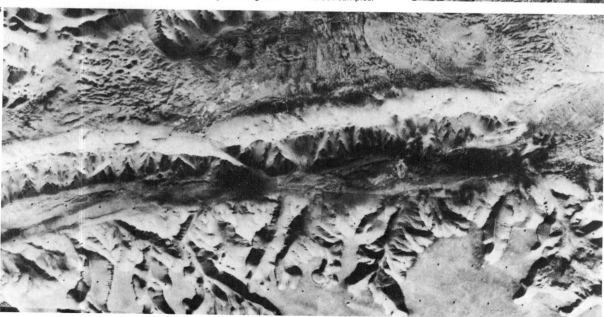

6 Valles Marineris is an enormous equatorial canyon that stretches nearly a third of the way round Mars. The far wall shows several large landslides that probably took place in series and perhaps were triggered by Marsquakes. Along the near wall another widening process seems to have occurred; a series of branch channels cuts into the plateau at the bottom. These may have been formed by slow erosion as a result of the release of ground water, or by mass wasting processes in which rock debris moves downhill as ground ice freezes and thaws. The photomosaic was made from pictures taken by Viking Orbiter 1 from a distance of some 4,200km (2,600 miles).

The moons of Mars

In 1877 Asaph Hall (1829–1907), using the 26in (66cm) Washington refractor in the United States, discovered two satellites of Mars; they were subsequently named Phobos and Deimos. Both are extremely small and are in no way comparable to the Moon. They had not been found before 1877, despite periodical searches, because they are both extremely faint.

Phobos and Deimos before Mariner
Telescopically, Phobos and Deimos appear as small, star-like points, but they caused a great deal of interest in the pre-Space Age period because of their unusual orbits [1]. Phobos moves round Mars at a mean distance of only 9,350km (5,800 miles) from the centre of the planet, so that the distance between Phobos and the Martian surface is about the same as that between London and Aden. The revolution period is only 7hr 39min; and since the rotation period of Mars is 24hr 37min, the "month", reckoning by Phobos, is shorter than the Martian day. In relation to Mars, Phobos rises in a westerly direction and sets toward the east; it is above

the horizon for only 4.5 hours at a time, during which it goes through more than half its cycle of phases and the interval between successive risings is a little more than 11 hours. The apparent diameter never exceeds 12.3 degrees, less than half that of the Moon as seen from Earth, and the amount of light sent down to the Martian surface is about the same as Venus sends to Earth. Phobos transits the Martian view of the Sun 1,300 times each year, taking 19 seconds or so to pass across the solar disc.

Even when above the Martian horizon, Phobos remains eclipsed by the shadow of the planet for long periods and from Martian latitudes greater than 69 degrees it never rises at all. The orbit of Phobos is practically circular and the inclination of the orbit to the equatorial plane of Mars is only a little more than one degree.

Deimos, smaller and more remote (23,500km [14,600 miles] from the centre of Mars), has a revolution period of 30hr 14min and remains above the Martian horizon for 2.5 days consecutively; but it sends less light to Mars than Sirius sends to Earth and to an

observer on Mars its phases would be almost imperceptible. Its maximum diameter is only about 12km (7.5 miles).

The nature of the two satellites is a matter for debate. They could be ex-asteroids that have been captured from the main minor planet zone. Some years ago it was suggested that Phobos was spiralling slowly down towards Mars and would collide with the planet in the foreseeable future; this led to a remarkable suggestion that it was being "braked" by the very tenuous Martian atmosphere. As its mass would have to be almost negligible for that to happen, the conclusion was reached that Phobos was a hollow space station built by the Martians. The idea came from Joseph Shklovsky (1916–), an eminent Soviet astronomer; but it received little support.

Mariner 9 discoveries
The first positive information about the satellites came from Mariner 9, which approached Mars in late 1971 and entered an orbit round the planet. During its approach, Mariner photographed both Phobos and Deimos and

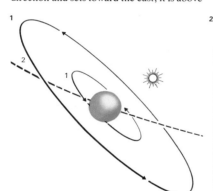

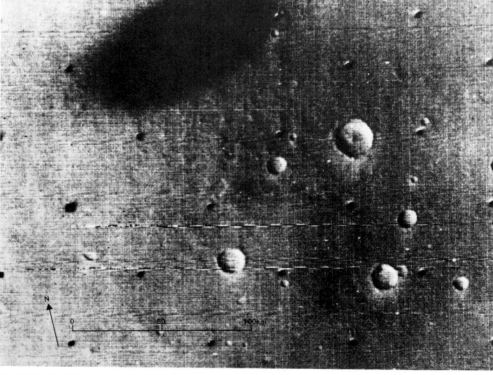

1 Both the satellites of Mars move in orbits that are practically circular and in the plane of the planet's equator. Phobos [1] is remarkably close to Mars and may approach to within 5,800 km (3,600 miles). It is the only known natural satellite with a revolution period shorter than the rotation period of its primary, and as seen from the planet it rises in the west and sets in the east. Deimos [2] is much farther out, with a period of 30hr 14min.

2 The Aethiopis region on Mars (lat. 14° N, long. 235°) is shown in detail in this Mariner photograph. The elliptical dark patch is the shadow of Phobos, measuring 130km (80 miles) by 280km (175 miles), so that an observer standing in the shadowed area would see a transit of Phobos across the Sun's disc. Because Phobos seems so much smaller than the Sun as seen from Mars, it could not produce a total eclipse.

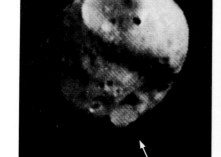

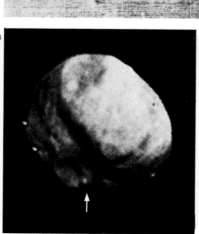

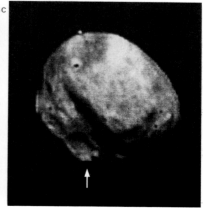

3 This series of photographs of Phobos [A–C] was taken by Mariner 9 in 1971. The approximate position of the south pole of the satellite is indicated by the arrow in each photograph. There is a pronounced surface bulge towards the top; this is the "synchronous" bulge permanently turned towards Mars, as the rotation period of Phobos is exactly the same as the time that it takes for the satellite to complete one journey round the planet – 7 hours 39 minutes.

showed that both are irregular in form. Phobos proved to be shaped rather like a potato, with a longest diameter of 28km (17 miles) and a shortest diameter of 20km (12 miles); its surface is pitted with craters, of which the largest formation (since named Stickney) has a diameter of 6.5km (4 miles) [4]. More than 50 features have now been charted [5], of which seven have been given official names: Roche, Wendell, Todd, Sharpless, D'Arrest, Stickney itself and the Kepler Ridge. Variations in surface height amount to as much as 20 per cent of the satellite's radius. The rotation is synchronous, so that Phobos always keeps the same face to Mars; the longest axis points to the planet.

Mystery craters
The origin of the craters of Phobos is not definitely known. Meteoritic impact has been suggested; the Japanese astronomer S. Miyamoto prefers the theory that the craters are of the blowhole variety, produced during the cooling period of the satellite. If impact is the cause, it must be agreed that Phobos has been severely battered; the diameter of

Stickney is approximately one-quarter the diameter of Phobos.

When Viking Orbiter 2 swung within 880km (545 miles) of Phobos in September 1976, it was able to photograph objects on the surface down to 40m (130ft) across. The features observed indicated that Phobos has the structural strength of solid rock; the main material is probably basalt. The escape velocity is a mere 20kmh (12mph), so that there can be no trace of atmosphere. The tiny moon was seen to be heavily cratered as expected but, surprisingly, showed striations (grooves) and chains of small craters. Similar crater chains appear on Earth's Moon. Mars and Mercury were formed by secondary cratering from a larger impact but craters are not so easily explained in the case of a small low-gravity body.

Deimos is of the same general type as Phobos, although smaller. It too is cratered, and the two main formations have been named Swift and Voltaire, after two writers who predicted in the eighteenth century that Mars would eventually be found to possess two satellites.

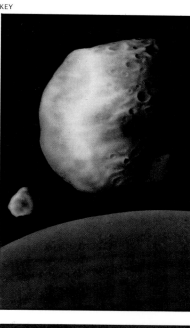

Phobos and Deimos the satellites of Mars, are much smaller than the Moon. Here they are shown as seen from the Martian equator while the Moon is shown as seen from Earth; all three are to scale. The nature of the satellites remains uncertain, but the Mariner 9 photographs suggest that both Phobos and Deimos are solid rock. They are quite different from our own Moon; it is possible that they are captured asteroids. Neither provides as much illumination as the Moon at night. Phobos gives about as much light to Mars as Venus does to Earth, while Deimos is still more faint. The surface of both satellites is exceptionally dark.

4 These Mariner 9 photographs of Phobos are the clearest pictures of the satellite so far obtained. Again, the arrow indicates the approximate position of the south pole. The largest crater, 6.5km (4 miles) in diameter, is Stickney. The apparent indentation [upper right] in pictures A and B shows up as a crater in the third [C]. Phobos is not even approximately spherical; its form is quite irregular.

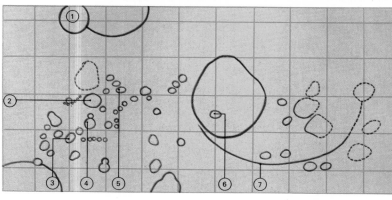

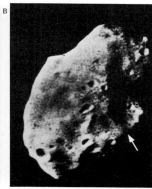

5 This Phobos map was compiled from the Mariner 9 photographs. Fifty craters have been charted, of which six have been given official names: Roche [1], Todd [2], D'Arrest [3], Sharpless [4], Wendell [5] and Stickney [6]. The ridge [7] has been named the Kepler Ridge. There is no doubt that the whole surface of Phobos is heavily cratered, and fractured, which is probably the result of large meteoroid impacts, although this is not certain.

6 This closest ever view of Phobos may help to resolve the question of how the tiny moons of Mars originated. Clues are a crater with a central peak, crater chains (indicated by arrows) running parallel to the equator, and striations (grooves) covering more than half the surface. One theory supposed that the grooves were caused by Phobos passing through a swarm of smaller bodies. The smallest visible object is 40 metres (130 ft) in diameter.

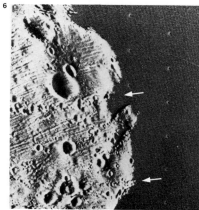

7 Deimos was viewed by Viking Orbiter 1 from a distance of 3,300km (2,050 miles); Mars is to the left. About half the side facing the camera is illuminated and the lighted portion measures about 12 by 8km (7.5 by 5 miles). While Mariner 9 photographs taken from greater distances showed only a few large craters, at least a dozen are prominent here; the largest have diameters of 1.3km (0.8 miles) and 1km (0.62 miles). A linear feature appears near the top.

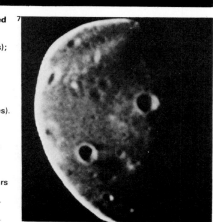

Minor planets

The Solar System is divided into two main parts by a wide gap between the outermost of the inner planets, Mars, and the first of the giants, Jupiter. A numerical relationship known as Bode's law, discovered by Titius of Wittenberg (1729–96), but popularized by Johann Bode (1747–1826) in 1772, led astronomers to suppose that there might be an extra planet there.

At the end of the eighteenth century, a group of astronomers headed by Johann Schröter (1746–1826) and Baron von Zach (1754-1832) formed themselves into what became known as the "celestial police", with the express intention of searching for the new planet. In fact they were forestalled.

New discoveries: the minor planets
On 1 January 1801, Giuseppe Piazzi (1746–1826), at Palermo in Sicily, discovered a star-like body that moved perceptibly from night to night and that proved to be a planet moving in the gap between Mars and Jupiter. It was named Ceres, in honour of the patron goddess of Sicily.

During the next few years the "celestial police" found three more planets: Pallas, Juno and Vesta. Together with Ceres, they became known as the minor planets, asteroids or planetoids. All are small and, apart from Ceres, less than 500km (300 miles) in diameter. Only Vesta is ever visible to the naked eye.

No more asteroids were evident and the "police" disbanded, but in 1845 a German amateur, Karl Hencke (1793–1866), discovered a fifth asteroid, Astraea. Since 1850 no year has passed without the discovery of several asteroids; the swarm may well exceed 50,000.

The diameters of some asteroids have recently been re-measured and in general earlier measurements underestimated them [Key]. Ceres is now thought to have a diameter of about 1,000–1,200km (600–750 miles), while the rest are smaller.

Irregular orbiters
Not all the asteroids keep strictly to their main zone [4]. In 1898 Carl Witt, at Copenhagen, discovered number 433, Eros, which can move well inside the orbit of Mars and occasionally approaches to within 24 million kilometres (15 million miles) of the Earth; that happened in 1931 and again in 1975. In 1931 Eros was extensively studied because exact calculation of its orbit can help in measuring the length of the astronomical unit or Earth–Sun distance. In shape Eros is elongated, with a long diameter of about 27 km (17 miles) and a short diameter of less than 16km (10 miles). Although small, Eros is larger than other so-called "Earth-grazers" such as Hermes, a mere 1km (0.6 mile) or so in diameter, which brushed past the Earth in 1937 only 780,000km (485,000 miles) away, less than twice the distance of the Moon. The Earth could be hit by such an asteroid and much damage would result, although the chances of a direct collision are extremely slight.

One asteroid, Icarus, actually approaches the Sun closer than does the planet Mercury. It must undergo some of the most extreme temperature changes in the Solar System. At the closest point in its orbit, only 28 million kilometres (17 million miles) from the Sun, the surface temperature on Icarus must be

CONNECTIONS

See also
276 Exploring the inner planets
218 Meteors and meteorites
176 Members of the Solar System

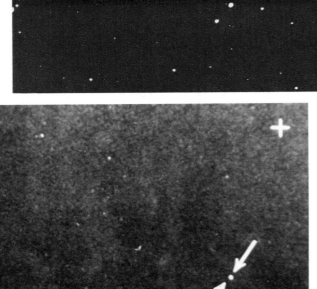

1 This photograph taken by Max Wolf (1863–1932) shows a star field together with two streaks that represent asteroid trails. The photograph is a time exposure; during it the driving mechanism of the telescope was adjusted so as to follow the stars (to compensate for the rotation of the Earth). The stars remained in the same relative positions while the asteroids shifted perceptibly against the background. Wolf was the great pioneer of this method of asteroid discovery. Previously the method had been to chart the same area of the sky for several consecutive nights, so that any star-like object that moved could be identified as an asteroid. Wolf's photographic method was far quicker and more efficient. Discovering asteroids is now very much easier, but keeping track of them and computing their orbits is time consuming.

2 Vesta, discovered in the early nineteenth century and shown here between the arrows, is the brightest but not the largest of the minor planets. It looks exactly like a star although its night-to-night movements betray its true nature. The cross to the upper right shows Vesta's position 24 hours later. Photographs do not show surface detail – asteroids are too small. Their rotation periods can be determined by variations in brilliancy.

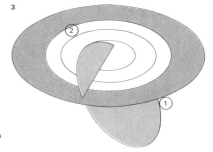

3 Icarus, about 1.5 km (1 mile) in diameter, was discovered by Walter Baade (1893–1960) in 1949. The orbit of Icarus [1] is highly inclined at 23°. It is the only asteroid that is known to have its perihelion inside the orbit of Mercury [2].

British scientist first to see Eros

John Innes

A BRITISH scientist was the first person to see the rocky surface of the asteroid Eros last night, as the American spacecraft Near touched down there.

Dr Louise Prockter, 36, was the first to see the potted surface of the asteroid some 196 million miles away from Earth.

It was the first time a man-made spacecraft had touched down on such an object.

Dr Prockter, a planetary geologist, was an advisor to the camera teams beaming out pictures taken from the surface of the asteroid, which is the size of the Isle of Wight.

Over the next few years Dr Prockter, originally from north London, will study the 160,000 photos Near took of the asteroid to see what can be learned from it.

Speaking from mission control in Columbia, Maryland, Dr Prockter said: "I was the first person to see the images as they

were beamed back. It was very exciting and very nerve-wracking. We were really surprised by how well it went today. We were amazed because we thought we would get no good pictures back at all.

"It's too early to say what the pictures will show and it will take years of study. But it was so exciting to be involved."

The 495kg spacecraft was not designed to land on the asteroid, which is near Mars.

Mission control was unsure whether they could slow its descent enough to prevent it from bouncing off the asteroid surface and back into orbit.

But shortly after 8pm British time it was confirmed that Near had successfully touched down on the surface and was continuing to beam back information to scientists at mission control.

Dr Prockter said: "We thought when it landed we wouldn't get anything back but it is still sending information back such as the temperature.

The surface of the asteroid pictured by the Near craft

We don't know how long that will go on for but I would imagine it won't be taking off again. It will probably stay there now."

Near began its 15-mile descent on to the asteroid surface yesterday morning, five years after it took off from earth. The difference in gravity fields

between Earth and Eros means that a coin dropped from head height on the asteroid would take five seconds to hit the ground.

Near has spent the last year taking photos of the asteroid, gathering information about its composition, size, shape and structure.

more than 500°C (900°F); at its aphelion (farthest point), only 200 days later, it has moved out to a distance of 295 million kilometres (183 million miles) – well beyond that farthest orbital point of Mars.

On the other hand, number 944, Hidalgo, has an eccentric path that takes it out almost as far as the orbit of Saturn, while the members of the Trojan group move in the same path as Jupiter. One group keeps approximately 60 degrees ahead of Jupiter and the other group 60 degrees behind; there is no danger of a collision. Although the Trojans are large by asteroid standards, their distance from Earth makes them faint objects.

Through a telescope, asteroids look exactly like stars and the only way to identify them is by checking their movements from night to night. Modern discoveries are made photographically. During a time exposure, an asteroid will often move enough to leave a trail on the plate rather than a point of light. Asteroids can therefore be a nuisance to astronomers. Photographic plates exposed for quite different reasons are often found to be dotted with asteroids and each has to be individually identified, which wastes time.

It is not yet known what asteroids are made of, but Mariner 9 photographs of the two dwarf satellites of Mars (Phobos and Deimos), which may well be captured asteroids, suggest that the surfaces of many may be pitted with craters. Some of the smaller satellites of the giant planets – the outer members of Jupiter's family, Phoebe in Saturn's and Nereid in Neptune's – also may be captured asteroids.

Origin of asteroids
The origin of the asteroids is still uncertain. According to one theory, they represent the debris of a former planet (or planets) that used to orbit the Sun beyond the path of Mars and met with some disaster in the remote past; but on the whole it seems more likely that they never formed a large body. The immensely powerful gravitational effect of Jupiter could prevent a large planet from forming in the region of the asteroid zone. It is also worth noting that all the asteroids combined would still not make up one body as large or as massive as the Moon.

The sizes of the first four asteroids to be discovered, Ceres [C], Vesta [D], Pallas [E] and Juno [F], together with the irregularly shaped Eros [B], are here compared with the Moon [A]. Their diameters are difficult to measure, being so small. Earlier measurements of Ceres gave 685km (426 miles), but new methods show that it is much larger – between 1,000 and 1,200km (600 and 750 miles).

4 Most of the orbits of the minor planets lie well beyond the orbit of Mars [1] and well inside that of Jupiter [2]. Within this distribution there are certain regions, known as Kirkwood gaps, in which there are fewer asteroids. These gaps lie at particular orbital distances from the Sun where Jupiter's gravitational field has forced the asteroids into different orbits.

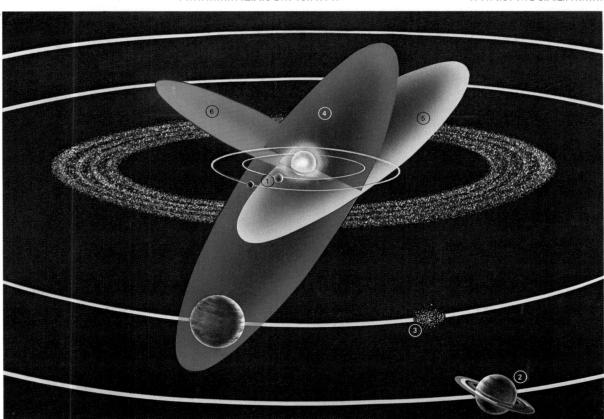

5 The orbits of the planets are shown from Earth [1] out to Saturn [2], together with those of some asteroids which are of particular interest (the illustration is not to scale). Most of the asteroids move in the region between the orbits of Mars and Jupiter; the Trojan asteroids [3] move in the same orbit as Jupiter but keep their distance and collisions are unlikely to occur; one group moves 60 degrees ahead of the planet and the other 60 degrees behind, although they move around for some distance to either side of their mean positions. Hidalgo [4] has a path which is highly inclined and so eccentric, much like a comet, that its aphelion is not far from the orbit of Saturn. Amor [5] and Apollo [6] belong to the so-called Earth-grazing asteroid group. All the Earth-grazers are very small; Amor has a diameter of 8km (5 miles) and Apollo about 2km (1.25 miles).

The planet Jupiter

Far beyond the main asteroid belt lies Jupiter, the largest of the planets. It has a greater mass than the other planets combined and it has been said that "the Solar System is made up of the Sun, Jupiter and assorted debris". The mean distance of Jupiter from the Sun is 778,300,000km (483,600,000 miles); the revolution period is 11.86 years and the synodic period (that is to say, the mean interval between successive oppositions) is 399 days. Thus Jupiter is well placed for observation for several months in each year and is surpassed only in brilliance by Venus and, on rare occasions, by Mars.

Jupiter's huge globe could swallow up 1,300 bodies the volume of the Earth but its mass is only 318 times that of the Earth because Jupiter is much less dense [1]. From the outer layers and possibly as far as the centre, the main constituent is hydrogen. The quick rotation period (less than ten hours) means that the equator tends to bulge out, and any casual glance through a telescope is enough to show that the planet is very much flattened at the poles. Jupiter's equatorial diameter is 143,000km (89,000 miles),

whereas the polar diameter of the planet is less than 135,000km (84,000 miles).

Telescopic observations

Through a telescope, the yellow disc of Jupiter is seen to be crossed by dark streaks that are known as cloud belts. Normally there are two prominent belts, one on either side of the equator, while others may also become conspicuous. With high magnifications, the details are complicated – and they are also continually changing, because Jupiter is a highly active world.

The quick rotation means that the various features can be seen to shift across the disc of the planet even over periods of a few minutes. Indeed, the rotation periods have been deduced from observations of this kind. When a feature reaches the central meridian as seen from Earth it is said to be in transit; and successive transit-timings provide all the information needed for working out the period of axial rotation. Jupiter does not spin in the way that a solid body would; different regions of latitude have different periods – thus the mean period in System I (between

the two equatorial belts) is five minutes shorter than that of the rest of the planet, referred to as System II. Moreover, various definite surface features also possess periods of their own and so drift around quite independently in longitude.

Jupiter and the Great Red Spot

Spots are often seen on Jupiter but most of them are short-lived. The exception is the Great Red Spot, which has been under observation now for more than 300 years; it sometimes disappears for a while, but always returns. This spot became prominent in 1878 when it developed into a brick-red elliptical patch 48,000km (30,000 miles) long by 11,000km (7,000 miles) broad, so that its surface area was greater than that of the Earth. It has been prominent again since the mid-1960s.

For many years it was believed that the Red Spot might be a kind of "island", floating in Jupiter's outer gas; variations in level could cause its occasional disappearances, so that when it sank it would be covered up. Alternatively it was attributed to

CONNECTIONS

See also
208 Jupiter panorama
212 The moons of Jupiter and Saturn
278 Exploring Jupiter and Saturn
176 Members of the Solar System

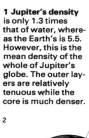

1 Jupiter's density is only 1.3 times that of water, whereas the Earth's is 5.5. However, this is the mean density of the whole of Jupiter's globe. The outer layers are relatively tenuous while the core is much denser.

1   5·5   1·3

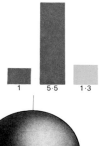

23·5   3·1

2 The Earth's axis is tilted at an angle of 23.5 degrees from the perpendicular to the plane of the orbit [1]. In the case of Jupiter [2] the tilt is only 3.1 degrees; in fact Jupiter is practically "upright". Of the other principal planets, only Mercury's axial inclination is like this.

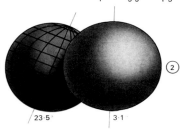

3 Jupiter comes to opposition at mean intervals of 399 days, so that astronomers can observe it clearly for several months each year. The diagram shows the opposition positions between 1960

and 1975. Because Jupiter's orbit, like that of the Earth, is eccentric, the opposition distance ranges from 589–669 million km (366–416 million miles). In 1975 Jupiter was at its closest to Earth.

4 The surface of Jupiter shows bright zones and dark belts. The nomenclature always used by observers is given here (south is at the top): [1] South Polar Zone; [2] South South Temperate Belt; [3] South Temperate Belt; [4] and [5] South Equa-

torial Belt, frequently seen to be divided into two well-marked components; [6] Equatorial Band; [7] and [8] North Equatorial Belt, also frequently divided into two components; [9] North Temperate Belt; [10] North North Temperate Belt; [11]

North North North Temperate Belt; [12] North Polar Zone; [13] South South Temperate Zone; [14] South Temperate Zone; [15] South Tropical Zone; [16] Equatorial Zone; [17] North Tropical Zone; [18] North Temperate Zone; [19]

North North Temperate Zone; [20] the Great Red Spot, together with its associated Hollow. The region between the south edge of the North Equatorial Belt and the north edge of the South Equatorial Belt is System I; the rest of

the planet (System II) has a rotation period that is on average five minutes longer. The belts show marked variations in intensity; for instance the South Equatorial Belt is sometimes as broad and dark as the North Equatorial.

a "Taylor column", that is, the top of a column of stagnant gas produced by the interruption of the atmospheric circulation by some large topographical feature on the surface of Jupiter. But results from the American Pioneer probes indicate that the spot is nothing more than a whirling storm; the reason for its colour remains a mystery.

What Jupiter is made of

The internal constitution of Jupiter has been investigated theoretically. One theory that was popular for many years suggested that there was a rocky core overlaid by a thick layer of ice, which was in turn overlaid by the atmosphere, but this idea has now fallen into disfavour. Spectroscopic work has shown that the outer gases are rich in hydrogen (together with hydrogen compounds, such as ammonia and methane) and hydrogen, possibly in a liquid state, is believed to be the main constituent throughout Jupiter because of its low mean density. Near the core, however, where pressures and temperatures are high, the hydrogen would start to assume some of the characteristics of a metal.

The temperature at the centre of Jupiter may reach several thousand degrees – much higher than that of the Earth, although there is no doubt that Jupiter must be regarded as a true planet and not a small star; the core temperature is much too low for nuclear reactions to begin. Yet Jupiter does seem to send out more energy than it should do if it depended entirely upon the Sun and this may be due to a very slight, steady contraction (much too small to be observable) that would release an appropriate amount of gravitational energy. Jupiter is also characterized by a powerful magnetic field and strong radio emissions; but astronomers have so far failed to provide a wholly plausible explanation of their origins.

Jupiter has a gaseous or possibly liquid surface and as a result no landing can take place there. Some people believe that life may exist below the outer clouds, where all the necessary ingredients are found and the temperatures may be tolerable, but this idea is highly speculative and confirmation of such a theory will be extremely difficult to obtain within the near future.

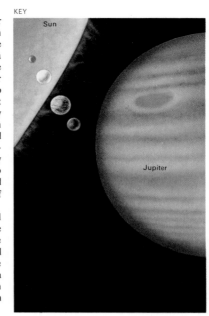

KEY

Jupiter was aptly named after the ruler of the Roman gods by the ancient astronomers, although they had no idea at the time of the planet's dimensions or the number of its satellites. It is larger than all the other planets combined, although it has a diameter about one-tenth of that of the Sun. Two of Jupiter's satellites are approximately the same size as the smallest of the principal planets, Mercury, and it has more satellites (13) than any other planet. This drawing shows the comparative sizes of the four inner planets together with segments of the Sun and Jupiter. Despite its great distance, Jupiter is a brilliant object in the sky.

5 One of the best colour photographs of Jupiter ever taken from the Earth was secured by G. P. Kuiper with the 61in (155cm) reflector at the Cataline Observatory, Texas. South is at the top. At this time the Great Red Spot was prominent and to the south of it was a well-marked white spot. The structure of the belts is seen to be decidedly complex, although through a small telescope they give the impression of being straight and uniform. Ganymede's shadow is also seen.

6 This series of Jupiter photographs was taken from the Lowell Observatory at Flagstaff in Arizona on 4 June 1972 [A and B] and 25 July 1973 [C]. They were made possible by combining images which were originally recorded at the telescope on black and white film, through colour filters. At that time the whole of System I, between the two equatorial belts, was of an unusual orange hue, although the effect had lessened by mid-1973 when the last photograph was taken. The Great Red Spot is not shown in the middle picture because it was then on the far side of the globe. But there is a comparatively good view of one of the shorter-lived bright white spots that appear in the South Temperate Zone.

207

Jupiter panorama

It takes a spacecraft a few days to reach the Moon. To reach Mars or Venus takes a few months. Jupiter, however, is so remote that a journey to it must take nearly two years. The difficulties of guidance are increased enormously by distance and there is also the problem of receiving the information sent out from on-board transmitters: an extremely small amount of energy reaches Earth from a probe at the distance of Jupiter, 629 million kilometres (390 million miles).

The first Jupiter vehicle
Pioneer 10, the first Jupiter vehicle, was launched in March 1972 and it was not until December 1973 that it reached its target. Its main task was to study the conditions in the region around Jupiter and to take photographs. Radio emissions from Jupiter (which had been picked up, purely by accident, in 1955 by B. F. Burke and W. Franklin in America) had indicated the presence of a very powerful magnetic field. It was thought likely that there would also be zones of intense radiation of the same basic type as the Van Allen belts encircling the Earth. Scientists in general were apprehensive about the effects of Jovian radiation upon the instruments carried in the spacecraft, particularly since Pioneer 10 was scheduled to pass over Jupiter's equatorial regions where the intensity of radiation would be greater than the radiation at the poles.

In fact, Pioneer 10 functioned excellently. It passed within 132,000km (82,000 miles) of Jupiter and sent back data about the magnetic field – which proved to be powerful yet different in structure from that of the Earth – and also about the radiation zones. The instruments were almost saturated; had Pioneer 10 approached much closer, the radiation would have put the equipment out of action altogether. After its rendezvous Pioneer 10 moved away from Jupiter and began a never-ending journey into space. It will escape from the Solar System in the 1980s and there seems no reason to doubt that for many millions of years to come it will continue to travel silently in the space between the stars.

Pioneer 11 followed a year later. It was launched in March 1973 and reached Jupiter in December 1974. This time the approach was from the pole of the planet and the spacecraft passed relatively quickly across the equatorial regions in a successful attempt to avoid the worst of the radiation. Further data were obtained, confirming the results from Pioneer 10. Later, after bypassing Jupiter, Pioneer 11 was put into an orbit that will take it out to a rendezvous with Saturn sometime in 1979.

The achievements of two Pioneers
The two Pioneer probes have answered some questions about Jupiter, although many puzzles remain. First, there is the question of the Great Red Spot, which is unique on Jupiter both because of its size and colour and because it is so long-lived. The "floating island" theory has proved to be wrong; the spot is not a semi-solid body floating in Jupiter's outer atmosphere. The Great Red Spot must be classed as a phenomenon of Jovian meteorology. Definite structure of some kind in it is shown on some of the Pioneer pictures [1, 4].

The bright zones on the surface of the

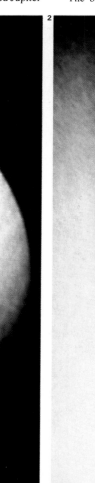

1 This photograph of Jupiter was taken on 1 December 1973 as the space probe Pioneer 10 neared the planet; the distance was then about 2,500,000km (1,550,000 miles). The Great Red Spot is well shown, with indications of structure; it lies in a bright zone. The irregular outline of the belts is also obvious. The black disc is the shadow of Io, the innermost of the large satellites, which moves in the outer part of Jupiter's magnetosphere; it is larger than our Moon.

2 This part of the surface of Jupiter was photographed from Pioneer 10 in December 1973 at a distance of 1,804,000km (1,121,000 miles). This section provides a view of one of the famous "plumes". It is thought that near the nucleus of the plume, cloud particles are forming from below and are then spreading into the nucleus. The "tail" of the plume is more than 64,000km (40,000 miles) long. The plume itself is higher than the surrounding clouds.

planet are at a higher level than the dark belts and they are also colder by several degrees. That was to be expected, but it has also been found that the surface temperature at the poles is the same as at the equator. If Jupiter depended upon heat received only from the Sun, the poles would be the coldest regions; there seems no doubt that there must be an internal source of heat. If this internal heat is more effective at high latitudes on Jupiter there should be a noticeable effect upon the structure of the gaseous layers, with the setting up of turbulence and convection currents. Photographs of the poles taken by both Pioneers [5] show that this is what happens; the change is conspicuous.

Further discoveries
Investigations of the precise structure of the layers had not previously been possible because from Earth it is hopeless to try to see details as delicate as those revealed by photographs taken from comparatively close range. The stable belt/zone structure breaks up at about 45 degrees Jovian latitude and increasingly towards the poles the regions are more unstable, with many disturbances within the cloud belts detectable [5].

Among other interesting features are so-called "plumes", which have a superficially cometary appearance [2]. Pioneer 10 recorded one plume that was still in existence when Pioneer 11 flew past a year later. Records of Jupiter made by Earth-based observers (mainly amateurs) showed that this particular plume had been in existence since 1964, so that it had lasted for virtually one Jovian year (12 Earth-years). Other plumes may have lasted longer but have not been recorded as continuous: from 1973 to 1975 the equatorial zone of the planet was unusually dark and its plume could be easily seen.

Experiences with the first two Pioneers will be put to good use with future probes. Spacecraft sent out to the more remote planets will, in general, pass by way of Jupiter so as to make use of the strong gravitational field of the planet in speeding them on their way, and the opportunities for carrying out further studies of Jupiter itself will not be missed. A Jupiter/Saturn Mariner is scheduled to be launched in 1977.

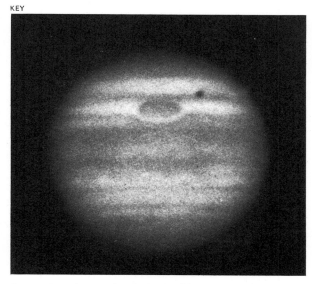

KEY

A comparison of this photograph of Jupiter taken from Earth (1964) with any of the Pioneer photographs vividly illustrates how much significant detail can be picked up by a space probe.

3 The two Pioneers bypassed Jupiter at an interval of one year (Pioneer 10 in December 1973, Pioneer 11 in December 1974) and it was obviously important to note any major changes in the surface structure in the interim period. This photograph was taken on 6 December 1974 by Pioneer 11 and the equatorial plume may be compared with the view as shown in illustration 2. The white plume is still easily recognizable and is in fact a long-lived feature; observational records of it from Earth go back for a Jovian year and in form it seems to have altered little.

4 This Pioneer 11 photograph was taken on 6 December 1974 at a distance from the planet of 1,100,000km (682,000 miles). The Great Red Spot is prominently displayed and close inspection shows that there is definite inner structure – it has even been compared with a "Cyclopean eye". There now seems no doubt that it is a kind of whirling storm, although whether or not it will gradually decay remains to be seen; certainly it is as conspicuous now as it used to be in the 17th century, when it was first observed. Note the marked phase of Jupiter.

5 Taken on 12 December 1974, when Pioneer 11 was 1,207,000km (750,000 miles) from Jupiter, this photograph shows the north pole, at a latitude of about 50 degrees. The pole itself is roughly on the line of the terminator across the top of the planet. This is one of the most significant views obtained, as it shows the obvious difference in surface structure between the polar regions and the equatorial zone. The convection cell structure at the pole is well displayed and the atmospheric circulation is different in high latitudes. There is a bluish cast to the pole (noted by Earth-based observers).

The planet Saturn

Saturn, the outermost of the planets known in ancient times, is a conspicuous naked-eye object, although in pre-telescopic times there was no means of distinguishing its rings, which are one of the most beautiful sights in the entire sky.

Saturn's mean distance from the Sun is 1,427 million kilometres (88 million miles) and its revolution period is 29.46 years. It comes to opposition once in approximately 378 days, so that it is well placed for observation for several months in every year.

Physical characteristics

Saturn is the second largest of all the planets [1]. Its equatorial diameter is 120,000km (75,000 miles), but the polar diameter is considerably less, because the planet is strongly flattened. This is partly because of its low density [2] (less than that of water, making it unique among the principal planets) and partly because of its rapid axial rotation. The rotation period at the equator is 10hr 14min and at the poles about 26 minutes longer, but precise measurements are less easy to make than with Jupiter; the

surface markings are less complex and, on the whole, they are also less distinct.

Saturn is a gas-giant whose main constituent is hydrogen. There is more detectable methane and less ammonia than in Jupiter, because a lower temperature has frozen more of the ammonia out of the planet's atmosphere. Although Saturn's mass is 95 times that of Earth, scientists believe that its surface gravity is only slightly greater than that of the Earth.

Saturn is made mainly of hydrogen. Near the core, the temperature is probably high, the pressure considerable and the hydrogen metallic. As yet there is no evidence for the existence of a magnetic field.

Observed through a medium-powered telescope, Saturn shows a yellowish disc, crossed by cloud belts that are basically the same as those of Jupiter, but considerably less active. Spots are relatively rare, but they do occur occasionally – the best modern example was that discovered in 1933 by the British amateur astronomer Will Hay. Hay's white spot [10] became conspicuous for a short period, but soon spread out and disap-

peared. It was undoubtedly due to the upflow of gaseous material from beneath the visible surface. Saturn shows no spots comparable with the famous Great Red Spot on Jupiter. Apart from the belts, every one of Saturn's surface features is comparatively short-lived and subject to change.

Saturn's bright ring system

The main glory of Saturn is its ring system [5]. There are two bright rings, A and B, separated by the wide gap known as Cassini's Division in honour of its discoverer, Giovanni Cassini (1625–1712). Closer to the planet there is a fainter semi-transparent ring discovered in 1850 by William Bond (1789–1859) at Harvard and independently by W. R. Davies in England, and known generally as the Crêpe or Dusky Ring. Other faint rings have been reported from time to time; in 1909 French astronomers reported an extra dusky ring outside Ring A (the outermost ring). It was known as Ring D. Some astronomers, however, are doubtful about these extra rings, whose existence has yet to be confirmed. There is also the

CONNECTIONS

See also
212 The moons of Jupiter and Saturn
278 Exploring Jupiter and Saturn
176 Members of the Solar System

1 Saturn is a giant planet, although not as large as Jupiter. Its volume is 744 times that of Earth. In this scale diagram the Earth would fit neatly into the gap between the Crêpe Ring and the surface of the planet.

Saturn —

Earth to same scale

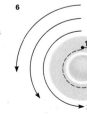

2 The mean density of Saturn's globe is only 0.7 that of water. This is much less than that of any other principal planet. It has been said that if Saturn were dropped into an ocean, it would float. The low density is due to the preponderance of the very light elements, hydrogen and helium. The columns show the mean densities of the Earth and Saturn respectively compared with the density of water.

1   5·5   0·7

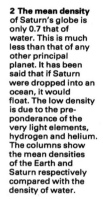

3 The inclination of Saturn's axis to the plane of the orbit is 26° 44', only slightly

greater than that of the Earth. The rings lie exactly in the plane of the planet's equator.

26° 44'

4 The aspects of Saturn's rings vary considerably as seen from Earth. At regular intervals the rings lie in the plane of the Sun and Earth [A]. The rings then seem to open out, until they are shown to maximum advantage [D], after which they close up again. When the south pole of Saturn is tilted towards the Sun, the southern ring-face is displayed and at such times part of the northern hemisphere of the globe is obscured [A–G]. In these diagrams south is to the top. Subsequently it is the northern ring-face which is displayed [H–L].

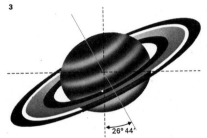

Ring B

Ring A

5 There are three principal rings. The outermost, Ring A, is 16,000km (10,000 miles) wide; Ring B is brighter and is 27,000km (17,000

miles) wide. Between it and the planet is the Crêpe Ring [1], which is less easy to observe. Rings A and B are separated by Cassini's Division [2].

6 The only prominent division in Saturn's rings was discovered by Cassini in 1675. It is caused by the gravitational effects of three of Saturn's inner satellites. A particle moving in Cassini's Division [1] will have a period half that of Mimas [2] so when the particle has completed

two revolutions, Mimas will have completed one [3]. When the particle has completed three revolutions, Enceladus will have completed one [4] and for four revolutions Tethys will have completed one [5]. These consistent perturbations will move the particle out of the division.

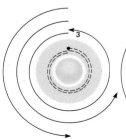

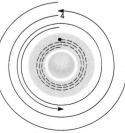

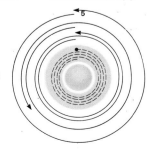

problem of the extra ring divisions, of which Encke's Division is the best known. It may well be that Encke's and other reported minor divisions are mere "ripples" or irregularities in the ring structure rather than true gaps such as Cassini's Division.

Edouard Roche (1820–83) gave his name to the distance between the centre of a planet and its satellite, within which the satellite cannot approach without being broken up. Saturn's rings lie inside the planet's Roche limit and cannot therefore be solid or liquid sheets. The rings are made up of relatively small particles, icy or ice-covered, with each particle moving round Saturn in its own independent orbit like a dwarf satellite. Cassini's Division is the result of the gravitational effects of Saturn's inner satellites [6], which may be said to keep the area of the division clear of ring particles.

The ring system is easily seen through a small telescope suitably placed. The rings lie in the plane of the planet's equator, so that they can often be seen at a suitable angle [4]. When edge-on to Earth, as happened in 1966 and will happen again in 1980, they appear as a thin line of light and can be traced only with powerful telescopes. This shows that although the ring system measures 272,000km (169,000 miles) from one side to the other, it is also extremely thin, with a thickness of not more than a few kilometres. If Saturn were represented as being the size of a tennis ball, the rings would be thinner than a sheet of tissue paper.

Information from space probes

Because Saturn is so far from the Earth, a space probe will inevitably take years to reach it. The first Saturn probe was Pioneer 11, which bypassed Jupiter in December 1974, sending back excellent pictures as well as miscellaneous information; it was then sent on for a rendezvous with Saturn, scheduled for 1979. It is even planned to send Pioneer through the ring system. Whether the instruments on board will still be operational afterwards is uncertain, but at least a Mariner probe to Jupiter and Saturn will then be on its way, so that in the foreseeable future some really detailed, close-range information about the ringed planet may be acquired.

The southern hemisphere of Saturn is displayed in this photograph. Part of the northern hemisphere is covered by the rings. There is a marked difference in brightness between Ring A and Ring B: the latter is much more brilliant.

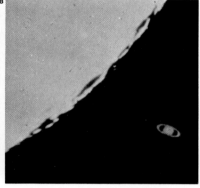

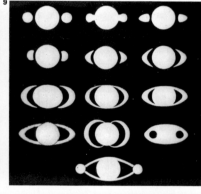

9 To early telescopic observers, Saturn was an observational enigma. The first recorded drawings seem to have been made by Galileo (1564–1642), whose feeble telescope was not strong enough to show the ring system in its true guise. He believed Saturn to be a triple planet and after two years' observation he lost sight of the rings altogether because during that time they had turned edge-on to the Earth.

7 Although Saturn is bright to the naked eye, it is quite impossible to detect any trace of the intricate ring system that forms part of the planet. The only way to distinguish Saturn from the stars is by its slow movement from one night to the next. Saturn is seen against its star field in this photograph.

8 Saturn seems to be close to the Moon. The limb of the Moon, to the upper left, is necessarily overexposed. The difference in apparent size between Saturn and the Moon is quite striking. The maximum diameter of Saturn as seen from the Earth is only 20.9 seconds of arc.

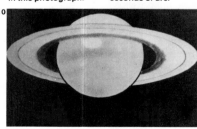

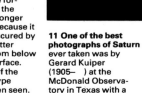

10 This drawing was made in August 1933 by Will Hay, a well-known British amateur astronomer whose private observatory was set up in outer London. It shows the white spot that suddenly appeared in the equatorial region of the planet and became prominent for a few weeks. The spot gradually lengthened and the portion of the disc following it darkened; subsequently the forward edge of the spot was no longer identifiable because it had been obscured by a mass of matter thrown up from below the visible surface. Other spots of the same basic type have also been seen.

11 One of the best photographs of Saturn ever taken was by Gerard Kuiper (1905–) at the McDonald Observatory in Texas with a 61in (155cm) reflector. The rings were opening out (edgewise presentation had occurred in 1966). It is clear that Ring B (inner) is much brighter than Ring A and indications of the so-called Encke's Division can be seen in Ring A itself. Kuiper himself stated that Encke's Division was a ring irregularity rather than a true gap (viewed visually with the Palomar 200in [508cm] reflector). The shadow of the ring on the disc is prominent – there is a bright region at the planet's equator – and the polar zones are rather dusky.

The moons of Jupiter and Saturn

Both the giant planets have extensive satellite families. Jupiter has 13 known attendants, Saturn 10. They seem to fall into two distinct categories: not all are large, but four – Io, Ganymede and Callisto in Jupiter's system, and Titan in Saturn's – are bigger than our Moon [2].

The Jovian satellites
The four bright Jovian satellites – Io, Europa, Ganymede and Callisto – were discovered by Galileo (1564–1642) in the winter of 1609–10, with one of his first telescopes. All would be naked-eye objects were they not overpowered by the brilliance of Jupiter itself. They were seen at about the same time by Simon Marius (1570–1624) and there was a dispute between Marius and Galileo over priority – which may be the reason why Marius' names for them were not generally accepted until quite recently.

Any telescope will show the "Galilean" satellites. Since their orbits lie virtually in the plane of Jupiter's equator, they tend to line up. Their phenomena are easy to observe. A satellite may pass in transit across Jupiter's disc [4]; it may pass behind the planet and be occulted; or it may be eclipsed by Jupiter's shadow [5]. Shadow transits are also seen. All the Galileans show perceptible discs and large telescopes can pick up surface details. Io and Ganymede were also photographed from the Pioneer probes of 1973 and 1974.

Ganymede is the largest and brightest of the Galileans; its diameter is about 5,000km (3,100 miles) according to recent measurements, which makes it larger than the planet Mercury. Callisto is nearly as big, but much less massive, so that its density is lower. Io and Europa are more like the Moon in size and density; Io was found by Pioneer 10 to have a tenuous atmosphere and also an ionosphere which affects the radio emissions from Jupiter because it moves through the outer part of the Jovian magnetosphere.

The remaining satellites are much smaller. Number 5, discovered by Edward Barnard (1857–1923) in 1892, is closest to the planet, with an orbit inside that of Io; its mean distance from the centre of Jupiter is only 181,000km (112,000 miles) and it has a period of only 11hr 57min. The estimated diameter is 200km (124 miles), so that it is not visible with small telescopes; it has been named Amalthea, although this name seems to be unofficial. All the other small, asteroidal-type satellites are farther out than Callisto; four of them move round Jupiter in a retrograde direction, which tends to support the idea that they have been captured from the minor planet zone. Moreover, they are so far from Jupiter, out to almost 24 million kilometres (15 million miles), that they are strongly perturbed by the Sun and their orbits are far from circular.

Titan: the unique satellite
The satellite family of Saturn is rather different from that of Jupiter. It contains only one satellite that is of planetary size (Titan) and only one that is definitely asteroidal in nature (Phoebe); the rest are of intermediate type.

Titan was discovered by the Dutch astronomer Christiaan Huygens (1629–95) in 1655 and is an easy object to pick up with a small telescope. It moves round Saturn at a distance of 1,220,000km (760,000 miles) in a circular orbit at half a degree in inclination

1 The satellites of Jupiter fall into several well-defined groups. In the first group [A] the satellite Amalthea [1] seems to be in a class of its own; the diameter is only about 200km (124 miles), so that in size it is asteroidal. Then come the four satellites discovered by Galileo from 1609–10 – Io [2], Europa [3], Ganymede [4] and Callisto [5] – with mean distances from Jupiter ranging between 422,000km (262,000 miles) for Io out to 1,880,000km (1,170,000 miles) for Callisto. The next group [B] consists of three satellites [6, 10, 7] plus a 13th, recently discovered, and the third [C] has four retrograde satellites [12, 11, 8, 9].

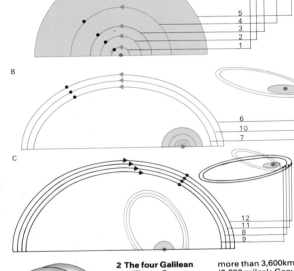

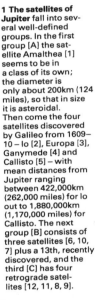

2 The four Galilean satellites – Europa [A], Io [B], Ganymede [C] and Callisto [D] – are compared in size with the Moon. Europa is smaller, but Io is slightly larger than the Moon, with a diameter of more than 3,600km (2,200 miles); Ganymede and Callisto are more nearly comparable with Mercury – Ganymede's diameter is 5,000km (3,100 miles) but Mercury's is 4,880 km (3,032 miles).

3 Surface details on the Galilean satellites may be discerned with very large telescopes. [A] Io, rather orange in colour, has a bright equatorial region and darker poles. This has been confirmed by the Pioneer probes. [B] Europa has the brightest surface of the four main satellites; unlike Io, Europa apparently has a darkish equatorial region, and its poles are brighter. The surface may well be ice covered. [C] Ganymede is the easiest of the Galilean satellites to study, and it has been photographed by the Pioneer probes. Bright areas, together with some darker areas that may be compared with the lunar seas, are visible. [D] Callisto has a relatively low reflecting power, and details are not easy to make out. This drawing, like the others, was made by Dollfus from observations with the Pic du Midi 24in refractor.

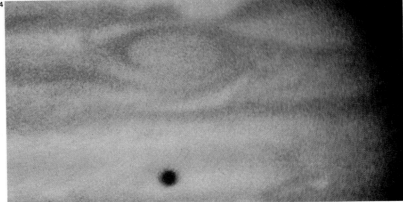

4 When a satellite passes in transit across Jupiter it may be seen as a bright spot. The two inner large satellites (Io and Europa) are easier to see in transit than Ganymede or Callisto because of their higher albedoes, or reflecting powers. Shadow transits are more striking. In this photograph, taken with the 61in (155cm) reflector at the Catalina Observatory in Texas, the shadow of Ganymede can be seen as a prominent black spot.

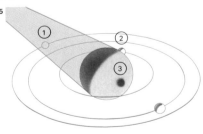

5 The Galilean satellites' orbits make them easy to observe. They may be eclipsed by the planet [1]; they may be occulted [2]; or their shadows may be seen in transit across the planet [3]. Observations of Callisto, the outermost of the large satellites, are less common because of its greater distance from Jupiter itself.

from the plane of the rings; it has a period of 15 days 22.5 hours. The diameter is given as 4,800km (3,000 miles), although this may be an overestimate, and some official lists reduce it to as little as 4,300km (2,700 miles). Titan, however, is decidedly larger than the Moon, being about the same size as the planet Mercury.

Titan is unique in being the only satellite known to have an appreciable atmosphere. The main constituent is methane; the ground pressure is about 100 millibars, which is ten times that on the surface of Mars. The atmosphere, however, is cyclic, inasmuch as its molecules escape continuously, although at a relatively slow rate in the extremely cold surface temperatures. The molecules cannot escape from the pull of Saturn itself and remain in the same orbit as Titan, so that the satellite picks them up again, and the overall atmospheric density remains almost constant. There may be clouds in the atmosphere of Titan and, as a result, surface details are difficult to see.

Titan is now of such interest to astronomers that there is talk of sending a special probe to it and certainly it should be photographed from the Mariners that are due to bypass Saturn within the next few years.

The other satellites of Saturn

Saturn's other satellites are much smaller. The inner four (Janus, Mimas, Enceladus and Tethys) are of low density and have been described as large ice balls; only Tethys is as much as 1,000km (600 miles) in diameter. Janus, the closest-in, was discovered by Audouin Dollfus (1924–) in 1966 and is visible only when the rings are edge-on. The fourth and fifth satellites, Dione and Rhea, are denser and more massive, although still smaller than the Moon. Titan comes next in order, then a very small satellite, Hyperion, and then Iapetus, which is much brighter west of Saturn than when to the east. Presumably it has a synchronous rotation and it must either be irregular in shape or else have a surface of unequal reflecting power. Finally comes Phoebe, up to 13 million kilometres (8 million miles) from Saturn; it is extremely small and its retrograde motion indicates that it is probably a captured asteroid.

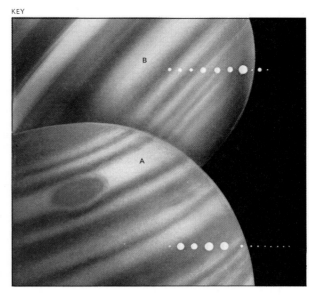

Jupiter [A] and Saturn [B] have 13 and 10 satellites respectively. Four of Jupiter's satellites and one of Saturn's are large.

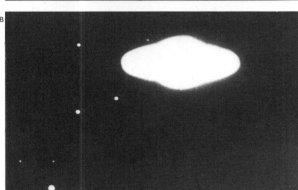

6 The inner satellites of Saturn [A] from right to left are Dione, Tethys, Mimas, Enceladus, Rhea and Titan. The image of Saturn is necessarily overexposed, otherwise the faint inner satellites would be lost. The same satellites were photographed on 24 March 1948 [B]. At that time the ring-system was fairly wide open. The two faint objects just above Titan are background stars.

7 Janus [arrowed] is visible only when the ring-system is edge-on. The brighter satellite to the left-hand side of Saturn is Mimas. This photograph was taken by Dollfus at the Pic du Midi Observatory.

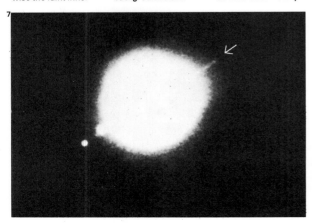

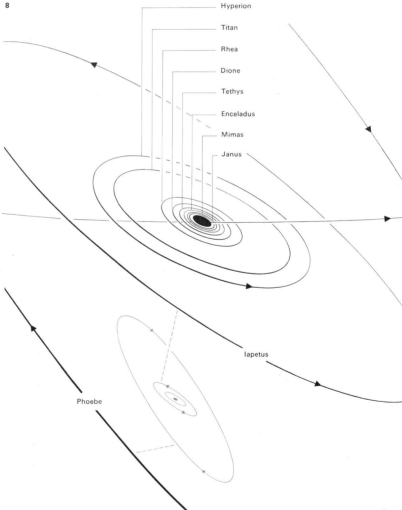

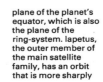

8 The orbits of Saturn's satellites, shown to scale, are varied. The orbits from Janus to Hyperion are almost circular, they move practically in the plane of the planet's equator, which is also the plane of the ring-system. Iapetus, the outer member of the main satellite family, has an orbit that is more sharply inclined. Phoebe is shown to the lower left, together with Hyperion and Iapetus; it has retrograde motion, and may be a captured asteroid. In 1905 William Pickering (1858–1938) reported another satellite between Titan and Hyperion, but it has not been seen since; he possibly mistook a star for a satellite.

The outer planets

In ancient times Saturn was the outermost of the planets known to man. There were seven main bodies in the Solar System (the five planets visible to the naked eye, plus the Sun and the Moon) and because seven was the mystical number of the astrologers no more planets were expected. Then in 1781 William Herschel (1738–1822) was mapping stars in the constellation of Gemini when he came upon an object that showed a disc and that moved perceptibly from night to night. Herschel thought that it must be a comet, but when its orbit was worked out he found that it was a planet much more remote than Saturn.

Uranus and its strange tilt
Uranus is dimly visible to the naked eye, but it is not surprising that it was overlooked until Herschel's fortuitous discovery. The planet is a giant, with a diameter of 51,800km (32,375 miles) – less than half that of Saturn. Its outer layers, at least, are gaseous and the surface temperature is extremely low. When seen through a telescope, Uranus shows a decidedly greenish disc; bright and dark zones may be made out with difficulty.

Uranus has a mean distance from the Sun of 2,869,600,000km (1,780 million miles) and a revolution period of 84 years. The axial rotation period is about 11 hours [2]. The axial tilt is very strange: it amounts to 98 degrees, which is more than a right-angle, so that from Earth Uranus is sometimes seen pole-on and sometimes with the equator crossing the centre of the disc. The reason for this peculiar inclination is unknown. It leads to an unfamiliar situation on the surface of the planet for each pole has a "night" lasting 21 Earth-years.

Our knowledge of Uranus is not at all complete and unfortunately, with the cancellation of the 1979 fly-by mission to Jupiter, Uranus and Neptune, there are no plans to send a probe to the major planets in the immediate future. Uranus has five satellites [3], all of which are smaller than our Moon and all of which revolve in the plane of the planet's equator, so that their movements are technically retrograde. There is some disagreement about the sizes of the satellites, but Miranda, the smallest, is about 550km (340 miles) in diameter; Umbriel about

1,000km (620 miles); and Ariel about 1,500km (930 miles). The two largest, Oberon and Titania, have diameters of approximately 1,600km (995 miles) and 1,800km (1,120 miles), respectively.

The discovery of Neptune
With the discovery of Uranus the Solar System seemed to be complete, but after a few years a strange problem arose. Uranus did not move as expected – it persistently wandered away from its calculated path. The only logical solution was that the action of a more distant, still-unknown planet was pulling Uranus out of position. This idea was proposed in 1834 and was communicated to George Biddell Airy (1801–92), who became Astronomer Royal at Greenwich the following year, but he showed little interest. John Couch Adams (1819–92), working at Cambridge in 1843, resolved to tackle the problem. He thought that using the perturbations of Uranus it would be possible to find the position of the unknown planet and after some months of hard work he felt that he had fixed the position accurately. He too con-

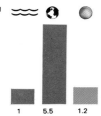

1 Uranus has a density 1.2 times that of water – rather more than that of Jupiter and much more than that of Saturn, although less than that of Earth (5.5).

2 The axial rotation of Uranus [A] lasts about 11 hours at its equator, although the rotation in the polar zones is slightly longer. The tilt of the axis as compared to the Earth [B] is 98 degrees – unique in the Solar System. Because this is more than a right-angle the rotation is technically retrograde (although not usually reckoned as such). All five satellites of Uranus move almost in the plane of the planet's equator.

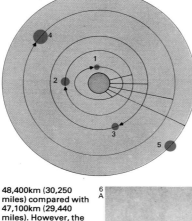

3 Orbits of the five satellites of Uranus are shown in this diagram as they would be seen looking down on the planet's pole: Miranda [1], Ariel [2], Umbriel [3], Titania [4] and Oberon [5]. When the planet is seen pole-on from Earth the orbits appear circular. But when Uranus is seen equator-on, as occurred in 1945, the orbits appear almost linear. Miranda is the smallest and most recently discovered of the satellites.

4 Uranus, was photographed with all five of its satellites by G. P. Kuiper, with the 82in (208cm) reflector at the McDonald Observatory, Texas, in 1948. That was the year in which Kuiper discovered Miranda, innermost and faintest of the satellites. Shown here are Ariel [1], Umbriel [2], Titania [3], Oberon [4] and Miranda [5]. The image of Uranus is overexposed and the ring effect is a purely photographic phenomenon.

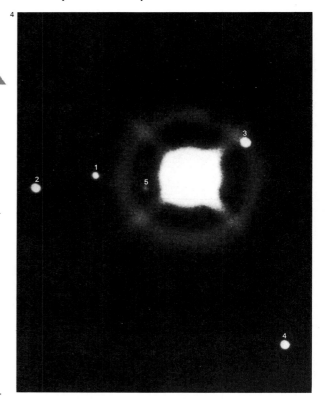

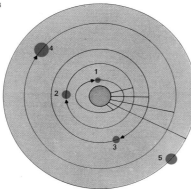

5 Uranus and Neptune are similar in size. Neptune is decidedly the more massive (17 Earth masses, as against only 15 for Uranus) and until recently it was thought to be larger,

48,400km (30,250 miles) compared with 47,100km (29,440 miles). However, the occultation of a star by Neptune in 1968 was estimated at 49,500 km (30,940 miles) and in 1970 an experiment with a balloon-borne telescope at Princeton University in the United States gave a new value for Uranus of 51,800km (32,375 miles). The density of Neptune (1.67) is greater than that of Uranus, and in fact it is the densest of the outer planets, with the exception of Pluto.

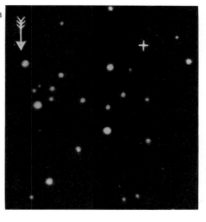

6 When Galle and d'Arrest set out to look for Neptune in 1846, using the calculations by Leverrier, they were able to make use of a new star map of the area, a corner of which is shown in diagram A. Challis, searching at Cambridge, had no such map and his task was thus much more laborious. B shows the corresponding portion of the sky; Leverrier's estimated position for Neptune is shown by a cross and the arrow indicates the planet's actual position.

tacted Airy, but again the Astronomer Royal took no action (although, to be fair, this was partly due to a series of misunderstandings). In the meantime, similar calculations made by the French mathematician Urbain Leverrier (1811–77) were sent to the Berlin Observatory where two observers, Johann Galle (1812–1910) and Heinrich d'Arrest (1822–75), quickly located the planet in almost exactly the position indicated by Leverrier. The discovery was made in 1846 [6]. Shortly afterwards James Challis (1803–82), working at Cambridge on the basis of Adams's work, located the planet; nowadays Adams and Leverrier are recognized as co-discoverers of Neptune.

Neptune is almost a twin of Uranus. It is slightly smaller, with a diameter of 49,500km (30,940 miles), and more massive; it does not have Uranus's remarkable axial tilt. The revolution period is 164.8 years and the mean distance from the Sun is 4,497 million kilometres (2,794 million miles). No telescope will show definite surface details on Neptune; all that can be seen is a pale bluish disc. Of the two satellites [8], Triton is larger

than our Moon and has a circular orbit, but moves round Neptune in a retrograde direction; the other satellite, Nereid, is small and has an eccentric orbit like that of a comet.

The enigma of Pluto

Even with the discovery of Neptune there were still discrepancies in the movements of the outer giants, and Percival Lowell (1855–1916) undertook new calculations with the aim of finding yet another planet. In 1930, at the Lowell Observatory in Arizona, the planet was duly located [9], although by then Lowell had been dead for 14 years.

Pluto, as the new planet was named, has set astronomers problem after problem. It is believed to have a diameter of about 3,500km (2,175 miles) which is about the size of our Moon. It has an eccentric, inclined orbit than can bring it closer than Neptune to the Sun [10]; and its lack of mass means that it cannot produce marked perturbations in the motions of Uranus or Neptune – yet it was because of such perturbations that Pluto was found. The revolution period is 247.7 years and the axial rotation 6.4 Earth-days.

The sizes of the outermost planets are compared here with Jupiter and Saturn.

Uranus and Neptune are giants with gaseous surfaces; Pluto is smaller than the

Earth and is looked on as a "terrestrial" planet although its nature is uncertain.

7 This photograph shows Neptune together with both its satellites, Triton and Nereid. Triton, near the lower right of the picture, is relatively bright (it is brighter than any of the satellites of Uranus) and was discovered by the English astronomer William Lassell only a few weeks after the discovery of Neptune itself. Nereid is very faint and can be photographed only with giant telescopes.

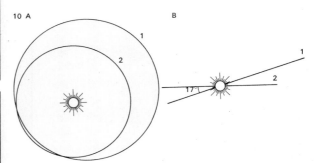

8 Orbits of Triton [1] and Nereid [2], Neptune's satellites, are quite different. Triton's is almost circular but it has retrograde motion; it is the only large satellite in the Solar System to behave in this way. Nereid has direct motion but an eccentric orbit like that of a comet.

9 These photographs show the discovery of Pluto in 1930 by Clyde Tombaugh at the Lowell Observatory on the basis of Percival Lowell's calculations. A on 2 March; B on 5 March – the shift of Pluto, indicated by the arrows, is very noticeable. The overexposed image is that of the third-magnitude star Delta Geminorum. Pluto is now of about magnitude 14 and may thus be seen with a telescope of moderate size.

10 Pluto has an exceptional orbit [A] that is both relatively inclined and decidedly eccentric. Here [1] it is compared with that of Neptune [2]. At perihelion Pluto may come within Neptune's path but the inclination of 17° [B] means that there can be no collision. The next perihelion passage is due in 1989. Its peculiar orbit has given rise to some doubt as to whether Pluto should be considered a true planet.

Comets

A great comet, with a brilliant head and a tail stretching half-way across the sky, is a spectacular object – and it is easy to understand why comets of this kind caused such terror in ancient times. Comets have always been regarded as unlucky and fear of them is still not dead in some primitive societies.

Yet a comet is not nearly as important as it may look. It is made up of small particles (mainly icy in nature) and tenuous gas. On several occasions the Earth has been known to pass through a comet's tail without suffering the slightest damage.

A comet's anatomy

A large comet is made up of three principal parts: a nucleus (containing most of the mass); a coma or head; and a tail [2]. The coma and tail appear only when the comet approaches the Sun and solar radiation vaporizes the icy nucleus. As the comet recedes the tail disappears. Small comets, however, are often devoid of tails and look rather like small patches of faintly luminous cotton wool in the sky.

The tails of comets are of two main kinds: gas and dust. Generally the gas tail is relatively straight, while the dust tail is curved – because it lags behind as the comet moves forwards. One remarkable feature of tails is that they always point more or less away from the Sun, so that when a comet is receding it travels tail-first. The causes of this are not known for certain, but it is thought that the tiny particles in the tail are repelled by the so-called "solar wind", a stream of electrified, low-energy particles constantly flowing outward from the Sun in all directions.

Comets are members of the Solar System but their paths, in most cases, differ from those of the planets in that they are much more eccentric. Dozens of comets with short-period orbits are known; the period of Encke's comet, for instance, is only 3.3 years, so that it is seen regularly and has been observed at more than 50 returns to perihelion (the point in its orbit when it is closest to the Sun – aphelion is the point farthest from the Sun) from when it was first sighted in the eighteenth century. Comets are not visible through the emission of light of their own, but through the reflection of the Sun's radiation, which causes the cometary material to fluoresce. Thus most comets cannot be followed throughout their orbits and can be seen only when relatively close to the Earth and the Sun.

Short- and long-period comets

All the short-period comets are faint and many of them are difficult objects to view telescopically. A few (notably Schwassmann-Wachmann I and the more recent Gunn's comet) have more circular paths [3] and can be followed throughout their orbits.

Other comets take decades to travel once round the Sun. The most famous of these is Halley's comet [5], which is a bright naked-eye object and is seen every 76 years or so (the period of a comet is not constant, because of the perturbations caused by the planets). Halley's comet last returned to perihelion in 1910 and is due back once more in 1986. Records of it go back well before the time of Christ.

Other great comets have much longer periods – so long, indeed, that we cannot measure them accurately. Thus comets of this

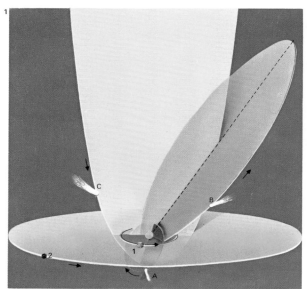

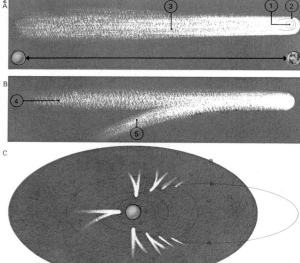

2 The anatomy of a large comet [A] contains the nucleus [1], which may or may not be a conglomerate; the coma [2] made up of small particles and tenuous gas; and the tail [3] extending away from the coma. The two types of tail shown in B are a gaseous tail [4], which is generally straight, and a dust tail [5], which lags behind the moving comet, so that the tail appears curved. The tail of a comet, like Halley's [C], always points approximately away from the Sun whatever its orbital position. A comet develops a tail only as it nears perihelion, losing it as it recedes from the Sun.

1 There are three main classes of comet. Short-period comets [A] often have their aphelia at approximately the distance of Jupiter's orbit [1]. Their periods amount to a few years and all short-period comets are faint. Long-period comets [B] have aphelia near or beyond Neptune's orbit [2] – Halley's is the only conspicuous member of the class. Comets with very long periods [C] have such great orbital eccentricities that the paths are almost parabolic. Because only a short arc can be measured, it is impossible to calculate the periods of these comets accurately. All the really brilliant comets, apart from Halley's, are of this type.

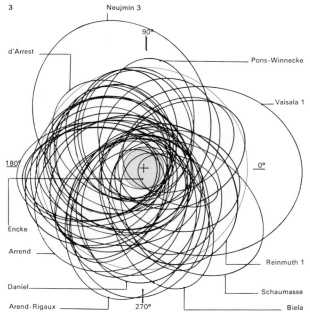

3 Some short-period comets, whose aphelia lie near Jupiter's orbit, are said to belong to Jupiter's family. Their orbits are shown in relation to those of Earth – the inner blue ring – and Jupiter the outer. Encke's comet has the shortest period (3.3 years); that of Schaumasse's is more than 8 years. It was thought that comets came from interstellar space and were captured by the planets, but this theory is no longer accepted as valid.

4 Donati's comet of 1858 was generally thought to be the most beautiful ever observed. It was a brilliant naked-eye object and had tails of both gas and dust. This picture is taken from an old woodcut.

kind cannot be predicted and are always apt to take astronomers by surprise. Such was the great comet of 1843, whose coma was larger than the Sun, even though its mass was, by astronomical standards, negligible. Other great comets appeared in 1811, 1882 and 1910. The Daylight comet of 1910 (not to be confused with Halley's) was probably the brightest to have been seen during the present century. Kohoutek's comet of 1973 was expected to be brilliant but proved, in the event, a great disappointment [9].

Short-lived comets
Some great comets approach the Sun very closely and are termed "Sun-grazers". As the comet passes perihelion the tail swings round, and it often happens that the original tail is destroyed and a new one forms. The tails of comets are produced from evaporation from the nucleus and there must be a steady wastage of material, so that by cosmic standards comets are short-lived. We even know of comets that have disappeared. Westphal's comet of 1913, which had a period of 62 years, faded out as it approached

perihelion and was never seen again. Biela's periodical comet, which took 6.75 years to complete one orbit, split in two in 1846; the "twins" were seen again in 1852 but that was their last appearance as comets. In 1872, when due to return, a bright meteor shower was seen in the region from which they ought to have come. This emphasizes the close connection between comets and meteors. Meteors may in fact be authentically regarded as cometary debris.

There is considerable uncertainty about the origin of comets. But according to J. H. Oort, a Dutch astronomer, there is a vast "comet cloud" at a great distance from the Sun; sometimes a comet will swing inwards towards the Sun, when it can be observed.

Although professional astronomers are engaged in comet-hunting, many discoveries are made by amateurs. G. E. D. Alcock, an English schoolmaster, has now discovered four, while the bright comet of 1970 was found by a South African, J. Bennett.

Comets are of great scientific interest and there is serious talk of sending a space probe to rendezvous with a suitable comet.

This brilliant comet is Bennett's comet, photographed on 12 March 1970. The tail is quite long and its fine, gaseous nature is clearly seen. The nucleus is not shown because of the over-exposure of the coma, or head, in order to bring out the structure of the tail. The coma is made up of material from the nucleus which is vaporized by solar radiation as the comet approaches the Sun, when its temperature may rise to several thousand degrees. A comet's emission spectrum reveals the presence of such elements as iron, calcium, sodium, potassium, copper, chromium, nickel and traces of several other metals.

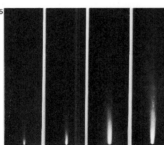 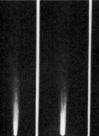 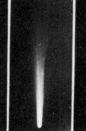 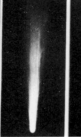    

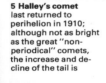    

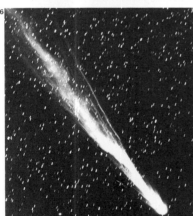

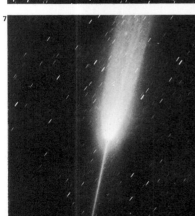

5 Halley's comet last returned to perihelion in 1910; although not as bright as the great "non-periodical" comets, the increase and decline of the tail is clearly shown in this sequence. As it approached perihelion the tail developed enormously; after the closest approach to the Sun the tail contracted, so that when the comet was last seen the tail had disappeared altogether. The seventh picture shows the tail shortly before perihelion passage.

6 Morehouse's comet of 1908 had a complex tail, the structure of which changed rapidly. Great disturbances must have been taking place, but the comet was not bright enough for the details of these changes to be seen on Earth.

7 Comet Arend-Roland of 1957 was one of the most interesting comets of recent times. The apparent "forward spike" is not an extra tail but is illuminated meteoritic debris lying along the comet's orbit.

8 Humason's comet of 1961, shown in this photograph taken with the 48in (121cm) Schmidt telescope at Palomar, USA, was one of the first comets to be photographed in colour. Because the telescope was tracking the comet, surrounding star images appear as short trails.

9 Kohoutek's comet of 1973 was not as spectacular as was hoped. There will be no opportunity to see it again since it will not return to perihelion for 75,000 years.

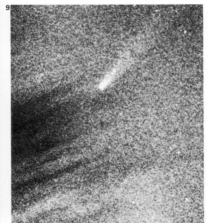

Meteors and meteorites

Meteors or shooting stars can usually be seen on clear August nights in the Northern Hemisphere. They are rapidly moving points of light, often with luminous tails, caused by objects travelling quickly across the sky. Such shooting stars have been known since antiquity but it was not until the beginning of the nineteenth century that their true nature became fully understood.

Meteorites are not as numerous as meteors and are an entirely different kind of heavenly body. They are not merely large meteors, nor are they related to comets, which they resemble visually as streaks in the sky. Rather they are considered to be much more closely associated with the asteroids or minor planets.

High-velocity particles

A meteor is a tiny particle, usually smaller than a grain of sand, moving around the Sun. It is so small that it can be seen only when it enters the Earth's upper atmosphere. With a velocity of entry possibly as high as 72km (45 miles) per second, the meteor sets up friction with the air molecules, which causes it to destroy itself well before it reaches the ground. The resulting luminous streak in the sky, which is characteristic of the shooting star, is not produced by the meteor itself but by its effect on the atmosphere through which it is falling.

Meteors are of two main kinds: shower and sporadic. Sporadic meteors may appear from any direction at any time. Shower meteors, on the other hand are associated with comets. The famous Leonid shower meteors of November, for example [3, 4], are linked with the faint periodical Tempel's comet and move in the same orbit as the comet itself. It has been said that meteors are mere cometary debris. This may be something of an oversimplification, but it is certainly true that one periodical comet, Biela's, was seen to disrupt and has now been replaced by a meteor shower [1, 2]. And there can be no doubt that as a comet moves along it "sheds" meteoric material.

The richness of a meteor shower is measured by its so-called zenithal hourly rate (ZHR). This is a measure of the number of meteors that would be seen by a watcher observing under ideal conditions with the shower radiant at the zenith. The most reliable annual shower, that of the Perseids, has a ZHR of about 70. Meteors below naked-eye visibility are not included, so that in fact there are many more meteors than might be thought. Those that are too small to produce any luminous effects are known as micrometeors and are extremely numerous.

Because the meteors in a shower are travelling through space in parallel paths, they seem to radiate from one particular point in the sky, which is known as the radiant. The principle is analogous with the view from a bridge overlooking a motorway. The parallel lanes of the motorway will seem to meet at a point near the horizon, which may be termed the apparent "radiant" of the lanes. Thus, on the same principle, the November Leonids have their radiant in Leo, the August Perseids in Perseus and so on.

Regular annual showers

Showers of meteors occur regularly on an annual basis. These include the Quadrantids (1–6 January, maximum 3–4 January);

1 Biela's comet once had a period of 6.75 years. In 1846, as shown in a contemporary drawing by Angelo Secchi (1818–78), it divided into a pair of comets. The division may have been caused initially by a close approach to Jupiter in 1842 with the pull of the Sun accounting for the rest of the change. The two comets, separated by over 2 million km (1.25 million miles), returned in 1852. Because of their unfavourable positions they were not seen in 1858 and did not appear in 1866. They have not been seen again.

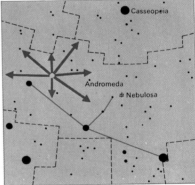

2 A brilliant meteor shower [red] from a radiant in Andromeda [blue] where Biela's comet should have been was seen in 1872. It was probably the comet's debris. The shower is now extremely feeble.

Casseopeia

Andromeda

※ Nebulosa

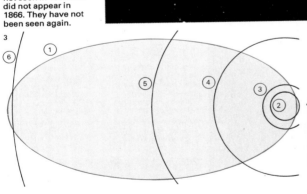

3 The orbit of the Leonid meteor stream [1] intersects the orbits of the Earth [2], Mars [3], Jupiter [4], Saturn [5] and Uranus [6].

Because the meteors are not distributed evenly, major meteor showers appear only occasionally. The average interval was once just over 33 years, but the expected showers of 1899 and 1933 were missed because the shower orbit had suffered planetary perturbations.

4 A splendid Leonid meteor shower was photographed from Arizona on 17 November 1966. The shower was not visible from Europe.

5 A meteor trail is seen here near the cluster Praesepe; the meteor was of about the second magnitude and lasted 1.5secs.

Lyrids (19–24 April); Eta Aquarids (1–8 May, associated with the famous Halley's comet); Perseids (25 July–18 August); Orionids (16–26 October); Taurids (20 October–30 November); the far southern Phoenicids (4–5 December); the Geminids (7–15 December); and Ursids (17–24 December). The Leonids, which are at their peak on 17 November, are less reliable because the meteors are clustered rather than spread along the cometary orbit. Thus a major shower cannot be seen until the Earth passes through the main swarm. This happened in 1799, 1833, 1866 and again in 1966, so that there could possibly be another display in 1999. In the years between these major showers the Leonids are sparse.

The history of meteorites

A larger body encountering the Earth may survive the journey to the ground without being destroyed. These meteorites may be of several kinds. Aerolites [10] are mainly of stone white siderites [12] have a high percentage of iron. There are various intermediate types. Etching a meteorite with acid will show

the characteristic forms, known as the Widmanstätten patterns, and this is one sure way of telling whether a piece of material is of meteoritic origin or not.

Meteorites have been known for many centuries – for instance, the Sacred Stone at Mecca is certainly a meteorite – but for a long time their cosmic origin was doubted. In 1795, when a 25kg (56lb) meteorite fell at Wold Cottage in Yorkshire, it was suggested that the object was a stone hurled out of the Icelandic volcano Hekla. However, in 1803 a meteorite group fell at L'Aigle in France and the famous astronomer Jean-Baptiste Biot (1774–1862) was able to demonstrate that the objects definitely came from the sky. Most museums have meteorite collections; the largest meteorite on display, at the Hayden Planetarium in New York, weighs about 31 tonnes. Fortunately, major meteorite falls are rare. The best example of craters produced by large meteorites are the Canyon Diablo Crater in Arizona [6, 7] and the Wolf Creek Crater in Australia. There is no known case, to date, of any human being having been killed by a falling meteorite.

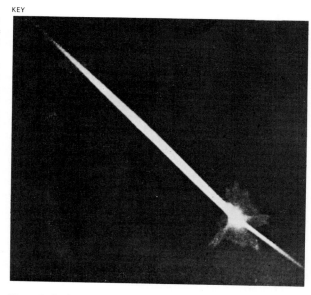

The exploding Andromedid meteor was photographed on 23 November 1895. This is one of the finest meteor photographs.

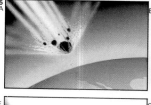

Wind direction

6 The Arizona meteorite crater was formed by several hickel-iron meteorites. Burning up as they plunged through the atmosphere [A], they shattered the Earth's outer layer of rock on impact [B]. Because of their high speed they burrowed, causing friction, heat, compression and shock waves [C], culminating in a violent explosion [D] that left a crater. Areas of meteorite fragments [E] show those that were unaltered by heat [blue], small heat-affected fragments [yellow], heavy heat-altered boulders [black] and metallic spheroids formed by condensation [red].

7 The impact crater near Winslow, Arizona, is more than 1km (0.6 mile) in diameter. It may well be over 10,000 years old. Many meteorite fragments have been found in the area.

8 The Hoba West meteorite, near Grootfontein in South West Africa, is the largest known meteorite. It weighs over 60 tonnes and its weight before entering the Earth's atmosphere may have been 20 tonnes more. The meteorite still lies where it fell in prehistoric times. No crater was produced. A meteorite of this kind could be highly destructive but fortunately major falls are extremely rare.

9 The Orgueil-meteorite (1864) is carbonaceous, having organic compounds.

10 The Norton-Furnas aerolite of 1948 is the heaviest (over 1 tonne) of its type of stony meteorite.

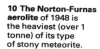
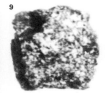
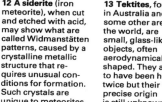

11 The most destructive fall of modern times was that of 1908 in the Tunguska region of Siberia. A meteorite came down in forested country, flattening pine trees for several miles around.

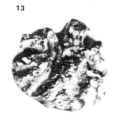

12 A siderite (iron meteorite), when cut and etched with acid, may show what are called Widmanstätten patterns, caused by a crystalline metallic structure that requires unusual conditions for formation. Such crystals are unique to meteorites.

13 Tektites, found in Australia and in some other areas of the world, are small, glass-like objects, often aerodynamically shaped. They appear to have been heated twice but their precise origin is still unknown.

The Sun and the solar spectrum

The Sun is a star; one of 100,000 million stars in our Galaxy. In the universe as a whole it is insignificant and is classed as a yellow dwarf star with a spectrum of type G; but in our planetary system – "the Solar System" – it is the all-important controlling body.

Immensely larger than the Earth, the Sun is made principally of hydrogen and helium and has a diameter of 1,392,000km (865,000 miles). Although it is big enough to contain more than a million bodies the volume of the Earth, its mass is only 1.990×10^{33} grammes – that is to say approximately 333,000 times that of the Earth. The reason why it is not as massive as might be expected is that its density is lower than that of an Earth-type planet. The mean value for the specific gravity is 1.409 (that is to say, 1.409 times that of an equal volume of water) but the Sun is not homogeneous and the density increases rapidly beneath the brilliant outer surface.

The Sun lies some 32,000 light-years from the centre of our Galaxy and takes approximately 225 million years to complete one journey round the galactic nucleus. It has an axial rotation period of 25.4 days at the equator, but this period is considerably longer near the solar poles because the Sun does not rotate in the manner of a solid body.

The photosphere

The bright outer surface of the Sun is known as the photosphere and has a temperature of 5,500°C. On it may be seen darker patches, which are known as sunspots [Key]. These are not truly black, but appear so by contrast; if a spot could be seen shining on its own its surface brilliancy would be greater than that of an arc-lamp.

To look at the Sun through any telescope or binoculars will almost certainly blind an observer permanently and dark filters are unreliable safeguards. The only sensible method is that of projection – using a telescope to throw the solar image on to a screen held or fastened behind the eyepiece. The Sun is not as generally smooth and feature-less as might be thought; granules exist on it, each of which is about 1,500km (1,000 miles) in diameter. Convection currents occur below the Sun's outer layer, and it is the rising gas columns they generate that cause the granules, whose dark edges show cooler gases dropping downwards [7].

A typical large sunspot consists of a central dark umbra surrounded by a lighter area of penumbra, although the shapes are usually very irregular and spots tend to occur in groups – generally with two main spots, one "leader" and one "follower" [3]. Some groups may be immensely complex and of tremendous area, but they are not long-lived. Even a large group will generally persist for only a few months at most, while smaller spots may last for only a few hours. As the Sun rotates the spots may be seen to be carried slowly across the disc from one side to the other. It takes about a fortnight for a spot to make the full crossing. After a similar interval, it will reappear on the opposite side of the disc – provided that it still exists.

Regular cycles

The cycle of solar activity is fairly regular and has a period of 11 years; thus there were maxima in 1957-8 and again in 1969–70 when groups were plentiful [4B]. At the intervening spot-minima the disc may re-

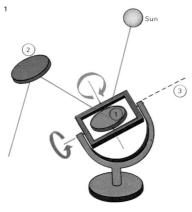

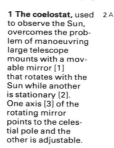

1 **The coelostat,** used to observe the Sun, overcomes the problem of manoeuvring large telescope mounts with a movable mirror [1] that rotates with the Sun while another is stationary [2]. One axis [3] of the rotating mirror points to the celestial pole and the other is adjustable.

2 **the heliostat** [A], an elaborate version of the coelostat, is installed in the 500ft (152.4m) telescope at Kitt Peak, Arizona, [B], to track the Sun. Sunlight falls on a rotating mirror [1]. It reflects down a tube to a concave mirror [2], focuses on to a plane mirror [3] and passes through a spectrograph [4].

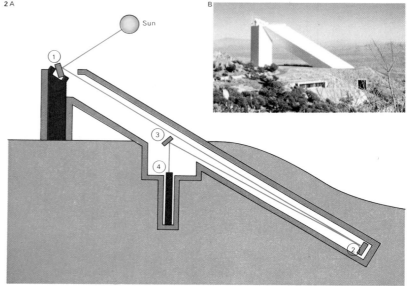

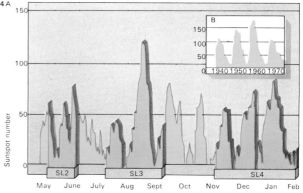

3 **The sunspot observations of 1947** show that on 11 February [A] the identities of the leader and follower were still in doubt. But as the lines indicate, the magnetic polarities were clear. From 9 March [B] to 7 April [C] the leader [1] and follower [2] are distinct. By 5 May [D] activity has ceased.

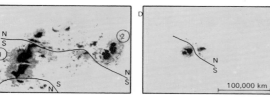

4 **Sunspot numbers** fluctuated widely [A] during the Skylab mission between 14 May 1973 and 8 February 1974. The periods when the Skylab was manned (28, 59 and 84 days respectively for the three crews) are shown in orange on the large graph [SL2, SL3 and SL4]. The inset graph [B] charts the sunspot cycle from 1935 to 1973. The 1969–70 maximum was much less intense than that of 1957–8.

100,000 km

5 **The penumbra of sunspots** [1] near the Sun's rim (the limb) seems to narrow on the side nearest the centre of the disc, indicating that the spots are depressions in the photosphere. This effect was first noted by A. Wilson in 1769.

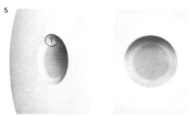

6 **A sunspot near the limb** illustrates the Wilson effect. A narrowing of the penumbra towards the centre of the Sun's disc is clear. The bright streaks are called faculae and are usually associated with major sunspot groups.

main featureless for many days on end.

Sunspots are associated with very strong magnetic fields and this has led to the modern theory of spot formation laid down by Harold Babcock (1882-1968) in 1962. The Sun has an overall magnetic field and it may be assumed that the lines of magnetic force run from one pole to the other below the bright surface. Owing to the difference in rotation period between the equatorial and the polar zones, the magnetic lines become distorted over an interval of some years and are "pulled out" along the equator, while the polar magnetic field is reinforced and becomes unstable. Eventually a loop of magnetic energy breaks through the surface producing two spots, one with north polarity and the other with south. Because of the magnetic linking, the polarities for the leading and following spots are opposite in the two hemispheres of the Sun. After about 11 years the "knots" in the lines break and the Sun reverts abruptly to its original state. But for the following cycle the polarities of spots in the two hemispheres are reversed.

Visual studies of the Sun's photosphere give us only limited information to draw on and most of our knowledge comes from instruments based on the principle of the spectroscope. According to the laws of spectroscopy as laid down by G. Kirchhoff (1824–87) in 1859, an incandescent solid, liquid or high-pressure gas will produce a continuous or rainbow spectrum, while a low-pressure gas will yield an emission spectrum consisting of isolated bright lines.

New elements found

Gaseous elements in the Sun's atmosphere absorb light at specific frequencies from the continuous spectrum emitted by the photosphere, producing gaps or dark lines on the spectrum. These dark lines are called, in the case of the Sun, Fraunhofer lines. The elements present in the Sun's atmosphere can be identified from the positions (that is, the frequencies) and intensities of the lines on the spectrum.

In this way more than 60 elements have already been found in the Sun. One of them – helium – was even identified in the Sun before it was found on Earth.

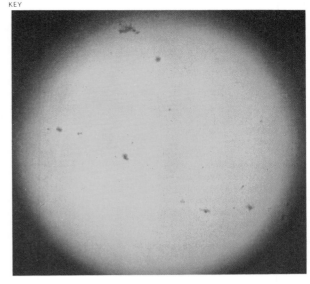

The solar maximum of 1958 was the most energetic ever recorded. The photograph shows a heavily spotted disc.

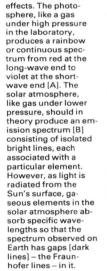

7 Solar granulation results from convection effects. These gaseous columns occur all over the Sun's disc. Their average diameter is about 1,500km, although their size range is fairly wide.

8 Large and complex sunspot groups were still common when the Sun was well past the peak of its 1947 cycle of activity. This group was photographed at Mount Wilson on 17 May 1951.

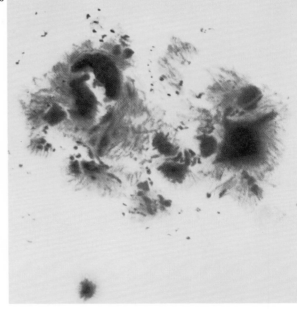

9 The solar spectrum combines two effects. The photosphere, like a gas under high pressure in the laboratory, produces a rainbow or continuous spectrum from red at the long-wave end to violet at the short-wave end [A]. The solar atmosphere, like gas under lower pressure, should in theory produce an emission spectrum [B] consisting of isolated bright lines, each associated with a particular element. However, as light is radiated from the Sun's surface, gaseous elements in the solar atmosphere absorb specific wavelengths so that the spectrum observed on Earth has gaps [dark lines] – the Fraunhofer lines – in it.

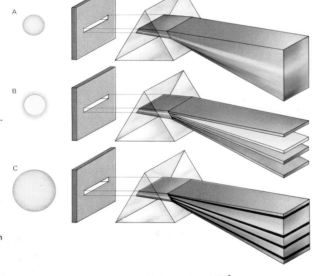

10 Lines in the solar spectrum can be accurately mapped with the 13ft (4m) spectrograph at the Mount Wilson observatory. The range illustrated is from 3,900 to 6,900 ångströms, ie from violet through to red. (One ångström equals one hundred millionth of a centimetre.) Each line can be identified; thus the D lines in the centre of the spectrum are due to sodium, the H-Alpha line to hydrogen.

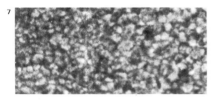

| K | H 4,000Å | Hδ | g | Hγ | 4,500Å | Hβ (F) | 5,000Å | bbb |

| Calcium | Iron | Hydrogen | Calcium | Hydrogen | | Hydrogen | Magnesium |

| Hα (C) 6,500Å | 6,000Å | D₁D₂ | 5,500Å |

Hydrogen | Sodium

The Sun's atmosphere and radiation

Ordinary telescopes show only the bright surface or photosphere of the Sun and features such as the spots, the granulation and the bright faculae (temporary patches on the Sun's surface) which lie above the photosphere itself. More complicated methods are needed to study the solar atmosphere because the Sun's surroundings can be seen with the naked eye (or with an ordinary telescope) only during the rare intervals when the Moon covers the Sun completely and produces a total eclipse.

Prominences and flares
The part of the solar atmosphere lying immediately above the photosphere is called the chromosphere ("colour sphere") because it has a characteristically reddish appearance. This is also the region of the large and brilliant prominences. To observe the prominences, instruments based on the principle of the spectroscope are used. There are two main types of prominences: eruptive [2] and quiescent [Key]. Eruptive prominences are in violent motion and have been observed extending to more than 50,000km (312,500

miles) above the Sun's surface. Quiescent prominences are much more stable and may hang in the chromosphere for days before breaking up. Both are most common near the peak of the solar cycle of activity.

Prominences are often associated with major spot-groups. Active groups also produce "flares", which are not usually visible, although a few have been seen. The flares are short-lived and emit streams of particles as well as short-wave radiation. These emissions have marked effects upon the Earth, producing magnetic storms or disturbances of the Earth's magnetic field that affect radio communications and compasses. They also produce the beautiful solar lights or aurorae [3, 4]. The Sun also sends out a constant stream of low-energy particles in all directions, making up what is now known as the solar wind. It is this emission that has a strong effect upon the tails of comets, forcing them to point away from the Sun.

In addition to sending out light, the Sun is an important source of infra-red (heat) and ultra-violet radiation, as well as radio waves, X-rays and gamma-rays. Studies are difficult

to carry out from Earth because of the screening effect of the atmosphere, but knowledge has been greatly increased as a result of work carried out by satellites and by the Skylab astronauts in 1973–4. It was fortunate that the Sun was reasonably active [5, 6] while the astronauts were in orbit, for many of the results could not possibly have been duplicated in ground-based observations.

The powerhouse
Although astronomers cannot prove most of their theories about the nature of the Sun, they have a good idea of its composition. The temperature increases toward the core, until at the centre of the globe it is estimated at about 10 million degrees centigrade. It is here, in what is called the Sun's "powerhouse", that the energy is being generated.

It is erroneous to suppose that the Sun is burning in the same way that a fire burns. A Sun made up entirely of coal, and radiating as fiercely as the real Sun does, would not last long on the cosmic scale and astronomers believe that the Sun is at least 5,000 million years old. (It is certainly older than the Earth,

CONNECTIONS

See also
230 The Sun and the solar spectrum
224 Solar eclipses
226 Star types
176 Members of the Solar System

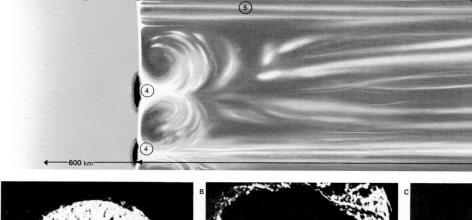

1 The main structure of the Sun cannot be drawn to an accurate scale. In the solar interior [1], nuclear transformations create energy. The convective zone [2] leads out to the relatively rarefied photosphere [3], which is surprisingly narrow and has sharp boundaries. Spots [4] lie in the photosphere and associated with them are the flares [5] and the prominences, which lie in the chromosphere [6]. The temperature of the chromosphere rises from 6,000°C at the bottom to more than 50,000°C near the upper portion. (Temperature here is purely a measure of the speeds at which the atomic particles are moving and does not indicate extra "heat".) In the chromosphere there are spicules [7] – masses of high-temperature gases shooting up rapidly into the immensely rarefied corona [8]. The corona is large and streamers [9] issue from it.

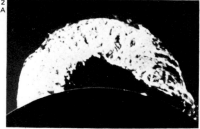

2 A large eruptive prominence occurred on 4 June 1946 at 16.03 hours [A]. It took the form of an arch. By 17.03 hours [B], it had been blown to 322,000 km (200,000 miles) above the Sun's surface. Little remained of the great arch by 17.23 hours [C]; the prominence is shown dispersing.

3 Aurorae, frequently associated with flares, are due to charged particles sent out by the Sun. The particles come from the Sun [1] and enter the Van Allen belts [2] which surround the Earth [3]. The Van Allen belts become overloaded and particles cascade down into the upper air, producing auroral glows.

4 Aurorae may take many forms such as curtains, arches and draperies, all with varied and lovely colours. They are aptly called "flaming surges". The electrified particles tend to spiral towards the Earth's magnetic poles, which is why aurorae are best seen from high latitudes; in low latitudes they are rare. Aurorae are commonest when the Sun is active, near the maximum of its 11-year cycle.

which has an age of about 4,600 million years.) The source of solar energy is to be found in nuclear transformations. Hydrogen is the main constituent and near the core, where the temperatures and pressures are so extreme, the second lightest element, helium, is formed from hydrogen nuclei by nuclear fusion. It takes four hydrogen nuclei to make one nucleus of helium; in the process a little mass is lost, being converted into a large amount of energy. The energy produced keeps the Sun radiating: the loss of mass amounts to four million tonnes per second. This may seem a lot but it is negligible compared with the total mass of the Sun; there is enough hydrogen available to keep the Sun shining in its present form for at least another 5,000 million years, perhaps longer.

Eventually the hydrogen will start to become exhausted and the Sun will change its structure drastically. According to present theory, it will pass through a red giant stage when it will have a luminosity at least 100 times as great as it is today; it will then collapse into a small dense star of the type known as a white dwarf.

The Earth also has a limited life-span. It cannot survive the red giant stage; along with the other inner planets it will certainly be destroyed eventually.

Solar research
Knowledge of the Sun has been drawn from many different areas of research. Radio astronomy is of especial importance. This is a method of studying astronomy in the long-wavelength region of the electromagnetic spectrum. The Sun is a strong radio source, a fact known since the early days of radio astronomy. The study of X-rays and gamma-rays from the Sun is much more recent because it depends upon instruments operating from above the Earth's hampering atmospheric layers.

There has been much discussion about the extent of the solar atmosphere. Beyond the chromosphere lies the corona, which is immensely rarefied and has no definite boundary. It merely "thins out" to become the solar wind. The Sun is a much more complicated and varied body than was originally thought, but it is probably a typical star.

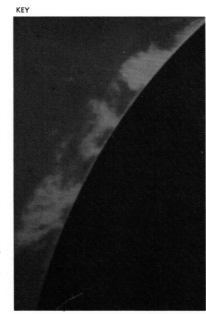

The most spectacular solar features are the streams of hot gas called prominences. Quiescent prominences may hang in the chromosphere for days or weeks, bulging out from the surface about 32,000km (20,000 miles). This example was photographed with a Lyot filter on a 4in (10cm) refractor. Eruptive prominences, the other main type of prominence, are thin flames of gas often reaching heights of 400,000km (312,500 miles); they are formed more frequently in those areas containing sunspots. The invention of the coronagraph in 1930 enabled continuous photography of prominences, which otherwise could be seen only during a total eclipse.

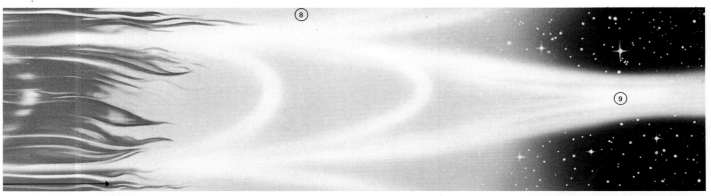

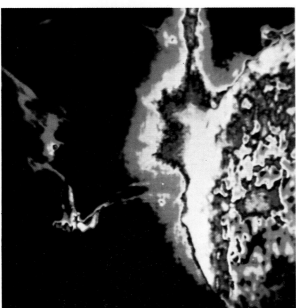

5 This solar prominence was photographed by the astronauts on board Skylab. The colours, in this extreme ultra-violet shot, are false. They represent the degree of radiation intensity from red, through yellow and blue, to purple and white where the activity is most intense. This picture could be taken only with equipment carried above the layers of the Earth's atmosphere.

6 This Skylab photograph depicts an eruptive solar prominence, which is seen rising to a great height. Matter at the apex of the arch seems to be reflected back to the Sun's surface.

Solar eclipses

The Skylab missions in 1973 and 1974 considerably improved man's knowledge of the Sun because they made possible extended observations of features not visible from the Earth's surface. Before the missions the best views of those features had been obtained during total eclipses.

Features of eclipses
The Moon is so much closer to the Earth than the Sun that despite its small size it looks just as big in our sky. The coincidence is fortunate: it means that when the three bodies are exactly lined up, the Moon can just blot out the Sun's brilliant photosphere, leaving the chromosphere and the corona to shine out unhindered [1B]. The spectacle is always brief, however, because the Moon's shadow only just touches the Earth [1A]: the track of totality can never be more than 269km (167 miles) wide and the maximum duration in any one spot is less than eight minutes. Hence astronomers have always done their best to take advantage of these opportunities. It was only after the famous eclipse of 1842 that most astronomers came to believe that the prominences belonged to the Sun, rather than to the Moon.

Because the Moon has an orbit that is not circular, its apparent size varies. At apogee (the point of greatest distance from the Earth), the full Moon looks ten per cent smaller than at perigee (the point nearest the Earth). The apparent diameter of the Sun also changes, being greatest in December and least in June because of the Earth's varying distance from the Sun. If the Moon appears smaller than the Sun, it is unable to cover the photosphere completely and the result is an annular (ring-shaped) eclipse, leaving a ring of sunlight showing round the dark mass of the Moon [1D]. There are also partial eclipses [1C], when not all of the Sun is hidden. Annular and partial eclipses are relatively unimportant because the Sun's surroundings do not come into view.

Eclipse records date back many centuries; there are records of an eclipse seen from China as long ago as 2136 BC. Eclipses do not occur at every new Moon because the lunar orbit is appreciably inclined to that of the Earth [1E]. Any eclipse, however, may be followed by another similar eclipse (total or partial) 18 years 10.3 days or 11.3 days with five leap years later when the Sun, Moon and Earth then return to almost the same relative positions. This period is known as the Saros (although there will be other eclipses in the intervening period). The Saros [Key] is not exact, but it is better than nothing at all as far as predictions are concerned and the ancients made extensive use of it.

Observing the corona
The main glory of a total eclipse lies in the view of the corona [5, 6]. It has been found that the shape of the corona varies according to the state of the solar cycle. Near spot-minimum the corona is fairly symmetrical, whereas near spot-maximum there are long streamers. The sky is dark enough for planets and bright stars to be seen and on several occasions unexpected comets have been found close to the hidden Sun. It is a pity that total eclipses are so rare in any particular place on Earth. In England, for example, the last total eclipse occurred in 1927; the next will not be until 1999.

CONNECTIONS

See also
220 The Sun and the solar spectrum
222 The Sun's atmosphere and radiation
268 A history of space achievements
270 Stations in space
212 Myths of autumn

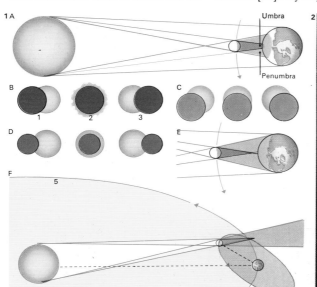

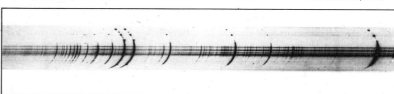

1 In a total eclipse [A] the main cone of shadow or umbra (not to be confused with the umbra of a sunspot) reaches the Earth's surface while to either side of it, in the partial shadow or penumbra, a partial eclipse is seen. The Moon and Sun [B] approach totality [1], arrive at totality [2] and leave totality [3]. C shows a partial eclipse that is not total anywhere on Earth. An annular eclipse [D] occurs when the umbra stops short of the Earth [E]. F shows how the tilt of the Moon's orbit [4] to the plane of the Earth's orbit [5] prevents an eclipse from occurring every month of the year.

2 As the Sun's disc reappears from behind the Moon after totality, there is a glorious "diamond ring" effect, as occurred in the total eclipse of 21 November 1966. It lasts for only a few seconds.

3 Just before totality begins, or just after it ends, at the moment of the "diamond ring" effect, the Sun's atmosphere is seen without the background photosphere. The dark absorption lines become suddenly bright emission lines, producing the "flash spectrum", a negative of which is shown here. The effect is brief, but it has been photographed many times and much information, particularly about the solar atmosphere, has been acquired.

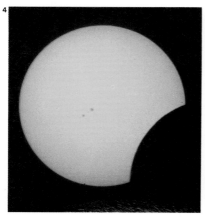

4 During the partial eclipse of 21 Nov 1966 several sunspots were seen on the disc.

5 Photographing an eclipse from an aircraft has advantages – there are no clouds and the aircraft can follow the Moon's shadow. Here the chromosphere and inner corona are clear.

6 The 1961 eclipse, taken with a long exposure, shows more of the outer corona, extending over a wide area. The inner corona and prominences are overexposed as a result.

With the development of spectroscopy as a research method, it became possible to study the chromosphere and the prominences at any time. The corona, however, is a more difficult feature to investigate because even its inner part is much fainter than the chromosphere. The French astronomer Bernard Lyot (1897–1952) developed a device known as a coronagraph, which can be used from high-altitude observatories to observe the inner corona. From Earth, the outer part remains unobservable except during totality. Knowledge of the Sun is therefore incomplete and attempts to increase this knowledge are not helped by the fact that certain radiations in the electromagnetic spectrum, including X-rays, can never penetrate to the surface of the Earth.

In the past, various methods have been worked out to overcome the screening effects of the Earth's atmosphere. For instance, balloons have been used, although they cannot fly as high as astronomers would like. The final solution was achieved when sophisticated equipment for studying solar X-rays was taken up in Skylab and used to good effect [7]. The corona was examined at all wavelengths and it is probably true to say that future total eclipses will be less important than those of the past. The first Skylab crew returned with exposed film of the corona, representing more hours of observations than had been acquired in the millennia of man's observations during natural eclipses.

Future studies of the Sun

Despite the newly acquired knowledge, many problems remain. A more complex coronal structure than was previously known has been revealed through Skylab ultraviolet experiments. X-ray pictures have also revealed low-density coronal regions, or coronal holes, which could be the source of disturbance in the solar wind.

Undoubtedly the future of the study of the Sun's outer surroundings lies in space-research methods. Skylab has shown the way and future orbiting stations, as well as equipment operated from the surface of the Moon, will tell man much more about his own particular star than he can claim to know at the present time.

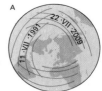

The 12 solar eclipses whose tracks of totality are shown here [A] belong to the same Saros family. The eclipse of 1991 will be total over Central America, but the next one in the cycle will be only partial there – that is, the "returns" are not exact. The eclipse of 7 March 1970 [B] was total over Mexico and Florida but the partial eclipse covered almost all North America. This made observation convenient.

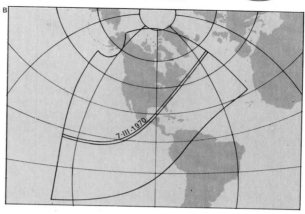

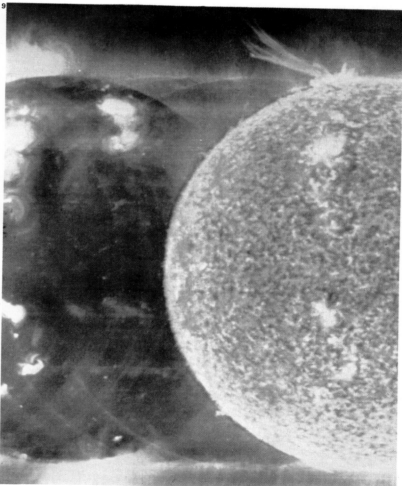

7 This X-ray image of the Sun's corona was taken by Skylab at the total eclipse of 30 June 1973. The Sun was near the minimum of its cycle of activity, and the corona was fairly symmetrical. The dark lane near the top is a "coronal hole".

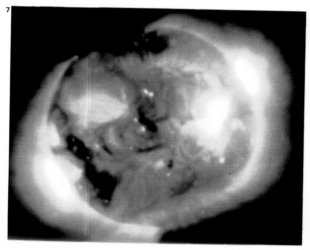

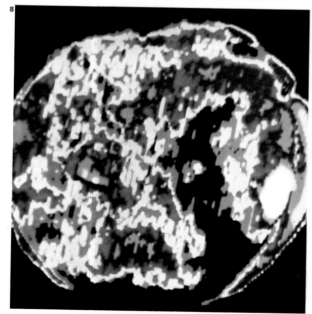

8 A large coronal hole is indicated by the dark stream in this "false colour", extreme ultra-violet picture of the Sun taken from Skylab. It illustrates that the structure of the corona itself is far from being perfectly uniform.

9 One of the most impressive events recorded during the studies of the Sun carried out from Skylab is shown in this picture. Film taken on 10 June 1973 showed a huge "blob" of tenuous material the size of the Sun. It was moving outwards through the corona at a velocity approaching 400km (250 miles) per second. The size can be appreciated from a comparison with the solar disc itself. During the mission more than 40 similar event were studied, but not all of them were the same size. Although the sizes of the "blobs" were so great, the actual amount of material involved was relatively slight because the material was so rarefied. It was the first time this type of event had been observed.

Star types

No star, apart from the Sun, is close enough to Earth to appear as anything but a point of light; studies of the Sun are therefore all-important in stellar astronomy. Moreover, the value of the telescope on its own is limited. Astrophysics mainly relies on instruments based on the principle of the spectroscope, which splits up light and gives information about the substances present in the light source.

Stellar spectra

The spectrum of the Sun was first studied by Isaac Newton (1642–1727) in 1666, but real progress was not made until the nineteenth century, mainly by Joseph von Fraunhofer (1787–1826), who mapped the dark absorption lines in the solar spectrum, still often called Fraunhofer lines. The lines were correctly interpreted by Gustav Kirchhoff (1824–87) and Robert Bunsen (1811–99) in 1859. Stellar spectroscopy, however, was a much more difficult matter, because so little light was available and spectroscopic equipment had to be used in conjunction with powerful telescopes.

Pioneer work, carried out largely by Angelo Secchi (1818–78) in Italy and William Huggins (1824–1910) in England, established that the stars may be divided into several reasonably well-defined spectral types. The system now adopted [3] is that drawn up at the Harvard College Observatory under the direction of Edward Pickering (1846–1919). The spectral types are given letters of the alphabet. In order of decreasing surface temperature the six main types are B, A, F, G, K and M; the complete sequence includes five more groups of rarer type, W, O, R, N and S, which denote somewhat different spectral characteristics. The sequence is alphabetically chaotic because there were several major revisions during the research period; types C and D, for instance, were found to be unnecessary.

The colour of a star is a key to its spectral type. Stars of types O, B and A are white or bluish-white; F and G, yellow; K, orange; and the rest orange-red. Subdivisions are given by figures; thus G0 is the hottest G-type star, G5 is midway in the sequence between G and K, and G9 is only slightly hotter

than K0 (the exact spectral class of the Sun is G2). Conventionally stars at the beginning of the sequence are referred to as "early" type stars and those near the end (types K, M, R, N and S) "late" type, although the Harvard sequence is no longer thought to be truly evolutionary. The situation is far more complicated than was once thought.

The Hertzsprung–Russell classification

In 1908 the Danish astronomer Ejnar Hertzsprung (1873–1967) drew up a diagram in which he plotted the stars according to their luminosities and their spectral types. Research of similar kind was being carried out in the United States by Henry Russell (1877–1957) and the diagrams produced are now known as Hertzsprung–Russell or H–R, diagrams [6]. They have proved to be immensely informative. Even a casual glance at an H–R diagram is enough to show that the stars are not randomly distributed all over it, although the H–R diagram does not, as was once supposed, mark a strict evolutionary sequence. Most of the stars in the diagram lie in a well-

1 If a camera is pointed at the night sky and a time exposure made without the camera being moved, stars will appear as trails because of the rotation of the Earth. The longer the exposure, the longer the trail. By making long exposures, such as the one shown, the different colours of the stars are more easily resolved. The hotter stars will appear blue or white, cooler stars yellow and still cooler stars as red trails.

2 The famous constellation Crux (the Southern Cross) was photographed from Rhodesia by J. McBain. The camera was attached to a driven telescope, so that the stars are shown as hard points and not as trails. Of the four main stars of the Cross, three are hot and white, but the fourth – Gamma Crucis – is a red giant and its colour is clearly shown here. The colours can be seen well through binoculars.

3 The Harvard classification of spectral type is illustrated for the six principal classes of stars: B, A, F, G, K and M. The spectrum for each category is shown together with the colour symbol that is repeated on the Hertzsprung–Russell diagram opposite. An example of a star in each class is given. B-type stars (Rigel): helium lines are prominent; 25,000°C surface temperature. A-type stars (Sirius): hydrogen lines are prominent; 10,000°C surface temperature. F-type stars (Polaris): calcium lines are prominent; 7,500°C surface temperature. The giant and dwarf division is appearing. G-type stars (the Sun): 5,700°C (giants) and 6,400°C (dwarfs) surface temperature. The giant and dwarf division is clear. K-type stars (Arcturus): 4,100°C (giants) and 5,100°C (dwarfs) surface temperature. M-type stars (Betelgeuse): 3,100°C (giants) and 3,500°C (dwarfs) surface temperatures; many are variable and advanced in evolution.

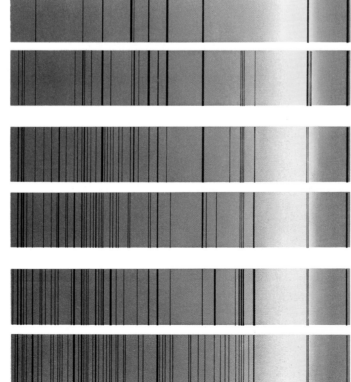

4 Magnesium lines in the green region of the Sun's spectrum were photographed by H. R. Hatfield with his spectrohelioscope. These magnesium lines lie at a wavelength of 5,170 Ångströms.

5 The double D line of sodium, also in the solar spectrum, was photographed by H. R. Hatfield. These are among the most prominent of all spectral lines and are well shown in many stellar spectra.

defined belt extending from the upper left to the lower right; this belt has become known as the main sequence; the Sun is a typical main sequence star.

It is also obvious that with the red and orange stars, and to a lesser extent with the yellow – that is to say, from types G to the end of the sequence – there is a sharp division into giants and dwarfs. Consider, for instance, two stars of type M: Betelgeuse in Orion and our nearest stellar neighbour Proxima Centauri. The surface temperatures are much the same, but this is the only point of similarity. Betelgeuse has a variable diameter of 420–560 million kilometres (260–350 million miles) – large enough to hold the entire orbit of the Earth – and a luminosity more than 10,000 times that of the Sun. The diameter of Proxima is less than one million kilometres (about 620,000 miles) and it has only one ten-thousandth of the Sun's luminosity [Key]. M-type stars with luminosities about the same as that of the Sun do not exist, as the H–R diagram shows. The discrepancy between giant and dwarf is rather less with earlier spectral types and

beyond type F it is more difficult to distinguish. (The white dwarfs, shown at the lower left of the H–R diagram, come into an entirely different category.)

The rarer classifications
Most of the stars lie in that part of the Harvard sequence from B to M. Stars of type W have high surface temperatures, of the order of 80,000°C and their spectra show bright emission lines, produced in the star's gaseous atmosphere. W-stars, also known as Wolf–Rayet stars, are rare; about 150 are known in our Galaxy and another 50 in the Large Magellanic Cloud. Allied to them are the O-type stars, with lower surface temperatures (about 35,000°C) and both bright and dark spectral lines. Zeta Orionis or Alnitak, for example, in Orion's belt, is of type O9.

At the other end of the sequence come stars of types R, N and S. All of them are remote, so that they appear faint, and almost all are variable. They are often called carbon stars because lines caused by molecules containing carbon are so prominent in their spectra. The reddest are of type S.

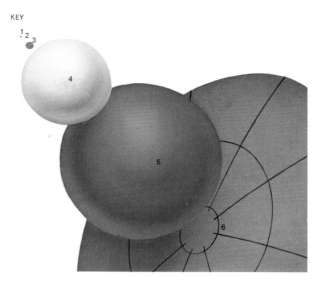

KEY

Stars vary in size, temperature and luminosity. Wolf 359 [1] is a faint red dwarf and Epsilon Eridani [2] is smaller and cooler than the Sun [3], and Rigel [4] is 50,000 times brighter. Aldebaran [5] is a red giant and Antares [6] is the largest red giant known.

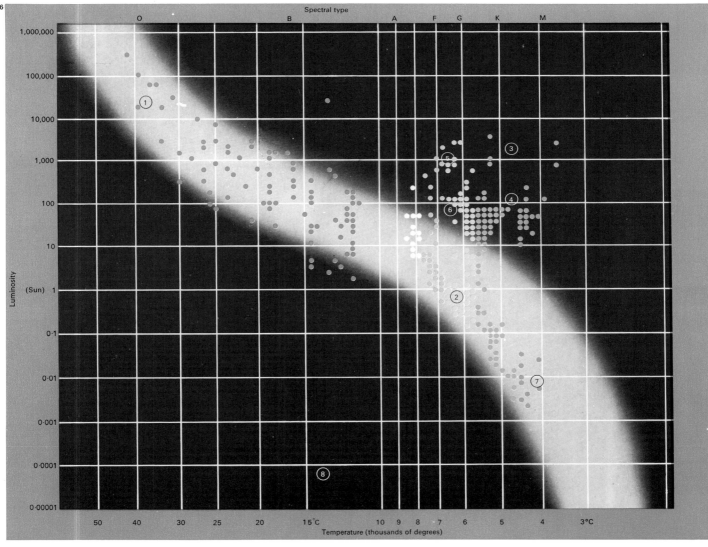

6 The Hertzsprung–Russell or H–R diagram is of fundamental importance. The stars are plotted on the graph according to their luminosities relative to that of the Sun, their spectral types and their surface temperatures. Most of the stars lie along the well-defined belt known as the main sequence. The main sequence extends from the upper left, with extremely hot O-type stars [1], through to G-type stars [2] such, as the Sun, and the red dwarfs of type M [7] of low luminosity. To the upper right lie the red supergiants [3] and the giant branch [4]. Also shown are the Cepheid variables [5] and the RR Lyrae variables [6]. To the lower left of the diagram are the white dwarfs [8]. Stars of types K and M are divided into giant and dwarf groups which can be clearly seen.

Stellar evolution

In the early years of the twentieth century, many astronomers assumed that stars evolved strictly along the course plotted on a Hertzsprung–Russell diagram [Key], starting as luminous white stars and ending as dim red ones. According to this theory, a star would begin by condensing out of interstellar dust and gas. Gravitational forces would shrink it so that the interior would heat up. The star would start shining as a large, diffuse red giant of type M1. It would continue to contract and heat up until it joined the top of the main sequence, moving down the main sequence until it became a faint M-type red dwarf. Eventually it would be transformed into a cold, dead globe.

Evolution of a star of solar mass

That plausible sequence of stellar evolution is now known to be completely wrong. A red giant such as Betelgeuse is not young. It is very old, has used up most of its energy reserves and is in an advanced stage of evolution. The stars are now known to shine because of nuclear reactions taking place inside them and the course of stellar evolu-

tion is known to depend largely upon the initial mass of a star when it is formed from the nebular material – a massive star evolves differently from a star of much lesser mass. The only common factor is that all stars begin their careers in gaseous nebulae of which the Orion Nebula M42 is unquestionably the best-known example.

As an embryo star shrinks it heats up, but if the mass is extremely low no nuclear reactions are able to start and the star never joins the main sequence. Instead it radiates feebly until its energy has been dissipated. In the case of a star with a mass about that of the Sun [3], a stage is reached, as the gravitational shrinkage continues, when heat is carried from the interior to the surface by convection. In a short time (perhaps only a century or so) the star becomes from 100 to 1,000 times as luminous as the Sun is today. After this initial burst of glory it continues to shrink and also becomes fainter – it is approaching the main sequence. Then when the core temperature has risen sufficiently, nuclear reactions begin. Hydrogen nuclei combine to form helium nuclei, resulting in a

loss of mass and the release of energy, and the star settles down on the main sequence to a long period of stable existence, lasting for perhaps as long as 10,000 million years. (The Sun, which is about 5,000 million years old, has thus reached the half-way stage in its main sequence career.)

Eventually the supply of available hydrogen "fuel" begins to run low and the star has to rearrange itself. The helium core contracts rapidly and is heated up once more, enabling hydrogen nuclei to "burn" in a shell surrounding the core while the outer layers expand and cool. The star swells out to become a red giant. The central temperature rises to about 100 million degrees centigrade, although the outer layers are cool and extremely rarefied.

White and black dwarfs

Further types of reactions follow, but at last there is no nuclear energy left and the star collapses into a small, dense white dwarf. Because the component atoms are crushed and broken they can be tightly packed and the star's density may reach more than

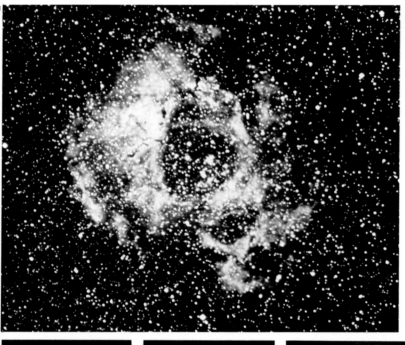

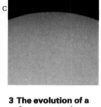

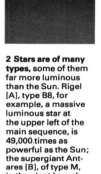

1 The Rosette Nebula, NGC 2237, is shown here in a photograph taken with the 48in (122cm) Schmidt telescope at Palomar, USA. The nebula lies in the constellation of Monoceros but is not a bright object. It is a typical emission nebula; the brightest star in it has a spectral type of O9. There is no doubt the nebula represents a region where fresh stars are being formed. Although the gas is so prominent, it is extremely rarefied and the formation of new stars within the region is by no means a rapid process.

2 Stars are of many types, some of them far more luminous than the Sun. Rigel [A], type B8, for example, a massive luminous star at the upper left of the main sequence, is 49,000 times as powerful as the Sun; the supergiant Antares [B], of type M, in the giant branch, has a diameter of 420 million km (260 million miles) and its luminosity is 3,400 times that of the Sun. Aldebaran [C], type K, 90 times more luminous than the Sun, is a less extreme red giant, with a diameter of 67 million km.

3 The evolution of a polar-type star is shown in this series of graphs. The star contracts out of the interstellar material [A]. It then joins the main sequence [B]. After perhaps 10,000 million years it leaves the main sequence and moves into the giant branch [C], increasing in luminosity to 1,500 times that of the Sun and expanding in diameter to 50 times that of the Sun. It then becomes unstable and matter is ejected [D]. Subsequently the star collapses into a small, extremely dense white dwarf [E]. The red line follows the star's evolutionary track.

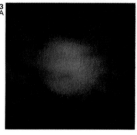

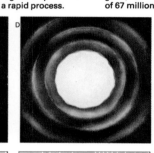
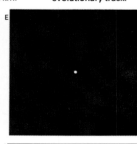
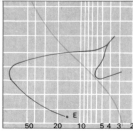

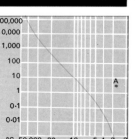
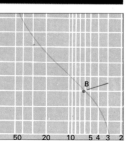
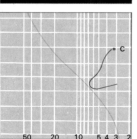
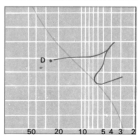

100,000 times that of water. After another long period all light and heat leaves the star and it becomes a dead black dwarf.

There is no positive information about black dwarfs, which send out no detectable radiation and we can only hazard guesses about their numbers. White dwarfs, however, are known to be common. The first to be identified (by Walter Adams [1876–1956] in 1916) was the companion of Sirius, which had been discovered by Alvan Clark (1832–97) more than half a century earlier, but had always been assumed to be cool and red. The surface temperature of the companion is greater than that of the Sun, but its diameter is only three times that of the Earth so that an immense amount of matter – almost as much as is in the Sun – is packed into a relatively small globe. Other white dwarfs since found are even more dense.

Evolution of a massive star
A star with a mass much greater than the Sun's [6] evolves much more rapidly. The luminous S Doradûs in the Large Magellanic Cloud, for example, cannot go on pouring

forth energy at its present rate for much longer than a million years, whereas the Sun will not leave the main sequence for at least another 5,000 million years and stars of lower mass change even more slowly.

Very massive stars do not merely collapse into white dwarfs. When the core temperature has reached about 5,000 million degrees centigrade there is a catastrophic change in structure; the core collapses and the outer layers of the star, in which nuclear reactions are still going on, are abruptly heated to about 300 million degrees centigrade. The result is a supernova outburst in which the star emits as much energy in a few seconds as the Sun does in millions of years. Material is ejected and when the convulsions are over all that remains is a cloud of expanding gas together with a neutron star or pulsar, even smaller and denser than a white dwarf. The Crab Nebula is a supernova remnant; the outburst was watched by Chinese observers in 1054. Of the two nebulae shown here, the Rosette [1] represents a stellar birthplace while the Crab [4] shows the death of a once glorious star.

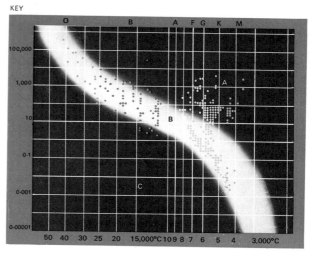
KEY

This Hertzsprung–Russell diagram in simplified form shows the main classes into which stars are grouped during their evolution. Typical stars in the giant class [A] have evolved out of the main sequence [B]. At the lower left [C] are white dwarfs. Luminosity is measured vertically on the diagram and spectral type across the top.

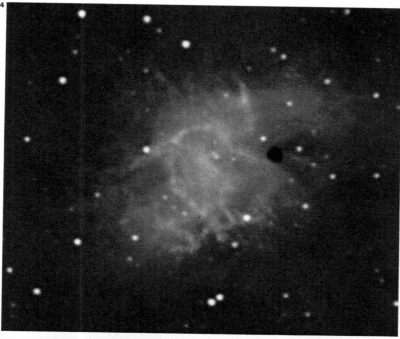

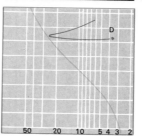

4 The Crab Nebula in Taurus, number one in Messier's catalogue, is an exceptional object. The supernova of 1054 became so bright that it could be seen in daylight with the naked eye, but after it faded below the sixth magnitude it was inevitably lost. The association between the 1054 supernova and the modern Crab Nebula has been questioned but seems to be no longer in doubt. The Crab contains the only pulsar to have been identified with an optical object; the pulsar has been termed the Crab's "powerhouse".

5 Stars on the main sequence, such as the Sun, are classed as dwarfs to distinguish them from members of the giant branch. The Sun [A] is a typical G-type main sequence dwarf. Capella is also of type G, but on the giant branch, with a luminosity 150 times that of the Sun. It is quite different from a white dwarf such as the companion of Sirius [B] which has collapsed, presumably from the giant stage, or a red dwarf such as Wolf 359 [C], one of the feeblest stars known, with a luminosity 0.00002 times that of the Sun.

6 The evolution of a massive star – that is, one with an initial mass of more than three times that of the Sun – is shown. The star contracts out of the interstellar material [A] and joins the main sequence [B]. After a period that is much shorter than for a solar-type star, it moves into the giant region of the H–R diagram [C], "burning" first helium and then heavier elements. Eventually it experiences a supernova explosion [D] and sends most of its material outwards, leaving a neutron star or pulsar [E].

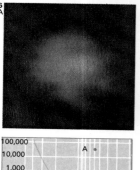

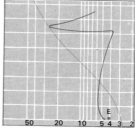

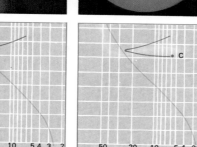

Galactic nebulae

Nebulae (clouds) are of various kinds and have proved to be of the utmost importance in modern astronomical theory. They appear in several parts of the sky as luminous patches that look like shining mist.

Catalogues of nebulae

Several catalogues of nebulae have been drawn up by astronomers over the years. One of the most famous is that published by the French astronomer Charles Messier (1730–1817) in 1781. It contains more than 100 objects. Ironically, Messier was not in the least interested in nebulae; he was a comet-hunter who compiled his catalogue of nebular objects in order to avoid confusing them with possible new comets. In the late nineteenth century an extensive catalogue based on the observations of William Herschel (1738–1822) and his son John (1792–1871), was compiled by the Danish astronomer Johan Dreyer (1852–1926). It is known as the New General Catalogue (NGC). Today the NGC numbers and Messier's numbers (M) are still used.

Messier catalogued all the nebulous objects, from star clusters to gaseous nebulae and to the systems such as the Andromeda Spiral, M31, that we now know to be galaxies. Astronomers have now agreed that the term "nebula" should be confined exclusively to clouds of gas or dust to avoid confusion.

Galactic nebulae are of two main kinds: emission and reflection. Both types occur not only in our own Galaxy, but in others. The so-called Tarantula Nebula lies in the Large Cloud of Magellan – 30 Doradûs (listed by Dreyer as NGC 2070) – and is much larger than the Orion Nebula, M42 [Key, 8], which is the most famous nebula in our Galaxy. The main constituent of all nebulae is hydrogen, which is the most abundant substance in the entire universe, but there is also a great deal of dust and it is this that absorbs starlight. Inside some of the nebulae are objects that cannot be seen, but can be detected by infra-red photography; Becklin's Object in the Orion Nebula is an excellent example. It may well be a star of tremendous luminosity, but it is permanently concealed from view.

Vast though they are, the nebulae are made up of extremely tenuous material. The gas is many million times less dense than the air we breathe. It has been calculated that if a 2.5cm (1in) core sample could be taken right through the Orion Nebula the total weight of material collected would be no more than that of one small coin.

Luminosity of nebulae

A nebula depends for its luminosity upon the presence of stars that are either close to it or are contained in it. If the stars are extremely hot, the hydrogen in the nebula is ionized and emits a certain amount of light of its own [Key, 2, 3, 8]. (Certain spectral lines in nebulae were once thought to indicate the presence of an unknown element, but it was subsequently found that the lines are due to familiar elements, such as oxygen, produced under unfamiliar conditions.) If the stars are less hot, the nebula shines only by reflection [5, 6]. If there are no suitable stars, the nebula does not shine at all; it remains dark and can be detected only because it blots out the light of stars beyond [5, 7]. There are various galactic nebulae within the range of small telescopes, although the vivid colours

CONNECTIONS

See also
226 Star types
228 Stellar evolution
232 From nebulae to pulsars
234 Pulsars and black holes

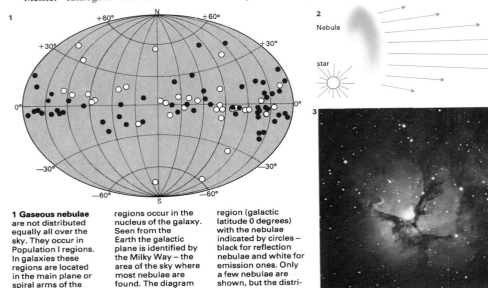

1 Gaseous nebulae are not distributed equally all over the sky. They occur in Population I regions. In galaxies these regions are located in the main plane or spiral arms of the galaxy. Population II regions occur in the nucleus of the galaxy. Seen from the Earth the galactic plane is identified by the Milky Way – the area of the sky where most nebulae are found. The diagram shows the Milky Way region (galactic latitude 0 degrees) with the nebulae indicated by circles – black for reflection nebulae and white for emission ones. Only a few nebulae are shown, but the distributions are clear.

2 An emission nebula emits light characteristic of the substance of which it is made up, if the nebulosity is illuminated by a suitably placed star. A star of the type W, O or B can cause ionization of the hydrogen atoms out to about 500 light-years, which is known as an H-II region. If the temperatures of the associated stars are too low there will be no emission from the nebula.

3 A famous emission nebula is the Trifid in Sagittarius, M20 (NGC 6514), shown in this Palomar photograph. Visible in a small telescope, it is 30 light-years across and more than 3,000 light-years away.

4 The main galactic nebula in Sagittarius is M8 (NGC 6523), known as the Lagoon Nebula. Described by John Flamsteed (1646–1719) in 1680, it is easy to view telescopically because the integrated magnitude is 6.0. M8 is a dense nebula with 10^3–10^4 atoms per cubic centimetre in the central region. It lies some 4,850 light-years from Earth. Associated with it is the galactic star cluster NGC 6530. M8 contains a number of T Tauri variables and also some dark globules that may eventually start to shine. Each globule has a diameter of about one light-year. Flare stars in M8 are also known. Moreover, M8 is a source of radio emission.

of the photographs below cannot be seen visually. The colours themselves are genuine, but are too faint to impress the eye.

According to a classification produced by Walter Baade (1893–1960), there are two kinds of regions in our Galaxy (and in other galaxies): he called these Populations I and II. In Population I areas [1] there is a great deal of interstellar material and the brightest stars are hot and white. In Population II areas the interstellar material has largely been used up in star formation and the brightest stars are red giants. Because these stars are well advanced in their evolutionary careers, Population II regions seem to be relatively old. Gaseous nebulae occur in Population I areas, so their stars are presumably young by cosmic standards.

The formation of stars
The most important feature of Population I areas from the theoretical point of view is that they are apparently regions in which star formation is in progress. According to current ideas, a star begins its career by condensing out of interstellar material. Nebulae are obvious sites for such activity because the material in other regions of space between the stars is much too tenuous. On average, interstellar space contains one atom of matter per cubic centimetre; nebulae, although rarefied, are more condensed than this. Objects such as the Orion Nebula, the Lagoon Nebula [4] and the Trifid Nebula [3] are in fact stellar birthplaces. The same is true of galactic nebulae in other systems, such as the Large Cloud of Magellan and the nebulae observable in the Andromeda Spiral. Dark patches in nebulae, known as globules, may well be embryo stars.

Nebulae also contain many stars that are variable in light and are unstable; these are known as T Tauri variables and are thought to be stars at an early stage in their careers that are still contracting towards the main sequence. Some stars have even been seen to increase in luminosity over a period of years, presumably because they have blown away their original dust clouds. One of these is FU Orionis, in the Orion Nebula, which became brighter in 1936 and must be one of the youngest stars known to us.

The Sword of Orion [M42] is the most spectacular of the gaseous nebulae.

Within the nebula the famous multiple star, the Trapezium, responsible for the nebula's luminosity, is clearly visible. The brightest star is Theta Orionis.

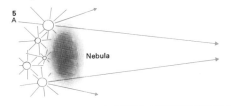

5 A dark nebula [A] cuts out the light of stars that lie at a greater distance from Earth. The Coal-sack [B] in the Southern Cross is the best example. The light of more distant stars is completely obscured, because light is absorbed by the nebula's solid particles, not by interstellar gas.

6 A bright nebula [A] shines by reflecting the light of a suitable star. The Pleiades cluster [B] in Taurus is an example of a reflection nebula. Such nebulae usually have a high dust content. The nebulosity in the Pleiades is best examined by using long-exposure photographs.

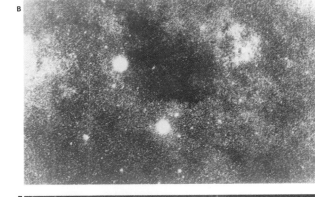
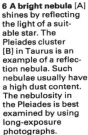

7 The Horse's Head Nebula in Orion lies close to Zeta Orionis, the most southerly star of the Belt. This dark nebula, shaped rather like the head of a knight in chess, is No 2024 in the New General Catalogue. It can be seen with a small telescope, although it is best studied in photographs such as this Palomar picture.

8 The bright Orion Nebula to the south of the Belt in the "Hunter's Sword" can be seen with the naked eye. Its luminosity is due chiefly to the multiple star, Theta Orionis, on the side of the nebula turned towards the Earth. If this star did not exist, the nebula would be just as dark as the Horse's Head Nebula.

231

From nebulae to pulsars

Until fairly recent times it was not generally realized how different the various types of nebulae are. Looking at the Omega Nebula in Sagittarius and then the Crab Nebula in Taurus, for instance, leads to the conclusion that they are much the same; but in fact they represent diametrically opposite ends of the stellar evolutionary sequence. In Omega – a diffuse nebula – stars are being forced out of the interstellar material; the Crab represents supernova explosion debris at the centre of which is a pulsar – a collapsed dense star.

Nebulae: the early stages

One interesting nebula is associated with the young star T Tauri, which is irregularly variable and is still contracting towards the main sequence. The nebula was discovered in 1852 by J. R. Hind, an English amateur who was using a 7in (17.8cm) refractor to hunt for asteroids and merely noted the T Tauri nebulosity in passing. Nine years later it was found that the nebula had disappeared. It has since been seen again and is within range of large telescopes (its official designation is NGC 1554), but it is not nearly as prominent

as it was when Hind discovered it. Moreover T Tauri itself is not hot enough to excite the nebular material to self-luminosity; it is, however, an infra-red source and there is no doubt that it is associated with the nebulous material around it from which it has been formed. There are other variable nebulae similarly associated with young stars; for instance, R Monocerotis in the Orion area and R Coronae Australis in the southern sky.

These, then, are nebulae involved in stellar birth. So are the familiar gaseous or galactic nebulae, such as M42 in Orion. Deep inside M42, permanently hidden from Earth by the nebular material, is an infra-red source known as Becklin's Object. It may be either a very young star or else an extremely powerful object at least a million times as luminous as the Sun; there is, however, no way of knowing because it is possible to study only its infra-red radiation, which can pass through the nebula and reach the Earth. In any case, star formation is in progress in the Orion cloud.

A. Blaauw and W. W. Morgan, in America, have studied an O-type star, AE

Aurigae, which has the high velocity of about 130km (80 miles) per second. It looks faint only because it is remote; it is in fact a luminous young star. Tracing its path "backwards", so to speak, indicates that about two and a half million years ago it was in the region of the Orion Nebula [1]. Moving in the opposite direction is another O-type star, Mu Columbae, which has similar velocity and is about equally far from the nebula. It has been suggested that a colossal disturbance probably hurled these two stars violently outwards from their place of birth.

Planetary nebulae

Other nebulae represent later stages in the evolution of a star; in particular there are the planetaries, which look like small, feebly luminous discs or rings, not unlike those of planets. The planetaries, like the diffuse nebulae, are gaseous, but they are neither planets nor nebulae so their popular name could hardly be less apt. The best known is the Ring Nebula, M57 Lyrae [4], which was discovered in 1779. It is made up of a central star surrounded by a spherical gaseous shell

1 The Orion Nebula, a stellar birthplace, is the most famous of all gaseous nebulae. This photograph was taken with the 200in (508cm) Hale reflector at Palomar, USA. The "hollow" in the border to the right-hand side is due to the presence of the multiple star Theta Orionis, which lies close to the Earth-turned edge of the nebula. There is no doubt that fresh stars are condensing out of the nebula.

2 The gaseous nebula M16 (NGC 6611) lies near the boundary between Sagittarius and the small constellation of Scutum, which adjoins the "tail" of Aquila. M16 has an integrated magnitude of 6.4, so that it is visible with binoculars; the distance from Earth is 5,900 light-years. Seen through a foreground of stars the nebula, photographed in red light, shows both bright and dark nebulosity.

3 The Omega Nebula M17 (NGC 6618), sometimes known as the Horseshoe Nebula, was discovered by the French astronomer L. de Chéseaux in 1746. It is an easy binocular object, 1.5°N and 2°E of the 5th-magnitude star Gamma Scuti. It is more massive than the Orion Nebula; like so many other diffuse nebulae it has bright areas as well as signs of dark obscuring material. M17 lies on the borders of the constellations of Sagittarius and Scutum.

that is incredibly tenuous. Looking at it, more is seen of the glowing material at the edge than at the centre so that the nebula looks like a ring. The diameter of M57 is almost one light-year but the gaseous surround is immensely rarefied, millions of times less dense than the Earth's air at sea-level. Some planetary nebulae are larger; for instance NGC 7293 Aquarii [5] is twice the size of M57. Other planetaries are asymmetrical; these include the Owl Nebula, M97 Ursae Majoris, and the Dumbbell Nebula, M27 in Vulpecula.

All planetaries are expanding and their age can hardly be more than a few tens of thousands of years; it has been estimated that if the gaseous shell is ejected from an old star – as is likely to happen – the material cannot continue to shine for more than 100,000 years or so. According to one theory, a planetary is produced by a red giant star "puffing off" its outer layers, so that the central stars in planetary nebulae represent the cores of old giants. These stars have high surface temperatures of about 50,000°C and have completed their main nuclear burning;

they are well on the way to becoming white dwarfs. The "puffing out" theory fits in well with the proposed evolutionary sequence, although it is by no means certain that every normal star inevitably becomes a planetary nebula at a late stage in its life.

Supernovae and pulsars

Finally there are nebulae that represent the end products of stellar evolution. Although the Crab Nebula is the best known example, there are others but almost all are much older than the Crab and so their forms are not as well marked. (In any case, the Crab, with its unusual central pulsar, seems to be an exceptional object.) With the Veil Nebula [6], in Cygnus, the arched shape of the luminous material is plain and all the evidence points to the conclusion that it is the debris of a supernova outburst that took place in prehistoric times. The present rate of expansion is 120km (75 miles) per second. There is thus a full sequence of nebulae, from those of the T Tauri type associated with stellar birth through to the stellar remnants of supernovae explosions.

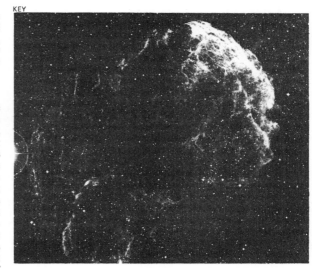

This gaseous nebula in Gemini is probably associated with the death of a star, because it gives every impression of being the remnant of a supernova outburst. Other types of nebulae are the birthplace of stars.

4 M57 (NGC 6720), the Ring Nebula in Lyra – the most famous of the planetaries – lies midway between the two naked-eye stars Beta and Gamma Lyrae. The variation in colours is due to temperature differences. The integrated magnitude is 9.3 and the distance from Earth 1,400 light-years. The central star, although clearly shown, is by no means bright; the second star [upper right] is in the foreground and not connected with M57.

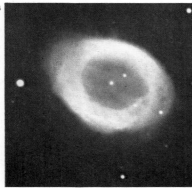

5 The planetary nebula NGC 7293, in Aquarius, was surprisingly left out of the famous catalogue of nebula objects drawn up by Messier; yet it is actually the brightest of all planetary nebulae and is definitely superior to the Ring Nebula in Lyra. It is 600 light-years from Earth. The central star is of magnitude 13.3 and is clearly shown in this photograph taken with the 48in (122cm) telescope at Palomar.

6 The Veil Nebula in Cygnus, NGC 6992 – sometimes called the Cirrus Nebula – may be a supernova remnant. This photograph (48in [122cm] Schmidt telescope, Palomar) shows the arched shape, which is significant. The nebula is 2,500 light-years away and from the motions in the gas it has been calculated that the supernova outburst occurred 50,000 years ago. In 25,000 years' time the nebula will cease to be luminous.

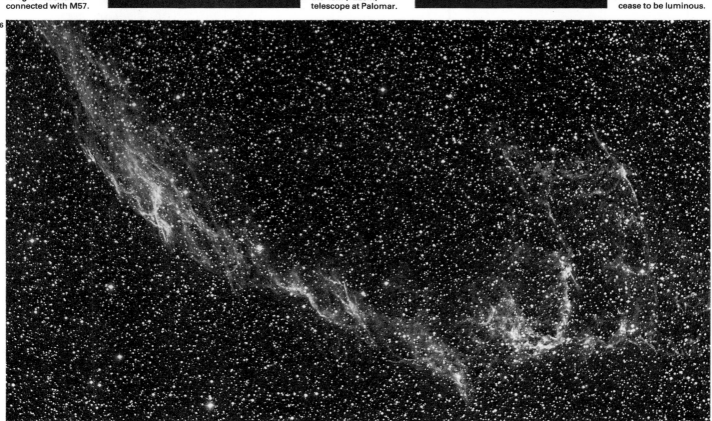

Pulsars and black holes

To explain the nature of a black hole, it is necessary to examine stellar evolution [1]. A star such as the Sun contracts towards the main sequence. When the core temperature has risen sufficiently, nuclear processes begin. After the supply of available "fuel" has run low, the star swells to become a red giant, after which it collapses to the white dwarf condition.

A more massive star will behave differently and when its nuclear reserves fail it will explode as a supernova, ending its luminous career as a neutron star, or pulsar, together with a cloud of expanding gas.

From white dwarf to black hole
In a white dwarf the atoms are crushed, broken and packed together so tightly that there is little waste space. In a neutron star the gravitational field is so intense that the protons and electrons are forced to combine with each other to form neutrons; the density of neutron star material far exceeds that of a white dwarf. There is now little doubt that the radio sources known as pulsars are really neutron stars. The pulsar in the Crab Nebula

has been identified with an optical object [Key, 3], and in 1977 the Australians identified another pulsar – in the constellation of Vela – with an extremely faint object of magnitude 26.5.

If a massive star collapses, it can pass through densities that correspond to the white dwarf and neutron star stages and still continue to contract, becoming steadily smaller and denser and entering a state of gravitational collapse where no known physical process can halt the contraction. Light will find it more difficult to escape and soon the body will contract within a critical radius (known as the Schwarzschild radius) at which point its gravitational field will become so strong that not even light will be able to move away from it. The star is then surrounded by what might be termed a "forbidden zone" from which nothing can escape. This is a black hole – a region that acts as a centre of gravitational attraction.

Inside a black hole, all the normal laws of physics break down. It has been suggested that the collapsed star may eventually be crushed out of existence altogether. And

there have been sensational predictions that black holes may extend until they swallow everything in the universe. But ideas of this kind are, at best, highly speculative.

In search of black holes
The most promising place to look for a black hole is in a binary system. Close beside the brilliant yellow star Capella is a small triangle of stars, known popularly as the Haedi or Kids. At the apex of the triangle is Epsilon Aurigae, which is always clearly visible to the naked eye although it is by no means brilliant. In 1821 it was discovered to be variable, with a magnitude range of 3.3 to 4.2. Later it was found that Epsilon Aurigae is an eclipsing binary of unusual type, for the eclipses take place only once in 27 years and last for more than 700 days.

The brighter member of the pair is a highly luminous yellow supergiant, 60,000 times as powerful as the Sun. The fainter component, which causes the eclipses, has never been seen; it radiates only in the infra-red and until recently all astronomers thought that it must be a large, cool star, still

1 The relative sizes of a giant, the Sun, a white dwarf, a neutron star and a black hole are shown. The ratios are given for each diagram; thus the diameter of the Sun is approximately 100 times that of a white dwarf. The neutron star has the same mass as that of the Sun.

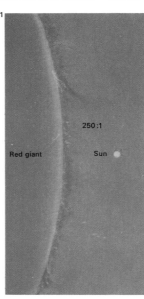

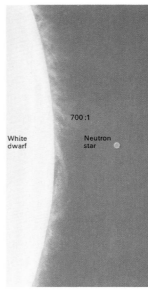
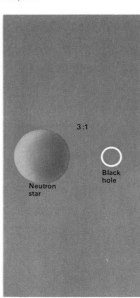

Red giant — Sun — 250:1 — Sun — White dwarf — 100:1 — White dwarf — Neutron star — 700:1 — Neutron star — Black hole — 3:1

2 A pulsar's radiation varies. The red on the diagram indicates the ends of the magnetic axis of a pulsar. As the pulsar rotates, the signal strength varies according to the position of the axis. When one end faces Earth [1], the intensity is at its maximum. When the other end faces Earth [2], intensity is at its minimum.

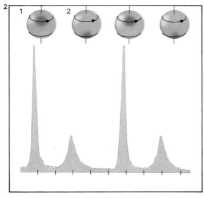

3 The Crab Nebula pulsar is of great importance to astronomers, for so far it is one of only two to have been identified with an optical instrument. These photographs were taken at the Lick Observatory, California, with its 120in (305cm) reflector. The pulsar is easily identifiable in A, but in B it is almost completely invisible. The whole pulse cycle amounts to only 33 milliseconds. There is now little doubt that pulsars are, in fact, neutron stars.

Time (seconds) — 0·010 — 0·020 — 0·030

200MHz — 400MHz — Optical — X-ray

4 Radiation at all wavelengths is emitted by the Crab pulsar. The pulse cycle at a wavelength of 200MHz in the radio range is shown [A] and at 400 MHz [B]; in the optical range [C] and in the X-ray range [D]. The beam of radiation has similar characteristics over the whole electromagnetic spectrum.

shrinking after its condensation from interstellar material, but not hot enough to shine by nuclear power. Now, however, there are suggestions that the infra-red member of the Epsilon Aurigae system may be a black hole.

This infra-red component seems to have a mass 23 times that of the Sun, which by stellar standards is high. It should therefore be luminous – but it is not. According to the American astronomers A. G. W. Cameron and R. Stothers it is a black hole, surrounded by a cloud of solid particles that are spiralling round the critical boundary – or "event horizon", as it is termed [7] – and are sending out the infra-red radiation detected from Earth. In time the particles will cross the event horizon and will enter the black hole, from which they never emerge.

X-ray sources
Another possible black hole is the companion of a supergiant star in Cygnus, known by its catalogue number, HDE 226868. The companion is a source of X-rays and it has been suggested that these are produced by material falling in towards the black hole and being accelerated to extremely high velocities.

X-ray astronomy is a recent development, because it involves sending equipment above the shielding layers of atmosphere. The technique dates back to the 1960s. Many X-ray sources have been found, one of which is the Crab Nebula [Key]. Apparently most galactic X-ray sources are members of binary systems and are neutron stars associated with giants. There are also what are termed X-ray novae, which flare up, last for weeks or months, and then fade away.

Most of the X-ray sources are members of our Galaxy and lie reasonably near the main plane of the Milky Way, but some other galaxies also emit X-rays, notably the massive system in Virgo known as Messier 87. It is also a source of radio emissions.

Progress in astronomy has been amazingly rapid during the past few years. In 1960 quasars and pulsars had not been detected and black holes were only of theoretical interest; X-ray studies had scarcely begun and even radio astronomy was primitive by modern standards.

The Crab Nebula (M1, NGC 1952) is the remnant of a supernova in 1054. It contains a pulsar, said to be the "powerhouse" of the Crab, which is 6,000 light-years distant.

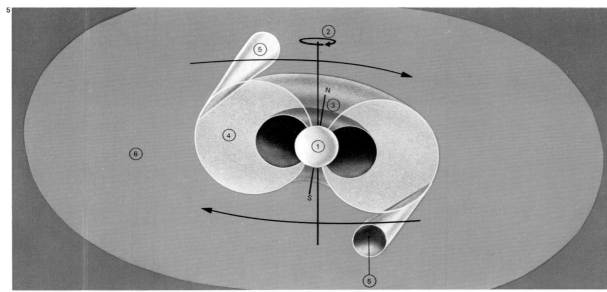

5 **When pulsars were discovered**, it was thought that the signals might come from rotating white dwarf stars. Now it is certain that a pulsar is a rotating neutron star [1], whose axis of rotation [2] does not coincide with the magnetic axis [3]. Near the star the plasma rotates [4], sending out radio waves in beams [5]. Beyond this the plasma is stationary [6]. It is now agreed that it is the magnetic field of the rotating neutron star that generates the pulses as it turns over and over. The mechanism is related to the region some distance from the neutron star where the magnetic field would have to travel at the speed of light to keep up with the rotation.

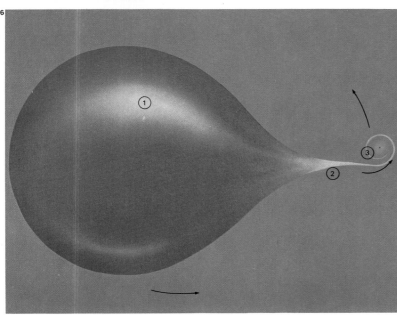

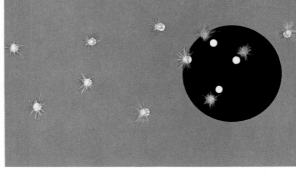

6 **The blue supergiant** [1], it is believed, has material pulled away from it in a jet [2] by the black hole [3], which is so small, it can be regarded as a point mass. Because of its small size, the black hole takes in material only slowly. Before this, the material is compressed and subject to great heat, so emitting powerful X-rays. The invisible companion of HDE 226868 in Cygnus may be a black hole with a diameter of about 100km (60 miles). The supergiant has a diameter 300,000 times larger.

7 **The "event horizon"** is the boundary of a black hole [black area]. The origins of some light sources are shown with locations of each light a moment later [white circles], depicting how gravity effectively bends light as it "falls" into the black hole.

Double stars

Our Solar System is centred on a single star, the Sun, but the universe contains many stars that appear to be close pairs or even members of complicated systems.

These double stars are surprisingly common, but are not always exactly what they seem. Some are indeed binary, or physically associated systems; the association of others is an illusion, the result of a mere line-of-sight effect. If two stars, as seen from Earth, happen to lie in much the same direction, they will appear to be side by side in the sky, even though there is no real connection between them. An example is Vega, the brilliant blue star in Lyra. It has a twelfth magnitude companion that is much farther away, although, from a terrestrial point of view, it is apparently close.

Binary stars and their structure

It was originally thought that all double stars must be the result of a line-of-sight effect. Not until 1793, and the observations of William Herschel (1738–1822), were true binary pairs discovered. In a binary system, the two components move around their common centre of gravity. For some pairs, the period of revolution is short – in extreme cases, less than 20 minutes – while for others it is long.

Gamma Virginis, not far from Spica, is made up of two exactly equal components with a revolution period of 180 years. The angular separation is now less than it was earlier in this century because the two stars are moving closer to the same line of sight. The pair used to be separable with any small telescope, but by AD 2016, when the apparent distance between the two will be at its minimum, Gamma Virginis will appear to be single, except in giant telescopes.

Mizar [Key], and its companion Alcor, in Ursa Major, is a particularly easy binary to spot – it was the first double star discovered by telescope. Like Alpha Centauri, it has two rather unequal components, one of magnitude 2.4, the other of 3.9.

Some pairs, such as Gamma Arietis, have the same spectral type for both components, but others are distinguished by their beautifully contrasting colours. Antares, the brilliant red star in the Scorpion, has a faint green companion and the same is true of the red giant star Alpha Herculis [3, 4]. But perhaps the best example is Beta Cygni [8, 9], or Albireo, which has a golden-yellow primary and a greenish-blue companion.

Spectroscopic and eclipsing binaries

If the separation between the components is slight, the binary will appear single. However, the revolution of the two components round their common centre will show up in the spectroscope [10]. The brighter component of the Mizar pair is a spectroscopic binary system.

There are also systems made up of more than two stars. Alpha Centauri [1, 2], for instance, the closest of all the bright stars, is made up of two rather unequal components (of magnitudes 0.0 and 1.7) and has a revolution period of 80 years. Closely associated with it is Proxima Centauri, making Alpha into a triple star. Proxima is the nearest star to Earth, but it is much less bright than Alpha. Epsilon Lyrae, near Vega, is an example of a wide pair, each component of which is again double. Castor, in Gemini, is a sixfold system, in which four of the components are bright

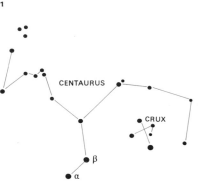

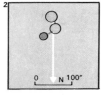

1 Alpha and Beta Centauri are the pointers to the Southern Cross, but are too far south to be seen from Europe.

2 Alpha Centauri is a triple star, visible through a telescope.

3 Hercules is not a brilliant constellation. However it contains three interesting features: the great globular cluster M13, just visible to the naked eye, and the two binaries, Rasalgethi and Zeta.

4 Alpha Herculis is a red giant (magnitude 3-4); its companion is greenish.

5 Zeta Herculis is another binary with unequal components (magnitudes 3.1, 5.6); its period is 34 years.

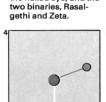

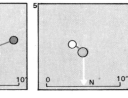

6 The pattern of Orion [A, B] shows two brilliant stars, Betelgeuse and Rigel. Rigel is one of the many double stars in Orion, which also contains the nebula M42, visible to the naked eye.

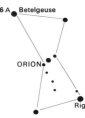

7 The multiple star Theta Orionis, nicknamed the Trapezium, is contained in the outer part of he Great Nebula. All the components are of spectral type O and presumably had a common origin. They are visible with a small telescope.

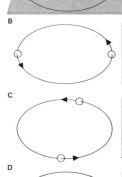

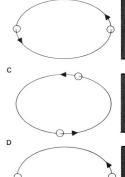

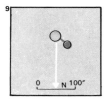

8 Cygnus is one of the richest of all constellations because it lies in the Milky Way. Its leading star, Deneb, is a highly luminous giant. The constel-lation takes the form of a cross, but the symmetry is spoiled by Albireo (Beta Cygni), a double with a yellow primary and also a greenish companion.

9 The golden-yellow primary of Beta Cygni is of type K, with an absolute magnitude of −2.2. Its companion is of magnitude 5.4. Despite their great separation these are still binary stars.

10 Analysis of a spectroscopic binary assumes that the stars are equal in mass, thus moving in circular orbit round their common centre of gravity [A]. The Earth, many light-years away, lies in the plane of their orbit. The stars move transversely to the line of sight from Earth [B]. Then the lower star moves towards the Earth and its spectral lines shift to the blue (or violet); the upper star shows a red shift and moves away and the lines in the combined spectrum appear double [C]. The stars again move transversely and the lines merge [D]. The sets of lines again shift in opposite directions [E]. Periodic doubling of lines indicates a binary.

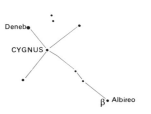

and the other two are dim red dwarfs; it is made up of two spectroscopic binaries and a third, much fainter companion star that is also a binary.

During the revolution of the two stars of a binary system, one component may pass behind the other, either totally or partially. When this happens, the light visible from Earth will be reduced and the star will seem to give a long, slow "wink". The prototype of these eclipsing binaries is Algol (Beta Persei) [11, 12], in which eclipses occur every 2.87 days and the magnitude drops from 2.2 to 3.5. Minimum magnitude lasts for 20 minutes and each fade – and recovery – takes five hours. Many stars of the Algol type are known. With Beta Lyrae [13, 14], near Vega, the components are close and less unequal. As a result there are two well-marked minima during the total period, which amounts to 12.9 days. Some eclipsing binaries have short periods – Delta Librae's, for example, is only 2.3 days. Others have long periods – 972 days for Zeta Aurigae, near Capella, and as much as 27 years for Epsilon Aurigae in the same region [15, 16].

It is evident that there is no essential difference between an eclipsing binary and an ordinary system; everything depends on the angle from which observations are made. Seen from a different angle, Algol would not exhibit any eclipses at all and would appear constant in light.

The importance of binary stars

Binary pairs were formerly thought to be the result of the fission or breaking up of a single star that rotated so quickly as to become unstable. However, it is now thought more likely that the components of the binary stars are formed separately in the same region of space and at the same time.

Binary stars make an important contribution to the total fund of astronomical knowledge. Measuring the mass of a single star is difficult, but observation of the components' orbital movements enables astronomers to estimate the combined mass of binary systems. Eclipsing binaries provide further opportunities for collecting data; study of their light-curves gives valuable insights into the diameters of the components.

KEY

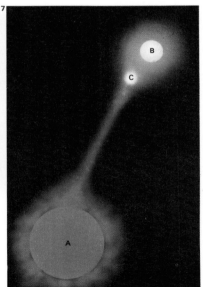

The 7 stars of Ursa Major [A] – the star pattern nicknamed the Plough – include Mizar, magnitude 2.4, or Zeta Ursae Majoris, in the Bear's tail. Naked-eye observation on a clear night will reveal Alcor, magnitude 3.9, apparently close beside Mizar. Through a telescope the binary system is seen clearly: there is an obvious difference between the primary and secondary components [B]. The system is further complicated – as the results of spectroscopic research show – for Mizar itself is another binary. Alcor is a true member of the system, but it is so far from Mizar that orbital revolution takes millions of years.

11 Perseus contains some fairly bright stars and has a distinctive shape. Algol lies in the southern part of the constellation and has much fainter stars (Kappa and Rho) on either side of it.

12 The Algol binary system has a small, bright star of type B8 (yellow) and a large, fainter star of type K (orange). When the brighter star passes in front of the fainter one [1], the total light re-ceived drops imperceptibly. When the two stars shine together [2], the light is constant. The main minimum occurs when the fainter star passes in front of the brighter one [3].

13 The brilliant blue star Vega, 26 light-years away, lies in the small constellation of Lyra. Fifty times as luminous as the Sun, Vega dominates the whole region. Close by is the quadruple star Epsilon Lyrae and a red, semi-variable, R Lyrae. Beta Lyrae or Sheliak, the eclipsing binary, makes up a pair with its neighbour, the third-magnitude Gamma Lyrae.

14 The components of Beta Lyrae are so close that they almost touch. Although they cannot be seen separately and are surrounded by complicated gas-clouds, it is known that they must be egg-shaped because of their nearness. Unlike Algol, Beta Lyrae is always varying – there are two minima, one of magnitude 3.8 and the other 4.3, taking place alternately. The maximum magnitude is 3.4 and the period is 12.9 days.

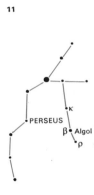

11

13

ε
Vega
γ
LYRA
β

PERSEUS
β Algol
κ
ρ

12
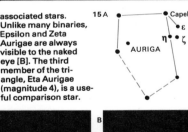
Hours
5 0·3 5   68·8   5 0·3 5

14
Magnitude
3·4
3·6
3·8
4·0
4·2
0   2   4   6   8   10   12   14
Days

15 The pattern of Auriga [A] is dominated by the bright yellow Capella. The two eclipsing binaries Epsilon and Zeta Aurigae lie in the small triangle close by Capella, but are not genuinely associated stars. Unlike many binaries, Epsilon and Zeta Aurigae are always visible to the naked eye [B]. The third member of the triangle, Eta Aurigae (magnitude 4), is a useful comparison star.

15A
Capella
ε
η ζ
AURIGA

B

17
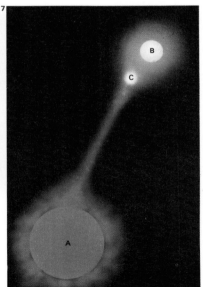

16
A
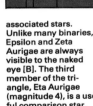

B
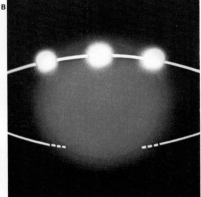

16 Zeta Aurigae is made up of an orange supergiant star (type K9, diameter 290 million km) and a hot white star (type B7, diameter 4 million km). When the bright star is eclipsed [A] (every 972 days), there is a three-week period when it shines through the supergiant's tenuous outer layers, producing informative spectral changes. When the bright star passes in front of the supergiant [B] there is no marked decrease in light.

17 Dwarf nova binaries or U Geminorum stars are close binaries, in which one star is a member of the main sequence, the other a white dwarf. Material is pulled off the larger star [A] across to the white dwarf [B] in a jet, striking the gaseous surround and producing a spot [C] brighter than the stars themselves. Variations in this jet produce rapid flickerings in the light, detectable only with electronic equipment. Periodic outbursts also come from the white dwarf.

Pulsating stars

Pulsating stars are variables whose brightness wanes with time due to cycles of expansion and contraction. The variations may be regular or irregular, varying from a few minutes to many centuries. Astronomers have been continually engaged in monitoring and searching for new variables [2, 3]. A notable observer, John Goodricke (1764–86), was the first to realize that the curious "winking" behaviour of Algol, in Perseus, is due to the periodical eclipse of the bright star by a darker companion; and he discovered the variability of Delta Cephei, which has proved to be one of the really important members of the Galaxy as far as theoretical research in astronomy is concerned.

Delta Cephei [5], in the far north of the sky, has a fairly small magnitude range of 3.6 to 4.3, so that it is never conspicuous and yet never becomes dim enough to be hard to see with the naked eye. Its period – the time between one maximum and the next – is 5.366 days and is absolutely regular, so that the brightness for any particular moment can always be predicted. Subsequently other stars of the same kind were discovered: Eta Aquilae in the Eagle, with a period of 7.17 days; Zeta Geminorum in the Twins, 10.2 days; and Kappa Pavonis [6B] in the southern constellation of the Peacock, 9.1 days. With modern methods many similar variables have been discovered – the known number now totals many thousands – and they have become classified as Cepheids.

The period-luminosity law

Cepheids are giant stars of high luminosity. Being well advanced in their careers they have become unstable. However they are quite unlike the explosive stars, whose behaviour cannot be predicted. But the Cepheids are of vital importance for one main reason: their changes in output provide a key to their real luminosities and hence to their distances.

The key was discovered in 1912 by Henrietta Leavitt, who was working on some photographs of the external system known as the Small Cloud of Magellan [Key, 2]. The Cloud contains Cepheids, and Leavitt found that the stars of longer period looked brighter than those of shorter period. For all practical purposes the stars in the Cloud can be regarded as being equally distant from Earth – just as, two men in New York City, one standing in Times Square and the other by the Statue of Liberty, are equally distant from London or Paris – and so it followed that the brighter Cepheids were genuinely the more luminous. If a star's real power and apparent brightness are known, then its distance can be worked out. Naturally, many corrections had to be made (notably for the absorption of light in space), but the principle was clear and the Cepheid period-luminosity law has provided the main method of gauging distance in the Galaxy.

Beyond the Galaxy

In 1923 Edwin Hubble (1889–1953), at Mount Wilson, found Cepheids in some of the "starry nebulae", including M31 in Andromeda. As soon as he had found their periods, he could obtain their distances. He realized that the Cepheids – and hence the spirals themselves – lay far beyond the limits of our Galaxy. Without Hubble's discovery of those convenient Cepheids, proof would

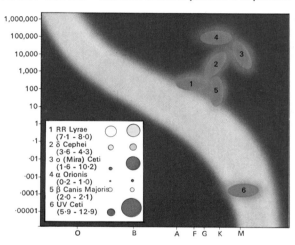

1 The Hertzsprung–Russell diagram shows a star's luminosity (or absolute magnitude) against its spectral type. Here the main classes of pulsating stars are plotted – RR Lyrae stars [1], Cepheids [2], the long-period Mira stars [3], red variables [4], Beta Canis Majoris variables [5] and flare stars [6]. The inset gives the variation in magnitude from minimum to maximum of the main members of each class. The Cepheids are of early spectral type, in a stage through which many stars are thought to pass.

1 RR Lyrae (7·1 - 8·0)
2 δ Cephei (3·6 - 4·3)
3 o (Mira) Ceti (1·6 - 10·2)
4 α Orionis (0·2 - 1·0)
5 β Canis Majoris (2·0 - 2·1)
6 UV Ceti (5·9 - 12·9)

2 The refracting telescope formerly in Peru and now at the Boyden Observatory in South Africa was used to obtain the photographs of the Small Cloud of Magellan from which, in 1912, Henrietta Leavitt (1868–1921) established the period-luminosity law of the Cepheid variables.

3 The 100in (254cm) Hooker reflector at Mount Wilson, USA, was completed in 1918. For 30 years the most powerful telescope in the world, it was used by Hubble in studies of short-period variables in external galaxies; at that time, the 1920s, no other instrument was powerful enough for research of this kind. The 100in is still in full operation. The mounting is of the English type, so that the telescope can never be pointed towards the celestial pole. The driving mechanism is powered by falling weights.

4 Trying to find short-period variables in external galaxies, Hubble concentrated on the Andromeda Spiral – not then known to be beyond our Galaxy. As the most impressive of the spirals, it seemed likely to be relatively near by cosmic standards. Hubble failed to find RR Lyrae variables in it, but he was able to locate Cepheids. As soon as he had observed their periods he was able to use the period-luminosity law to show that the Cepheids – and hence the Spiral itself – must be beyond the Milky Way. His estimate was 750,000 light-years but the real distance is 2.2 million light-years.

have been extremely difficult to obtain. It is true that his original estimates were found to be too low because of an error in the Cepheid scale that did not come to light until the work of Walter Baade (1893–1960) in 1952. Hubble believed the Andromeda Spiral to be 750,000 light-years away; but the real distance is more than 2 million light-years [4].

Because Cepheids are so powerful, they can be seen over immense distances and even at about 40 million light-years they are still detectable. There are also some associated variables of shorter period (less than a day) all of which appear to be of about the same luminosity, roughly 90 times that of the Sun. They are known as RR Lyrae stars, after the best-known member of the class.

Long-period stars
Cepheids and RR Lyrae variables [7] are pulsating stars, alternately swelling and shrinking. There are also stars that pulsate in much longer periods, of from a few weeks up to a year or more. These are the long-period variables, often called Mira stars after Mira [10, 11, 12], the "Wonderful Star" in Cetus,

the Whale. Virtually all stars of this kind are old red giants of tremendous size and high luminosity; they have used up their available hydrogen "fuel" and are unstable. There is no Cepheid-type period-luminosity law and indeed the periods and the amplitudes are not constant. Mira itself has a period of 331 days, but this may vary by a week or so either way from one cycle to another. At some maxima Mira may become as bright as the Pole Star (magnitude 2), while at other maxima it is no brighter than the fourth magnitude. When at its faintest the magnitude is about 10, so that ordinary binoculars will not show it. Another Mira star is Chi Cygni, in the Swan, which ranges between magnitudes 3.3 and 14.2.

There are also semi-regular variables, such as Betelgeuse in Orion, with small amplitudes and periods that are very irregular indeed. Most, though not all, are red giants and they too swell and shrink, changing their output of energy as they do so. Stars such as Betelgeuse are extremely large, with diameters of up to 580 million kilometres (360 million miles).

KEY

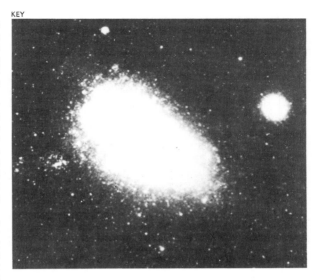

The Small Cloud of Magellan is in the southern sky, inaccessible from the great northern observatories. Photographs taken from Peru before World War I confirmed that it contains Cepheids, on which the period-luminosity law was based.

5 Delta Cephei lies in the far north of the sky. It forms a triangle with Epsilon and Zeta Cephei, which act as convenient comparison stars; the fluctuations of Delta are obvious.

7 The periods of RR Lyrae stars, formerly called cluster-Cepheids, are much shorter than those of the classical Cepheids; they are all of about the same luminosity.

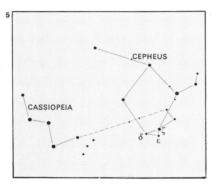

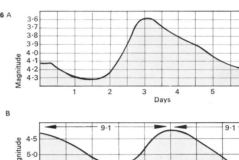

6 Cepheid variables are notable for their regularity. Delta Cephei itself has a period of 5.3 days [A]; in comparison the light-curve of the southern Cepheid variable Kappa Pavonis has a period of 9.1 days [B]. The shapes of the curves are not identical. Since Kappa Pavonis has the longer period, it is the more luminous. Other famous Cepheids are Zeta Geminorum (10.2 days) and Eta Aquilae (7.17 days).

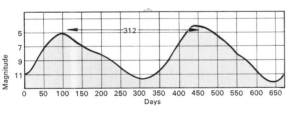

8 Eta Geminorum is a semi-regular variable with a small magnitude range; neither the period nor the amplitude is constant from one period to another but the fluctuations are relatively slight.

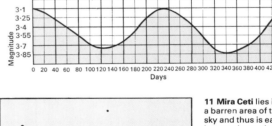

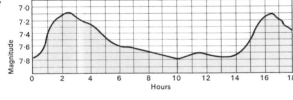

9 The light-curve of R Leonis is typical of long-period variables of the Mira type. As with all Mira stars, both period and amplitude are subject to fluctuation. At its brightest, about magnitude 5, R Leonis is a naked-eye object.

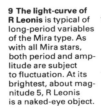

10 Mira Ceti is another long-period variable that is visible to the naked eye. The average period is 331 days; the magnitude ranges from 1.7 to 4 at maximum and down to 10 at minimum.

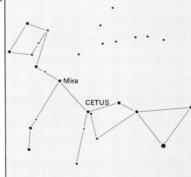

11 Mira Ceti lies in a barren area of the sky and thus is easy to locate when at its brightest. But it is visible to the naked eye for only a few weeks each year, although not regularly in each period.

12 The size of Mira is here compared with the Sun; it has a diameter, like all red giants, of more than 160 million km (100 million miles). The diameter changes as the star's output of energy varies.

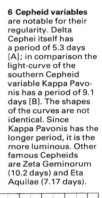

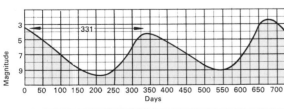

Irregular stars

Not all variable stars are predictable. Like the regular pulsating stars, the many irregular variables are categorized into groups of similar type. For instance all semi-regular stars, such as Betelgeuse, have only approximate periods – the time elapsed between a star's magnitude at maximum and minimum. R Coronae stars [2] remain normally at maximum and suffer sudden, unpredictable drops to minimum. U Geminorum stars [3, 4] or "dwarf novae" are normally at minimum, but increase abruptly to maximum before fading away again. RV Tauri stars [5, 6] are G- to K-type giants and have alternate deep and shallow minima, superimposed upon which are periods of total irregularity. Flare stars [Key], for example the M-type giant UV Ceti, show sudden increases over periods of minutes and remain at maximum only briefly so that their changes can actually be watched. Recurrent novae show sudden, violent outbursts over periods of years; thus T Coronae exploded in 1866 and again in 1946. Normal novae [7–10] show one outburst only and then return to their former obscurity. The exceptional highly brilliant star Eta Carinae [11, 12] is classed as a pseudo-nova.

It is customary to draw light curves of irregular variables and novae in the same way as those for regular pulsating stars, relating apparent magnitude to time. (Apparent or visual magnitude is the apparent brightness of a celestial body as seen with the eye. The brighter the object, the smaller the numerical value of the magnitude.) It must be emphasized that the apparent magnitude of a star is its brilliancy as seen from Earth; it is not a reliable guide to a star's real luminosity. Only the variable stars and novae show short-term changes in apparent magnitude.

Semi-regulars and irregulars

Most of the semi-regular stars are red giants. They are regarded as unstable because they swell and shrink. Betelgeuse in Orion is one such star. Sometimes it will almost equal Rigel in brilliance; its mean magnitude (0.85) is comparable to that of Aldebaran. It has an approximate period of five to six years between maximum and minimum, but the irregularities are very marked. Rasalgethi, or Alpha Herculis, another semi-regular variable, is easily visible to the naked eye. Semi-regular stars that can be detected only by telescope are also common. Generally the variations in the magnitude are not great.

Most irregular variables are telescopic objects; however, Gamma Cassiopeiae [1] can rise to almost the brilliance of Castor in the Twins – as it did in 1936. The spectral changes were of great interest and apparently the star was throwing off a shell of material.

Probably the most erratic variable in the sky is Eta Carinae, in the southern hemisphere. During the middle of the nineteenth century it shone more brightly than any star in the sky apart from Sirius, but since 1867 it has been too faint to be seen with the naked eye, although binoculars will pick it up. It is orange-red and surrounded by nebulosity; when seen through a telescope it appears as a small patch rather than a sharp point like a normal star. It is luminous and remote.

R Coronae and U Geminorum stars

R Coronae Borealis [2], in the Northern Crown, is the prototype of one class, of which fewer than 50 members are known. R

1 The W of Cassiopeia contains the irregular variable star Gamma. Alpha Cassiopeiae or Shedir (spectral type K) is also suspected of slight variability, although with a small range of magnitude.

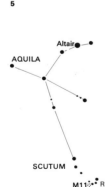

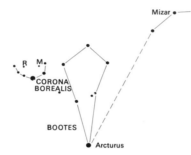

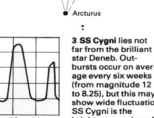

2 R Coronae lies in the bowl of the Northern Crown (Corona Borealis) not far from Arcturus. It is normally about the sixth magnitude, so that, together with the most useful comparison star (M) with a magnitude of 6.6, it can clearly be seen with binoculars. At some minima R Coronae drops to 15 and cannot be seen through moderate-sized telescopes.

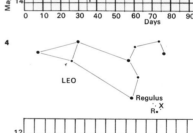

3 SS Cygni lies not far from the brilliant star Deneb. Outbursts occur on average every six weeks (from magnitude 12 to 8.25), but this may show wide fluctuation. SS Cygni is the brightest member of U Geminorum variables.

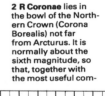

4 X Leonis, near the long-period R Leonis, is of the U Geminorum class. Normally of the 15th magnitude, it brightens up to magnitude 12 about every 22 days.

5 The brightest of the RV Tauri variables is R Scuti in the tiny constellation of Scutum, the Shield, near the tail of Aquila, the Eagle. It is easy to find because it is one of four stars making up a quadrilateral and is not far from the beautiful open cluster of M11, which is nicknamed the Wild Duck. The light curve shows magnitude plotted against time; but the curve is only an average because all RV Tauri stars are erratic in their behaviour. The range of R Scuti is between magnitudes 5 and 8.6 so that at maximum it is visible to the naked eye. It always stays within the range of binoculars, appearing reddish in colour. It is therefore a favourite object for amateur astronomers.

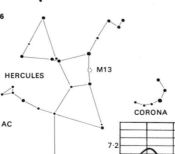

6 All the RV Tauri stars are highly luminous and are among the most massive variables known; some are at least 25 times as massive as the Sun. Unfortunately, all are remote and few are within binocular range. The alternate deep and shallow minima may sometimes be replaced by spells of total irregularity. AC Herculis, whose light-curve is shown here, is a case in point; it has a magnitude range from 7 to 8.5.

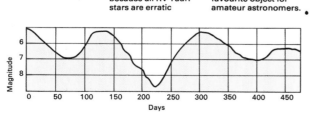

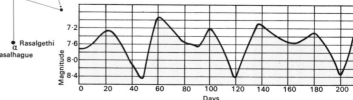

Coronae stars stay at their maxima for most of the time but suffer sudden, unpredictable falls to minima. They are poor in hydrogen, but rich in carbon; it has been suggested that the minima may be due to the accumulation of carbon particles in the star's outer atmosphere, which causes the radiation from the star itself to be temporarily shut in. R Coronae at maximum is on the fringe of naked-eye visibility.

The U Geminorum or SS Cygni stars normally stay at minimum, but undergo periodical outbursts. The average interval between outbursts of SS Cygni [3] is about six weeks. It is now known that all SS Cygni or U Geminorum stars are close binaries, with one white dwarf component together with a late-type red dwarf.

Normal and recurrent novae
A nova is not a new star; it is a formerly obscure star that has suddenly increased in brilliancy. Some past novae have been brilliant; for instance both Nova Persei (1901) [7] and Nova Aquilae (1918) [8] exceeded the first magnitude at maximum. Once a nova

has passed its peak it fades back to its original brightness, although it may take years to do so. It is thought that the outburst affects only the star's outer layers – whereas in a supernova explosion the star is destroyed in its old form. Many, perhaps all, novae are spectroscopic binaries.

HR Delphini [10], one of the most interesting novae of modern times, was discovered in July 1967 by the English amateur George Alcock. It never became brighter than magnitude 3.6, but it was slow to fade and remained a naked-eye object for a year. By 1975 it had fallen to below magnitude 11, but it may not fade much more. It is one of the few novae whose pre-outburst magnitude of 12 is well known. Since it is about 30,000 light-years away, we are watching the results of an explosion that must have happened some 30,000 years ago.

A few stars have been known to suffer more than one outburst; T Coronae blazed up from magnitude 9 to 2 in 1866 and from magnitude 10 to above 3 in 1946. Such stars are the recurrent novae, but not many of them are known.

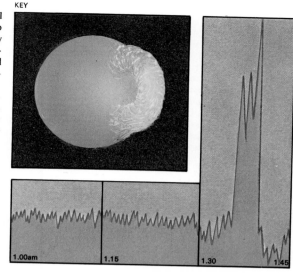

KEY

Radio intensity

1.00am   1.15   1.30   1.45

Flare stars, such as UV Ceti, are near-by dwarfs and, alone among variable stars, change in brilliancy so quickly that they can be watched as their luminosity in- creases over a period of minutes. The sudden increases are due to intense flare activity.

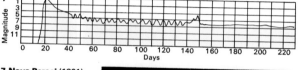

7 Nova Persei (1901) became very brilliant, but declined rapidly. The decrease was attended by marked fluctuations. Nebulosity round the star was illuminated, giving the false impression of an expanding cloud. Today Nova Persei is a faint object.

8 Nova Aquilae (1918) appeared with dramatic suddenness. At its maximum, on 9 June, its magnitude was −1.4, equal to Sirius. It soon faded, but remained visible to the naked eye until November 1918. A nebulous cloud round it steadily expanded and became dimmer, finally disappearing from view in 1940. The old nova is still visible, although extremely faint. Nova Aquilae was the brightest exploding star in modern times.

9 Nova (DQ) Herculis (1934) exceeded the 2nd magnitude and was unusual because it showed a temporary recovery after its drop from maximum. These photographs taken in 1951 show that the nova is a close binary

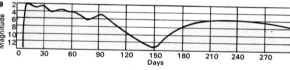

with associated nebulosity. It is classed as a "slow nova".

losity, but the star itself must also be intrinsically variable. In the telescope it appears strongly red and seems to be much less well defined than a normal star.

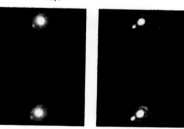

11 Eta Carinae is not a normal nova. It is intensely luminous and is associated with complex nebulosity. The fluctuations may be due in part to variations in density of the nebu-

10 HR Delphini (1967) has the distinction of being the slowest true nova on record. The maximum lasted for six months, although with definite fluctuations, and the subsequent fall was gradual. It is likely that it can be observed through a small telescope.

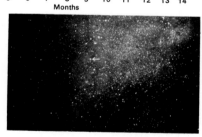

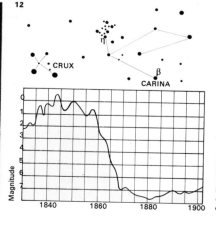

CRUX

β
CARINA

12 In the Keel of the Ship, near the Southern Cross and the 2nd-magnitude star Beta Carinae, lies Eta Carinae. It is the most erratic of all variables. It reached its maximum brightness in the 1840s when it was exceeded only by Sirius; it then declined and since about 1867 has been invisible to the naked eye, although binoculars will show it. It is surrounded by nebulosity and is in every way an exceptional object of high luminosity.

Stellar clusters

Man lives in a part of the Galaxy in which the density distribution of the stars in space is about average. His nearest neighbour, Proxima Centauri, is more than four light-years away and there are not many stars within a radius of ten light-years from the Sun. Here and there in the Galaxy, however, are groups of stars making up genuine clusters. The best-known example is the cluster of the Pleiades, or Seven Sisters [5], in Taurus; several others are easily visible to the naked eye and the number of known clusters visible through a telescope is immense.

Open clusters
Clusters are essentially of two kinds: open and globular. The open or loose clusters are to be found in the spiral arms of our Galaxy and are irregular in shape; they may be rich with thousands of members, or poor with only a dozen or two. There is no possibility of their being caused by line-of-sight effects.

There are wide differences between the various open clusters. In the Pleiades cluster the brightest stars are hot and white and there is a large reflection nebula indicating the presence of a great deal of interstellar material; by cosmic standards the group is extremely young. Several of its chief stars are known to be in rapid rotation and one of them, Pleione, is so unstable that it periodically sheds some of its material, and produces a shell or gaseous ring. This can be studied only by spectroscopic methods, round its equator.

In the second of the Taurus clusters, the Hyades (round Aldebaran) [3], the star density is smaller; the principal members are not so energetic and the amount of material spread between the stars is less. The Hyades are not as spectacular as the Pleiades because they are largely overpowered by the brilliant orange light of Aldebaran. Yet Aldebaran is not a true member of the cluster at all and here a line-of-sight effect is found, because Aldebaran, in fact, lies midway between Hyades and ourselves.

Other naked-eye clusters are Praesepe, the Beehive [4], in Cancer and the lovely southern cluster round Kappa Crucis, known as the Jewel Box because it contains stars of varied colours. In Perseus, not far from the W of Cassiopeia, is the Sword Handle [2], which is made up of two rich clusters in the same telescopic field.

Open clusters are not stable associations and must eventually be disrupted by the gravitational pulls of field stars in our Galaxy. It has been estimated that most of them have life-spans of no more than 1,000 million years before being scattered to the extent of losing their separate identity. One of the oldest clusters known, M67 in Cancer, which is easily visible with binoculars near the star Alpha Cancri, may be more than 4,000 million years old, but is located well away from the galactic plane and as a result is less liable to disruption.

Globular clusters
Globular clusters are of a different type altogether. Only about 120 are known to exist in our Galaxy. They are symmetrical and may contain hundreds of thousands of stars; as seen from Earth they are so condensed towards their centres that they are difficult to resolve into individual stars. Even so, the danger of stellar collisions occurring remains slight; but to an inhabitant of a

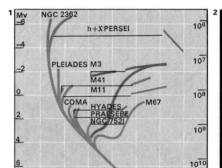

1 The colour-magnitude graph of open clusters shows absolute magnitude (Mv) against colour index (B−V=photographic magnitude minus visual magnitude). The age (years) is given on the right-hand scale.

2 The twin clusters in Perseus were catalogued by Dreyer as NGC 869 and 884; they are also known as Chi and h Persei. Each cluster is 75 light-years across and contains about 350 stars.

3 The Hyades, round Aldebaran (not itself a cluster member), are easily visible to the naked eye. The group extends in a kind of V formation away from Aldebaran. The individual stars are easy to make out and one of them, Theta Tauri, is a naked-eye double. The Hyades are so scattered that the group is not nearly as spectacular as the Pleiades. The best view is obtained with binoculars or with a telescope of low magnification and a wide field.

4 Praesepe, in Cancer, is an example of an open cluster visible to the naked eye. It is 525 light-years away and has long been known. It is not a condensed cluster, and lies well away from the galactic plane.

5 The Pleiades cluster (member 45 in the Messier catalogue) is 410 light-years away. At least seven of its stars are visible to the naked eye, and the total number of members is about 500. The leading stars are hot and white.

planet that was moving round a star in such a region, the night sky would be scattered with thousands of stars that would be shining more brightly than Sirius or even the full Moon as it appears to us.

The distribution of the globular clusters is not uniform over the whole sky; they surround the galactic centre, so that from Earth they are only seen in the direction of the centre. The distances of globular clusters have been measured by using the RR Lyrae variables contained in them. Because all RR Lyrae stars have similar periods and luminosity, their distances can be calculated more easily, enabling more effective measurements of the globular clusters themselves. It was by this method that Harlow Shapley, more than 60 years ago, was able to work out the size of our Galaxy from his studies of globular clusters. Because of their remoteness they form an "outer framework" to the main galactic system [Key].

The brightest globular clusters, Omega Centauri and 47 Tucanae, lie so far south in the sky that they cannot be seen from Europe or most of North America. Both are easily visible with the naked eye and Omega Centauri in particular is a superb sight through a telescope; it is resolvable to its centre. In the north, the best example is M13 in Hercules [9], which lies at a distance of 26,700 light-years and is about 100 light-years in diameter. It is just visible to the naked eye and binoculars show it well.

Globular clusters belong to the galactic halo and move round the nucleus in highly inclined, eccentric orbits. They are stable and are not subject to disruption in the same way as open clusters; they are made up of Population II objects, so that their brightest stars are mostly of late spectral type.

Moving clusters

In addition to the open and globular clusters, there are the so-called moving clusters, whose members are widely separate but move through space in the same direction and with the same velocity. Hot, luminous O- and B-type stars form "stellar associations", of which nearly 100 are now known. One association of this kind is centred on the Orion Nebula.

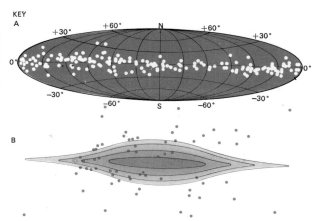

Two main types of stellar clusters are found in and around our Galaxy. Open or loose clusters [A] are composed of Population I stars and they lie near the main plane of the Galaxy, although there are a few exceptional clusters that are well away from it. An example is the old cluster M67. The open clusters are therefore a part of the general rotation of the Galaxy round its nucleus. Globular clusters [B], which are made up of Population II stars, are distributed throughout the galactic halo. The position of the Sun is indicated by the small open circle.

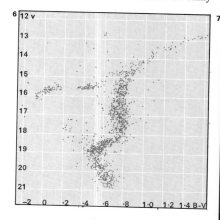

6 The globular cluster M3 in Canes Venatici (the Hunting Dogs), is 48,500 light-years from us. In this colour-magnitude diagram apparent magnitude, v, is plotted against the colour index (B–V). The RR Lyrae variables, used to measure the distance of the cluster, lie at v=15.7 – the gap in the horizontal branch.

7 The globular cluster M3 in Canes Venatici contains more than 44,000 stars of magnitude 22.5 or brighter within an 8' radius of the centre. It is typical of its kind and quite symmetrical.

8 The globular cluster M5 in Serpens was discovered in 1702 and is a bright telescopic object. It is exceptionally rich in RR Lyrae variables and more than 100 have been discovered in it.

9 M13 in Hercules is the finest globular cluster in the northern sky and is found between the stars Zeta and Eta Herculis. It is 26,700 light-years away from Earth. Surprisingly, it is poor in RR Lyrae variables and fewer than 20 have been discovered, as against over 100 in other clusters.

10 Globular clusters attend other galaxies as well as our own; there are many, for instance, round the Andromeda Spiral. This photograph shows the giant elliptical galaxy M87 around which some 1,000 clusters have been detected. These appear as small dots on the picture. M87 is at a distance of 40 million light-years from Earth.

Our Galaxy

The Solar System, with the Sun at its centre, is an insignificant part of a local system of about 100,000 million suns called the Galaxy. In relation to those other stars, the Sun is neither an exceptionally luminous nor an exceptionally dim star.

The Sun is without doubt older than the Earth and its age cannot be less than 5,000 million years. The Galaxy itself is presumably much older, although it cannot be claimed that we yet have any clear knowledge of its early history. The Galaxy is a flattened system [1] and when we look along its main plane we see many stars in almost the same line of sight. That is why the stars in the Milky Way seem so closely crowded together that they appear almost to touch [Key, 6, 7].

Early observations
The Milky Way is a dominant feature of the night sky although unfortunately city-dwellers never see it clearly because its soft radiance is apt to be swamped by the glare of street lights. One of the earliest and best descriptions of it was written about AD 150 by Ptolemy of Alexandria, the last great classical astronomer and mathematician.

"The Milky Way", he wrote, "is not a circle but a zone, which is everywhere as white as milk and this has given it the name it bears. Now this zone is neither equal nor regular anywhere, but varies as much in width as in shade or colour, as well as in the number of stars in its parts and in the diversity of positions; also in some places it is divided into two branches, as is easy to see if we examine it with a little attention."

Most countries have legends about the Milky Way, but its nature was not known until Galileo (1564–1642) first examined it with a telescope in the winter of 1609–10. He found it to be made up of "a mass of innumerable stars", an observation that may be checked even with binoculars.

The first man to record the approximate shape of our Galaxy was William Herschel (1738–1822), who compared it to "a cloven grindstone". Herschel is best remembered for his discovery of the planet Uranus in 1781 although his major contributions were in the field of stellar astronomy. In particular he considered whether the starry or resolvable nebulae, such as that in the constellation of Andromeda, might be separate star systems well beyond our own.

We can never see the centre of the Galaxy because of the obscuring effect of the interstellar materials. Present knowledge is derived mainly from radio astronomy, by which the centre can be located. It lies beyond the glorious star clouds in Sagittarius where the Milky Way is particularly thick with stars [4]. It has been suggested that there may be a quasar there or even a black hole, but these speculations rest on uncertain evidence. Certainly radio waves come from the galactic centre and it was the source of the first radio waves from the sky ever to be detected – by Karl Jansky (1905–50) – in the early 1930s.

The form of our Galaxy
During World War I, Harlow Shapley (1885–1972) in America measured the size of our Galaxy (from his studies of RR Lyrae variable stars in globular clusters). He also proved that the Sun, together with the Earth and the other members of the Solar System,

1 The shape of the Galaxy would look different if viewed from separate vantage points well out in space. In an edge-on view [A] the shape is flattened, with a pronounced nucleus. (S indicates the position of the Sun.) As seen from an angle [B], the general form of the Galaxy remains clear, but the spiral arms are now displayed. The Galaxy is a rather loose spiral, but the arms can be clearly seen.

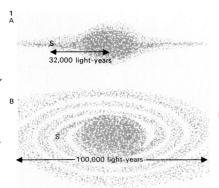

32,000 light-years

100,000 light-years

2 Stars within a radius of ten light-years from the Sun make up only a small area of the Milky Way. The two ellipses around the Sun indicate the possible maximum distances of orbiting comets. Most of the nearby stars are faint red dwarfs; there are also some white dwarfs (such as the companion of Sirius). The only stars more luminous than the Sun are Sirius, Procyon and Alpha Centauri.

BD12° 4523
Munich 15040
Alpha Centauri
ε Indi
Proxima Centauri
Σ 2398
61 Cygni
OA (N) 17415
Krüger 60
Ross 248
Groombridge 34
Corboda Vh 243
Sirius
Lalande 8760
ε Eridani
Procyon
BD 51° 658
4   8   12   16 light-years

3 Galaxy NGC 7331 in Pegasus is similar in size and mass to our own, although its spiral arms are wound more tightly. It contains some 100,000 million stars. Only when the existence of galaxies outside our own was established could a true idea of the status of our Galaxy be worked out; even in 1920 it was believed that our Galaxy was unique and that the other spiral galaxies were contained inside it.

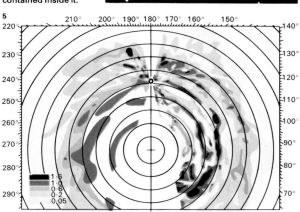

4 The centre of the Galaxy lies beyond the star clouds in Sagittarius which are seen in this photograph taken with the 48in (122cm) Schmidt telescope at Palomar. In addition to the star clouds, the photograph shows a considerable amount of dark nebulosity, stellar matter which betrays its presence by blotting out the light of stars beyond. This is the richest part of the entire Milky Way area.

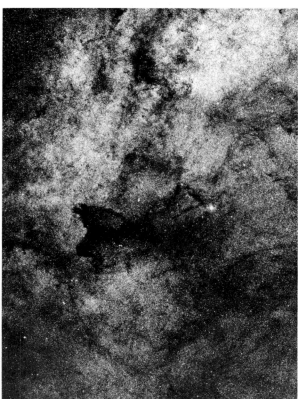

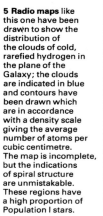

5 Radio maps like this one have been drawn to show the distribution of the clouds of cold, rarefied hydrogen in the plane of the Galaxy; the clouds are indicated in blue and contours have been drawn which are in accordance with a density scale giving the average number of atoms per cubic centimetre. The map is incomplete, but the indications of spiral structure are unmistakable. These regions have a high proportion of Population I stars.

lies well away from the centre; the modern estimate of its distance from the galactic nucleus is 32,000 light-years (rather than was believed until recently). However, the comparative size and structure of the Galaxy was uncertain and became clear only with the work of Edwin Hubble (1889–1953), who showed in the 1920s that the spiral galaxies [3] are external systems, of the same basic type as our own.

If so, then there seemed no reason to doubt that our own Galaxy must also be spiral (although because the Earth is situated inside it, the spiral effect is naturally lost). The distribution of bright stars ("Gould's Belt") provided some sort of confirmation, but the final proof came from radio astronomy. During World War II Hendrik van de Hulst (1918–) and his colleagues in The Netherlands calculated that the clouds of cold hydrogen spread through the Galaxy should radiate at a wavelength of 21cm and in 1951 E. Purcell and H. Ewen, in the United States, showed that this is what does happen. When the positions and the movements of these hydrogen clouds were worked out, an

unmistakable spiral structure was found [5]. It has also been established that the Galaxy is rotating, not as a solid body, at one rate, but showing differential rotation. In the neighbourhood of the Sun, the revolution period is approximately 225 million years – known unofficially as the cosmic year. One cosmic year ago, the Earth was at the beginning of the Triassic period, when the giant reptiles were replacing amphibians as the dominant life form.

The relative size of the Galaxy
It used to be thought that the Galaxy must be exceptionally large, but this illusion, too, has now been shattered. Its size is above average but some other known systems are decidedly bigger, including the Andromeda Spiral, M31, in our local group.

Nowadays the term "Galaxy" is taken to refer to the star system and Ptolemy's "Milky Way" to the luminous aspect in the sky. In appearance, the Milky Way is extremely beautiful; it is particularly rich in areas such as Crux, Cygnus and the Scorpio–Sagittarius region.

NGC 7000 has been given the unofficial title of the "North American Nebula" because of its resemblance to that continent. It is in the constellation of Cygnus, 1,000 light-years away from Earth. This photograph was taken with the 48in (122cm) Schmidt telescope at Palomar. The nebula is associated with the exceptionally luminous supergiant Deneb or Alpha Cygni. The comparatively dark areas seen in the photograph are caused by an intervening cloud of opaque dust, which cuts out the light of the nebula as well as that of the background stars. The nebula is one of the richest areas of the Milky Way – the galactic plane of our Galaxy.

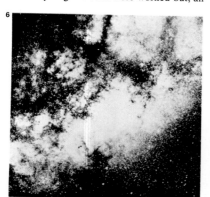

6 Myriads of stars can be seen in this photograph, taken at Palomar, although it covers only a small part of the Milky Way. Many of the stars are much more luminous than our Sun.

7 These star fields in the Milky Way were photographed at the Naval Observatory, Flagstaff, Arizona. The trail of the balloon satellite Echo I is seen as it crossed the field of view during the exposure.

8 The Milky Way, as mapped by Martin and Tatiana Keskula at the Lund Observatory in Sweden, has co-ordinates referring to galactic latitude and longitude measured from the galactic plane. The zero point for longitude is the intersection between the galactic plane and the celestial equator near the borders of Aquila and Serpens. The north galactic pole is in Coma Berenices, the south in Sculptor.

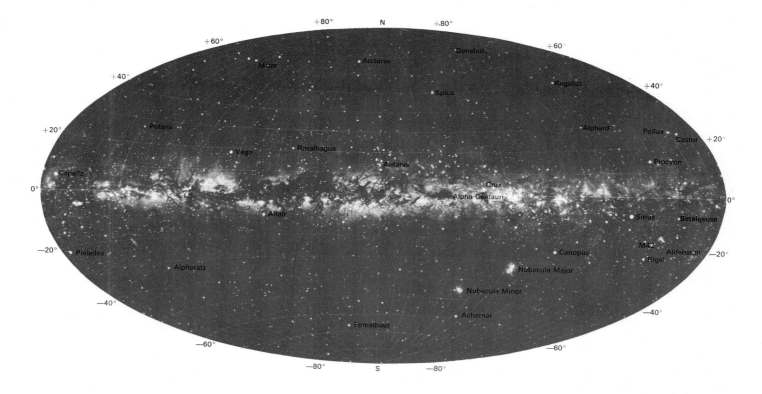

The local group of galaxies

Galaxies tend to occur in groups that are usually termed clusters – although there must be no confusion with the open and globular clusters of stars that occur in our own Galaxy and others. Many clusters of galaxies are known, some of which include hundreds of members. It is therefore not surprising to find that our Galaxy is a member of such a system, known generally as the local group.

The local group [Key] also includes the Andromeda Spiral, the Triangulum Spiral, the two Clouds of Magellan and more than two dozen smaller systems that are classed as dwarf galaxies. There may also be two large extra members, known as Maffei 1 and Maffei 2 in honour of their Italian discoverer Paolo Maffei, but neither of these systems can be seen properly because they lie inconveniently close to the plane of the Milky Way and are heavily obscured by dust inside our Galaxy. Their nature is thus uncertain.

A stable system

It is generally agreed that the universe is expanding and that all the galaxies outside the local group are receding at various rates.

The members of the local group, however, are not moving away from our own Galaxy (indeed, the Andromeda Spiral actually shows a movement of approach, although for the most part this is because of the Sun's own motion around the galactic centre). The local group makes up a stable system.

In attempting to determine the membership of the group, theorists face a serious initial difficulty. The distance of the Andromeda Spiral [1, 2] was originally estimated as being 750,000 light-years. But this estimate produced a number of anomalies; the globular clusters surrounding the spiral were calculated, at that distance, to be systematically different in size from similar globular structures in our Galaxy. Equally unexpectedly, RR Lyrae stars could not be found inside the Andromeda Spiral. Although RR Lyrae stars are less luminous than Cepheid variables, they should still have shown up quite well had the spiral been only 750,000 light-years away. Even a subsequent increase in the estimated distance of the Andromeda Spiral to 900,000 light-years did not resolve these anomalies.

The distances of the galaxies had been calculated by reference to the period and luminosity of the Cepheid variables. In 1952, Walter Baade (1893–1960) announced that the accepted Cepheid scale was inaccurate. There are, in fact, two kinds of Cepheids, one much more powerful than the other. The Cepheids in the Andromeda Spiral had been taken as being less luminous but they were really members of the highly luminous class, which meant that they were much more remote than had been believed. Today, the Spiral is estimated to lie at a distance of 2.2 million light-years.

Distance and relative size

The Andromeda Spiral is the largest member of the local group, with our own Galaxy a rather poor second. Next comes the Triangulum Spiral [4] and then the two Clouds of Magellan or Nubeculae, which are too far south to be visible from Europe or the United States. The remaining galaxies in the local group are much smaller and less rich.

The Clouds of Magellan [5, 6] look rather like detached portions of the Milky Way, but

1 The great spiral in Andromeda has been known for many centuries and was noted by the Arab astronomer Al-Sufi in the tenth century. It was first described telescopically by Simon Marius (1570–1624), a contemporary of Galileo, as looking "like the flame of a candle seen through horn". It is distinctly visible to the naked eye under good conditions, but even through a telescope of considerable size it is disappointing, appearing to be nothing more than an elongated blur of light. Part of the reason for this is that the system lies at a difficult angle from Earth. Were it face-on, as is the Whirlpool Galaxy (M51) for instance, it would be more imposing. As well as containing clusters, gaseous nebulae and variables of all kinds, novae have also been seen frequently. In 1885, one supernova, S Andromedae, could just be seen with the naked eye.

2 The nucleus of the Andromeda Spiral, M31, is usually overexposed in photographs to bring out the structure of the spiral arms. Where the exposure is correct (as in this picture) the arms do not show up. The scattered stars are members of our own Galaxy and simply happen to lie in the foreground. The nucleus of M31 contains mostly Population II stars, as do many galactic nuclei.

3 The dwarf galaxy in Sextans [centre] in the local group has relatively few stars. The bright star that appears below it is a foreground star in our own Galaxy.

each lies at a distance of about 150,000 light-years from Earth. They are irregular in form (suggestions of spirality in the Large Cloud do not seem convincing) and have been regarded as satellite galaxies of our own, although whether they are moving around our Galaxy is extremely problematical. The richer Large Cloud is about 40,000 light-years in diameter and the Small Cloud 20,000 light-years, so that both are much smaller than our Galaxy. The distance between their centres is 75,000 light-years and their genuine association is shown by the fact that both seem to be contained in a common envelope of rarefied hydrogen. Population I characteristics are evident; novae and huge gaseous nebulae have been seen in them. Both contain Cepheid variables.

The Andromeda Spiral also has two companion galaxies, M32 (NGC 221) and NGC 205. Both are dwarf elliptical systems, made up of Population II stars. They are easily visible in small telescopes, but do not seem to be of the same class as the Clouds of Magellan. For a long time it was thought that the Clouds of Magellan, the spirals in Andromeda and

Triangulum and the companions to the Andromeda Spiral were the only members of the local group, but many smaller systems have now been discovered.

Sparsely populated systems

All these smaller systems are of relatively low mass; for instance, the two dwarf systems in Sculptor and Fornax make up a total of only about one per cent of the mass of our Galaxy. These and similar systems are so sparsely populated that they are not immediately recognizable as galaxies in their own right. They are made up of Population II stars, so their leading stars are old red giants; there is little or no interstellar material prominent enough to be detected from Earth, which suggests that star formation has ceased.

There seems to be no real difference between the local group and many other groups of galaxies known to us. However, the local group is much easier to study because it is in our own part of the universe. The really small galaxies, such as those in Sculptor and Fornax, would not be detectable at all if they lay at distances of millions of light-years.

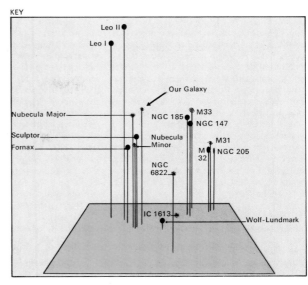

KEY

The local group of galaxies is small with fewer than 30 members. Only the Andromeda Spiral (M31) and the Clouds of Magellan (Nubecula Major and Minor) are naked-eye objects.

4 The Triangulum Spiral, M33, is the third largest member of the local group and the loose spiral form is well shown. At its distance of 2,350,000 light-years, it is slightly more remote than M31.

5 The Small Cloud of Magellan lies in the far southern constellation of Tucana. It is an easy naked-eye object in a dark, clear sky. It contains many variable short-period stars.

6 The Large Cloud of Magellan is irregular in shape and contains many stars. Part of it lies in the constellation of Dorado, part in Mensa. It is a bright naked-eye object even in moonlight.

7 In the Large Cloud lies the Great Looped Nebula surrounding the star 30 Doradûs. This star itself is variable and is the most luminous star known, about a million times as powerful as the Sun. But it cannot be seen without optical aid.

Types of galaxies

As can be seen from even a cursory examination of photographs, galaxies assume various forms [Key]. There are galaxies that show a spiral pattern, some of them loose, some tightly coiled. Among these are the barred spirals, in which the arms seem to protrude from the ends of a "bar" apparently through the centre of the system. Other galaxies appear elliptical, ranging from extremely long, narrow systems to shapes that are almost circular as seen from Earth. Finally, there are irregular galaxies, with no definite shape at all. Most of the dwarfs belong to this category, but there are larger irregular systems such as M82, a radio source lying within Ursa Major.

The Hubble classification
The study of galaxies entered its modern phase in the early 1920s, when work carried out by Edwin Hubble (1889–1953), using the 100in (254cm) Hooker reflector telescope at Mount Wilson in California, definitely confirmed the existence of external systems that were not outlying parts of our own Galaxy. Hubble established a method of classification

that has served as the basis of later, more complicated, classifications [1]. Hubble's system distinguished three basic types of galaxies: the spirals, the ellipticals and the barred spirals. The irregular galaxies were not classified separately although they were recognized. It was clearly tempting to regard the Hubble classification as an evolutionary sequence. However it is purely a classification intended to recognize increasing degrees of flattening; the elliptical systems are in fact spheroidal – they look elliptical only in projection. Present knowledge is so limited that most astronomers regard any overall evolutionary sequence with a certain degree of scepticism.

There is still little idea of how spiral arms are formed, although it has definitely been established that the stars in the disc of a spiral galaxy mostly revolve about the centre in approximately circular paths in the same direction. The spiral pattern appears to revolve also in the same direction with the arms "trailing". There is uncertainty about whether or not the spiral arms – apparently some sort of wave moving through the stars

and gas – are long-lasting on the cosmic scale. Spirals, with their numerous hot Population I stars and their interstellar matter, appear to be less advanced in their evolution than the elliptical galaxies, where the leading stars are red giants of late spectral type and where there is comparatively little nebular material to be observed.

In what are known as the Seyfert galaxies – first noted by Carl Seyfert (1911–60) in 1943 – the nuclei are almost stellar in appearance and the spiral arms are relatively obscure and tightly wound. Seyfert galaxies emit radio waves and great disturbances may well be in progress there. A particularly good example of a Seyfert galaxy is M77 in Cetus, which has a total mass estimated at about 800,000 million times that of the Sun.

The Hubble constant
Even before Hubble's researches had proved that the galaxies lie well beyond our own system, it had been found that the 40-odd galaxies for which suitable spectra had been obtained were apparently receding. This was established by the Doppler shift [11]. If a

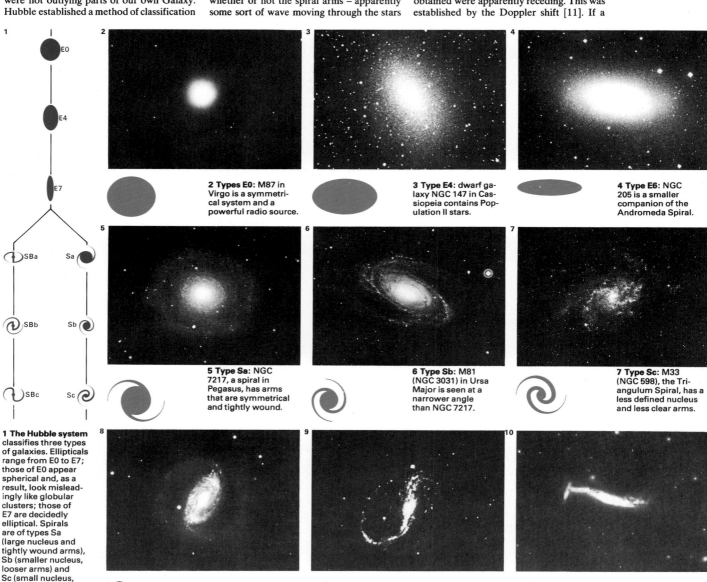

1 The Hubble system classifies three types of galaxies. Ellipticals range from E0 to E7; those of E0 appear spherical and, as a result, look misleadingly like globular clusters; those of E7 are decidedly elliptical. Spirals are of types Sa (large nucleus and tightly wound arms), Sb (smaller nucleus, looser arms) and Sc (small nucleus, loose arms); barred spirals are divided into SBa, SBb, SBc. Hubble's system is still used.

2 Types E0: M87 in Virgo is a symmetrical system and a powerful radio source.

3 Type E4: dwarf galaxy NGC 147 in Cassiopeia contains Population II stars.

4 Type E6: NGC 205 is a smaller companion of the Andromeda Spiral.

5 Type Sa: NGC 7217, a spiral in Pegasus, has arms that are symmetrical and tightly wound.

6 Type Sb: M81 (NGC 3031) in Ursa Major is seen at a narrower angle than NGC 7217.

7 Type Sc: M33 (NGC 598), the Triangulum Spiral, has a less defined nucleus and less clear arms.

8 Type SBa: NGC 3504 in Leo Minor has spiral arms extending from the ends of a bar.

9 Type SBb: NGC 7479 in Pegasus has a pronounced bar with arms clearly extending from its ends.

10 Type SBc: in the Hercules cluster has a dominant bar and the arms are little more than extensions.

galaxy is receding, the lines in its spectrum are shifted towards the red or long-wave end. The farther away a galaxy lies, the greater the speed at which it is receding.

Hubble himself established that there is a definite empirical relationship between distance and the speed of recession. In essence the speed of recession is proportional to the distance; the factor of proportionality is known as Hubble's constant.

Measuring distances

Measurement of the distances of galaxies cannot be precise. For nearby systems – those of the local group, and even beyond – the period-luminosity relation for Cepheid variables can be used. This method for establishing the distance of Cepheids, which can then be used as "standard candles" for other stellar distance measurements, seems to be reliable now that the difference between types of variable stars has been clarified. Cepheids are powerful stars that can be observed at distances of up to several millions of light-years.

Supergiant stars, however, are more powerful than the Cepheids and it seems probable that the brightest supergiants in our Galaxy are more or less equal to the supergiants in other galaxies, so that they too can be used as indicators – although the results are probably less accurate than with the Cepheids. This method works for distances up to about 40 million light-years. There, in the Virgo cluster of galaxies, are galaxies of all kinds, including spirals. Having calculated their sizes it is possible to use them in the same way as the Cepheids and the supergiants, although accuracy is again less. Supernovae can also be used when they appear in remote systems. For the extremely distant galaxies there is as yet no means of measuring a distance independently of the red shift, but if the red shift can be measured the distance may be inferred (assuming that Hubble's law is valid).

Galaxies that lie beyond the local group yield little detail even when viewed with large telescopes. Only detailed photography can reveal the diverse and fascinating nature of more distant star systems scattered throughout the universe.

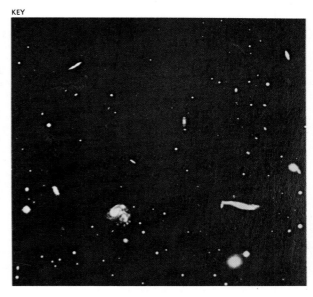

These galaxies in Hercules include spirals, barred spirals and elliptical systems. They form a genuine cluster.

11 The Doppler shift occurs when, from an approaching light source [A], more waves per second enter the eye than would enter from a source without relative motion [B], and the wavelength shifts towards the violet end of the spectrum. Without relative motion [B], the light is unaffected. Recessional velocity [C] produces an apparent increase in wavelength and a spectral red shift.

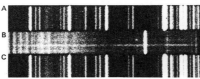

12 One of the most remote galaxies is 3C-295 [arrowed]. Its spectrum [B] is compared with laboratory spectra [A] and [C]. The position of the white line, shifted to the right, shows that 3C-295 is red-shifted.

13 NGC 6946 is a loose spiral galaxy. The nucleus is well marked but not large compared with the spiral arms, which are much less pronounced and tightly wound than those of galaxies classed Sa.

14 The Whirlpool galaxy, M51 in Canes Venatici, is 37 million light-years away. It is face-on to us and therefore is excellently displayed. This was the first spiral galaxy to be recognized (by the Earl of Rosse in 1845).

15 The Sombrero Hat galaxy, M104, part of the Virgo cluster, is at a distance of 41 million light-years.

16 Spiral galaxy NGC 253, in Sculptor, is almost edge-on to Earth. As a result, the spiral form is not well displayed.

Radio galaxies and quasars

Some galaxies are powerful sources of radio waves as well as light. These are known as "radio galaxies". No doubt all galaxies send out long-wavelength radiations because of the supernova remnants and other discrete radio sources inside them, but the energy of radio galaxies is of an entirely different order of magnitude.

Problems of radio galaxies
A typical galaxy with a strong radio source is M87 in the famous Virgo cluster, approximately 60 million light-years from Earth. A curious jet issues from it which looks as though it may be composed of material being ejected at high velocity. M87 sends out about 10,000 times as much energy at radio wavelengths as it might be expected to do. Some other radio galaxies are much more remote; for instance Cygnus A, the first radio galaxy identified optically (in 1954), is 700 million light-years distant, assuming a Hubble constant of 50km/sec/megaparsec (a parsec is the distance at which a star would show a parallax of one second of arc; it is equal to 3.26 light-years).

Many theories have been proposed to explain the radio emissions from these exceptional galaxies. Originally it was thought that there might be not one galaxy, but two – in fact, two separate systems passing through each other in opposite directions. If so the individual stars would seldom collide, but the interstellar matter would be in collision throughout the encounter and would – it was suggested – produce the radio emission that is recorded. Certainly radio galaxies such as Centaurus A give the impression of being compound. Cygnus A, like many other radio galaxies, manifests two powerful centres of radio emission straddling the optical image with a weak radar source coinciding with the optical object.

Further research showed that collisions could not produce nearly enough energy to explain the observations and the theory of colliding galaxies was abandoned. It now seems that the radio emissions are the result of tremendous explosions inside the galaxies themselves. One excellent example of this is M82, an irregular galaxy in Ursa Major [3]. It has been found that there are huge, intricate

gas structures inside the galaxy, moving about at speeds of up to 160km (100 miles) per second and from the present movements it seems that an outburst took place near the centre of M82 about one-and-a-half million years ago (although since the distance of the galaxy is 10.5 million light-years, the explosion occurred 12 million years ago in our time frame). The radio emission from objects like M82 is believed to be generated by the synchroton process – that is, energy radiated by the acceleration of high-energy electrons in a strong magnetic field. Unfortunately the cause of these outbursts in radio galaxies is not yet understood.

A new object in the sky
In 1960 the search for distant objects in the universe led to a surprising sequence of events. A few objects, which from their radio properties were believed to be distant, were identified with blue star-like objects on photographs obtained with the Palomar telescope. Until 1963 they were believed to be a hitherto unidentified type of star in the Milky Way. In March of that year M. Schmidt,

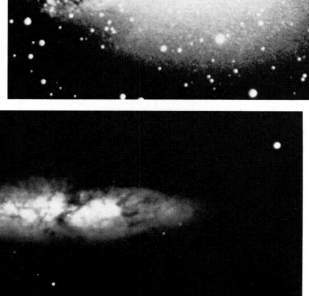

1 Galaxies NGC 4038 and 4039 in Corvus are classified Sc because each galaxy is a loose spiral. Each is also a radio source. There is no doubt that the two systems are genuinely associated and that they lie at the same distance from the Earth. They give the appearance of being interlocked and it was this particular aspect that led to the theory that radio emissions from galaxies are due to collisions. If this theory had been valid, NGC 4038 and 4039 would have been spectacular examples. However it is now known that the theory of colliding galaxies is wrong. The star to the lower right is in our own Galaxy and hence shows up in the foreground.

2 The radio galaxy Centaurus A (NGC 5128) is now thought to be a single system, although it was once believed to be a collision of two galaxies. At its distance of 12 million light-years, it is one of the closest of the radio galaxies and seems to contain an unusual amount of diffused dusty material. The radio sources do not coincide with the centre of the optical object, but lie on either side.

3 The irregular galaxy designated M82 is a radio source 10,500,000 light-years away. There seem to be intricate hydrogen gas structures of immense size moving at velocities of up to 160km (100 miles) per second; and all the indications are that a tremendous explosion took place inside the nucleus of the system 1.5 million years before our present view of it. M82 is therefore the best-known example of an exploding galaxy.

working with the 200in telescope, succeeded in identifying the spectrum of the radio object known as 3C 273 (that is, the 273rd object in the third Cambridge catalogue of radio sources) [4] and at the same time Greenstein and Matthews published their red shift measurements on another blue object – 3C 48. The red shifts were extremely large and as more objects of this type were identified it became clear that a new class of object had been discovered, more remote than any hitherto recognized and in many cases receding with velocity greater than half the speed of light.

Problems associated with quasars

These distant objects, known as quasars, a contraction of their original name, quasistellar objects, have provided astronomers with a series of baffling problems. Assuming that the estimates of their distances are correct, a powerful quasar may far outshine a whole galaxy such as our own – and it is difficult to know how a relatively small object can emit so much energy. Radio measurements of the angular diameters of quasars

and the rapidity of the light variations found in some of these objects imply that the main output of energy may be located in a region of space that is only a few light-years in width.

None of the processes of energy production encountered in normal stars or galaxies seems adequate to account for these phenomena and many theories about quasars have been put forward over the years. It was suggested that a quasar might be produced by many supernovas exploding in quick succession, but there seems no reason why such an event should happen; and theories involving the possibilities of anti-matter or black holes are entirely speculative.

It is possible that quasars and certain types of radio galaxies (notably the Seyfert systems – those having bright central nuclei with emission line spectra) are different stages of evolution of the same class of object, but there is still no reliable information. Recently two American astronomers, J. Oke and J. Gunn, have found that the peculiar object known as BL Lacertae [8] may turn out to be a quasar embedded in a normal galaxy.

KEY

Optical telescopes reveal objects in the heavens that emit light, whereas radio telescopes detect their longer wavelength radiation. Each technique can highlight a different feature of the same object. Here the radio map of the Andromeda galaxy [red lines] is superimposed on its optical image. The galaxy is a weak radio emitter, but the map reveals radio sources that are invisible because they emit no light, such as the one near the lower right corner. The wavelengths detected by radio telescopes range from an upper limit of about 30m down to 1cm, below which any incoming radio waves from space are blocked by the atmosphere of the Earth.

4 The quasar 3C 273, photographed with the 200in (508cm) Palomar reflector, lies in Virgo and with a magnitude of 13 is visually the brightest of the quasars. It was one of the earliest to be identified, in 1963.

5 Quasar 3C 147 has also been photographed from Palomar with the 200in reflector. The quasar [arrowed] looks remarkably like the object just below it, which is an ordinary star in our Galaxy.

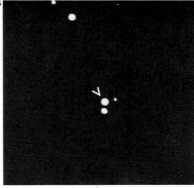

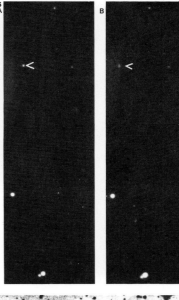

6 These two photographs of the quasar 3C 345 were taken at the Royal Greenwich Observatory in August 1966 [A] and in September 1971 [B]. The quasar is arrowed and the decrease in brightness compared with the other stars can be seen quite easily. Quasars show fluctuations over short periods and this is evidence that they must be extremely small compared with galaxies.

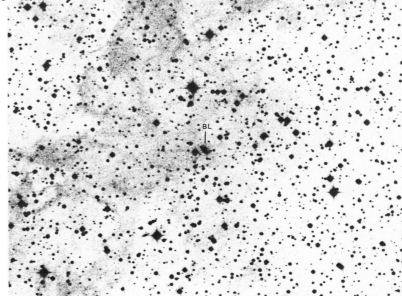

7 The galaxy marked Gal A is the closest of the three radio galaxies in this photograph and may be associated with the elliptical galaxy to the upper left. 3C 390.3 and Gal B are also radio sources identified optically in this photograph.

8 The extraordinary object BL Lacertae, which superficially looks like a star, is variable and near minimum a slight fuzziness is recorded around it. The spectrum is quite featureless, with no bright or dark lines. It has been found

that the outer "fuzz" shows a spectrum resembling that of an elliptical galaxy. BL Lacertae is not a star; it is remote and luminous – possibly intermediate in luminosity between a galaxy and a quasar. There is a strong infra-red emission.

The expanding universe

Of all the unanswered questions facing mankind, that of the origin of the universe is one of the most fascinating – and certainly one of the most enigmatic.

The Doppler significance

An analysis of the structure of a luminous body will show whether it is approaching or receding. If it is approaching, the wavelength of the light appears slightly shortened and the body appears "too blue". If it is receding, the apparent wavelength is lengthened and the body appears "too red". This is known as the Doppler effect after Christian Doppler (1803–53), the Austrian physicist who first drew attention to it in 1842. The Doppler effect shows up in the spectrum of any body that is self-luminous. If the spectral lines are shifted toward the red or long-wavelength end, then this indicates that the body is receding.

The spectrum of an external galaxy is made up of the combined spectra of millions of stars, but the main lines are identifiable and it has been found that apart from the members of our local group of galaxies [1], all

the shifts are to the red. If these shifts are Doppler effects, it follows that the whole universe is in a state of expansion. It has also been found that the farther away a galaxy lies, the greater its red shift – and hence the greater its recessional velocity. This was shown by the work of Edwin Hubble (1889–1953) at Mount Wilson during the years following 1923 when he first showed conclusively that certain nebulae were in fact external galactic systems rather than objects in our Milky Way system.

Theories of the universe

Several years before Hubble discovered the observational evidence for the expansion of the universe, a Dutch astronomer, Willem de Sitter (1872–1934), found a solution to the cosmological theory published by the mathematical physicist Albert Einstein (1879–1955) in 1917. Soon afterwards the Russian scientist A. Friedmann (1888–1925) discovered a whole range of solutions to the Einstein equations in which the radius and mean density of the universe varied with time. But many problems remain.

By adjusting the parameters in the equations, the theoretical models predicted either a universe expanding indefinitely as time advanced or one that would eventually collapse. Many distinguished theorists such as Arthur Eddington (1882–1944) and George Lemaître (1894–1966) developed variations of the models of the expanding universe all of which had in common an initial "beginning" in time when the primeval material was compressed into an infinitely small space.

In 1946 George Gamow (1904–68) developed the idea (popularized as the "big bang" theory) that this initial state was at an extremely high temperature, resulting in a primordial explosion. He also proposed that the common elements were formed from the primeval hydrogen in the first minutes after the beginning of the expansion.

The difficult conceptions involved in a beginning of time (and a comparison of the predicted age of the universe and that of the Earth) led Fred Hoyle (1915–) and T. Gold to propose in 1948 that the universe never had a beginning, but was in a steady state of continuous creation. That is, hyd-

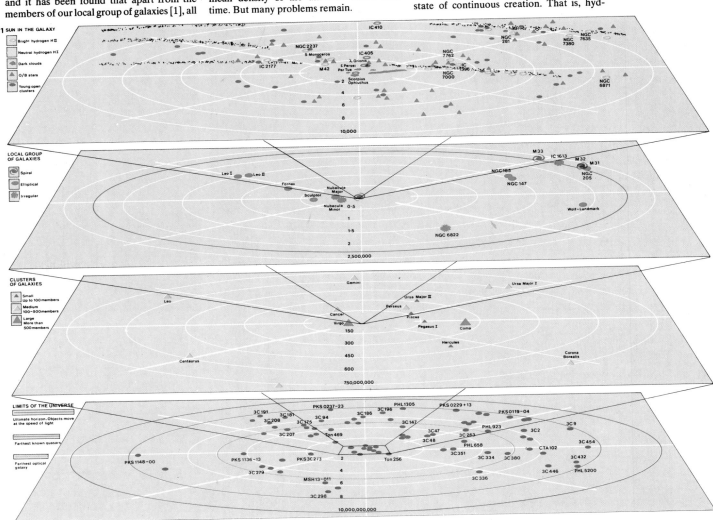

1 The immense scale of the universe is shown in these diagrams. The region of our Galaxy which can be examined optically is shown first. On this scale the entire Solar System would be a microscopic dot. In addition to the visible stars there are clusters, stellar associations and gaseous nebulae such as the Rosette Nebula NGC 2237 and the Orion Nebula M42. Distances are given in thousands of light-years, so that the outermost white line represents a distance of 10,000 light-years from the Sun. The local group of galaxies contains more than 24 members, of which the largest are the spiral galaxy M31 in Andromeda, our own, the Triangulum Spiral M33 and the Clouds of Magellan, the companions of our Galaxy; the other members such as Leo I and Leo II are dwarf galaxies. Distances are given in millions of light-years. Two recently discovered galaxies, Maffei 1 and 2, may be members of the local group but are so heavily obscured by dust in the plane of our Galaxy that they are hard to study. The galaxies in the local group are not receding from us. The area out to 750 million light-years contains many clusters of galaxies – such as the rich Virgo cluster. The region out to 10,000 million light-years cannot yet be studied by either optical or radio methods. Even the most remote objects, the quasars, are not as far away as this.

rogen atoms were being continually created and forming into stars and galaxies at a sufficient rate to replace the galaxies that were moving out of the field of view because of the expansion of the universe. During the postwar epoch of radio investigations of the distant parts of the universe great arguments arose about whether the measurements supported the steady state or evolutionary models of the universe.

In 1965 scientists at the Bell Telephone Laboratories, New Jersey, discovered by accident a radiation from the sky with maximum intensity at a wavelength of 7cm distributed quite uniformly over the heavens. In subsequent years measurements of the spectrum of this microwave radiation seem to have confirmed their initial claim that this is the "relic" radiation from the initial high temperature-high density phase of the universe envisaged by Gamow.

Questions about evolution

Since the discovery of the relic microwave radiation the concept that the universe is in an evolutionary, as distinct from a steady, state has been widely accepted. Areas of doubt persist, however. One concerns the future behaviour of the universe. Is there enough material in the universe to overcome by gravitational attraction the forces of expansion? The critical density is 2×10^{-29} gm/cc but it is unlikely that measurements of this density can settle the issue because of the uncertainty about how much unobservable material exists in space. It seems more probable that the issue will be settled by observations of the remote objects in the universe such as quasars; for example, how do their recessional velocities vary with distance?

At present the results are too scattered to enable any conclusions to be made. A far more pressing question concerns the state of the universe at the beginning of time. The theories propose a singular condition of infinite density at time zero. The microwave measurements probably refer to an epoch only a minute or so after the beginning of the expansion. Present physical theories can envisage a much earlier phase only a fraction of a second after the beginning of the expansion of the universe.

M101 in Ursa Major is a typical spiral galaxy. This is one of the systems that is close enough to be studied in detail.

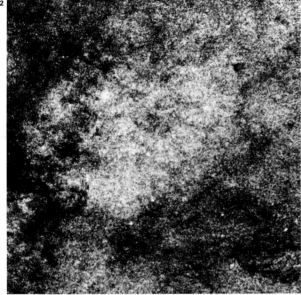

2 Star-clouds in the Galaxy, shown here in the region of Sagittarius and photographed with the Schmidt telescope, indicate the direction of the centre of our Galaxy (illustration 1). All galaxies beyond the local group are receding from our Galaxy which, apart from the Andromeda system, is the largest member of the local group.

3 M33, the Triangulum Spiral, is the most distant known member of the local group of galaxies, shown in illustration 1, and is not receding from us. It has only about 1/25 the mass of our Galaxy. M33 is a normal open-type spiral (Sc). The distance is 2.35 million light-years from Earth.

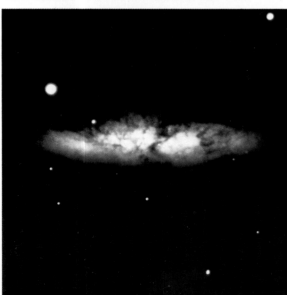

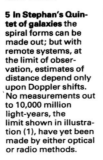

4 M82 in Ursa Major is an irregular galaxy 10.5 million light-years away, beyond the local group. Movements of the gas within it indicate that a tremendous explosion took place there 1.5 million years before our present view of it. M82 is also a strong emitter of radio waves.

5 In Stephan's Quintet of galaxies the spiral forms can be made out; but with remote systems, at the limit of observation, estimates of distance depend only upon Doppler shifts. No measurements out to 10,000 million light-years, the limit shown in illustration (1), have yet been made by either optical or radio methods.

Mapping the constellations

It requires little imagination to make patterns out of the stars in the sky. There are some groups that beg for treatment: for instance the seven stars making up the so-called Plough or Big Dipper and those within the pattern of Orion. In the far south the four stars of the Southern Cross are equally distinctive, even though they look more like a kite than a cross.

At an early stage the stars were divided up into constellations, each of which was given a name. It is, however, important to remember that these visual categories are created simply by line-of-sight and that the stars in any particular constellation are not necessarily associated; thus in the Plough, the "end" star, Alkaid, is more than twice as far away from Earth as the second star, the celebrated binary Mizar. Nor are the constellations permanent: patterns change with time as the stars move.

Early star maps
Ancient star-gazers, who had no idea of the construction of the universe, believed the stars to be equally distant. If they were, the

constellation patterns would presumably be of real importance. The zodiacal constellations are probably the oldest: the Babylonians traced the Zodiac and divided it into 12 constellations [1], which led to our present division of the year into 12 months. Both the Chinese [6] and the Egyptians [4] drew up maps of the sky, showing named constellations; and so, rather later, did the Greeks. It is the Greek system that has survived.

Naming the constellations
About 150 BC Hipparchus, one of the greatest astronomers of the Classical period, compiled a star catalogue. It was this catalogue that was used by Ptolemy of Alexandria as a basis for his own work, undertaken around AD 150. Ptolemy gave a list of 48 constellations, all of which are still to be found on astronomical maps even though their boundaries have been modified. Ptolemy also divided the stars up into magnitudes, or grades of apparent brilliancy, from 1 (very bright) down to 6 (the faintest stars normally visible with the naked eye on a clear night).

The "Ptolemaic" constellations were named after mythological figures, living creatures, or – a relatively few – inanimate objects. The list included virtually all the famous constellations that can be seen from the latitude of Alexandria where (as far as is known) Ptolemy spent all his life. His Latin names are still used: Ursa Major (the Great Bear), Aries (the Ram) and Aquarius (the Water-bearer), and those of mythological characters such as Perseus, Cepheus, Cassiopeia and Andromeda.

It has been claimed that the sky is a complete picture-book [1] in which the classical legends are illustrated and preserved. There is, for instance, the tale of how some tactless boasting by a proud queen, Cassiopeia, about the beauty of her daughter Andromeda led to the princess being chained to a rock on the sea-shore, there to await the coming of a monster sent by the sea-god. This is one of the legends that has a happy ending, for Andromeda was rescued by the gallant hero Perseus, who was returning from an expedition during which he had killed a hideous creature named Medusa, the Gorgon, who

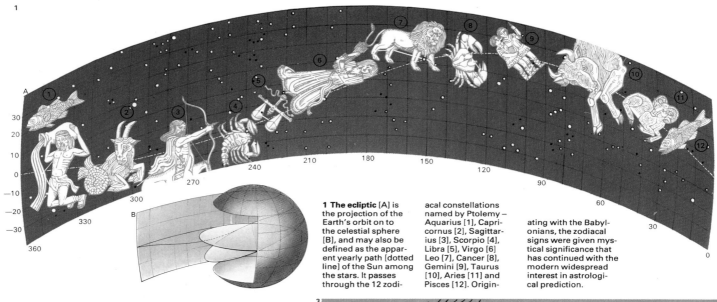

1 The ecliptic [A] is the projection of the Earth's orbit on to the celestial sphere [B], and may also be defined as the apparent yearly path [dotted line] of the Sun among the stars. It passes through the 12 zodiacal constellations named by Ptolemy – Aquarius [1], Capricornus [2], Sagittarius [3], Scorpio [4], Libra [5], Virgo [6], Leo [7], Cancer [8], Gemini [9], Taurus [10], Aries [11] and Pisces [12]. Originating with the Babylonians, the zodiacal signs were given mystical significance that has continued with the modern widespread interest in astrological prediction.

2 This early Arab water clock is surrounded by an Arab zodiac, with the constellation figures clearly represented. The Arab star catalogues were more accurate than any previously made.

3 Indian constellation patterns included Makara [A], the Sea-monster, and Kumbha [B], the Waterpot, containing an elixir which the gods took to heaven – an occasion celebrated by an Indian festival.

4 The impact of Greek astrology on the Egyptians can be clearly seen in these two zodiacal figures (Scorpio and Sagittarius), found inside an Egyptian mummy case, dating from the second century AD.

had snakes instead of hair and whose glance would turn any onlooker to stone. All the main characters in the story can be seen in the Northern Hemisphere: Andromeda, her parents Cassiopeia and Cepheus, the sea-monster (Cetus) and of course Perseus himself, with the Gorgon's head marked by the "demon star", the eclipsing binary Algol. Another legend concerns Orion, the great hunter. A scorpion was sent by Apollo who wished to protect his sister Artemis from Orion. Orion escaped from the scorpion but was killed by one of Artemis' arrows, she being unaware of whom she had shot. Artemis then set Orion's image among the stars, but far away from the Scorpion.

Extending Ptolemy's list

Few of the constellation patterns bear the slightest resemblance to the objects after which they are named; Scorpio is one of the exceptions. However, the system has been followed for so long that it will certainly never be altered now.

Ptolemy's list did not include all the stars visible from Alexandria and he could know

nothing about the stars to be seen from places farther south. Later astronomers extended his list, even to the extent of stealing stars from the original 48 groups; there was a time when no astronomer seemed to feel happy without making at least one addition to an already crowded sky. The climax came with the maps by J. E. Bode, in the late eighteenth century, which included groups with barbarous names such as Sceptrum Brandenburgicum (the Sceptre of Brandenburg), Lochium Funis (the Log Line) and Officina Typographica (the Printing Press). Subsequently the list was reduced to a more manageable 88 groups and the boundaries were rigidly defined by a commission of the International Astronomical Union in 1934.

The constellations are unequal in size and in importance. One (Argo Navis, the Ship Argo) was so huge that it has been split up into several parts; the largest constellation now recognized is Hydra (the Watersnake), which sprawls its way across the sky but has only one bright star. It is interesting to note that the famous Southern Cross, Crux, is the smallest constellation; it was named in 1679.

The 12 constellations of the Zodiac are shown in relation to the Sun during its apparent movement along the ecliptic – a result of the orbital motion of the Earth. An observer on Earth will see the Sun in Aries in March, but as the constellation is only above the horizon during daylight it will be unobservable.

5 Japanese zodiacal signs are animals. The Dog [A], Cock [B], Snake [C], Tiger [D], Rat [E] and Owl [F] are shown in the form of netsuke or buttons worn on a kimono.

6 Chinese zodiacal signs are also animals, some of which are depicted here on a vase [A] of the 5th or 6th century AD. This Chinese horoscope [B] has an intricate layout.

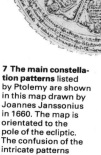

7 The main constellation patterns listed by Ptolemy are shown in this map drawn by Joannes Janssonius in 1660. The map is orientated to the pole of the ecliptic. The confusion of the intricate patterns led the 18th-century astronomer William Herschel to state that they were created to "cause as much inconvenience as possible". The patterns were subsequently modified and laid down by the International Astronomical Union in 1934. The revisions are now universally accepted.

255

Star guide: northern sky I

The far northern sky is dominated by the constellation of Ursa Major, the Great Bear, probably the most famous constellation in the sky. Its seven bright stars make up the pattern that is commonly known as the Plough or, in the USA, the Big Dipper.

Ursa Major – the key constellation

The shape of the Plough is so distinctive that it cannot be overlooked. Six of the stars are of about the second magnitude, but the seventh, Delta Ursae Majoris or Megrez, is below the third. Its relative faintness cannot be missed – and yet astronomers of ancient times ranked it equal with its companions, so that if their descriptions are accurate (which is by no means certain) Megrez has faded appreciably. Alpha or Dubhe, the brighter of the Pointers to the Pole Star, is orange; the rest of the stars are white or bluish-white. Mizar or Zeta Ursae Majoris has a fainter star, Alcor, close beside it. Through a telescope, Mizar can be seen to be a fine double.

From Britain and the northernmost part of the USA, Ursa Major is circumpolar – that is to say, it never sets; and it is extremely useful as a guide for locating other stars and constellations. Ursa Minor, the Little Bear, is easily found [Key]. It resembles a faint and distorted version of the Great Bear, but it has two stars of the second magnitude – Polaris itself and the orange Beta, or Kocab, sometimes nicknamed the Guardian of the Pole. Between the Bears there sprawls the long, dim constellation of Draco, the Dragon, whose brightest star, Gamma or Etamin, is of about the second magnitude. Alpha Draconis, or Thuban, between Kocab and Alkaid in the Plough, used to be the pole star in ancient times.

Map 1 – from Hercules to Virgo

In the large but rather ill-formed Hercules the brightest star, Beta, is above the third magnitude. Alpha, or Rasalgethi, is a semi-regular variable, with a range of between magnitudes 3 and 4; it is a huge red giant star with a small greenish companion visible in small telescopes. Close by it is the second-magnitude Rasalhague, the brightest star in another large, rather dim group, Ophiuchus, the Serpent-bearer. However, the most interesting objects in Hercules are the globular clusters M13 (NGC 6205) and M92 (NGC 6341). M13, the finer, is just visible with the naked eye and a fairly small telescope will resolve its outer portions into stars.

Following round the "tail" of the Great Bear will lead to Arcturus, the brilliant orange star in Boötes, the Herdsman. Arcturus is the brightest star in the northern hemisphere of the sky. Its magnitude is −0.06 and it is light orange in colour, with a K-type spectrum. It is 36 light-years away from Earth and 100 times more luminous than the Sun. The rest of Boötes is not notable, although Epsilon (Izar) is a beautiful double star. Close to Boötes is the conspicuous little semicircle of stars making up Corona Borealis, the Northern Crown, which contains the celebrated variable star R Coronae. This is normally of about the sixth magnitude, but suffers sudden, unpredictable drops to minimum. R Coronae is the prototype star of its class and is much the brightest example. Also in this constellation is T Coronae, which is normally of the tenth magnitude, but flared up to naked-eye visibility

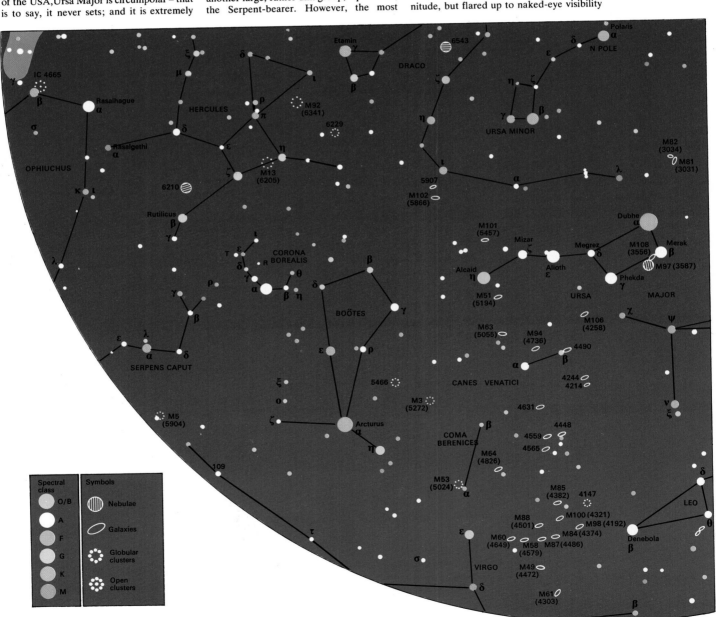

in 1866 and once again 80 years later.

Part of Virgo is also shown, although its leading star, Spica, is in the southern hemisphere. This region is particularly rich in faint galaxies and between Virgo and Ursa Major lies the constellation of Coma Berenices (Berenice's Hair), which looks almost like a large, dim cluster. Canes Venatici, the Hunting Dogs, has only one star as bright as the second magnitude.

Map 2 – from Leo to Canis Minor

Leo, the Lion, is the most easily visible of the constellations during spring in the Northern Hemisphere. Its leading star, Regulus, lies at one end of a curved line making up the pattern known as the Sickle and is of magnitude 1. Gamma, or Algieba, is a fine double star with rather unequal components. Beta or Denebola, on the other side of Leo, is now of the second magnitude, but in ancient times it was ranked as being of the first and there is a chance that it, too, has faded appreciably. Adjoining Leo is Cancer, the Crab, which contains the famous open clusters M44 (NGC 2632) or Praesepe, easily visible with the

naked eye on a dark night, and M67 (NGC 2692), which is visible with binoculars and is thought to be one of the oldest of the loose star clusters.

The brilliant constellation of Orion is cut in half by the celestial equator, so that part of it is shown on this map; the leader is the orange-red Betelgeuse. Not far from Orion are the Twins, Castor and Pollux, in the constellation of Gemini; Pollux is of the first magnitude and Castor between the first and the second – another possible case of fading. Pollux has a K-type spectrum and is clearly orange; Castor is a multiple star, made up of two main components, each of which is a very close binary, together with a fainter companion that is also a binary. The other bright star in Gemini is Gamma or Alhena (magnitude 2). Adjoining Gemini is Canis Minor, the Little Dog, with one brilliant first-magnitude star, Procyon, which is one of our nearest stellar neighbours and has a white dwarf companion. The Milky Way flows through Gemini and its neighbour Auriga, the Charioteer, resulting in a region of many rich star-fields.

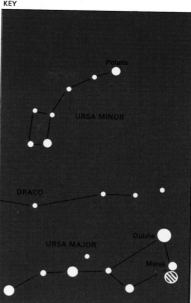

KEY

The north celestial pole is easy to find because Polaris lies within one degree of it. Polaris can be found by following the line of the Pointers, Merak and Dubhe, in the Great Bear. As the Earth rotates from west to east, the celestial pole remains stationary while all other objects appear to circle it slowly. Depending on the position of the observer, the Pole Star will be in a different area of the sky; only at the North Pole is it seen directly overhead. As the celestial pole is thus an extension of the Earth's axis, so also is the celestial equator a projection of the Earth's Equator. A star at the celestial equator will rise in the east and set in the west 12 hours later.

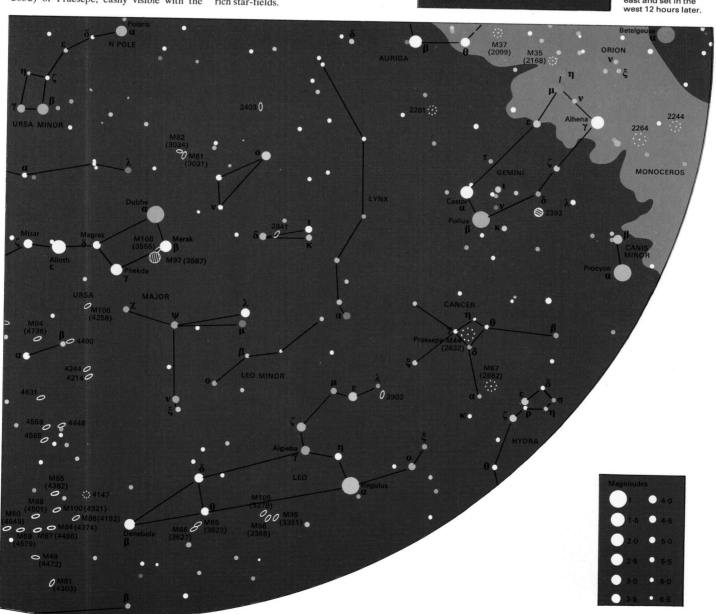

Star guide: northern sky II

Cassiopeia, whose leading stars make up a well-marked W or M formation, is second in importance only to Ursa Major among the constellations of the far north.

Cassiopeia – the key constellation
Like Ursa Major, Cassiopeia is circumpolar over Britain and the northern USA. The best way to find it is to extend a line from Mizar, the second star in the tail of the Great Bear, through Polaris and along a line for an equal distance in the opposite direction. Gamma Cassiopeiae, the middle star of the W, is an irregular variable. It is usually just below the second magnitude, but sometimes, as it did in 1936, it can flare up to 1.6. Its spectrum is peculiar and it is a highly unstable star. Alpha or Shedir is of type K and is suspected of slight variability; Beta is invariable at magnitude 2.3. Tycho Brahe's supernova of 1572 flared out in Cassiopeia; it is still a distinctive source of radio signals.

Two of the stars in the W point to Perseus, which has one second-magnitude star, Alpha or Mirphak, as well as the celebrated eclipsing binary Algol (Beta) which has a mag-nitude range of 2.2 to 3.5. Perhaps the most notable object in Perseus is the Sword Handle, Chi-h (known officially as H.Vi.33–4), which consists of two beautiful open clusters in the same telescopic field, making up a glorious spectacle. Each cluster has a diameter of 75 light-years and lies at a distance of 7,000 light-years from Earth.

Map 1 – from Auriga to Triangulum
The Milky Way flows through Cassiopeia and Perseus and on into Auriga, where the leading star is Capella – almost exactly equal in brilliancy to Vega and on the opposite side of the celestial pole. From Britain it is almost overhead during winter evenings. Capella is 45 light-years away and is of the same spectral type as the Sun, although it is a giant star and much more luminous. The small triangle beside it makes up the Haedi, or Kids. Epsilon Aurigae, the apex of the triangle, is a remarkable object; it is a binary and the secondary is either a young star or a black hole. Zeta Aurigae, close beside it, is also an eclipsing binary of long period (972 days). Auriga itself is made up of a quadrila-teral of stars, easy to identify; the constella-tion includes several bright, open clusters.

Taurus, the Bull, adjoins Auriga. Aldebaran, orange and of the first mag-nitude, lies in line with the three stars of Orion's belt, which are just in the southern hemisphere. Taurus contains the two most famous open clusters in the sky, the Pleiades and the Hyades, as well as the Crab Nebula, M1, near the third-magnitude Zeta Tauri.

Cassiopeia can also be used to locate the Square of Pegasus, which is prominent during autumn evenings in the Northern Hemis-phere. Of the four leaders, Beta, or Scheat, is semi-regular with a period of approximately 35 days; it is visible as a huge, red giant. The line of stars leading off from Pegasus make up Andromeda – most celebrated because of the presence of the spiral galaxy M31. Alpheratz, or Alpha Andromedae, is included in the Square of Pegasus. Of the other second-magnitude leaders of Andromeda, Beta is orange-red and Gamma, also orange, is a fine binary easily separable with a small tele-scope. Pegasus contains one bright star away from the Square, Epsilon or Enif. Adjoining

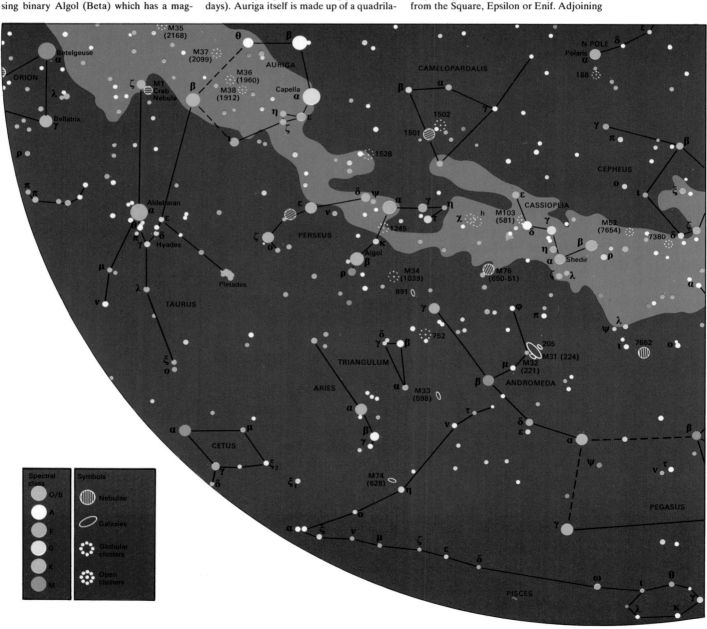

the Square is the dim zodiacal constellation of Pisces; and in this region also lie Aries, with the second-magnitude Hamal (Alpha) and Triangulum. Triangulum contains the loose spiral galaxy M33, a member of our local group of galaxies, and visible with binoculars on a clear night.

Map 2 – from Lyra to Delphinus

Vega, in Lyra, is one of three brilliant stars making up what has been unofficially called the Summer Triangle. Vega is the fifth brightest star in the whole sky and is easy to identify. From Britain and the northern United States it is almost overhead during summer evenings. Lyra is a small constellation, but contains a great many interesting objects. Epsilon Lyrae is a quadruple star, while Beta is the famous eclipsing variable. M57 (NGC 6720) is the most famous of the planetary nebulae, which are surrounded by gaseous shells, and lies between Beta Lyrae and the third magnitude Gamma.

The other two members of the Summer Triangle are Deneb in Cygnus and Altair in Aquila. Cygnus is a superb group. It is shaped

rather like an X and among its many interesting objects are the beautiful double Beta or Albireo, in which the primary is golden-yellow (magnitude 5) and the companion blue (magnitude 3), and the long-period variable Chi, between Albireo and the centre of the X, which has a great magnitude range (3.3–14.2) and a period of 407 days. Like most of its kind it is red and at maximum is easy to locate. The Milky Way is particularly rich in this area, so that Cygnus is well worth looking at through binoculars.

Altair, in Aquila, is recognizable, partly because of its brightness and partly because it has a fainter star to either side of it. Close to it is a line of stars, of which the central member, Eta Aquilae, is a typical Cepheid, with a period of 7.17 days.

In the general area of Cygnus and Aquila there are some small but quite distinctive constellations – notably Delphinus, the Dolphin. It was here that the famous slow nova, HR Delphini, appeared in 1967; it rose to magnitude 3.6 and is still visible with a small telescope. Sagitta is another small but easily identified group.

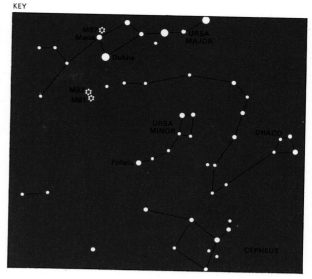

The area of the North Pole is marked by Polaris, which is within one degree of the pole itself, with Ursa Major to one side and the rather formless Cepheus to the other.

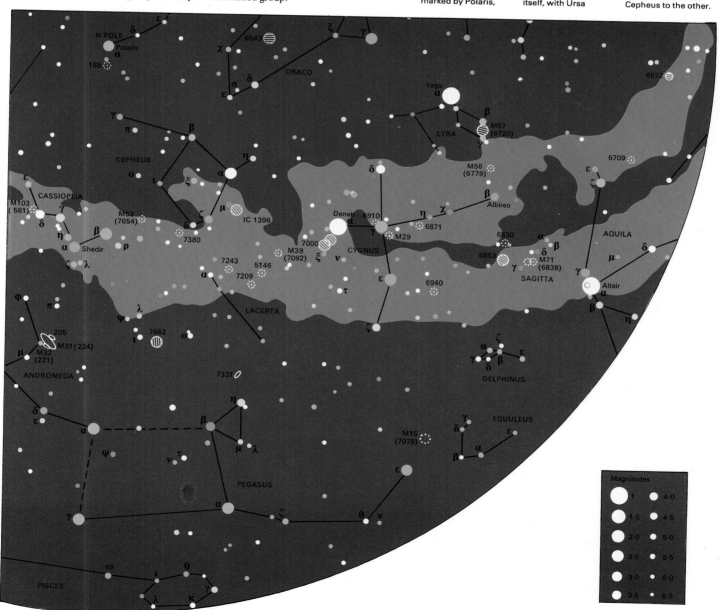

Star guide: southern sky I

There can be little doubt that the stars of the far south are more splendid than those of the far north. Brilliant constellations such as Centaurus, Carina and above all Crux are not visible from Europe or most of the United States, and the same is true of the bright external systems known as the Nubeculae or Magellanic Clouds. Other stars, such as Sirius, which are visible from the Northern Hemisphere, are much brighter seen from south of the Equator.

Crux – the key constellation

Crux, the Southern Cross, is the most famous of all the far southern groups and is, incidentally, the smallest of all the recognized constellations. Since it is not at all visible from the Northern Hemisphere it is not one of the ancient constellations, and it was not added to star maps until the seventeenth century. It is not like a cross because there is no central star to make up the X as there is with Cygnus in the northern sky; Crux more nearly resembles a kite but it is so compact that it cannot be mistaken. Acrux or Alpha Crucis, the leader, is a fine binary 270 light-

years from Earth. The magnitudes of the components are 1.6 and 2.1, giving a combined naked-eye magnitude of 0.8. The pair may be seen well with a small telescope. Beta Crucis (magnitude 1.3) is a very luminous B-type star and Gamma Crucis (magnitude 1.6) is a red giant. The fourth star of the pattern is much fainter at magnitude 3 and thus spoils the symmetry. Crux also includes the Jewel Box cluster, Kappa Crucis (NGC 4755), and the dark nebula that is usually termed the Coal-sack – a huge mass of dust and gas cutting out the light of stars beyond. Most astronomers consider this to be the finest of the dark nebulae.

Map 1 – from Carina to Hydra

Carina is part of the old constellation of Argo Navis, the ship that carried Jason and his companions in search of the Golden Fleece – but Argo was so large that it was cut up into a Keel, a Poop and Sails (Carina, Puppis and Vela). Carina contains many bright stars including Canopus, an F-type supergiant of magnitude –0.7, and also the extraordinary object Eta Carinae, which is wreathed in

nebulosity and is variable. For a period between 1834 and 1844 it ranked as one of the brightest stars in the sky at magnitude –0.7, but for almost a century now it has been below naked-eye visibility at magnitude 7.7.

The so-called False Cross is made up of two stars in Carina and two in Vela, all of about the second magnitude. The Milky Way flows through the Ship and the region abounds with clusters and rich star fields. By contrast Canis Major is relatively barren of interesting telescopic objects, but it contains several bright stars as well as Sirius. Although it is at least 20 times more luminous than the Sun, Sirius owes its apparent eminence to its closeness (8.7 light-years from Earth) rather than to its power. It is feeble compared with Canopus which appears appreciably fainter (magnitude –0.7 as against –1.47), but is much farther away and extremely luminous.

Map 2 – from Hydra to Scorpio

Hydra, the Watersnake, is a barren area of the sky. However, it does contain the reddish second-magnitude star Alphard, the solitary

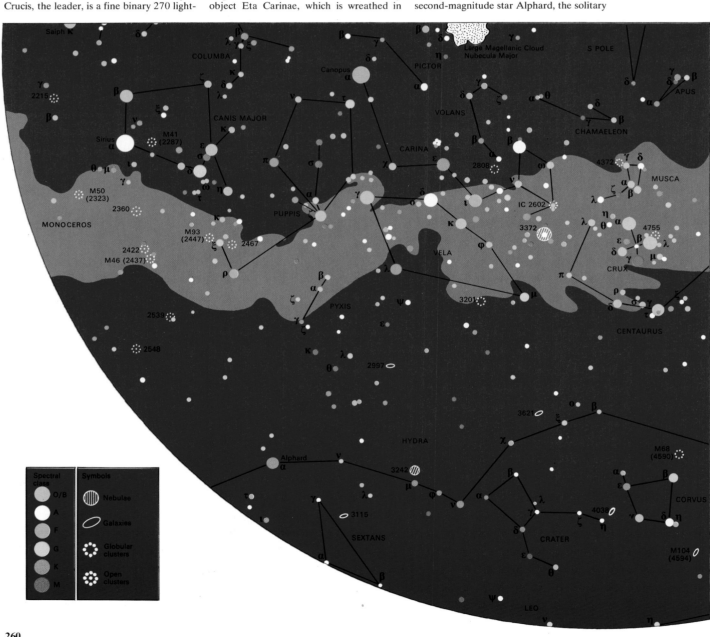

one, which appears distinct against its isolated background. The conspicuous quadrilateral of Corvus, the Crow, is also shown, and part of Virgo, including the first-magnitude Spica. Gamma Virginis, or Arich, is a fine binary with a period of 180 years.

Crux is more or less surrounded by Centaurus which is yet another magnificent group with many brilliant stars. Alpha Centauri, sometimes called Rigil Kent or Toliman, is a binary with a period of 80 years; any small telescope will separate its components of types G and K. At its distance of 4.3 light-years it is the nearest of all the bright stars, and its faint red dwarf companion Proxima, the nearest star to Earth, is only slightly closer. Adjoining Alpha is the remote Beta Centauri; at magnitude –4.3 it is more than 4,000 times as luminous as the Sun and bluish-white in colour. Also in Centaurus lies Omega (NGC 5139), the finest of all the globular clusters – conspicuous to the naked eye as a hazy patch and resolvable with a small telescope.

Adjoining Centaurus are Lupus, the Wolf, and Triangulum Australe, the Southern Triangle. Lupus is rather formless, but the Triangle is distinctive. Its leader, Alpha, is strongly orange in hue and is of magnitude 1.9. The other two members of the Triangle, Beta and Gamma, are of the third-magnitude.

Also shown on this map is Scorpio or Scorpius (the Scorpion), which is one of the most distinctive of the zodiacal constellations and one of the few which slightly resembles its namesake, as it consists of a long line of stars, many of them bright. Its brightest star, Antares, is a vast red giant, with a diameter of about 420 million kilometres (260 million miles); it is about 400 light-years away and its luminosity is almost 5,000 times that of the Sun. It has an apparent magnitude of 1, and rises well above the horizon over most parts of Europe.

Next to Scorpius is the obscure zodiacal constellation of Libra. The brightest star, Beta Librae, is of magnitude 2.7. Delta Librae is an eclipsing binary, with a magnitude range from 4.8 to 5.9. It is of Algol type and bright enough to be seen throughout its period with a pair of binoculars.

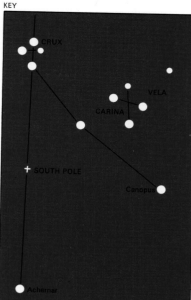

KEY

The south celestial pole is, compared to its counterpart in the north, more difficult to locate. The north celestial pole is conveniently marked by a bright star, Polaris, in the constellation of Ursa Minor, which lies within one degree of the polar point. Unfortunately there is no convenient south polar star; the pole lies in a barren region made up of the faint constellation of Octans. The best way to locate the pole is to follow the longer axis of Crux to the point about midway between Crux and the bright star Achernar, in the constellation of Eridanus. The nearest naked-eye star to the pole is the fifth-magnitude Sigma Octantis.

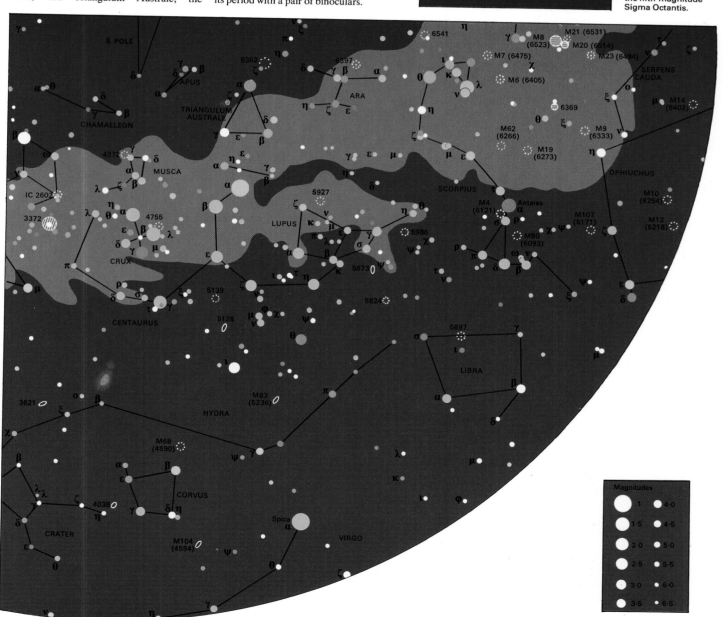

Star guide: southern sky II

Because of the great quantities of dust that lie in the main plane of the Galaxy, it is impossible to see through to the galactic centre, but at least its position is known. It lies beyond the star clouds of Sagittarius, at a distance of about 33,000 light-years from Earth.

Sagittarius – the key constellation
The constellation of Sagittarius (the Archer) is not hard to identify; it contains several reasonably bright stars, although there are no stars of the first magnitude. It is, however, the southernmost constellation of the Zodiac and so is never well seen from Europe. In shape Sagittarius is rather hard to define; some people with vivid imaginations have compared it to a teapot. It adjoins the Sting of Scorpius and between the Sting and the leader of Sagittarius (Epsilon Sagittarii, or Kaus Australis) there are two bright open clusters, M6 and M7. Sagittarius is also rich in globular clusters.

Strangely enough the star catalogued as Alpha Sagittarii is obscure; the Greek letters are not necessarily in sequence. Near Alpha is the circle of stars marking Corona

Australis, the Southern Crown, which is not nearly as prominent as the Northern Crown, but easier to identify.

If Sagittarius is an especially rich region, the south pole of the sky is particularly barren. There is no bright south polar star; the nearest naked-eye object is Sigma Octantis, which is a star of only the fifth magnitude and a rather poor substitute for the northern Polaris.

Map 1 – from Grus to Capricornus
The four "Southern Birds" are shown: Grus (the Crane), Pavo (the Peacock), Tucana (the Toucan) and Phoenix. This is admittedly a confusing area because only Grus has a distinctive form; it really does give the impression of a flying crane. The leaders, Alpha (Alnair, magnitude 2.1) and Beta (magnitude 2.2) are quite different. Alnair is bluish-white, but Beta is orange, and the difference is striking when the stars are seen through binoculars or any telescope. Adjoining Grus is Piscis Austrinus, the Southern Fish, which has as its leader the first-magnitude Fomalhaut – 23 light-years

distant and about 15 times as luminous as the Sun. Tucana is the most obscure of the four birds, but it contains a fine double, Beta, and 47 Tucanae, a fine globular cluster second only to Omega Centauri.

Finally there are two obscure constellations of the Zodiac – Aquarius and Capricornus. There are a few interesting objects in Aquarius, notably the bright globular cluster M2, while in Capricornus there is the naked-eye double Alpha or Al Giedi; Beta Capricorni is also a wide, easily seen double, readily separable with a small telescope or even with binoculars.

Map 2 – from Cetus to Orion
Cetus, the Whale or Sea-monster, is a long, rather faint constellation, most of which lies in the southern hemisphere although the head is just north of the equator. It has one second-magnitude star, Beta or Diphda, which is suspected of variability. Here too is Omicron Ceti or Mira, the most celebrated long-period variable in the sky. It has a mean period of 331.6 days and at some maxima it has been known to exceed the second mag-

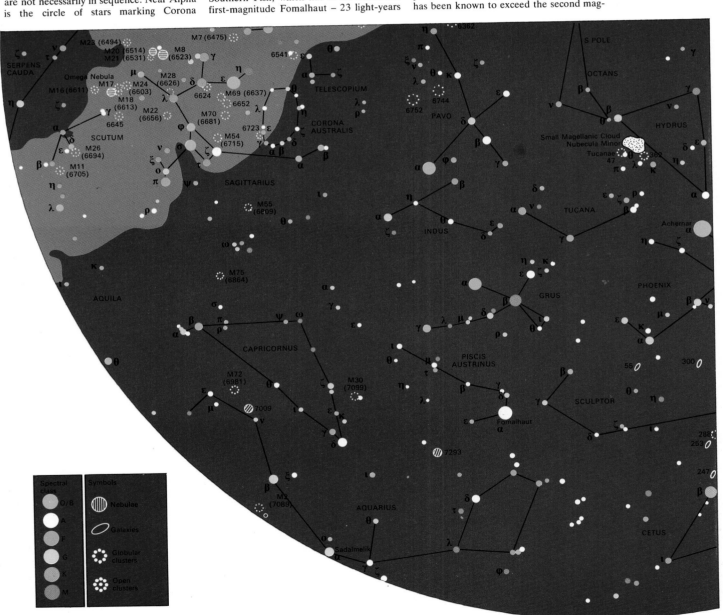

nitude; at minimum it descends to the tenth. Its variability was recognized as long ago as 1638. However, for much of the year it is below naked-eye visibility. Mira is a red giant and its colour is extremely prominent.

Close to Phoenix lies Achernar, the "End of the River" and the only really bright star in the long constellation of Eridanus, which winds its way from near the south pole as far as the boundaries of Orion. Achernar (magnitude −1.3) is 75 light-years from Earth and has a luminosity 256 times that of the Sun. Farther along the line of Eridanus is Theta, or Acamar, which is a fine double, and – like Castor in Gemini and Megrez in the Great Bear – is suspected of having faded during historical times, because ancient astronomers ranked it of the first-magnitude and it is now below the third.

Not far from the south pole the two remarkable Clouds of Magellan can be seen. Unfortunately they can never be seen by observers in Europe or the United States. The clouds are external systems, even though they look at first glance like detached portions of the Milky Way. They are about 150,000 light-years away from the Earth and are therefore the most remote objects clearly visible with the naked eye apart from the spiral galaxy M31 in Andromeda. The Large Cloud is so bright that even moonlight will not hide it. Binoculars bring out its form well and telescopic research has shown that it contains objects of all kinds, including globular clusters and gaseous nebulae. One star, S Doradûs, is thought to be a million times more luminous than the Sun, and yet without optical aid it cannot be seen. The Large Cloud also contains the Great Looped Nebula which is visible to the naked eye.

Orion, divided by the equator of the sky, is visible from every inhabited part of the world. The equator passes near Delta or Mintaka in the Belt, so that the brilliant Rigel lies well in the southern hemisphere. Rigel (magnitude −7.0) is a highly luminous star, thought to be about 49,000 times as powerful as the Sun and lies at a distance of 850 light-years from Earth. Also in the southern part of Orion lies the Great Nebula, M42, which contains the multiple star Theta Orionis, often known as the Trapezium.

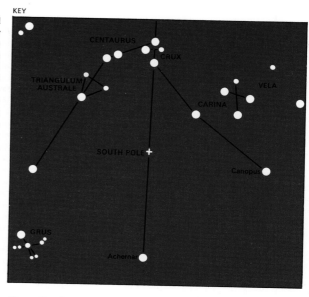

The south polar area contains no bright star, but it is surrounded by brilliant constellations such as Crux and Centaurus.

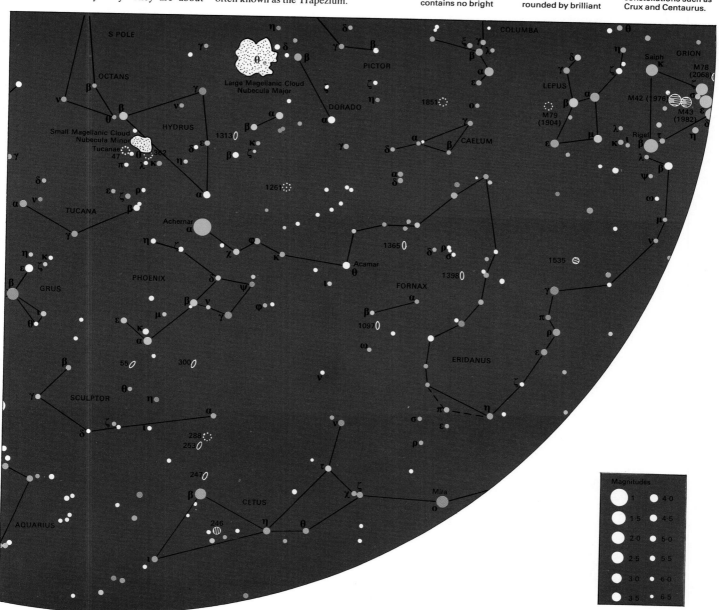

Seasonal star maps: northern

All the stars of the northern sky are visible to an observer in the Northern Hemisphere in the course of a year. The charts on these two pages are suitable for observers living between latitudes 30° and 50°N [Key]. The horizon is given by the latitude marks near the bottom of the charts. Thus for an observer who lives at latitude 30°N, the northern horizon on chart 1 [N] will pass just above Deneb and Deneb will be invisible.

Limits of visibility

A star rises earlier, on average, by two hours each month; thus the chart for 22.00hrs on 1 January will be the same as that for 20.00hrs on 1 February, 18.00hrs on 1 March and so on. All the times given in the charts are in Greenwich Mean Time with no allowance made for adjustments such as British Summer Time.

A star that never sets is said to be circumpolar; thus Ursa Major is circumpolar from England, while Arcturus is not. Limits on the visibility of a star for an observer at any latitude can be worked out from its declination (that is, the star's angular distance north

or south of the celestial equator). To an observer in the Northern Hemisphere, a star is at its lowest point when it is below the pole and therefore due north. The circumpolar region of the sky can be found by subtracting the observer's latitude from 90°. Suppose, for example, that an observer lives at latitude 51°N. Subtracting 51 from 90 gives 39: a star which is north of declination +39° will never set, while a star south of declination −39° will never rise. Thus, from latitude 51°N, which is approximately the latitude of London, the star Capella (declination +46° in round figures) will be circumpolar while Arcturus (declination +19°) will be well seen but will not be circumpolar; Canopus (declination −53°) will remain permanently below the horizon from this latitude.

Alkaid or Eta Ursae Majoris, the most southerly of the seven principal stars of the Great Bear, has a declination of approximately +50°. It will therefore be circumpolar from latitudes north of 40°N (90−50=40) and it will be invisible to an observer from latitudes south of 40°S.

The charts show the southern [S] and the

northern [N] aspects of the sky from the viewpoint of an observer in northern latitudes; the descriptions that follow are for the late evening (when the sky appears the same as it did three months previously). Inevitably, charts of this kind involve some distortion, but used solely as recognition aids they are quite satisfactory.

Stars of chart 1

In winter, the southern aspect is dominated by Orion and its retinue. Capella is almost at the zenith, or overhead point, and Sirius is at its clearest – although as seen from northern Europe or the United States, Sirius is always rather low down and so the effects of the Earth's atmosphere make it twinkle strongly. Other stars in the general region of Orion are Aldebaran, Castor and Pollux and Procyon, while Orion itself has two brilliant leaders, the white Rigel and the orange-red Betelgeuse. Orion is invaluable as a guide to other groups; for instance, Aldebaran can be found by following the upward direction of the three stars in Orion's belt.

The sickle of Leo is prominent in the east

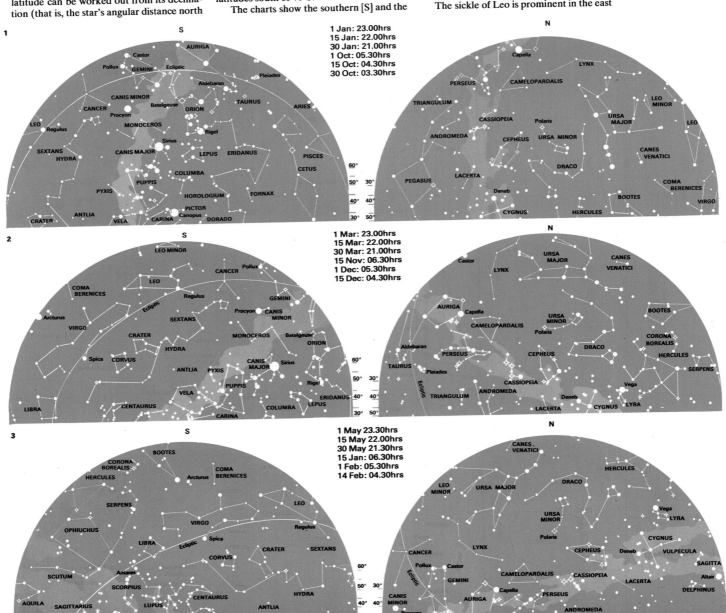

(as shown here it is cut by the two maps). In the northeast lies Ursa Major, the Great Bear – the other invaluable aid to recognizing various constellations. The two Pointers show the way to Polaris, the Pole Star (Alpha Ursae Minoris), which is of the second magnitude. It lies less than one degree away from the celestial pole, so that its declination exceeds +89°. Vega is at its lowest and is not shown on the first chart; it is just circumpolar from England, but not from the New York area. Vega, Polaris and Capella lie approximately in a line, with the Pole Star in mid position; thus when Capella is almost at the zenith, Vega is at the horizon (on winter evenings) and when Vega occupies the zenith, Capella can be seen at the horizon (on summer evenings).

Stars of charts 2–6
During spring evenings (chart 2), Orion is still above the horizon; Leo is high up with Virgo to the east; Capella is descending in the northwest and Vega rising in the northeast. In the west, Aldebaran and the Pleiades are still quite clearly visible.

By early summer (chart 3), Orion has disappeared, although Castor and Pollux remain in view. Vega has risen, Capella is descending and Ursa Major is not far from the zenith. This is the best time for evening viewing of Arcturus, which is of a light orange hue and is actually the brightest star in the Northern Hemisphere (fractionally superior to Vega or Capella).

During summer evenings (chart 4), Vega occupies the zenith and its brilliance and distinct bluish colour make it unmistakable. An excellent view is afforded of the so-called Summer Triangle (Vega in Lyra, Deneb in Cygnus and Altair in Aquila) and, in the south, of the brilliant Antares in the Scorpion and the star clouds of Sagittarius in the direction of the centre of our Galaxy.

In autumn (chart 5), the Square of Pegasus is high, the Summer Triangle is still apparent and Ursa Major is at its lowest.

By early winter (chart 6), Pegasus is still high and Vega and its companions are sinking, but Orion has returned once again and will dominate the evening sky until well into the following year.

From these northern latitudes the stars shown on these pages can be seen.

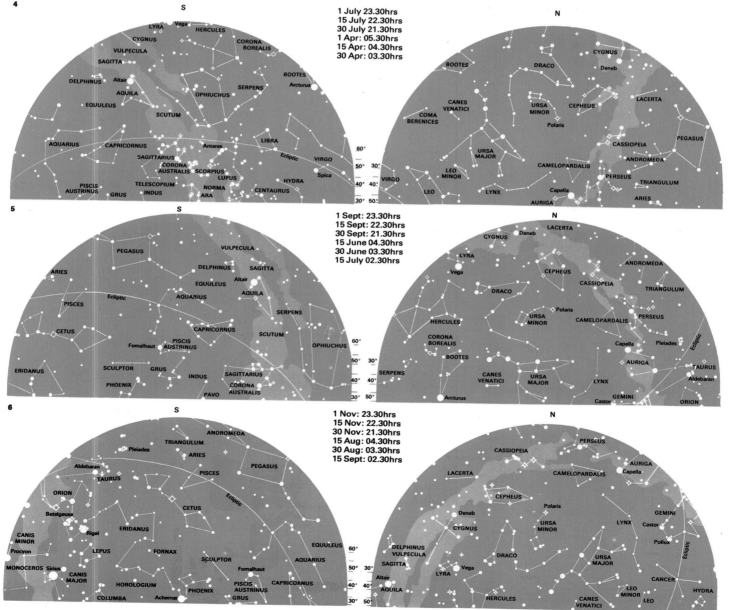

Seasonal star maps: southern

The far southern skies are much superior to those of the Northern Hemisphere for astronomical observation. They contain a number of brilliant groups that cannot be viewed by observers from either most of Europe or the United States.

Northern observers lack both the Southern Cross and the two Clouds of Magellan, which are of special importance astronomically. Other objects of the far south include Alpha Centauri, nearest of the bright stars; the globular clusters Omega Centauri and 47 Tucanae; and the remarkable irregular variable, Eta Carinae. The Milky Way is very rich in its southernmost portions and in the Southern Cross there is the so-called Coal-sack – the best example of a dark nebula – together with the glorious open cluster that is nicknamed the Jewel Box.

Features of chart 1
The six pairs of charts given on these two pages are arranged with the left-hand chart looking towards the northern horizon and the right-hand chart looking towards the southern. They are valid for observers living in South Africa, Australia, New Zealand and most of South America (allowing for differences in aspect) and calculations can be made by reference to a star's declination.

In the evenings during January, the southern summer, Orion is high up; from the Southern Hemisphere Rigel is higher than Betelgeuse, while the Belt stars point "up" towards Sirius and "down" towards Aldebaran. The whole of the Ship (Argo) is displayed; this huge constellation was regarded by astronomers as an unwieldy section for observation and was "cut up", the principal sections now recognized being Carina (the Keel), Vela (the Sails) and Puppis (the Poop). Carina contains Canopus, the second brightest star in the sky, which is very high during January evenings. It does not appear as brilliant as Sirius (its apparent magnitude is −0.7, as against −1.47 for Sirius) but in reality it is much more luminous and lies at a distance of hundreds of light-years from Earth.

Crux is rising in the southeast in chart 1. Strictly speaking it is shaped more like a kite than a cross; it is the smallest of the 88 constellations in the sky, but is very compact. Two of its stars (Acrux or Alpha Crucis and Beta) are ranked as being of the first magnitude. The third (Gamma Crucis) is ranked just below this, while the fourth star of the kite pattern is much more faint. Even a casual glance shows that of the four main stars, Gamma is orange-red (spectrum M) and the others white. The two Pointers to the Cross are Alpha and Beta Centauri; Alpha is the brightest star in the sky apart from Sirius and Canopus and lies at a distance of only 4.3 light-years. It is a fine binary, separable with a small telescope; Alpha Crucis is also a splendid binary.

Achernar in Eridanus (the River) is to the southwest. Probably the best way to locate the region of the south celestial pole is to look midway between Achernar and Crux, but there are no really bright stars to help in identification of the pole itself.

Charts 2 and 3
By March evenings [chart 2] Canopus is descending in the southwest and Crux is rising to its greatest altitude; together with

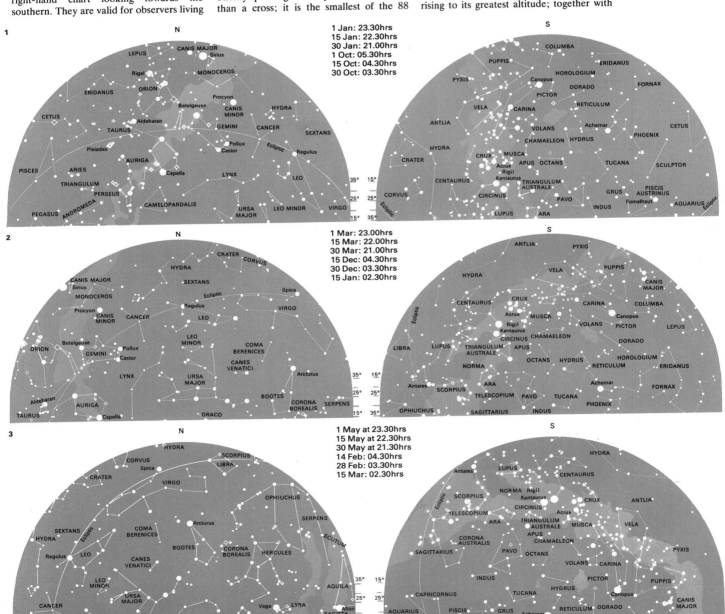

1 Jan: 23.30hrs
15 Jan: 22.30hrs
30 Jan: 21.00hrs
1 Oct: 05.30hrs
15 Oct: 04.30hrs
30 Oct: 03.30hrs

1 Mar: 23.00hrs
15 Mar: 22.00hrs
30 Mar: 21.00hrs
15 Dec: 04.30hrs
30 Dec: 03.30hrs
15 Jan: 02.30hrs

1 May at 23.30hrs
15 May at 22.30hrs
30 May at 21.30hrs
14 Feb: 04.30hrs
28 Feb: 03.30hrs
15 Mar: 02.30hrs

the Pointers they comprise a splendid group. Not far from it is the magnificent globular cluster Omega Centauri, much the finest of its type in the entire sky. The Milky Way is extremely rich in this whole area and even binoculars will give a good view of the Coalsack, which conveys the impression of being a virtually starless area (there are a few stars in the foreground). Scorpio (or Scorpius) is coming into prominence in the southeast, Orion is dropping and Leo can be seen high in the north region.

May evenings show Alpha and Beta Centauri very high up, with Crux; Canopus is visible in the southwest, but Orion and Sirius have set. Arcturus is prominent in the north, with Spica and Virgo not far from the zenith. The Scorpion is now dominant.

The Scorpion group is a splendid one, with a long chain of stars that cannot be mistaken – the red giant Antares is particularly striking. There are many bright, open clusters in the region, as sweeping the sky with binoculars will reveal. Adjoining the Scorpion is Sagittarius, which contains many moderately bright stars.

Charts 4, 5 and 6
The Scorpion is near the zenith in chart 4 while Crux and Centaurus remain prominent in the southern aspect of the sky. The brilliant northern stars Vega, Altair, Deneb and Arcturus are all visible but Canopus is at its lowest and is virtually absent from the sky for a short while.

Pegasus is high in the north in chart 5; Vega, Altair and Deneb can still be seen and Fomalhaut is almost overhead; the Scorpion is sinking in the southwest and Crux almost out of view.

Orion has returned in chart 6 and with it Sirius, Canopus and the other adjacent stars. When Crux and Centaurus also return the southern sky will again be at its brilliant best.

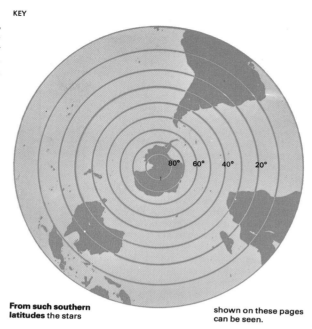

From such southern latitudes the stars shown on these pages can be seen.

March is the best time of year for seeing Ursa Major, although the constellation never rises to observers in the southernmost parts of the latitudes for which the maps have been drawn. (From New Zealand, for instance, the seven main stars of Ursa Major never rise, but from Rhodesia they can attain a considerable altitude.)

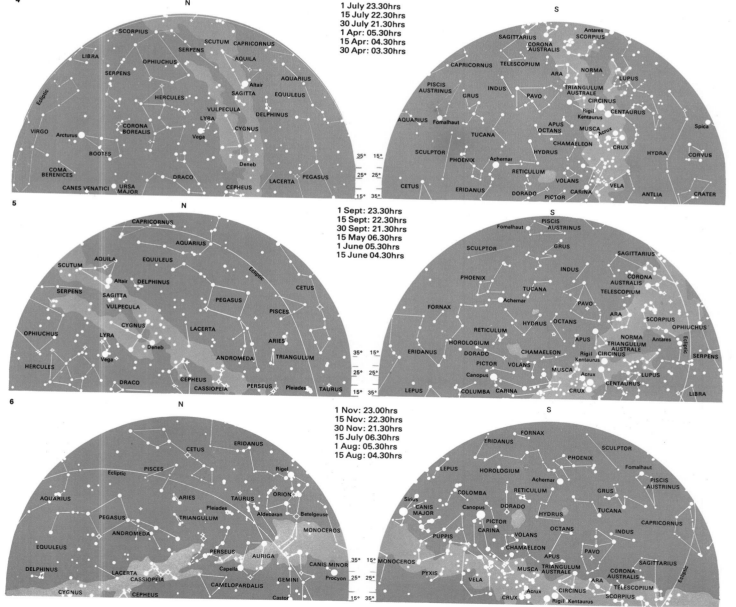

267

A history of space achievements

The idea of reaching other worlds is far from new [Key]. As long ago as the second century AD a Greek satirist, Lucian of Samosata, wrote a story about a journey to the Moon, although he did not intend it to be taken seriously; as he commented, his *True History* was made up of nothing but lies from beginning to end. Another early space-travel story was written by the German astronomer Johannes Kepler (1571–1630) and was published in 1634. In Kepler's version the "astronaut" was transported to the Moon by an obliging demon.

Early space rockets

It is only in modern times that space research has become a practical possibility. In the Earth's atmosphere winged aircraft can use conventional engines, such as gas turbines, which need a supply of air to burn their fuel. The atmosphere does not extend upwards for very far on a cosmic scale and in space, where there is no air, a vehicle needs an engine that does not "use" air, that is, a rocket motor. In a rocket, a stream of gas emitted from the exhaust of the vehicle provides the motive power needed; it has been said that a rocket "kicks against itself" and the presence of a resisting atmosphere is actually a hindrance.

The great theoretical pioneer of astronautics was a Russian, K. E. Tsiolkovsky [1], whose first papers on the subject appeared in 1903, although at the time they aroused little comment. Tsiolkovsky realized that solid fuels, such as gunpowder, are inadequate for space flight and he proposed the use of a liquid-propellant rocket motor. The first rocket to have an engine of this type was fired by R. H. Goddard (1882–1945) [2], in America, in 1926; although it achieved a maximum speed of only 97km/h (60mph) in its brief flight, it proved the principle to be valid. Subsequently a German team, including Wernher von Braun (1912–), developed liquid-propellant rockets with some success. The team was then taken over by the Nazi government for military purposes and transferred to the island of Peenemünde, in the Baltic, where the V2 weapon was developed in time to be used in the last stages of World War II.

The V2 was the direct ancestor of the space probes of today, as after the end of the war most of the German researchers went to the United States, where progress continued steadily. In 1949 a two-stage rocket [3], launched from the United States, reached the altitude of almost 400km (250 miles).

From Earth satellites to Moon landing

By the early 1950s, scientific rockets were proving useful in studying the upper atmosphere and what may be called "near space", but it was realized that an artificial satellite would achieve the same purpose much more efficiently. A scientific vehicle carried up in a rocket and set in a stable orbit round the Earth will behave in the same way as a natural astronomical body, provided it is clear of the resisting atmosphere.

The first artificial satellite [4] was launched by the USSR on 4 October 1957, marking the real start of the Space Age. The vehicle, Sputnik 1, was only about the size of a football and carried little apart from a radio transmitter, but it paved the way for all future research. Other Soviet satellites followed and in 1958 the United States put into orbit their

1 1903: Konstantin Tsiolkovsky (1857–1935) founded astronautics when he published a series of papers in Russia that laid down the fundamental principles with remarkable foresight and accuracy.

2 1926: R. H. Goddard sent up the first liquid propellant rocket in the United States of America. Although his rocket was small and feeble, Goddard's experimental work was of immense significance.

3 1949: An early step rocket was launched from White Sands, USA. It consisted of a V2 carrying a WAC Corporal rocket which reached an altitude of 393km (244 miles), then a record height.

4 1957: Sputnik 1, the first artificial satellite, was launched by the USSR.

5 1959: Luna 1 was the first successful lunar probe, bypassing the Moon at 6,400km (4,000 miles).

6 1961: Yuri Gagarin of the USSR became the first man in space – launched on 12 April, his craft, Vostok 1, made a complete circuit of the Earth at a height ranging from 180km (112 miles) to 301km (187 miles). His flight of 1hr 29 min proved that man could invade space and return safely to a prearranged position. Gagarin made only one space flight. With tragic irony, he was killed in 1968 in the crash of an ordinary aircraft.

7 1961: Alan Shepard was the first US space traveller, making a 15-minute sub-orbital "hop" on 5 May. A decade later, he commanded the Apollo 14 Moon flight. He is now a US Navy admiral.

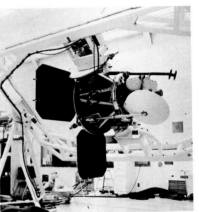

8 1962: Telstar, the first active communications satellite, provided a transatlantic television relay. Only 876mm (34.5in) in diameter, it is still in orbit but all track of it has now been lost.

9 1965: Two US space vehicles launched on 5 and 15 December, Gemini 6 and 7, were brought to within 30cm (12in) of each other in a rehearsal of the docking procedure essential to the success of the Apollo Moon landing project.

10 1966–7: Orbiters 1 to 5 were put into closed orbits around the Moon. All these space vehicles were operated successfully and sent back thousands of photographs covering the whole of the lunar surface.

11 1968: US astronauts Frank Borman, James Lovell and William Anders made the first manned flight around the Moon in Apollo 8 from 21 to 27 December. This was an essential preliminary to the actual Moon landing in 1969. Having reached the vicinity of the Moon, Apollo 8 was put into a closed path and completed ten circuits before blasting back into Earth transfer orbit.

first Explorer 1, satellite. It provided the first information about the radiation zones that surround the Earth, which are now known as the Van Allen belts.

The first probes to the Moon were also Russian. In January 1959 Luna 1 [5] by-passed the Moon and later in the year two more probes were launched. One of them hard-landed on the Moon while the other made a round trip, sending back photographs of the Moon's far side. In 1961 came the first manned satellite, Vostok 1, in which Yuri Gagarin (1934–68) [6] made a full circuit of Earth. By this time the whole concept of research was being taken very seriously although less than a decade before it widely ridiculed.

the early 1960s Earth satellites oped that were able to send back tographs of the Earth, as well as cellaneous information. They for communication purposes. 8], the first active television launched.

les became capable of three men instead of only

one and space-docking [9] manoeuvres were carried out. Meanwhile the US Apollo prog-ramme to land a man on the Moon was being developed, culminating with the Apollo 11 flight in July 1969, when Neil Armstrong (1930–) and Edwin Aldrin (1930–) made the first Moon walk.

Exploring the Solar System

The first successful planetary probe was Mariner 2, which bypassed Venus in 1962 and sent back the first reliable information about that peculiar world. Mars was con-tacted in 1965 by Mariner 4 and in 1971 Mariner 9 [14] entered a closed orbit round Mars, sending back thousands of high-quality pictures. In 1974 the US Mariner 10 by-passed both Venus and Mercury and the first Jupiter probe, Pioneer 10 [15], reached its target in 1973. By 1975, Pioneer 11 was on its way to Saturn. In the same year the Soviet Veneras 9 and 10 [18] sent back photographs from the surface of Venus that showed sur-face details to be markedly different from those anticipated. The US Vikings 1 and 2 successfully landed on Mars in 1976.

KEY

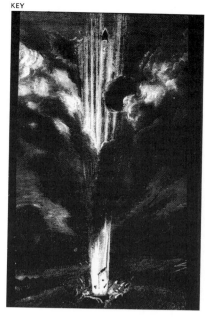

In his science-fiction story *From The Earth To The Moon* (1865), the French novelist Jules Verne (1828–1905) sug-gested that men could be sent to the Moon in a projectile fired from a huge cannon he called "Columbiad" The principle is not practical and in his story Verne had to invent a second natural satellite to swing the projec-tile back to its starting-point, Earth. But Verne's use of scientific detail to give an air of realism to fanciful adventures prompted others, including H. G. Wells (1866–1946), to write of space travel. Their fantasies were only a few generations away from the actuality of the first Moon landing.

14 1971: Mariner 9 became the first vehicle to orbit Mars. The cameras shown in this detail operated from December until late 1972, taking photographs that greatly changed ideas about the planet.

12 1969: Neil Arm-strong and Edwin Aldrin made the first lunar landing during the Apollo 11 mission with Michael Collins piloting the orbiting command module. Their walk was in Mare Tranquillitatis.

13 1970: Lunokhod 1, a Soviet "crawler", moved around the lunar surface sending back valuable infor-mation. Carried by Luna 16, which was launched in Sept-ember, it operated until October 1971.

15 1972: Pioneer 10, the first Jupiter probe, by passed the planet in December 1973 at a distance of about 31,400km (82,000 miles), send-ing back detailed information and pictures of the planet.

16 1973: Skylab, the first American space station, was manned by three successive crews, the last spending 84 days in space. Owen Garriott of the second crew is seen here at the tele-scope mount console.

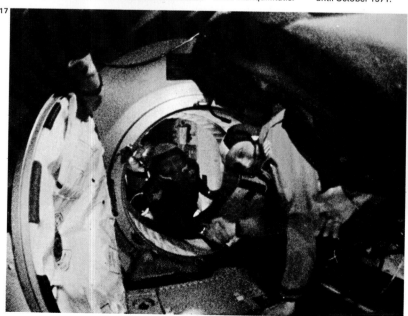

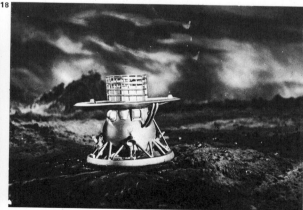

17 1975: Soyuz-Apol-lo, the first joint US-USSR space pro-ject, enabled the crews of the Soyuz Soviet craft and a modified Apollo cap-sule to enter each-other's vehicles after docking in space. The

American commander was Thomas Stafford, with Vance Brand and Donald Slayton. The Rus-sians were comman-ded by Alexei Leo-nov. It was a valu-able exercise in joint scientific research.

18 1975: Veneras 9 and 10, two Soviet probes (this is a model), made soft land-ings on Venus, oper-ating for about an hour. Each sent back a picture of an unex-pectedly rock-strewn landscape.

Stations in space

The idea of a space station or artificial manned satellite permanently circling the Earth outside its atmosphere was discussed by a Russian scientist Konstantin E. Tsiolkovsky (1857–1935) [Key] about the turn of the century, although at the time it could be regarded as little more than a fantasy. He looked upon the space station both as a stepping-stone refuelling base for spaceships visiting other planets and as a laboratory in which scientists could carry out experiments that would be impossible to perform on Earth. He even proposed growing plants in space stations to provide their crews with alternative sources of food and oxygen.

Overcoming gravity
The effect that weightlessness would have on space travellers was unknown. In orbit – where gravitational attraction is opposed by equal and opposite inertial forces – a body experiences no mechanical stress, and astronauts and any loose objects float weightlessly. The same is true of an unpowered spaceship moving in frictionless space towards or away from the Earth.

It was widely believed that even short periods of weightlessness (or zero gravity) would have ill effects on space travellers and thoughts turned to creating artificial gravity. One of Tsiolkovsky's first designs showed a huge cylindrical space station that spun on its central axis. The crew had their feet firmly planted on the inside walls, with their heads pointed at the spin axis, by action of centrifugal force. Vegetation in a "cosmic garden" grew inwards towards the centre.

As late as 1952 Wernher von Braun (1912–) who worked on the V2 at Peenemünde and who was mainly responsible for the rockets that launched America's first artificial satellite and the Apollo mooncraft, proposed a space station that had the form of a huge rotating wheel [2]. The crew quarters were in the rim of the wheel, docking ports for visiting spacecraft were in the central hub and tubular "spokes" allowed people to move from one part of the space station to the other.

Yuri Gagarin's flight in Vostok 1 in 1961 showed that weightlessness is not uncomfortable and since then men have remained in orbit under conditions of zero gravity for nearly three months. However, it is still uncertain whether, after spending much longer periods in space, the human body can adjust to normal gravity conditions without ill effects and we may yet see attempts to create artificial gravity in large-scale vehicles, which would enable astronauts to spend possibly limitless periods in outer space.

Present and future space stations
True orbital stations were first launched during the early 1970s after experience had been gained with Soviet Soyuz and American Apollo spacecraft, which carried a variety of scientific instruments.

Although the Russians had considerable trouble with the first of their 19-tonne Salyut stations, they have since flown a number of highly successful missions. The Americans had problems, too. Skylab was damaged when it was launched and had to be repaired in orbit before space teams could begin their experiments. It was manned by three successive crews who spent 28, 59 and 84 days in space respectively [1].

CONNECTIONS

See also
268 A history of space achievements
224 Solar eclipses
158 Man in space

In other volumes
158 Man and Machines

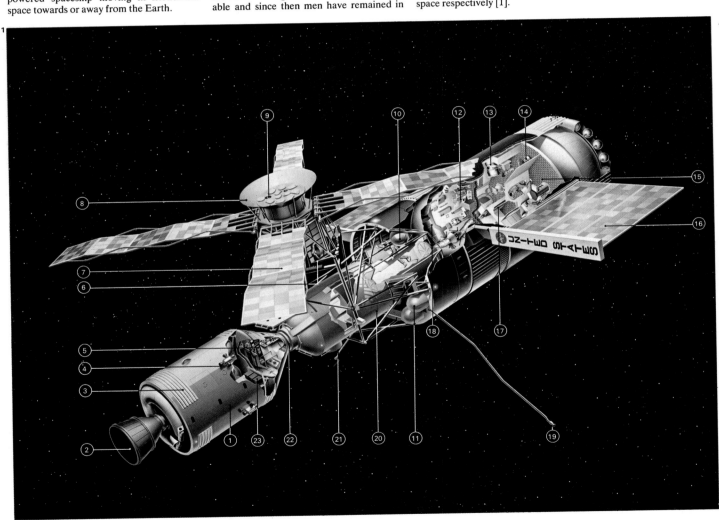

1 Skylab, the first American space station, was launched on 14 May 1973 and was manned by three successive crews for a total of 171 days. Its weight was about 82 tonnes and it measured 25m (82ft) long and 6.7m (22ft) across the workshop section. While in orbit, the instruments and systems were powered by solar cells. The various components are indicated by the following key: [1] modified Apollo spacecraft (command module plus service module) to take the crews to the space station; [2] service propulsion system engine, with a thrust of 9,100kg (20,000lb); [3] radiators; [4] attitude control jets used in docking, each nozzle with a thrust of 45kg (100lb); [5] crew station in command module; [6] Apollo telescope mount; [7] solar cells, converting sunlight to electricity to power the Apollo telescope mount; [8] sun shield (this gave trouble initially but was later rectified); [9] telescope apertures; [10] oxygen tank and [11] nitrogen tank, for the two-gas atmosphere inside Skylab; [12] manoeuvring unit; [13] lower body negative pressure device; [14] gravity substitute workbench; [15] food provisions; [16] solar cells (the foreground wing was torn off during launch); [17] sleep restraints; [18] water containers; [19] aerial; [20] multiple docking adapter; [21] alternative docking port; [22] atmosphere interchange duct; and [23] descent battery packs charged by solar energy from the telescope mount.

The first ferry craft to bring crew and cargo to space stations were versions of existing spacecraft launched by expendable rockets. To reduce the cost American scientists are now building the re-usable "space shuttle", which will take off vertically like a rocket, launch satellites or visit a space station and fly back to Earth like an aeroplane [3]. The winged orbiter is intended to be re-used at least 100 times.

When not used for other purposes, the shuttle can become a miniature space station in its own right. Within its 18.3×4.6m (60×15ft) cargo bay will fit the European Space Lab in which up to four scientists can work without spacesuits for up to 30 days as the spaceplane swings round the Earth. After a mission Space Lab can be replaced by another cargo ready for another flight.

The uses of orbiting stations

The value of orbiting stations is to carry out work that cannot be done more cheaply by unmanned satellites. The first space stations were used for research in biology, chemistry, physics, observations of the Earth's natural resources and studies of the Sun and other stellar objects.

Skylab and the Apollo spacecraft that took part in the "space handshake" with the Soviet Soyuz in 1975 each carried small electric furnaces to melt various metal samples under weightless conditions. In the future it may be possible to make ultra-lightweight foamed steels with many of the properties of solid steel, to combine dissimilar materials such as steel and glass and to grow crystals of great purity for the electronics industry.

The gravity-free environment should also be ideal for isolating biological materials for the treatment of certain diseases and to purify vaccines.

Future space stations will be assembled from modules (units) ferried into orbit by the shuttle. Eventually, it may be possible to build orbiting power stations that have immense solar-cell collectors generating electricity from sunlight. The energy would be beamed to Earth by microwave antennae for use in factories and homes, and would provide a limitless supply of power.

KEY

Konstantin Eduardovich Tsiolkovsky was the "father of astronautics" and the man who laid down the principles of space stations. Tsiolkovsky was a shy, deaf Russian teacher whose first papers on astronautics, although written about 1897, appeared in 1902. Even then they aroused little comment and Tsiolkovsky did not become famous until near the end of his life. Today there is a Tsiolkovsky Museum in Kaluga, where he lived. Although he was not an experimenter many of his theories were correct and as a space pioneer he was decades ahead of his time. He was also the first to stress the importance of liquid propellants in rockets.

2 Early space-station design took the form of a wheel with the power supplies in the hub and the crew quarters arranged round the rim. This design was originally worked out to provide for "substitute gravity" by rotation of the wheel. It was then believed that zero gravity might be harmful to astronauts even over short periods, but after Yuri Gagarin's flight and subsequent experiences with Skylab and the Soviet Salyut stations this was disproved.

3 The NASA Space Shuttle has a winged Orbiter that lifts off [1] with a large external tank containing the ascent propellants and two solid rocket boosters which separate [2] when the craft has climbed 45km (28 miles). The tank is discarded [3] just before the Orbiter goes into orbit. Typical payloads are the European Space Lab, space probes for release in orbit [4], or modules for assembly into a space station. The Orbiter uses its manoeuvring engines as retro-rockets to re-enter the atmosphere [5], when it endures high temperatures, for which it is reinforced with heat-resistant materials. It lands back at base [6].

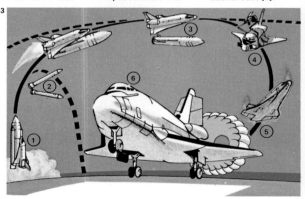

4 Future orbiting space stations will be different from the neat wheel design of pioneer days. A station, assembled in space, is designed to move round the Earth at a height of about 500km (300 miles) and will accommodate a crew of up to 100 members. In this picture, we see a space shuttle that has just delivered a propellant module for a spaceship bound for Mars. The nearer shuttle has just fired its retro-rockets to start its journey back to Earth. Below, the surface of the Earth itself is largely covered with cloud. Various other vehicles are shown in the black sky.

Colonizing the Moon

The Apollo missions to the Moon were essentially reconnaissances [Key]. All that Apollo could hope to do was to take three men to the neighbourhood of the Moon, land two of them on the surface for a brief period and then bring all members of the crew home safely. There was no provision for rescue in the event of a failure of the lunar module during the surface expedition and the time scale was very limited. Yet Apollo was an essential part of the main programme of lunar colonization and it showed that there is no reason why bases should not be set up on the Moon sometime in the future.

Problems encountered on the Moon
There is no question of turning the Moon into a kind of second Earth. The main problem is the lack of atmosphere. Unfortunately, the Moon is an airless world and there is not the slightest chance of providing it with a breathable atmosphere; the low escape velocity means that it is incapable of retaining a dense atmosphere similar to that of the Earth. Lack of atmosphere means a total lack of water and it now seems that – contrary to earlier

expectation – it will not be practicable to extract water from the lunar rocks for the simple reason that there is none to extract. Neither is there any hope of finding underground supplies of ice. Colonists of the future will have to take everything with them and it will be a long time before a lunar station can hope to become self-supporting.

The development of lunar bases
By 1990 or thereabouts, the first permanent lunar bases should have been developed and these will be a great advance on the chemically propelled vehicles of Apollo. By then, too, the space-station projects should be well under way and it will be practicable to consider going back to the Moon. Possibly the first step will be to send supplies to the surface, setting them down at a prearranged point to await the arrival of the explorers, so that when the astronauts land they will find supplies of various kinds ready for use.

The lunar modules themselves may be used as the first bases, but this pioneer phase should not last for long and more elaborate designs should be developed quite quickly.

One pattern, dating from the 1930s, is that of a series of domes, each kept inflated by the pressure of air inside it and equipped with a system of airlocks for the exit and entry of the crew members. This kind of design might be developed. Fortunately, it is now known that there is no danger from meteoritic bombardment so that the relative fragility of the domes should not prove a major problem. In the pre-Apollo days it was thought that it might be necessary to construct lunar bases underground for protection.

Even when space-shuttles have been perfected, the expense of travelling between the Earth and the Moon will still be considerable and every possible method will have to be used to cut supply journeys to a minimum. Everything (including human waste products) will have to be "recycled", particularly the atmosphere. The colonists will spend long periods on the Moon which involves making the conditions as comfortable as possible; inside the base it will be essential for the colonist to be able to take off his spacesuit and behave as naturally as is possible under conditions where gravity is

CONNECTIONS

See also
178 The Moon

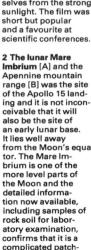

1 Science-fiction writers have always been attracted by the Moon. Jules Verne (1828-1905) described a circumlunar voyage more than 100 years ago and H. G. Wells (1866-1946) a fantastic world peopled by insect-like beings. In 1902 came the first famous Moon voyage film, produced by George Méliès (1861-1938). This frame from the film shows the arrival of the rocket (to the distress of the Moon), after which the space travellers go for a walk on the surface, not forgetting to put up their umbrellas to shield themselves from the strong sunlight. The film was short but popular and a favourite at scientific conferences.

2 The lunar Mare Imbrium [A] and the Apennine mountain range [B] was the site of the Apollo 15 landing and it is not inconceivable that it will also be the site of an early lunar base. It lies well away from the Moon's equator. The Mare Imbrium is one of the more level parts of the Moon and the detailed information now available, including samples of rock soil for laboratory examination, confirms that it is a complicated patchwork of lava flows.

3 The first lunar base is unlikely to be at all like the final elaborate stations. Here is the scene after the pioneer expedition has established itself. In the foreground is the basic space-station module which will serve as the centre of the future station. It can accommodate a crew of up to a dozen members and it can also provide all that is needed for a prolonged stay, although in an emergency the expedition can be halted and the crew returned to Earth by means of ferry vehicles. Also shown is a lunar Rover, similar to the vehicles used by the last three Apollo expeditions, which proved to be so successful; a cargo-landing craft with a separate cargo module; and a lunar drill.

only 17 per cent of the Earth's. There is also the question of recreation, which will be important in view of the relatively long tours of duty that are to be expected. Books, films and musical recordings present no problems; but what about physical recreation? No doubt various new sporting activities will be developed on the Moon, suited to conditions of low gravity.

Essential supplies of food

To send all food supplies from the Earth will be impracticable and efforts will be made to persuade edible plants to grow on the Moon. Of course this cannot be done in the open, but inside the domes the principles of hydroponic farming, whereby plants can be grown without soil, may be used. The plants are suspended in netting inside a tank and are fed by liquid nutrients circulated beneath them. The principle has been tested and excellent results have been achieved, so that there seems no reason why it should not work on the Moon.

The early lunar bases may be staffed entirely by scientists. There may be physicists, anxious to take advantage of the low gravity, the limitless hard vacuum and the chance to study all the radiations coming in from space; astronomers, thankful to escape from the restrictions imposed by the screening layers in the atmosphere of the Earth; chemists, biologists, medical men – in fact, scientists of all disciplines. The lunar base should add appreciably to the sum total of our knowledge.

This will be the second phase. Then, as the colony becomes more and more self-supporting, it may be able to take in non-scientists as well, at least for brief visits. The idea of "holidays on the Moon" may no longer be fantastic in 100 years' time. By then, there may be children who have been born on the Moon and who regard it and not the Earth as their home world. Before the end of the twenty-first century the Moon should support not one base, but many, used for a variety of purposes.

It is remotely possible that by the end of the twenty-first century, the indigenous population of the Moon may be demanding independence from the home planet Earth.

Astronaut James Irwin of the Apollo 15 mission is standing in the Hadley-Apennines region – one of the possible sites for the first lunar base. The peak of Hadley Delta is seen beyond.

4 A permanent lunar base will probably be set up in the far north of the Moon in order to avoid the intense daytime heat of lower latitudes. (The night temperature is virtually the same at all points on the surface.) The relatively low altitude of the Earth in the sky indicates the high altitude of the site. At the limb of the Earth-turned hemisphere, an observer on the Moon would see the Earth at the horizon (with slight variations due to the irregular orbit of the Moon) while from the Moon's far side the Earth would never rise at all. In this illustration the Earth is shown as full, with its surface details masked by cloud. It has just passed through the Milky Way into the constellation of Gemini, the Twins. The red star to the right is the semi-regular variable, Eta Geminorum. The base is made up of several domes, each with its separate system of airlocks. It will be essential to conserve atmosphere to the greatest possible extent and individual airlocks are necessary to guard against the sudden failure of pressure in one of the domes. Radio aerials and various kinds of instruments are also shown. The illumination is purely by Earthlight (because the Earth is full, the Sun is below the horizon; to a terrestrial observer, the Moon is new) and the radiance on the lunar rocks is bluish.

The Martian base

In the early part of the twentieth century it was generally supposed that Mars might well be able to support terrestrial life, and that it might even be inhabited. In 1877 the Italian astronomer Giovanni Schiaparelli (1835–1910) had set the world talking about intelligent beings on Mars who had constructed a vast network of canals to irrigate their dying planet. "All the vast extent of the Continent", he wrote, "is furrowed upon every side by a network of numerous lines of a more or less pronounced colour. . . . Some of the shorter ones do not attain 300 miles; others extend for many thousands." It did not seem to matter that, at the distance of Mars, such features would have to be tens of kilometres across to be seen at all.

Life on Mars
The myths were shattered once and for all when the first space probes reached Mars in the 1960s and 1970s. Instead of great Sahara-like deserts, there were thousands of Moon-like craters, huge volcanoes and immense canyons and features such as the dried-out beds of ancient rivers (or thin lava streams according to one theory). Whatever water exists on Mars may be locked up beneath the surface in the form of ice and permafrost. The glistening white polar caps appear to be mainly water ice. The climate is cold and the thin, mainly carbon dioxide atmosphere (5 to 8 millibars pressure) does not allow water to exist in a free state.

The space probes solved another mystery. The so-called "dark areas" on Mars, which some astronomers had linked with the growth and decay of vegetation, seem to have been caused by wind storms which redistributed light-coloured dust over a darker surface. In 1975 two Viking spacecraft, each carrying an automatic laboratory, were sent to see if the soil contained microbes. The quest for life unique to another planet will ensure that exploration continues.

The main problems in sending people to Mars are time and distance. Whereas astronauts can land on the Moon and return in less than two weeks, a Mars ship must leave the Earth-Moon system and embark on an immense journey which takes it right around the Sun. Before such a project can even begin it is necessary to build a space station in Earth-orbit which can be used as an assembly and refuelling depot. The men and women who make such journeys must learn to live and work in space for 1½ to 2½ years.

Landing on Mars
Opportunities to launch to Mars come round at intervals of about 25 to 26 months when Mars is at opposition. One American plan, shelved for the time being because of the enormous cost, involved a possible expedition this century. Two atomic-powered ships each 82.3m (270ft) long were each to carry six explorers. Work had already started on Nerva rocket engines which use nuclear heat to expand liquid hydrogen fuel into a powerful propelling jet. The journey was to take the explorers around the Sun to a point in space where Mars would be in nine months' time. For most of the flight the two ships would be docked together nose-to-nose, separating before they arrived. They were to orbit Mars for 80 days while three explorers from each ship descended to the surface in landing craft.

CONNECTIONS

See also
194 The planet Mars
202 The moons of Mars

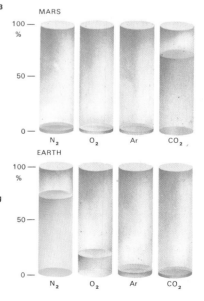

1 The engineering replica (simulator) of the Viking lander at the Jet Propulsion Laboratory (JPL) at Pasadena, California, was used by scientists to solve problems as they occurred with the actual spacecraft on Mars. They were thus able, for example, to free a locking pin on the Viking 1 soil scoop that had failed to eject. JPL commanded the scoop arm to shake out the pin after working out a series of movements with the simulator. The "repair" was effected across more than 340 million km (212 million miles) of space. The sampler scoop, cameras [upper left and centre] and meteorology beam [upper right] can clearly be seen in this photograph.

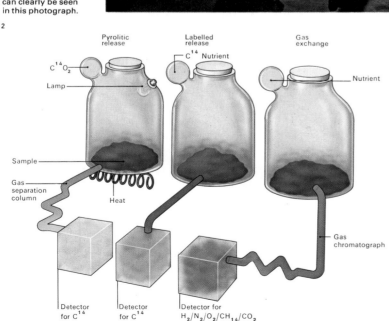

2 The search for life was an important task for Viking. Each lander has an automatic biology laboratory to which soil samples were delivered by a mechanical scoop in three experiments: Pyrolytic Release was designed to detect any micro-organisms that live by photosynthesis, taking CO_2 and using it to make organic matter in artificial sunlight (lamp). Labelled Release looked for signs of metabolism, evidence that organisms in the soil were maintaining and reproducing themselves. Gas Exchange looked for any exchange of gases between microbes in the soil and atmosphere of the chamber.

3 The main constituents of the Martian atmosphere are here compared with those of Earth's atmosphere.

Inside their pressurized, temperature-controlled landers, the astronauts would enjoy a "shirtsleeve" environment. They would have to don spacesuits before leaving the ship in order to set out their instruments, take samples and make excursions in a Martian roving vehicle. Then they would stow their samples and blast off from Mars, leaving the lower section of the landing craft behind, to rendezvous and dock with their orbiting mother ships. At a pre-calculated time, the ships were to blast out of Mars orbit and continue their long journey round the Sun, flying close to Venus and using its gravity to slow down and arrive back at the Earth station after a round trip of 21 months. The enormous cost and complexity of such missions – one 1969 estimate suggested $80,000 million – makes it unlikely that man will land on Mars this century.

Intelligent machines
A scientific base on Mars would need a much larger investment and its value must be weighed against the ability of intelligent machines to gather scientific information.

For example, computer-controlled roving vehicles have been devised to avoid the time-lag in radio communications (up to 23 minutes in the case of Mars) which rules out direct radio-control steering from Earth. The Russians have experimented with a six-legged vehicle with a computer "brain" and a laser "eye". The laser probes ahead for obstacles and the computer works out a suitable avoidance path.

It is even possible that unmanned spacecraft can be made to land on Mars, release an automatic rover and obtain soil and rock samples at different sites. After the samples have been placed into the return rocket, at the appropriate time it would be launched back to Earth. Any specimens taken from Mars must be kept in isolation in case they spread some unknown disease, the best place being a space laboratory in Earth orbit. Even the rocks and soil obtained from the Moon were kept in sterile conditions in a special receiving laboratory until they were declared safe by biologists. This underlines the many unknowns which face any expedition to other bodies of the Solar System.

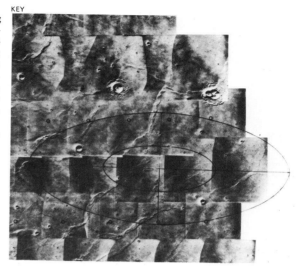

The Viking 1 landing site in Chryse Planitia – a channelled lowland some 2.5km (1.5 miles) below Mars' mean "sea-level" – lies northeast of the great Martian volcanoes and the Tharsis Plateau. The craft landed near the centre of the target ellipse in the picture.

4 The procedure for establishing a base on Mars will differ from that on the Moon. It will be necessary to build a complete base immediately since Earth is too far away in case of emergency. In this illustration a Mars Excursion Module (MEM), similar to the type needed to carry astronauts to the surface of the planet, has just arrived. Its cargo includes an inflatable pressure dome like the one already set up in the background. Another MEM is about to land a roving laboratory which the astronauts will use to explore the ground. The landing technique follows the practice of Viking. After release from the orbiting mothercraft, the MEM fires a braking rocket to descend. It enters the Martian atmosphere at about 16,000km/h (10,000mph) to be slowed first by aerodynamic drag on the blunt heat shield, then by parachutes. A few thousand metres up, the parachutes are discarded and retrorockets cushion the touchdown. Equipment on Mars must be prepared for the climate. Pressure domes are double-walled to protect the people inside from the cold. Radar dishes are mounted in radomes mainly for protection from wind-blown dust. Other supplies include nuclear generators for power supply and reserves of oxygen for life-support systems, food and water. Sources of water and oxygen must be established to make the base less dependent on Earth. The search will already have begun with projectile-like probes launched from orbit which penetrate many metres into the soil to determine its chemical composition and the presence of permafrost. In likely places astronauts will drive out in their roving laboratory to make a thorough study. In the picture, geologists are shown taking core samples for analysis.

Exploring the inner planets

Although Venus approaches the Earth more closely than any other planet, it took a long time to discover the true nature of its surface because it is perpetually shrouded by a cloudy white atmosphere. Only slightly smaller than the Earth, Venus orbits some 42 million kilometres (26 million miles) nearer the Sun and intercepts about twice as much heat and light. These simple facts led astronomers to some quite bizzare theories. The Swedish Nobel prizewinner Svante Arrhenius in 1918 imagined the planet was covered in seas, swamps and steamy jungles perhaps populated by primeval monsters.

Astronomers of the 1930s and 1940s had different ideas. With the spectroscope they had discovered that the chief constituent of the atmosphere was the heavy gas carbon dioxide. This suggested that radiation from the Sun would be trapped by a "greenhouse" effect creating high temperatures which, some believed, might attain the boiling-point of water. One theory was that Venus was in the throes of extensive vulcanism and that its thick atmosphere was volcanic dust suspended in a perpetually warm fog. The American astronomers F. L. Whipple and D. H. Menzle thought it possible that the planet was entirely covered by ocean. Others anticipated a hot, windy desert.

Venus: a hostile environment

Radar signals bounced off the planet from Earth in the 1960s indicated a rugged surface and possibly large craters. By finding a fixed point of reference these radar soundings showed that Venus turned on its axis – in a reverse direction to Earth – once in 243 days.

Space probes – beginning with the fly-by of America's Mariner 2 in 1962 and continuing with the Soviet Venera capsules which entered the atmosphere – produced clear evidence of surface temperatures far above the melting-point of lead and atmospheric pressures that would crush any normal type of spacecraft.

In 1975 Veneras 9 and 10 swung into orbit around Venus after releasing camera-equipped capsules which landed some 2,200km (1,370 miles) apart. Ruggedly built to withstand extremely high pressures, they descended with the help of parachutes and a circular air brake. They were also specially cooled and insulated to work for at least half an hour on the fiercely hot surface. The first capsule – transmitting to Earth via its orbiting mother craft – worked for 53 minutes. It sent a panoramic picture of its surroundings showing a scattering of sharp-edged rocks 30 to 40cm (12 to 16in) across which appeared to be little affected by heat or wind erosion. The rocks seemed to be comparatively young and of a kind which may have been produced by subsidence or a meteorite impact. Instruments showed the pressure of the atmosphere to be 90 times that on Earth, the temperature 485°C (900°F). The second capsule, which transmitted for 65 minutes, showed a different landscape typical of old mountain formations. The rocks resembled huge pancakes with sections of cooled lava or the debris of weathered rocks in between. Here the pressure was 92 atmospheres, suggesting that the capsule had come down in the area of a valley.

The Venera 9 and 10 mother craft, continuing the survey, built up a detailed picture of the planet's cloud cover adding to informa-

CONNECTIONS

See also
188 The planet Mercury
190 The planet Venus
204 Minor planets

1 The dense atmosphere of Venus led scientists to believe that it must be "super-refractive": that is to say, the rays of light would be bent to such an extent that an observer on the planet's surface would have the impression of being in a vast bowl, with the horizon curving upwards all around him. However, the Venera 9 and 10 pictures showed that this is not the case. The atmosphere of Venus does not show the super-refraction that was forecast.

2 Venus has proved to be a world quite different from anything that had been expected. The pictures from Veneras 9 and 10 show what the Russians have called a "stony desert", shown in this artist's impression. The rocks are relatively smooth, and it is thought that the erosion on Venus is less than on the Earth or even Mercury. The open sky can never be seen through the dense, corrosive clouds of acidic vapour that surround the planet, making Venus a gloomy, hostile world. As a potential colony, Venus has proved to be a disappointment. The surface features are almost certainly volcanic in origin. It is not known whether or not active vulcanism is in progress.

tion received from America's Mariner 10 in 1974. Clouds tending to spiral around the planet reached a height of 65km (40 miles) with atmospheric gases moving at different speeds at different heights. Near the surface wind speed was low, growing with altitude until at the cloud tops it was 60 times that of the planet's own slow rotation.

Mikhail Marov, project leader for the Soviet probes, agreed with American specialists that the brilliant white clouds might be laden with sulphuric acid and possibly small amounts of other acids. He thought the high surface temperature might be polluting the atmosphere with the vapours of low-melting-point metals. However, from the evidence of the surface cameras, the atmosphere was dust-free and a remarkable amount of light penetrated to the surface.

Mercury: solar observation point

Mercury, the nearest planet to the Sun, takes nearly 88 days to make one revolution at an average distance of 58 million kilometres (36 million miles). A little less than half Earth's diameter, the planet turns slowly on its axis,

completing a day in 58.5 Earth-days. As a result its surface is alternately baked by the Sun and frozen by the cold of outer space.

Mariner 10 – which flew past Mercury in 1974 after skirting Venus – discovered a Moon-like world with craters, mountains and valleys. There was barely a trace of atmosphere and no detectable magnetic field.

The asteroids: beacons in space

Mercury and Venus are the only planets moving at a distance from the Sun less than that of the Earth. Farther out, Mars is our nearest neighbour and then comes the main zone of asteroids or minor planets [Key]. The largest of these bodies, Ceres, is about 1,000–1,200km (600–750 miles) across, but most are so small that contacts with them would be more in the nature of docking operations than landings. Some asteroids wander far from the main belt. Icarus, for example, goes within 28 million kilometres (17 million miles) of the Sun, while Hidalgo swings out almost as far as Saturn. One day some of these bodies may be visited by spacecraft from Earth.

KEY

The asteroids or minor planets may one day be the object of an exploratory mission. In this artist's impression the astronauts have docked with Eros and are setting up an infla-table, semi-trans-parent dome; they are preparing to make a geological survey of the asteroid. Eros is about 27km (17 miles) in diameter, and it is irregular in shape, like most of the asteroids, so that its horizon slope seems strange. The orbit of Eros takes it away from the main belt of the minor planets and it may approach to within 24 million km (15 million miles) of Earth. Its surface is pitted with craters resulting from collisions with debris in the asteroid belt.

3 **Mariner 10** approached Mercury in March 1974, seven weeks after by-passing Venus, and sent back the first close-range pictures, showing that Mercury, like the Moon, is cratered. In this picture Mariner 10's two "paddles" spanning nearly 9m (29.5ft) are seen; these provide the solar power for the various on-board systems. The dish antenna transmits information to Earth, and there are devices to measure magnetic fields, charged particles and ultra-violet and infra-red radiation. Mariner passed Mercury three times and on each occasion produced valuable data. It will continue to orbit the Sun indefinitely, although its "useful" life was over after it passed Mercury in the spring of 1975.

4 **The planet Mercury** has virtually no atmosphere and it is just as hostile as the Moon. In this picture the Sun is hidden behind a plug of lava from a volcanic erup-tion millions of years previously and since worn down by alternate expansion and contraction as a result of Mercury's great diurnal temperature range. Craters are seen on the surface, and the bright star in the upper right is the Earth-Moon pair. Mercurian days and nights are long because the planet rotates so slowly on its axis, and it is a desolate, lifeless world.

Exploring Jupiter and Saturn

Chemically propelled vehicles have taken men to the Moon. Whether they will suffice for manned journeys to Mars is problematical and they will certainly be inadequate for manned missions to Jupiter. The transfer orbit, at present the only means of making an extended journey, allows a probe to utilize the gravitational pull of a planet to accelerate and direct the probe on its way to the target planet. With a probe moving in a transfer orbit almost all the journey is carried out in free fall, so that no propellant is expended. Using this technique a probe would take more than two years merely to reach the neighbourhood of Jupiter. Such a long journey is technically impractical. For missions to the giant planets we must await the development of efficient nuclear rocket motors and dispense with the idea of extended transfer orbits.

Jupiter and its satellites

Jupiter, the closest of the giants, has no solid surface and is surrounded by zones of intense radiation that would be lethal to any astronaut incautious enough to come within range. There is, too, the high escape velocity – 60.22km (37.4 miles) per second – which would make manoeuvring almost impossible. The only alternative, then, would be to land on some of the Jovian satellites.

Ganymede would probably be the first choice. It is of planetary size (slightly larger than Mercury, although not so massive) and is reasonably far out, with a mean distance from Jupiter of 1,000,000km (621,000 miles). From Ganymede the view would be spectacular by any standards: the giant planet would seem to spin rapidly, showing the panorama of its belts, its zones and its Great Red Spot. Like all large satellites, Ganymede has a captured or synchronous rotation. It takes rather more than 7.15 days to complete one revolution round Jupiter and its axial rotation period is the same, so that one side of Ganymede constantly faces Jupiter and to an observer on it the appearance of the planet would constantly change.

Like Ganymede, the smaller but denser satellite Europa has almost no atmosphere. It may well be ice-covered and the view of Jupiter from its surface must be imposing. The innermost large satellite, Io [1] is the smallest known body with an atmosphere (possibly of ammonia) and Amalthea [2] is a dwarf moon whirling round Jupiter at a distance of only 181,000km (112,000 miles) in a period of almost 12 hours. To an observer on Amalthea, Jupiter would fill a quarter of the sky and its surface features would seem to change much more slowly than from Ganymede or Europa because Amalthea has a rotation period only two hours longer than that of Jupiter itself.

It is tempting to picture an observatory on Amalthea, ideally placed to study events on Jupiter, but this may never be possible simply because Amalthea lies inside the zone of Jovian radiation, which may make it a highly dangerous world.

The problems of reaching Saturn

If manned expeditions go out to Jupiter within the next 200 years (and it may happen much sooner than this), then Saturn [3] would be the next target. Here the problems are somewhat different. The distance involved is greater; but in compensation

1 Jupiter would dominate the sky to an observer standing on Io, one of its four large satellites. He would clearly see the dark belts, the bright zones and the Great Red Spot. Io lies 422,000km (262,000 miles) from the centre of Jupiter, which is rather more than the distance of the Moon from the centre of the Earth. But the revolution period of Io is only one day 18.5 hours because Jupiter's powerful gravitational pull causes it to move much faster in its orbit. Io moves in the Jovian Magnetosphere and affects radio emissions from the planet. Parts of Io's surface may well be ice-coated.

2 Jupiter, as seen from Amalthea, would look magnificent, with the shadows of Io and Europa clearly visible. Officially known as Satellite 5, Amalthea is the innermost member of the Jovian family, lying only about 110,000km (70,000 miles) from the planet's surface. It moves at a mean distance of 181,000km (112,000 miles) from Jupiter's centre. Amalthea is only 200km (124 miles) in diameter and is possibly distorted in shape by the intensely powerful pull of gravity from the planet. Expeditions to Amalthea could prove highly dangerous because it lies within the radiation zone of Jupiter.

Saturn seems to lack the lethal radiation zones that characterize Jupiter. Of the numerous satellites of Saturn, Titan is far more interesting and important than any member of the Jovian family.

From the inner satellites Saturn would look magnificent, despite the edge-on presentation of the rings. The passage of a vehicle through the ring-system itself would be extremely hazardous. The danger would come not from radiation, but from solid particles of matter. Presumably the exploration of the regions in and near the ring-system would be left to automatic probes, certainly in the beginning.

Saturn, like Jupiter, lacks any solid surface, so that landings would have to be restricted to the satellites. All the inner members of the family move practically in the plane of the rings and, as a result, the rings appear edge-on. But this does not apply to Iapetus, which moves at a mean distance of 3.6 million kilometres (2.2 million miles) from Saturn; it completes one revolution in 79.33 days and is about 1,100km (700 miles) in diameter – about half the size of our Moon.

Because its orbit is appreciably inclined to the ring-plane, Iapetus would provide the best base for studying Saturn itself.

The lure of Titan

It is Titan that really captures the imagination. Here is a world of planetary size, 1,200,000km (760,000 miles) from Saturn, with an atmosphere that may contain clouds and with a density ten times greater than that of Mars at the surface. Unfortunately Titan's atmosphere is composed mostly of methane, which will not support life as it is known on Earth, and some hydrogen, but the satellite is regarded with special interest and there is even talk of sending a probe to study it.

If a base were to be built on Titan, it could be on the same lines as those for Mars where everything would have to be made self-supporting; and there could be no chance of a prompt rescue in an emergency.

The close-range view of Saturn would exceed by far in splendour and interest anything else in the inner part of the Solar System: it would be unequalled in the experience of mankind.

Saturn, surrounded by a host of stars, appears to the naked eye as a particularly bright star. Although remote, it is far closer than any star and lies within probe range.

3 Saturn, seen from Rhea, would display four of its inner satellites: Dione, Tethys, Enceladus and Mimas. With the Sun below the horizon, Saturn's strong yellow light would cast a bright glow over Rhea's surface. Rhea is the sixth farthest satellite from Saturn and moves round the planet at a distance of 527,000km (327,000 miles) from its centre, the distance from Rhea to Saturn's surface being about 467,000km (290,000 miles). Rhea takes four days 12.5 hours to complete one journey round Saturn. It is much smaller than the Moon and nothing positive is known about its surface except that it lacks an atmosphere and the temperature is extremely low. Like the other inner satellites (and indeed all of Saturn's attendants, apart from the two outermost members, Iapetus and Phoebe) Rhea moves virtually in the plane of Saturn's equator, which is also in the plane of the rings. Therefore, from Rhea, the rings would appear as a thin line of light and would always be seen edge-on. An observer would never be able to see the Cassini Division.

Exploring the outer planets

In the mid-1970s Jupiter was the most remote world to have been contacted by a space probe, although by 1979 the Pioneer 11 probe is expected to have reached Saturn. Plans are now being made to send vehicles out to the next giant planet, Uranus, which has features of special interest, but the journey will take much longer – a point that is not always easy to appreciate from a casual glance at a map of the Solar System (Uranus is 2,720 million kilometres [1,690 million miles] from the Earth). A spacecraft travelling to Uranus will have reached only the half-way mark by the time it begins to cross the orbit of Saturn.

It is much too early to speculate about when the first manned expedition will venture as far as Uranus. Vehicles much more sophisticated than those planned at the moment will be needed; even so, landing on the planet's surface would be impossible. Uranus, like Jupiter and Saturn has a surface of gas, although the constitution of the planet differs in various important details. There is no evidence so far to indicate the existence of dangerous radiation zones like those of Jup-

iter. Of the five satellites, Miranda is the closest to the planet (only 130,000km [80,000 miles] out) but it is likely that the first landing will be made on one of the larger satellites, such as Ariel [1]. Because of its strange axial tilt Uranus sometimes appears as a crescent with its horns extending from one side of the equator to the other rather than from pole to pole – a case unique in the Solar System. The strange greenish light of Uranus upon the rocks of one of its satellites will make an eerie picture, and one of emptiness and desolation.

Neptune and Triton

Beyond Uranus lies Neptune – but again, the distances involved are immense; the distance of Neptune from Earth is 1.5 times that of Uranus [2]. But at least there is a more promising satellite, Triton, which is much larger than any of the attendants of Uranus and may possibly have the same kind of atmosphere as Titan (Saturn's largest satellite) although no proof of this has yet been obtained. Triton is unique among large satellites in that it moves round its primary in a retrograde direction.

Neptune has a rotation period of 15hr 48min, while Triton has a revolution period of only 5 days 21 hours. Because these movements are in opposite senses, the drift of surface markings on Neptune will be rapid from the viewpoint of a Tritonian observer, providing a fascinating view of Neptune.

If a major outpost is to be established in these desolate regions of the Solar System it is most likely to be on Triton. The other satellite of Neptune, Nereid, is extremely small – less than 300km (200 miles) in diameter – and with its highly eccentric orbit would offer few advantages as an observation base. Even from Triton the other planets would not be seen to advantage; only Uranus would seem brighter than it does from Earth and it would be an inferior planet, keeping in the same area of the sky as the shrunken but still intensely brilliant Sun. Moreover, when Uranus and Neptune lie on opposite sides of the Sun an observer on Triton would find Uranus difficult to see over a period of years. Saturn would be even more elusive because the distance between Neptune and Saturn is much greater than that between Saturn and

1 Uranus is seen here from its satellite Ariel in this artist's impression. A probe visiting Uranus from Earth would be launched to the neighbourhood of Jupiter; the powerful Jovian gravity would then be used to pull the probe and accelerate it out beyond Jupiter and on towards Uranus. The gaseous nature of Uranus clearly prohibits any landing. But landings may be possible on some of its five satellites. Of these, the closest to Uranus, Miranda, is very small. The innermost of the main satellites is Ariel, which moves round Uranus at a distance of 192,000km (119,000 miles) from the centre of the planet in a period of 2 days 12 hours 29 minutes. Nothing is known about the surface of Ariel but its size seems to be considerably smaller than the Moon – about 1,500km (930 miles) in diameter. Ariel, like the other satellites, moves in the plane of Uranus's equator. In this view Uranus looks like a crescent, but the horns extend from one side of the equator to the other. With its remarkable inclination of 98°, in 1985 one pole of Uranus will face the Sun (so that there will be a "polar day" lasting 21 Earth years) and from Ariel or any other satellite Uranus will appear as a half-disc. The changing surface details will be displayed, although the pale greenish disc of Uranus is markedly less active than either Jupiter or Saturn and less interesting visually.

Earth and the inner planets would be virtually out of view. Yet from Triton there would be no interference from brilliant solar radiation and a base there could still make useful observations outside the Solar System and thus contribute substantially to man's knowledge of the universe.

The outermost planet

Little is known about Pluto, the outermost planet [3]. It is slightly smaller than Triton and has a surface of methane ice, which tends to confirm current theories explaining the origin of the solar system and the sequence of planetary formation. At perihelion, or closest point to the Sun, it comes within the orbit of Neptune; the next perihelion passage is due in 1989. At aphelion, Pluto recedes to more than 7,000 million kilometres (4,500 million miles) from the Sun.

Information sent back from a future space probe to Pluto would be of great interest. The mass of Pluto is still not accurately known but the effect it has on a fly-by probe should enable scientists to calculate its value. If astronauts ever land there, they will find that

the Sun looks no larger than Jupiter does as seen from Earth, although it will still shed a certain amount of light over the bleak Plutonian surface.

Communications with Earth will be slow. A radio wave would take about five hours to cross the space from Pluto to Earth – so that if a message is transmitted from Earth, there will be an interval of ten hours before any reply can be received.

Exploring the comets

Although Pluto is the outermost planet, there may well be opportunities for studying material from even farther out in the Solar System. Comets [Key], those wraith-like and insubstantial objects, mostly have very eccentric orbits and there is a real possibility of sending a probe through a comet that has come from the region beyond the orbits of Neptune and Pluto. This may be attempted in the near future – perhaps even with Halley's comet at its next return in 1986, although the retrograde motion of this comet presents a number of extra problems to space scientists and astronomers.

Comets, the erratic wanderers of the Solar System, can journey farther than the known outermost planet, Pluto. Shown here is a probe approaching a comet on its journey; the Earth and Moon appear in the upper right. Unlike a planet a comet is not a solid, massive body; it is composed of relatively small particles, mainly icy in nature, together with extremely tenuous gas. Hence, there is no reason why a probe should not be able to pass right through it. The tail is particularly tenuous, thus the background stars can be seen through it virtually undimmed. Some comets are known to travel a distance equivalent to a third as much again of Pluto's distance from the Sun.

2 Neptune, unlike Uranus, has a "normal" axial inclination of 29° – less than 6° greater than that of the Earth. Surface details on its bluish disc are difficult to make out, but there seems little doubt that Neptune and Uranus are similar. The illustration shows a view as it would appear from Nereid, the smaller of Neptune's two satellites, which has an eccentric orbit; at its closest to the planet it approaches to within 1,400,000km (870,000 miles) and this is depicted in the illustration. Nereid takes almost one Earth year to complete a full revolution around the planet Neptune.

3 Pluto at times comes within Neptune's path because of its relatively eccentric orbit. The next perihelion is due in 1989: for some years on either side of that date, Pluto will no longer be the outermost known planet. At aphelion it recedes to more than 7,300 million km (4,600 million miles) from the Sun. The temperature of Pluto is estimated to be as low as −230°C (−382°F). So far no atmosphere has been detected; it possesses a surface of methane ice. Its composition may well resemble that of the moons of the outer planets rather than that of the planets themselves.

Beyond the Sun's family

Exploration of the Solar System is progressing and, provided that the present rate of progress is maintained, all the planets will have been contacted by automatic probes within the next 50 years – probably well before – and manned expeditions will have been sent to those worlds like Mars that are not overwhelmingly hostile and are within reasonable range. Yet even when man has finished his exploration of the Solar System, he will hardly have begun to explore the vastness of the universe.

Problems of interstellar travel

The Solar System is only a small part of the universe. If the distance between the Earth and the Sun were represented by 2.5cm (1in), then the nearest star would be almost 7km (4.3 miles) away. Stellar distances are so great that man's present technology cannot yet begin to bridge them. Although two interstellar probes have been dispatched – Pioneer 10, which by-passed Jupiter in December 1973, and Pioneer 11, which did so about a year later – neither will approach a star for many thousands of years. Neither can

transmit signals to Earth. They are gambles, launched into space in the hope that they may reach a world of advanced beings who can communicate with Earth – and in the hope that when that happens mankind will have advanced sufficiently to be able to understand that communication.

Even if it travelled at the speed of light, a probe would take more than four years to reach Proxima Centauri – the nearest star that is sufficiently like the Sun to have a family of planets moving round it. According to the theory of relativity, which has so far survived every test, it is impossible for any material body to travel at the speed of light and any spacecraft that can be planned in the present state of technology is, by comparison, very slow indeed.

Journeys to other solar systems will have to be made by methods that are as yet unknown and will probably be as technologically advanced as television would have been in the days of Julius Caesar. All kinds of suggestions have been made. A favourite science-fiction idea is that of the space-ark, in which those who set out die at an early stage

of the voyage, leaving their descendants to finish the journey. Alternatively, it has been suggested that the travellers should be put into a state of suspended animation, being conveniently woken up just before arrival at a suitable planet. Then there are theories involving telepathy and teleportation – transporting matter through space in much the same way as television transmits pictures. They are all intriguing, but at present beyond man's powers to implement.

All that can really be said is that interstellar travel is impossible by any known method. However, a spectacular breakthrough may come eventually and it is conceivable that alien beings may visit the Earth before man has developed sufficiently to be able to visit them.

Finding new planets

What are the prospects of finding planets orbiting other stars? The Sun is a normal G-type dwarf and there is no reason to regard it as exceptional in any way. Moreover, G-type dwarfs are common and it thus seems that planetary systems are likely to be abundant in

CONNECTIONS

See also
176 Members of the
Solar System
226 Star types

1 The Moon can be reached from Earth in a few days and to send a rocket to Mars or Venus involves a journey of only a few months. But interstellar travel presents quite a different picture and poses many problems. The distances involved are millions of millions of kilometres and even light, moving at a speed of 300,000km (186,000 miles) a second, takes more than four years to reach the nearest star. Rockets powered by chemical fuels of the type in use today would be hopelessly inadequate for flight to any star. In the USA, much research has gone into the possibility of building what is termed a photon rocket, in which the gases emitted from the exhaust of a chemical rocket are replaced by a stream of photons – that is to say, a beam of light. The thrust produced would be low, but it could be maintained indefinitely and over a period of years the acceleration would build up until the velocity of the vehicle approached that of light. (According to relativity theory, the precise velocity of light can never be reached by a material body, since it would involve infinite mass.) In principle, the photon rocket can be compared to a gigantic electric torch which is driven forward by the light emitted. One possible design is shown here. The rocket would be more than 9.5km (6 miles) long with a crew of 300–500. Our Galaxy is in the background. In the inset, lower right, our Galaxy is shown first face-on and then edge-on, with the position of the Solar System indicated by a red circle. Even at the velocity of light, it would take a probe 100,000 years to pass from one side of the Galaxy to the other.

the Galaxy. Other kinds of stars are less promising. For instance, a red giant star that has left the main sequence and has swelled to many times its original size is likely to have swallowed up any planets it may once have had [2], while a hot, massive blue or white star will have run through the earlier part of its evolution so quickly that planets that resemble the Earth will hardly have had time to develop.

Then there are the faint red stars, so feeble that they have never joined the main sequence and are on their way to extinction. Barnard's Star, at a distance of only a little more than five light-years from Earth, is one of these. As it moves through space it "wobbles" slightly and there have been suggestions that it is being pulled out of position by an orbiting planet or planets. If so, what would such a planet be like?

An alien environment

By Earth standards, a planet associated with a star like Barnard's would be a dreary world. Its only light would come from a dim red sun and it would therefore be cold. Any life there

would have to contend with an environment that man would find intolerable. Yet it would be unwise to dismiss such planets as possible places for colonization.

Barnard's Star is not active by stellar standards. Stars that are much more violent become unstable as they use up their nuclear energy and some produce nova-like outbursts [3]. The effects of these upon any orbiting planets would be catastrophic. If our Sun became a nova, all life on Earth would be destroyed in a matter of hours. Luckily this cannot happen yet, because the Sun is a stable star and will not significantly change its structure for at least 5,000 million years. There must, however, have been planetary systems that were engulfed and destroyed by their dying central stars.

There would have been a period of warning in which the inhabitants of a threatened planet, if they were sufficiently advanced technologically, could have taken steps to save themselves. The obvious step would be into space – abandonment of the world about to die in favour of a new start on a new world, whatever the difficulties.

KEY

In this cluster of stars, NGC 5897 in Libra, photographed with the 200in (508cm) reflector at Palomar, each star is a sun in its own right. Many may have associated planet-families but there is no direct optical proof; no telescope yet built or planned can hope to show a planet of another star.

2 Zeta Aurigae is a binary system made up of a vast red supergiant together with a much smaller white star, about to be eclipsed. The red supergiant is an old star that has left the main sequence and has swelled out to a diameter greater than that of the orbit of the Earth around the Sun. Because it has become luminous, it will have raised the temperatures of its planets to intolerable levels and any inner planets of its original system will have been destroyed. No life can be expected in a system of this kind. The illustration shows a view from a hypothetical planet.

3 This barren, dead planet is on the outskirts of a hypothetical planetary system – the "Pluto" of the system. The central star, a binary system, has flared up as a nova, resulting in a tremendous, although temporary, increase in luminosity. The planet's surface has been scorched by the tremendous radiation; its water has evaporated and even its atmosphere has been driven away so that no life can possibly survive. The inner planets of the system have been completely destroyed. The sky is enriched by an aurora-like display, caused by a shell of gases released by the star as its outer layers expand. When the nova outburst is over, the planet will remain, cold and sterile, circling the feeble remnant of its once glorious sun.

Worlds of many kinds

Life on Earth has developed along its familiar lines because the conditions are suitable for it. If the Earth were smaller, colder or less massive, then life would have taken on different forms; and if the conditions were unsuitable, no living organisms would have developed at all. Life, wherever it is found, is suited to its environment. If a star like the Sun were attended by a planet the size and mass of the Earth orbiting it at a distance of 150 million kilometres (93 million miles), then Earth-type life might reasonably be expected. In 1972 Pioneer 10 was launched to probe beyond the Solar System. It carried a plaque [Key] to communicate with any intelligent life that might encounter it.

Alien life forms?
This does not mean that all forms of life must correspond with the terrestrial pattern. There is nothing, in theory, against an intelligent astronomer having six legs and two heads. If he were made up of the same materials as ourselves he would not, despite his appearance, qualify as one of those interesting creatures novelists call BEMs, or "bug-

eyed monsters". This term is reserved for entirely alien creatures, breathing pure methane, say, and able to survive in temperatures of −150°C or below. It cannot be said definitely that alien life does not exist; all that can really be done is to take the available facts and then put the most reasonable interpretation upon them. When this is done, the existence of bug-eyed monsters begins to appear improbable.

Rational discussion of life on other planets must be confined to "life as we know it". Any extension to include alien life forms means that speculation becomes not only endless, but also pointless.

An Earth-type planet may be expected to produce Earth-type life, essentially similar to our own and no doubt subject to the same weaknesses. For instance, the star Delta Pavonis, at a distance of 19 light-years from the Earth, is strikingly similar to the Sun; we have no idea whether or not it has a planetary system, but there seems no reason why it should not be attended by a world similar to the Earth, in which case its inhabitants may at this very moment be speculating about the

possibility of intelligence upon a planet orbiting a fourth-magnitude yellow star in their own sky. If this hypothetical planet lay farther from Delta Pavonis than the Earth does from the Sun, the colder climate would produce life forms more akin to those in our polar zones; if it lay closer, life would be more equatorial. Of course, this can be no more than speculation because nobody knows whether a planet capable of supporting life will actually produce it; but again there seems to be no valid reason why it should not.

Different perspectives
Many stars in the Galaxy are members of binary systems and it is fascinating to picture a planet lit by two suns – perhaps of different colours, one yellow and one blue, giving strange, spectacular colour effects. Then there are the variable stars, some of which are completely regular while others are violently explosive. It is hard to see how a variable star could be attended by a life-bearing planet because there would be extreme fluctuations in climate; but most variable stars are well advanced in their evolution, so that life on

CONNECTIONS

See also
252 The expanding universe

1 A planet of Proxima Centauri, the nearest star to the Earth, may exist, according to recent investigations. This picture of its surface is thus founded on something more than sheer imagination.

Proxima Centauri is a dim red dwarf that has never joined the main sequence nor passed through the giant stage. It does not cast as much light as the Sun on its attendants. The planet thought to be

moving round it is assumed to have an orbital period of between 10 and 12 days. The outer edge, or limb, of Proxima is not sharp, like that of our Sun, but diffuse, because the den-

sity of the outer layers is low. It is also assumed that the planet has a tenuous atmosphere. Because Proxima is a flare star, the climate of an inner planet will be unstable

and the landscape extremely desolate. No life can be expected there. But water, shown in the picture as a lake fringed by glittering crystals of ice, may survive. The black circle repre-

sents a possible satellite as it would be seen in silhouette against the red disc of Proxima. In the sky is seen a familiar constellation pattern: the W of Cassiopeia, all of whose stars are

remote and will therefore look much the same from Proxima as from the Earth. To the left of the W is another star – our own Sun – which will be easily visible with the naked eye.

any of their surviving planets may have died out in the remote past.

The Sun lies in a relatively sparsely populated region of the Galaxy, at the edge of one of the spiral arms (from which it emerged perhaps 5,000 million years ago). If the Sun lay in one of the rich globular clusters the sky would be ablaze; there would be many stars shining more brilliantly than Venus does to us and since globular clusters are "old" by stellar standards there would be many red stars which would have left the main sequence. There is no reason why stars inside globular clusters should not be attended by planetary systems. Astronomers there would be at something of a disadvantage, because it would be difficult for them to see clearly beyond the confines of their own dense cluster and they could know little about the greater universe beyond.

Interstellar communication
The only means of achieving communication with other planetary systems, using existing techniques, is by radio. Radio waves move at the same velocity as light and would take years to reach even the nearest star; nevertheless, the reception of a signal-pattern rhythmical enough to be classed as artificial would be of unparalleled significance. Attempts have already been made to distinguish such a pattern.

In 1960 radio astronomers at Green Bank, West Virginia, began an ambitious programme known officially as Project Ozma. With powerful equipment they concentrated upon the two nearest stars that are reasonably like the Sun – Tau Ceti and Epsilon Eridani, both of which are slightly smaller and cooler than the Sun and are more than ten light-years away. A wavelength of 21.1cm (8in) was selected because this is the wavelength of the radio signals emitted by the clouds of cold hydrogen spread throughout the Galaxy. It is logical to believe that other astronomers, wherever they may be, are also devoting their attention to this particular wavelength. The earliest that a signal from an alien world could be received by astronomers on Earth is 1980, but it may, and probably will, be many years after that before contact is made.

This plaque is carried on Pioneer 10 (launched in 1972, it will be the first vehicle to leave the Solar System). The radiating lines represent 14 pulsars; the binary notation gives their frequencies relative to the universal constant – the hydrogen atom [upper left]. The regular decrease in the pulsars' frequencies will give the time elapsed since launch. The Earth's position in the Solar System with Pioneer's path [bottom] and the male and female in proportion to Pioneer are shown.

2 Communication with alien civilizations could be achieved by means of mathematics – a system man discovered rather than invented. One suggested method is the transmission of two signals of different kinds (say dots and dashes). 209 signals are sent as 0's and 1's. A listener could represent the 0's by black squares and the 1's by white squares (or vice versa). 209 has only two factors: 11 and 19. Accordingly, a receiver has the choice of dividing the signals into 11 groups of 19 or 19 groups of 11. The second alternative gives the correct intelligible picture – that of a biped, as shown.

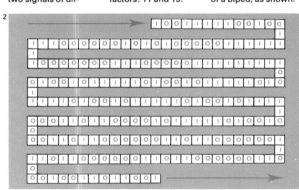

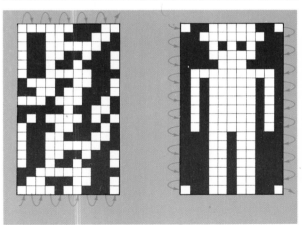

3 A radio telescope of the kind pictured here may well exist on a planet far away in the Galaxy and could be used to transmit codes in mathematical form. The parent star [bottom left] is like the Sun, and the planet itself is like the Earth, which means any forms of life found there might be similar to those now on Earth. A large nearby satellite is shown. The radio telescope is constructed along essentially the same lines as those already set up on Earth and it could establish contact with an instrument such as the Haystack radio telescope at the Massachusetts Institute of Technology.

Scientific and astronomical tables

SYMBOLS USED IN MATHS AND SCIENCE

Symbol	Meaning	Symbol	Meaning	Symbol	Meaning
$+$	Plus; positive	f, F, φ	A function of	GCD, gcd	Greatest common divisor
$-$	Minus; negative	$<$	[In geometry] an angle	LCD, lcd	Least common denominator
$\times$	Multiplied by	$\perp$	Perpendicular; is perpendicular to	LCM, lcm	Least common multiple
$\div$	Divided by	$\parallel$	Parallel; is parallel to	Σ	Sum of specified terms
$\pm$	Plus or minus	$\triangle$	Triangle	$\int$	Integral
$\mp$	Minus or plus	$\cong \equiv$	Congruent, is congruent to	$\lim y=b, \lim y=b$	The limit of y as x nears a is b.
$:$	Ratio; in proportion to	$\therefore$	Therefore	$x \to a \quad x=a$	
$=$	Equals	$\because$	Because	$\triangle y$	An increment [of y]
$\neq$	Does not equal	π	Pi, 3.14159	∂y	A variation or increment [in y]
$\equiv$	Is identical with	$x°$	Degree of angle [x degrees]	dy	Differential of y
$\not\equiv$	Is not identical with	x'	Minute of angle [x minutes]	$\ni$	[In logic] such that
$\approx$	Approximately equals	x''	Second of angle [x seconds]	$\sim p, -p, \bar{p}, p'$	Not p
$\sim$	Equivalent or similar to	sin	Sine	$p \wedge q, p.q, p\&q$	Both p and q
$>$	Is greater than	cos	Cosine	$p^2 \vee q$	At least one of p and q
$\not>$	Is not greater than	tan	Tangent	$p/q, p|q$	Not both p and q
$<$	Is less than	ctn, cot	Cotangent	$p{\downarrow}q, p{\triangle}q$	Neither p nor q
$\not<$	Is not less than	sec	Secant	$p{\Rightarrow}q, p{\Rightarrow}q, p{\subset}q$	If p then q; only p if q
$\geq$	Is greater than or equal to	cosec	Cosecant	V, 1	Universal class
$\leq$	Is less than or equal to	covers	Conversed sine or coversine	$\emptyset, \wedge, 0$	Null class
$\propto$	Is directly proportional to	exsec	Exsecant	$a \in M$	The point a belongs to the set M
$\rightarrow$	Approaches as a limit	e	Base of the natural logarithms	M = N	The sets M and N coincide
x^n	x to the nth power	$\log_a x$	Logarithm [base a] of x	$M \subset N$	M is a subset of N
$\sqrt{x}, x^{\frac{1}{2}}$	Square root of x	$\ln_a, \log_a, \log_a e$	Natural logarithm of a	$M \subseteq N$	M is a subset of N
$\sqrt[n]{x}, x^{1/n}$	The nth root of x	$\log a, \log_{10} a$	Common logarithm of a	$M \cap N, M N$	Intersection of M and N
$\frac{1}{x^n}, x^{-n}$	Negative exponent	antilog	Antilogarithm	$M \cup N, M + N$	The sum of M and N
$x!, \lfloor x$	Factorial of x [Here $1 \times 2 \times 3 \dots x$]	colog	Cologarithm	$M - N, M \sim N$	All points of M not in N
∞	Infinity	exp x	e^x	$\sim M, C[M], \bar{M}$	The complement of M

BASE SI UNITS

Physical quantity	SI unit	Symbol
length	metre	m
mass	kilogram[me]	kg
time	second	s
electric current	ampere	A
thermodynamic temperature	kelvin	K
luminous intensity	candela	cd
amount of substance	mole	mol

SUPPLEMENTARY SI UNITS

Physical quantity	SI unit	Symbol
plane angle	radian	rad
solid angle	steradian	sr

DERIVED SI UNITS WITH SPECIAL NAMES

Physical quantity	SI unit	Symbol
frequency	hertz	Hz
energy	joule	J
force	newton	N
power	watt	W
pressure	pascal	Pa
electric charge	coulomb	C
potential difference	volt	V
electric resistance	ohm	Ω
electric conductance	siemens	S
electric capacitance	farad	F

Physical quantity	SI unit	Symbol
magnetic flux	weber	Wb
inductance	henry	H
magnetic flux density	tesla	T
luminous flux	lumen	lm
illuminance	lux	lx
activity (radioactive)	becquerel	Bq
absorbed dose	gray	Gy

SI PREFIXES (DECIMAL MULTIPLES AND SUBMULTIPLES)

Submultiple	Prefix	Symbol	Sub multiple	Prefix	Symbol	Multiple	Prefix	Symbol	Multiple	Prefix	Symbol
10^{-1}	deci	d	10^{-9}	nano	n	10^1	deca	da	10^9	giga	G
10^{-2}	centi	c	10^{-12}	pico	p	10^2	hecto	h	10^{12}	tera	T
10^{-3}	milli	m	10^{-15}	femto	f	10^3	kilo	k	10^{15}	peta	P
10^{-6}	micro	μ	10^{-18}	atto	a	10^6	mega	M	10^{18}	exa	E

IMPERIAL UNITS OF LENGTH

4 lines	1 barleycorn
3 barleycorns	1 inch
12 inches	1 foot
3 feet	1 yard
5.5 yards	1 rod (pole or perch)
4 rods	1 chain
10 chains	1 furlong
8 furlongs	1 mile
3 miles	1 league
6 feet	1 fathom
2.5 fathoms	1 nautical chain
400 nautical chains	1 nautical mile

IMPERIAL UNITS OF CAPACITY

60 minims	1 fluid drachm
8 fluid drachms	1 fluid ounce
5 fluid ounces	1 gill
4 gills	1 pint
2 pints	1 quart
2 quarts	1 pottle
2 pottles	1 gallon
2 gallons	1 peck
4 pecks	1 bushel
4 bushels	1 coomb
2 coombs	1 quarter

UNITS OF MASS — AVOIRDUPOIS

16 drams	1 ounce
16 ounces	1 pound
14 pounds	1 stone
2 stones	1 quarter
4 quarters	1 hundredweight
20 hundredweights	1 ton
700 grains	1 pound

The hundredweight and ton are sometimes called the long hundredweight and long ton to distinguish them from the US units, the short hundredweight [100 pounds] and the short ton [2,000 pounds]

UNITS OF MASS — TROY

4 grains	1 carat
6 carats	1 pennyweight
20 pennyweights	1 ounce
12 ounces	1 pound [troy]
25 pounds [tr]	1 quarter
4 quarters	1 hundredweight
20 hundredweights	1 ton

UNITS OF MASS — APOTHECARIES'

20 grains	1 scruple
3 scruples	1 drachm
8 drachms	1 ounce
12 ounces	1 pound

INTERCONVERSION FACTORS

Length	metres	inches	yards
1 metre	1	39.3701	1.0936
1 inch	0.0254	1	0.2778
1 yard	0.9144	36	1
	kilometres	miles	nautical miles
1 kilometre	1	0.62137	0.53996
1 mile	1.60934	1	0.86898
1 nautical mile	1.852	1.1508	1

1 light-year	9.4607×10^{15} metres
1 parsec	3.0857×10^{16} metres
1 astronomical unit	1.495×10^{11} metres

Area	sq metres	sq inches	sq yards
1 sq metre	1	1.550	1.196
1 sq inch	0.000645	1	0.00077
1 sq yard	0.8361	1,296	1
	sq kilometres	sq miles	acres
1 sq kilometre	1	0.386	247.105
1 sq mile	2.58999	1	640
1 acre	0.00405	0.00156	1

1 hectare = 0.001 sq km = 2.47105 acres

Volume	cubic metres	cubic feet	gallons
1 cubic metre	1	35.3146	219.969
1 cubic foot	0.0293	1	6.2288
1 gallon [UK]	0.00455	0.160544	1
1 gallon [US] = 0.83268 gallon [UK]			

Mass	kilograms	pounds	tons
1 kilogram	1	2.2046	0.000984
1 tonne	1,000	2.204.6	0.98420
1 pound	0.45359	1	0.000446
1 ton [long]	1,016.047	2.240	1

Energy	joules	cals	K Wh	btus
1 joule	1	0.2388	2.778×10^{-7}	9.478×10^{-4}
1 calorie	4.1868	1	1.163×10^{-6}	0.00397
1 kilowatt hour	3.6×10^{-6}	8.598×10^{-5}	1	3,412.1
1 British Thermal unit	1,055.06	251.997	2.931×10^{-4}	1

BEAUFORT SCALE

			Speed [knots]
0	Calm	Smoke rises vertically	<1
1	Light air	Smoke or leaves indicates movement, otherwise almost calm	1-3
2	Light breeze	Wind felt on face, leaves rustle, etc.	4-6
3	Gentle breeze	Flag extended; leaves and twigs in constant motion	7-10
4	Moderate breeze	Small branches moved; dust and litter raised	11-16
5	Fresh breeze	Small trees begin to sway	17-21
6	Strong breeze	Large branches in motion; whistling in telephone wires	22-27
7	Near gale	Whole trees in motion; inconvenience experienced in walking	28-33
8	Gale	Twigs broken off; walking impeded	34-40
9	Strong gale	Slight structural damage	41-47
10	Storm	Widespread damage	48-55
11	Violent storm	Severe damage results	56-63
12	Hurricane	Severe damage results	>64

THERMOMETER COMPARISONS [CELSIUS/FAHRENHEIT]

C	F	C	F	C	F	C	F
−17.8	0	21.1	70	48.9	120	76.6	170
−6.7	20	25	77	50	122	80	176
−1.1	30	26.7	80	54.4	130	82.2	180
0	32	30	86	55	131	85	185
4.4	40	32.2	90	60	140	87.8	190
5	41	35	95	65	149	90	194
10	50	37.8	100	65.6	150	93.3	200
15	59	40	104	70	158	95	203
15.6	60	43.3	110	71.1	160	98.9	210
20	68	45	113	75	167	100	212

CHEMICAL ELEMENTS
*denotes the mass number of the most stable isotope

Element	Symbol	Atomic number	Atomic weight	Melting point [°C]	Boiling point [°C]	Relative density or density
actinium	Ac	89	*227	1050	3200 [est.]	10.07
aluminium	Al	13	26.9815	660.2	2467	2.699
americium	Am	95	*243	995	2607 [est.]	13.67
antimony	Sb	51	121.75	630.5	1640	6.684
argon	Ar	18	39.948	−189.2	−185.7	1.78 kg m^{-3}
arsenic	As	33	74.9216	817 [grey]	sublimes 613 [grey]	5.37 [grey]
astatine	At	85	*210	250	350	
barium	Ba	56	137.34	725	1140	3.5
berkelium	Bk	97	*247			14 [est.]
beryllium	Be	4	9.01218	1278	2970	1.85
bismuth	Bi	83	208.9806	271.3	1560	9.75
boron	B	5	10.81	2300	2550	2.34 [crystalline]
bromine	Br	35	79.904	−7.2	58.78	3.12 [liquid]
cadmium	Cd	48	112.4	320.9	765	8.65
caesium	Cs	55	132.9055	28.5	690	1.87
calcium	Ca	20	40.08	848	1487	1.55
californium	Cf	98	*251			
carbon	C	6	12.01115	sublimes 3500	4827	1.9-2.3 [graphite] 3.1-3.5 [diamond]
cerium	Ce	58	140.120	795	3468	6.7
chlorine	Cl	17	35.453	−100.98	−34.6	3.124 kg m^{-3}
chromium	Cr	24	51.996	1890	2482	7.19
cobalt	Co	27	58.9332	1495	2870	8.9
columbium	Cb		see niobium			
copper	Cu	29	63.546	1083	2595	8.96
curium	Cm	96	*247	1340		13.51 [est.]
dysprosium	Dy	66	162.50	1407	2335	8.56
einsteinium	Es	99	*254			
erbium	Er	68	167.26	1522	2510	9.045
europium	Eu	63	151.96	826	1439	5.25
fermium	Fm	100	*257			
fluorine	F	9	18.9984	−219.62	−188.14	1.696 kg m^{-3}
francium	Fr	87	*223	30	650	
gadolinium	Gd	64	157.25	1312	~3000	7.898
gallium	Ga	31	69.72	29.78	2403	5.91
germanium	Ge	32	72.59	937.4	2830	5.32
gold	Au	79	196.9665	1063	2660	19.30
hafnium	Hf	72	178.49	2150	5400	13.31
helium	He	2	4.0026	−272.2	−268.6	0.178 kg m^{-3}
holmium	Ho	67	164.9303	1461	2600	8.803
hydrogen	H	1	1.00797	−259.14	−252.5	0.0899 kg m^{-3}
indium	In	49	114.82	156.61	~2000	7.31
iodine	I	53	126.9045	113.5	184.35	4.93
iridium	Ir	77	192.22	2410	4130	22.42
iron	Fe	26	55.847	1539	2800	7.90
krypton	Kr	36	83.80	−156.6	−152	3.733 kg m^{-3}
lanthanum	La	57	138.9055	920	3454	6.17
lawrencium	Lr	103	*257			
lead	Pb	82	207.19	327.3	1750	11.3
lithium	Li	3	6.941	179	1317	0.534
lutetium	Lu	71	174.97	1656	3315	9.835
magnesium	Mg	12	24.305	651	1107	1.738
manganese	Mn	25	54.938	1244	2097	7.21-7.44
mendelevium	Md	101	*256			
mercury	Hg	80	200.59	−38.87	356.58	13.55
molybdenum	Mo	42	95.94	2610	5560	10.22
neodymium	Nd	60	144.24	1024	3127	6.80
neon	Ne	10	20.179	−248.67	−246.05	0.8999 kg m^{-3}
neptunium	Np	93	237.0482	640	3902 [est.]	20.25
nickel	Ni	28	58.71	1453	2732	8.90
niobium	Nb	41	92.9064	2468	4927	8.57
nitrogen	N	7	14.0067	−209.86	195.8	1.251 kg m^{-3}
nobelium	No	102	*256			
osmium	Os	76	190.2	~3045	5027	22.57
oxygen	O	8	15.9994	218.4	−182.96	1.429 kg m^{-3}
palladium	Pd	46	106.4	1,552	3327	12.02

CHEMICAL ELEMENTS

Element	Symbol	Atomic number	Atomic weight	Melting point [°C]	Boiling point [°C]	Relative density or density
phosphorus	P	15	30.9738	44.1 [white]	280 [white]	1.82 [white]
platinum	Pt	78	195.09	1769	3800	21.45
plutonium	Pu	94	*244	641	3327	19.84
polonium	Po	84	*209	254	962	9.40
potassium	K	19	39.102	63.65	774	0.86
praeseodymium	Pr	59	140.9077	931	3212	6.77
promethium	Pm	61	*147	1080	2460	
protactinium	Pa	91	231.0359	1200	4000	15.4 [est.]
radium	Ra	88	226.0254	700	1140	5
radon	Rn	86	*222	−71	−61.8	9.73 kg m^{-3}
rhenium	Re	75	186.2	3180	56.27 [est.]	21.0
rhodium	Rh	45	102.9055	1966	~3727	12.4
rubidium	Rb	37	85.4678	38.89	688	1.53
ruthenium	Ru	44	101.07	2310	3900	12.41
samarium	Sm	62	150.35	1072	1778	7.5
scandium	Sc	21	44.9559	1539	2832	2.99
selenium	Se	34	78.96	217 [grey]	684.9 [grey]	4.79 [grey]
silicon	Si	14	28.086	1410	2355	2.33
silver	Ag	47	107.868	961.93	2212	10.5
sodium	Na	11	22.9898	97.81	892	0.97
strontium	Sr	38	87.62	769	1384	2.54
sulphur	S	16	32.064	112.8 [rhombic] 119.0 [monoclinic]	444.6	2.07 [rhombic] 1.957 [monoclinic]
tantalum	Ta	73	180.9479	2996	5425	16.65
technetium	Tc	43	98.9062	~2200	5030	~11.5
tellurium	Te	52	127.6	449.5	989.8	6.24
terbium	Tb	65	158.9254	1360	3041	8.234
thallium	Tl	81	204.37	303.5	1457	11.85
thorium	Th	90	232.0381	1750	~3800	11.72
thulium	Tm	69	168.9342	1545	1727	9.31
tin	Sn	50	118.69	231.89	2270	5.75 [grey] 7.31 [white]
titanium	Ti	22	47.90	1675	3620	4.54
tungsten	W	74	183.85	3410	5927	19.3
uranium	U	92	238.029	1132	3818	~18.95
vanadium	V	23	50.9414	1890	3380	6.1
wolfram	W	74	see tungsten			
xenon	Xe	54	131.30	−111.9	−107.1	5.887 kg m^{-3}
ytterbium	Yb	70	173.04	824	1193	6.97
yttrium	Y	39	88.9059	1523	3337	4.46
zinc	Zn	30	65.37	419.58	907	7.133
zirconium	Zr	40	91.22	1852	4377	6.51
[Element 104]		104				
[Element 105]		105				

ALLOYS

Name	Typical composition	Properties	Uses
Brass	70% Cu 30% Zn	malleable, corrosion resistant, good conductivity	rust-free parts, electrical contacts, etc
Bronze	90% Cu 10% Sn	hard, corrosion resistant, good for casting	bearings, coins, tools, statues, etc
Phosporbronze	90% Cu 9.7% Sn 0.3% P	strong, hard ductile, corrosion resistant	bearings, ships' propellers
Gunmetal	88% Cu 10% Sn 2% Zn	strong, wear and corrosion resistant	bearings, gears, etc
Manganin	84% Cu 12% Mn 4% Ni	high resistivity, low expansion	resistance wire
Constantan	60% Cu 40% Ni	resistance insensitive to temperature	resistance wire
German silver	62% Cu 15% Ni 22% Zn	hard, corrosion resistant	cutlery, surgical instruments
Monel metal	67% Ni 29% Cu 1.7% Fe 1% Mn +C+Mg	high strength and corrosion resistance	pumps, propellers, chemical plant
Nichrome	77.3% Ni 21% Cr+Mn+Fe	high resistivity low expansion	heating elements
Nimonic	80% Ni 19.5% Cr+Ti+Al	high melting point	gas-turbine blades
Duralumin	94.3% Al 4% Cu+Mn+Mg	light, high strength and conductivity	aerials, aircraft parts
Carbon steel	98.4% Fe 0.8% C+Mn+Si+P	hard, high tensile strength	construction, cutting blades, wire
Stainless steel	85.1% Fe 13.7% Cr 0.3% C+Ni+Mn+Si	high strength and corrosion resistance	cutlery, chemical plant ball bearings

THE SOLAR SYSTEM

Object	Distance from Sun in millions of km [millions of miles]		Diameter in km	[miles]	Mass in Earth masses	Rotation period	Sidereal period	Number of satellites
Mercury	58	[36]	4,880	[3,032]	0.05	58.7 days	88 days	0
Venus	108.19	[67.20]	12,100	[7,500]	0.82	243 days	224.70 days	0
Earth	149.59	[92.950]	*12,756	[7,926]	1.00	23 hr 56 min 4 s	365.25 days	1
Mars	227.94	[141.32]	6,790	[4,210]	0.11	24 hr 37 min 23 s	686.96 days	2
Jupiter	778.38	[483.6]	*142,800	[88,700]	317.9	9 hr 51 min	11.86 years	13
Saturn	1,427	[887]	*120,000	[75,000]	95.2	10 hr 14 min	29.46 years	10
Uranus	2,869	[1,780]	51,800	[32,375]	14.6	10 hr 48 min	84.0 years	5
Neptune	4,496	[2,793]	49,800	[30,940]	17.2	15 hr 48 min	164.8 years	2
Pluto	5,900	[3,658]	5,998	[3,725]	0.08	6.39 days	247.7 years	0
Moon	[384,000 km from Earth (239,000 miles)]		3,477	[2,160]	0.012	27.30 days	—	—
Sun	—		1,392,300	[865,000]	333,000			9 planets

* denotes equatorial value.

1 earth mass = 590 × 10^{22} kg (1.3 × 10^{23} lb).

ASTRONOMICAL DISTANCES

Name	Nature	Distance [light-years]
Formalhaut	Star	23
Vega	Star	26
Aldebaran	Star	65
Pleiades	Globular cluster	410
Rigel	Star	850
Deneb	Star	1,500
M71	Globular cluster	8,000
Large Magellanic Cloud	Globular cluster	180,000
Palomar 3	Galaxy	180,000
Andromeda Spiral	Galaxy	2,200,000
M87	Galaxy	40,000,000
3C 273	Quasar	1,500,000,000

ASTRONOMICAL SYMBOLS

Symbol	Name
☉	Sun
☽	Moon
☿	Mercury
♀	Venus
♁	Earth
♂	Mars
♃	Jupiter
♄	Saturn
♅	Uranus
♆	Neptune
♇	Pluto

BRIGHTEST STARS

Name	Apparent magnitude	Absolute magnitude	Distance [light-years]
Sirius	−1.47	+0.7	8.7
Canopus	−0.71	−5.5	300
Alpha Centauri	−0.27	+4.6	4.3
Arcturus	+0.06	−0.3	36
Vega	0.03	+0.3	26
Capella	0.09	+0.1	45
Rigel	0.15	−8.2	850
Procyon	0.34	+2.8	11
Achernar	0.49	−1.3	75
Betelgeux	variable	variable	650
Hadar	0.61	−4.3	300
Altair	0.75	+2.1	16

NEAREST STARS

Name	Apparent magnitude	Absolute magnitude	Distance [light-years]
Proxima Centauri	10.7	15.1	4.3
Alpha Centauri	0.0	4.4	4.3
Barnard's Star	9.5	13.2	6.0
Wolf 359	13.5	16.5	8.1
Lalande 21185	7.5	10.5	8.2
Luyten 726-8	12.41	15.4	8.7
Sirius	−1.5	1.4	8.7
Ross 154	10.6	13.3	9.3
Ross 248	12.2	14.7	10.3
Epsilon Eridani	3.7	6.1	10.8
Ross 128	11.1	13.5	11.1
Layten 789-6	12.2	14.6	11.1

ASTRONOMERS ROYAL

Name	Dates
John Flamsteed	1675-1719
Edmund Halley	1720-1742
James Bradley	1742-1762
Nathanial Bliss	1762-1764
Nevil Maskelyne	1765-1811
John Pond	1811-1835
Sir George Biddell Airy	1835-1881
Sir William Henry Mahoney Christie	1881-1910
Sir Frank Watson Dyson	1910-1933
Sir Harold Spencer Jones	1933-1955
Sir Richard van der Riet Woolley	1955-1971
Sir Martyn Ryle	1972-

MILESTONES IN SPACE FLIGHT

Spacecraft	Launched		Achievements
USSR Sputnik 1	4 October 1957	Unmanned	1st artificial satellite
USSR Sputnik 2	3 November 1957	Unmanned	1st inhabited spacecraft [the dog, Laika]
US Explorer 1	31 January 1958	Unmanned	1st US satellite; discovered inner Van Allen belt
USSR Luna 1	2 January 1959	Unmanned	1st spacecraft to escape the Earth's gravitational pull
USSR Luna 2	12 September 1959	Unmanned	1st landing on the Moon
USSR Luna 3	4 October 1959	Unmanned	1st orbit of the Moon; 1st photographs of the far side
US Pioneer 5	11 March 1960	Unmanned	Interplanetary probe; studied Sun's magnetic field
US Tiros 1	1 April 1960	Unmanned	1st weather satellite
USSR Vostock 1	12 April 1961	Manned	1st manned space flight [Yuri A. Gagarin]
US Mercury-Redstone 3	5 May 1961	Manned	1st US manned space flight [Alan B. Shepard]
US Mercury-Atlas 6	20 February 1962	Manned	1st US manned orbit of the Earth [John H. Glenn]
US Telstar	10 July 1962	Unmanned	1st transatlantic relay of television transmissions
US Mariner 2	27 August 1962	Unmanned	1st fly-by of Venus; measured atmospheric and surface temperatures
USSR Mars 1	1 November 1962	Unmanned	1st fly-by of Mars
USSR Vostok	16 June 1963	Manned	Valentina Tereshkova becomes 1st woman in space
USSR Voskhod 1	12 October 1964	Manned	1st 3-man spacecraft
US Mariner 4	28 November 1964	Unmanned	Mars fly-by that 1st photographed surface and studied the atmosphere
USSR Voskhod 2	18 March 1965	Manned	1st spacewalk [for 10 minutes by Aleksei Leonov]
US Gemini 3	23 March 1965	Manned	1st manned orbital manoeuvres
USSR Venera 3	16 November 1965	Unmanned	1st craft to land on another planet [Venus]
USSR Luna 9	31 January 1966	Unmanned	1st lunar soft landing; photographed surface
US Gemini 8	16 March 1966	Manned	1st docking with another spacecraft
USSR Luna 10	31 March 1966	Unmanned	1st spacecraft to enter lunar orbit
US Surveyor 1	30 May 1966	Unmanned	1st US lunar soft landing; photographed surface
USSR Venera 4	12 June 1967	Unmanned	Landing on Venus; transmitted data on atmosphere before touchdown
US Apollo 8	21 December 1968	Manned	1st manned orbit of the Moon
US Apollo 11	16 July 1969	Manned	1st landing on the Moon [Neil A. Armstrong and Edwin E. Aldrin]
USSR Venera 7	17 August 1970	Unmanned	1st transmission from the planet's surface
USSR Luna 16	12 September 1970	Unmanned	Lunar soft landing; automatically took soil samples
USSR Soyuz 11	6 June 1971	Manned	1st inhabitation of Salyut space station [23 days]
US Pioneer 10	2 March 1972	Unmanned	1st satellite to relay data on Jupiter and leave Solar System
US Apollo 17	7 December 1972	Manned	Last and longest stay on the Moon [75 hours]
US Pioneer 11	6 April 1973	Unmanned	1st spacecraft enroute to Saturn
US Skylab 2	25 May 1973	Manned	Longest space flight [84 days]. Wideranging scientific experiment
US Mariner 10	3 November 1973	Unmanned	Relayed 1st close-up photographs of Mercury
US Skylab 3	16 November 1973	Manned	1st occupation of Skylab space station
Soyuz/Apollo	17 July 1975	Manned	Russian-American link-up in space
US Viking 1	20 August 1975	Unmanned	1st successful landing on Mars

SIGNS OF THE ZODIAC

Sign	Symbol	Period
Aries [Ram]	♈	21 March–20 April
Taurus [Bull]	♉	21 April–21 May
Gemini [Twins]	♊	22 May–21 June
Cancer [Crab]	♋	22 June–23 July
Leo [Lion]	♌	24 July–23 August
Virgo [Virgin]	♍	24 August–23 September
Libra [Scales]	♎	24 September–23 October
Scorpio [Scorpion]	♏	24 October–22 November
Sagittarius [Archer]	♐	23 November–22 December
Capricorn [Goat]	♑	23 December–20 January
Aquarius [Water-bearer]	♒	21 January–19 February
Pisces [Fishes]	♓	20 February–20 March

INDEX

Picture Credits

Every endeavour has been made to trace copyright holders of photographs appearing in *The Joy of Knowledge*. The publishers apologize to any photographers or agencies whose work has been used but has not been listed below.

Credits are listed in this manner: [1] page numbers appear first, in bold type; [2] illustration numbers appear next, in brackets, [3] photographers' names appear next, followed where applicable by the names of the agencies representing them.

16–17 Adam Woolfitt/Susan Griggs Picture Agency. **18** Fritz Goro/T.L.P.A. © Times Inc 1976/Colorific; **19** Paul Brierley. **20–1** [3A] Spectrum Colour Library; [3B] Spectrum Colour Library; [5] Spectrum Colour Library; [6] Michael Holford; [7] Ronan Picture Library. **22–3** [4A] Ronan Picture Library. **24–5** [Key] Ronan Picture Library/Royal Astronomical Society; [1A] Trustees of the British Museum; [1B] Ronan Picture Library/E. P. Goldschmidt & Co Ltd; [1C] Ronan Picture Library; [5A] Ronan Picture Library/Royal Astronomical Society; [5B, 6A, B] Ronan Picture Library. **26–7** [Key] Paul Brierley; [1] Mary Evans Picture Library; [2] Anthony Howarth/Susan Griggs Picture Agency; [5] Ken Lambert/Bruce Coleman Ltd; [7] Cooper Bridgeman Library; [8] David Levin. **28–9** [Key] Hans Schmid/ZEFA; [4] Gerry Cranham; [8] Barnabys Picture Library; [9] David Levin. **30–1** [Key] Sally & Richard Greenhill; [2] David Levin; [4] Mansell Collection; [6A] David Levin; [9A] Racing Information Bureau; [9B] IBM. **32–3** [Key] Dr D. E. H. Jones; [1A] Dr D. E. H. Jones; [1B] Dr D. E. H. Jones; [1C] Paul Brierley; [2] Paul Brierley; [4] Fritz Goro/T.L.P.A. © Time Inc 1976/Colorific; [6] Dr D. E. H. Jones; [7] Photri. **34–5** [Key] Spectrum Colour Library; [3] David Levin; [7] David Levin. **36–7** [Key] Dr D. E. H. Jones; [5] Dr D. E. H. Jones. **38–9** [Key] Paul Brierley; [2] Spectrum Colour Library; [5] David Levin; [6A] David Levin; [6B] David Levin; [7] David Levin. **40–1** [Key] Pictor; [1A] David Levin; [8A] Barnabys Picture Library. **42–3** [Key] The Royal Institution; [1] Dr D. E. H. Jones; [3] Dr D. E. H. Jones; [4B] Spectrum Colour Library; [5] Dr D. E. H. Jones; [6B] ZEFA. **44–5** [Key] Dr D. E. H. Jones; [7] Dr D. E. H. Jones. **46–7** [Key] Art & Antiques Weekly; [8A] David Levin; [8B] Brian Coates/Bruce Coleman Ltd. **48–9** [Key] Institute of Electrical & Electronics Engineers Inc; [5] Dr D. E. H. Jones; [7] Dr D. E. H. Jones; [8] Dr D. E. H. Jones; [10] William MacQuitty. **50–1** [Key] R. K. Pilsbury/Bruce Coleman Ltd; [4] Ron Boardman; [5] Escher Foundation, The Hague; [6A] David Strickland; [8A] Spectrum Colour Library. **52–3** [Key] Dr D. E. H. Jones; [7] Dieter Buslau/*Construction News*. **54–5** [Key A] National Gallery; [Key B] National Gallery. **56–7** [Key] Paul Brierley/S.T.L. Research; [7] Barnabys Picture Library. **58–9** [Key] David Levin; [5] David Strickland. **60–1** [Key] Photri; [2] CERN; [3] Dr A. M. Field, Virus Reference Laboratory, Colindale; [4] Spectrum Colour Library; [5] Scala; [6] Spectrum Colour Library; [7] Photri. **62–3** [1] C. M. Dixon; [2] Ron Boardman; [5] Ronan Picture Library; [6] Solvay & Cie; [7] Popperfoto; [9] Bettmann Archive. **64–5** [Key] Photri; [1] Spectrum Colour Library; [2] Spectrum Colour Library; [3] Ronan Picture Library; [4A] Ronan Picture Library; [4B] Cavendish Laboratory/Cambridge University; [7] Science Museum; [8] UK Atomic Energy Authority. **66–7** [1] F. Rust/ZEFA; [2] David Levin; [3] International Society for Educational Information, Tokyo; [4] American History Picture Library; [6E] Photri; [8] Photri. **68–9** [Key] Photri; [1] Photri; [2] David Levin; [3] Photri; [5] Photri; [6] David Levin; [7] ZEFA. **70–1** [Key] John Walmsley; [3] Spectrum Colour Library; [5D] London Transport Executive; [6] David Strickland. **72–3** [3] Photri; [5D] David Levin. **74–5** [Key] Spectrum Colour Library; [3] Popperfoto; [4] Camera Press; [6] Camera Press; [7] Ronan Picture Library. **76–7** [Key] Adam Woolfitt/Susan Griggs Picture Agency; [2] Spectrum Colour Library; [3A] Spectrum Colour Library; [4] Photri; [5] Spectrum Colour Library; [9] Institution of Civil Engineers. **78–9** [Key] Hawker Siddeley Aviation; [1B] Spectrum Colour Library; [4] Picturepoint; [6] Spectrum Colour Library. **80–1** [2] Lyn Cawley. **82–3** [5] David Strickland; [6] David Strickland; [7] Fabbri. **84–5** [Key] Picturepoint; [2] Picturepoint; [4B] Camera Press; [7] Photri. **86–7** [2] Shell Photographic Library; [5] Picturepoint; [6] CERN; [7] Graeme French; [8] David Levin; [9] David Levin. **88–9** [Key] Ron Boardman; [4] *Construction News*; [5] K. Helbig/ZEFA; [6] Picturepoint; [8] Gerry Cranham. **90–1** [Key] Mansell Collection; [8] B.O.C. Ltd. **92–3** [1] Photri; [2] Photri; [3] Paul Brierley. **94–5** [2A] Air Products & Chemicals Inc; [4A] Paul Brierley; [4B] Paul Brierley; [5A] CERN. **96–7** [Key] De Beers Industrial Diamond Division; [1A] Picturepoint; [1B] Paul Brierley/Daly Instruments; [2] Paul Brierley/British Aluminium Co; [3] Ford Motor Co; [4A] Joseph Lucas Ltd; [4B] Joseph Lucas Ltd; [5] Paul Brierley/RCA; [6] Paul Brierley/Southampton University; [7] Photri. **98–9** [Key] Ronan Picture Library; [1] Spectrum Colour Library. **102–3** [3] Bob Croxford; [6] David Strickland; [8A] David Strickland; [8B] David Strickland; [9] David Strickland. **104–5** [Key] Horst Munzig/Susan Griggs Picture Agency; [3] Victor Englebert/Susan Griggs Picture Agency; [4] Paul Brierley. **106–7** [Key] Ronan Picture Library; [5] Photri. **108–9** [Key] Science Museum; [5] Spectrum Colour Library; [7] Spectrum Colour Library; [8] Picturepoint; [10] Courtesy of the GPO. **110–11** [Key] Paul Brierley; [7A] Paul Brierley/Welding Institute; [7B] William Vandivert; [7C] William Vandivert. **114–15** [9] Central Electricity Generating Board. **116–17** [Key] The Royal Institution; [5] Mansell Collection. **118–19** [Key] Professor E. Laithwaite; [5] Spectrum Colour Library. **120–1** [1] Imperial War Museum; [10] Cubestore Ltd. **122–3** [4E] Otis Elevators Ltd; [7] Paul Brierley/Lintrol/Imperial College. **124–5** [Key] W. Canning & Co Ltd; [1] Paul Brierley; [2] A.S.E.A.; [5] A.S.E.A.; [8] Monitor. **126–7** [2A] David Levin; [2B] Paul Brierley/UKAEA Culham Lab; [6A] David Levin; [6B] Central Electricity Generating Board; [8] Spectrum Colour Library. **128–9** [8A] David Levin; [9A] Marshall Cavendish/Kim Sayer; [10A] Paul Brierley; [10B] Paul Brierley. **130–1** [Key] Mullard Valves Ltd; [5A] David Levin; [5B] David Levin. **132–3** [Key] Paul Brierley; [2] Chris Steele-Perkins/Science Museum. **134–5** [6] UK Atomic Energy Authority; [7] Paul Brierley/STL Research. **136–7** [Key] Cooper Bridgeman; [2] Picturepoint; [3] David Levin; [4] Picturepoint; [5] National Gallery; [6] Michael Holford; [9] Mary Evans Picture Library. **138–9** [1] Popperfoto; [7] Shell Photographic Library; [8] Photri; [9A] Kim Sayer; [9B] Kim Sayer. **140–1** [Key] David Strickland; [4] ZEFA; [6] Dr J. Holloway/Leicester University/courtesy Argonne National Laboratory, Argonne, Illinois, USA; [12] Source unknown. **142–3** [Key] David Strickland; [3] Radio Times Hulton Picture Library; [4] Spectrum Colour Library; [7] A. F. Kersting; [8] Citroen. **144–5** [Key] Picturepoint; [3] Spectrum Colour Library; [4] Paul Brierley; [6] Dead Sea Works; [7] Spectrum Colour Library. **146–7** [Key] Paul Brierley. **148–9** [Key] Mansell Collection; [1] Ronan Picture Library; [3] Vitatron UK Ltd. **150–1** [Key] Paul Brierley; [1] Paul Brierley; [2] Paul Brierley. **154–5** [Key] Colorsport; [4] P. H. Ward/Natural Science Photos. **156–7** [Key] Dr Robert Horne; [4A–C] Sir John Kendrew; [5] Dr Audrey Glavert; [8] Daily Telegraph Colour Library. **161** Photri. **162** NASA. **163** Photri. **164–5** [Key] Patrick Moore Collection. **166–7** [8] Patrick Moore Collection. **168–9** [6] Hale Observatories, Mount Wilson and Palomar; [7] Hale Observatories, Mount Wilson and Palomar. **170–1** [Key] Patrick Moore Collection; [1] Novosti Press Agency; [2] Australian Information Service; [3] Lick Observatory; [4A] Hale Observatories, Mount Wilson and Palomar; [5] Patrick Moore Collection; [6] US Naval Observatory. **172–3** [Key] Patrick Moore Collection; [2] J. Arthur Dixon/by courtesy of Sir Bernard Lovell; [3] P. Daly; [4] Hale Observatories, Mount Wilson and Palomar; [5] Lund Observatory; [7] US Naval Observatory. **178–9** [Key] Georgetown University Observatory; [4A] Ronan Picture Library; [4B] Patrick Moore Collection; [6] Picturepoint; [7] H.R. Hatfield; [8] NASA; [9] NASA; [10] Hale Observatories, Mount Wilson and Palomar; [11] NASA; [12] NASA; [13] H. Brinton. **180–1** [Key] Royal Astronomical Society; [11] NASA; [12] NASA; [13] NASA; [14] NASA; [15] NASA; [16A–E] NASA. **182–3** [7A] Lick Observatory; [8] Royal Astronomical Society. **184–5** [Key] Fairchild Space and Defence Systems. **186–7** [Key] Novosti Press Agency; [1] NASA; [2] NASA; [3] NASA; [4] NASA; [5] NASA; [6] NASA; [7] NASA; [8] NASA. **188–9** [Key] Patrick Moore Collection; [4] NASA/Courtesy of Dr John Guest; [8] NASA; [9] NASA; [10] NASA; [11] NASA; [12] NASA. **190–1** [4] H.R. Hatfield; [5] NASA; [9] NASA; [10] NASA; [11A] NASA; [11B] NASA; [12] NASA. **192–3** [1A] NASA; [3] NASA. **194–5** [Key] NASA; [5A–D] C.F. Capen. **196–7** [Key] NASA; [4] NASA; [8] NASA; [9] Photri; [10] NASA. **200–1** All photographs NASA. **202–3** [2] NASA; [3A–C] NASA; [4A–C] NASA; [6] Photri; [7] NASA. **204–5** [1] Max Wolf/Royal Astronomical Society; [2] F. C. Acfield. **206–7** [5] G. P. Kuiper; [6] Lowell Observatory, Arizona. **208–9** [Key] H. E. Dall; [1, 2, 3, 4, 5] NASA. **210–11** [Key] US Naval Observatory; [7] Patrick Moore Collection; [8] H. R. Hatfield; [11] Hale Observatories, Mount Wilson and Palomar. **212–13** [4] G. P. Kuiper; [6A, B] G. P. Kuiper; [7] G. P. Kuiper. **214–15** [4] G. P. Kuiper; [7] G. P. Kuiper; [9A, B] Patrick Moore Collection. **216–17** [Key] Source unknown; [5] Hale Observatories, Mount Wilson and Palomar; [6] E. E. Barnard/Royal Astronomical Society; [7] E. M. Lindsay/Royal Astronomical Society; [8] Hale Observatories, Mount Wilson and Palomar; [9] Royal Greenwich Observatory. **218–19** [Key] Butler/Royal Astronomical Society; [4] D. McLean/Royal Astronomical Society/Kitt Peak Observatory; [5] T. J. C. A. Moseley; [7] Patrick Moore Collection; [8] Source unknown; [9] Institute of Meteorites, New Mexico; [10] Source unknown; [11] Novosti Press Agency; [12] Source unknown; [13] Source unknown. **220–1** [Key] Royal Greenwich Observatory, Herstmonceaux; [2B] P. Daly; [6] Hale Observatories, Mount Wilson and Palomar; [7] Hale Observatories, Mount Wilson and Palomar. **222–3** [Key] W. M Baxter; [2A, B, C] Roberts/Royal Astronomical Society; [4] Patrick Moore Collection; [5] NASA; [6] NASA. **224–5** [2] NASA; [4] H. Brinton; [5] NASA; [6] A. Kung; [7] NASA; [8] NASA; [9] NASA. **226–7** [1] P. Gill; [2] J. McBain/Patrick Moore Collection; [4] H. R. Hatfield; [5] H. R. Hatfield. **228–9** [1] Hale Observatories, Mount Wilson and Palomar; [4] Hale Observatories, Mount Wilson and Palomar. **230–1** [Key] H. R. Hatfield; [3] Hale Observatories, Mount Wilson and Palomar; [4] Hale Observatories, Mount Wilson and Palomar; [5B] Source unknown; [6B] Hale Observatories, Mount Wilson and Palomar; [7] Hale Observatories, Mount Wilson and Palomar; [8] Hale Observatories, Mount Wilson and Palomar. **232–3** [Key] Hale Observatories, Mount Wilson and Palomar; [1] US Naval Observatory; [2] Hale Observatories, Mount Wilson and Palomar; [3] Hale Observatories, Mount Wilson and Palomar; [4] Hale Observatories, Mount Wilson and Palomar; [5] Hale Observatories, Mount Wilson and Palomar; [6] Hale Observatories, Mount Wilson and Palomar. **234–5** [Key] Hale Observatories, Mount Wilson and Palomar; [3A, B] Royal Astronomical Society. **236–7** [Key A, B, 6B, 15B] H. R. Hatfield. **238–9** [Key] Mount Stromlo Observatory, Australia; [2] Patrick Moore Collection; [3] Hale Observatories, Mount Wilson and Palomar; [4] Hale Observatories, Mount Wilson and Palomar. **240–1** [1] K. G. Malin-Smith; [7] Hale Observatories, Mount Wilson and Palomar; [8] Source unknown; [9] Hale Observatories, Mount Wilson and Palomar; [10] T. J. C. A. Moseley; [11] Patrick Moore Collection. **242–3** [2] US Naval Observatory; [3] K. G. Malin-Smith; [4] H. R. Hatfield; [5] K. G. Malin-Smith; [7] Royal Astronomical Society; [8] Hale Observatories, Mount Wilson and Palomar; [9] US Naval Observatory; [10] Hale Observatories, Mount Wilson and Palomar. **244–5** [Key] Carnegie Institute Washington/Hale Observatories, Mount Wilson and Palomar; [3] Carnegie Institute, Washington/Hale Observatories, Mount Wilson and Palomar; [4] Hale Observatories, Mount Wilson and Palomar; [6] Hale Observatories, Mount Wilson and Palomar; [7] US Naval Observatory; [8] Lund

Observatory. **246–7** [1] Carnegie Institute, Washington/Hale Observatories, Mount Wilson and Palomar; [2] US Naval Observatory; [3] Hale Observatories, Mount Wilson and Palomar; [4] Hale Observatories, Mount Wilson and Palomar; [5] Mount Stromlo Observatory, Australia; [6] Royal Astronomical Society; [7] Radcliffe Observatory. **248–9** [Key] Hale Observatories, Mount Wilson and Palomar; [2] Lick Observatory; [3] Hale Observatories, Mount Wilson and Palomar; [4] Hale Observatories, Mount Wilson and Palomar; [5] Hale Observatories, Mount Wilson and Palomar; [6] Hale Observatories, Mount Wilson and Palomar; [7] Hale Observatories,

Mount Wilson and Palomar; [8] Hale Observatories, Mount Wilson and Palomar; [9] Lick Observatory; [10] Hale Observatories, Mount Wilson and Palomar; [12] Hale Observatories, Mount Wilson and Palomar; [13] US Naval Observatory; [14] US Naval Observatory; [15] US Naval Observatory; [16] Hale Observatories, Mount Wilson and Palomar. **250–1** [1] Hale Observatories, Mount Wilson and Palomar; [2] Hale Observatories, Mount Wilson and Palomar; [3] US Naval Observatory; [4] Hale Observatories, Mount Wilson and Palomar; [5] Hale Observatories, Mount Wilson and Palomar; [6A, B] Royal Greenwich Observatory; [7]

Source unknown; [8] Source unknown. **252–3** [Key] Hale Observatories, Mount Wilson and Palomar; [2] Hale Observatories, Mount Wilson and Palomar; [3] Hale Observatories, Mount Wilson and Palomar; [5] Hale Observatories, Mount Wilson and Palomar. **254–5** [4] Photoresources; [5A–F] Photoresources; [6A] Photoresources; [6B] Snark International; [7] Source unknown. **268–9** [1] Patrick Moore Collection; [2] Patrick Moore Collection; [3] NASA; [4] Novosti Press Agency; [5] Novosti Press Agency; [6] Novosti Press Agency; [7] NASA; [8] NASA; [9] Photri; [10] Photri; [11] Photri; [12] NASA; [13] Novosti

Press Agency; [14] NASA; [15] Photri; [16] Photri; [17] Photri; [18] Novosti Press Agency. **270–1** [Key] Patrick Moore Collection. **272–3** [Key] NASA; [1] by permission of Madame Malthete Melies/Copyright S.P.A.D.E.M. Paris 1976; [2] Royal Astronomical Society. **274–5** [Key] NASA; [1] Photri. **278–9** [Key] Patrick Moore Collection. **282–3** [Key] Hale Observatories, Mount Wilson and Palomar.

Colour photographs credited above to Hale Observatories are copyright by the California Institute of Technology and the Carnegie Institute of Washington.

Artwork Credits